# New Frontiers in Regional Science: Asian Perspectives

Volume 77

This series is a constellation of works by scholars in the field of regional science and in related disciplines specifically focusing on dynamism in Asia.

Asia is the most dynamic part of the world. Japan, Korea, Taiwan, and Singapore experienced rapid and miracle economic growth in the 1970s. Malaysia, Indonesia, and Thailand followed in the 1980s. China, India, and Vietnam are now rising countries in Asia and are even leading the world economy. Due to their rapid economic development and growth, Asian countries continue to face a variety of urgent issues including regional and institutional unbalanced growth, environmental problems, poverty amidst prosperity, an ageing society, the collapse of the bubble economy, and deflation, among others.

Asian countries are diversified as they have their own cultural, historical, and geographical as well as political conditions. Due to this fact, scholars specializing in regional science as an inter- and multi-discipline have taken leading roles in providing mitigating policy proposals based on robust interdisciplinary analysis of multifaceted regional issues and subjects in Asia. This series not only will present unique research results from Asia that are unfamiliar in other parts of the world because of language barriers, but also will publish advanced research results from those regions that have focused on regional and urban issues in Asia from different perspectives.

The series aims to expand the frontiers of regional science through diffusion of intrinsically developed and advanced modern regional science methodologies in Asia and other areas of the world. Readers will be inspired to realize that regional and urban issues in the world are so vast that their established methodologies still have space for development and refinement, and to understand the importance of the interdisciplinary and multidisciplinary approach that is inherent in regional science for analyzing and resolving urgent regional and urban issues in Asia.

Topics under consideration in this series include the theory of social cost and benefit analysis and criteria of public investments, socio-economic vulnerability against disasters, food security and policy, agro-food systems in China, industrial clustering in Asia, comprehensive management of water environment and resources in a river basin, the international trade bloc and food security, migration and labor market in Asia, land policy and local property tax, Information and Communication Technology planning, consumer "shop-around" movements, and regeneration of downtowns, among others.

Researchers who are interested in publishing their books in this Series should obtain a proposal form from Yoshiro Higano (Editor in Chief, higano@jsrsai.jp) and return the completed form to him.

Tofael Ahamed
Editor

# Remote Sensing Application II

## A Climate Change Perspective in Agriculture

Springer

*Editor*
Tofael Ahamed
Institute of Life and Environmental Sciences
University of Tsukuba
Tsukuba, Ibaraki, Japan

*Editor-in-Chief*
Yoshiro Higano

ISSN 2199-5974 ISSN 2199-5982 (electronic)
New Frontiers in Regional Science: Asian Perspectives
ISBN 978-981-97-1187-1 ISBN 978-981-97-1188-8 (eBook)
https://doi.org/10.1007/978-981-97-1188-8



This Springer imprint is published by the registered company Springer Nature Singapore Pte Ltd.
The registered company address is: 152 Beach Road, #21-01/04 Gateway East, Singapore 189721, Singapore

# Foreword

I am honored to have the opportunity again to provide further insights into the impacts of this volume, *Remote Sensing Application II: A Climate Change Perspective in Agriculture*, which is edited by Dr. Tofael Ahamed, my former colleague at the University of Tsukuba. As readers may see by the title, this book is a sequel to *Remote Sensing Application: Regional Perspectives in Agriculture and Forestry* in the Series (published on 8th May 2022). I had an opportunity to provide a Foreword to the first book and emphasized the huge potential of GIS/Remote Sensing technologies in analyzing policy issues related to climate change and the global environment through interdisciplinary and multidisciplinary methods.

This book focuses more on the vulnerability of the agricultural production system caused by climate change and, therefore, the risk of food shortage, especially in south and southeast Asia and in Africa. It also provides leading-edge modeling techniques such as machine learning and fuzzy expert systems, which are useful for practical regional planning for the next 20 years in light of the emergence of issues under unavoidable uncertainty.

Contributors are internationally active experts in the fields, and readers will be re-confirmed how GIS/Remote Sensing is powerful and effective in the sustainability management of interphase interaction among ecological and social human systems. They will be motivated to further expand the frontier of GIS/Remote Sensing in regional science and related fields.

University of Tsukuba
Tsukuba, Japan

Yoshiro Higano

# Preface

Climate change has become inevitable due to global warming, which has resulted in extreme floods, drought, riverbank erosion, coastal area erosion, cyclones, and saline water intrusion. These climate extremes often affect the lives and livelihoods of people, mainly in low-lying deltaic regions. Countries in Southeast Asia and Africa are most vulnerable to the impacts of global climate change, which has become a great concern among researchers and policy-makers. In recent times, the application of remote sensing with other analytical methods, such as multicriteria analysis, machine learning, and cloud computation, has provided researchers with data exhibiting increased spatial and temporal coverage as well as higher computational capabilities and faster decision-making. In this regard, the application of remote sensing methods along with machine learning models can play a significant role in the adaptation of policies to the adverse effects of climate change through vulnerability analysis, land cover/land use change analysis across large spatial scales, and crop and orchard suitability analysis.

With the above concepts, this book encompasses 17 chapters and focuses on three main thematic points, namely, climatic change and impacts of droughts and floods, land use/land cover changes, climatic adaptation to mitigate the risk to food production, and climate risk mitigation. First, beginning with **Chap. 1,** remote sensing methods with different analytical approaches are applied to assess the vulnerability and impacts of climate extremes, such as riverbank erosion, floods, cyclones, and droughts. In this regard, at the first stage, multicriteria-based erosion vulnerability analysis is performed by applying a fuzzy expert system to remote sensing datasets. At the second stage, a flood vulnerability analysis is presented using the same approach. At the third stage, damage area assessment is conducted using the normalized difference vegetation index (NDVI), modified vegetation condition index (mVCI), and damage difference vegetation damage index (DVDI) considering cyclones, where the occurrence of cyclones and their frequency are examined by analyzing global datasets. At the fourth stage, drought severity is measured using several satellite remote sensing indices, such as the normalized difference vegetation index (NDVI), vegetation condition index (VCI), temperature condition index (TCI), and standard precipitation index (SPI). Finally, an inventory of knowledge from indigenous people with spatial diversity is presented as a source of indigenous knowledge for addressing climate change impacts. These combined approaches may help subsequent studies related to the effects of climate change.

In **Chap. 2**, the impact of drought stress was determined in tea estates using a remote sensing technique with the standardized precipitation index (SPI). The land surface temperature (LST) and soil moisture index (SMI) were measured from Landsat 8 OLI/TIRS images. Maps of the normalized difference moisture index (NDMI), normalized difference vegetation index (NDVI), and leaf area index (LAI), as well as yield maps, were developed from Sentinel-2 satellite images. The drought frequency was calculated from the drought classification utilizing the SPI. It was suggested that the SPI could represent an important index for measuring drought stress in tea estates.

In **Chap. 3**, land use/land cover (LULC) change assessment was performed over a 26-year time span to determine the most common land use class, and inundation vulnerability was analyzed to identify vulnerable agricultural land use areas within a 25-km zone of the main river systems in Bangladesh using the Google Earth Engine (GEE) platform and ArcGIS®. The distributed river basin encompasses eight administrative divisions from North to South Bangladesh covering an area of 64,556 $km^2$ (north: Rangpur, northwest: Rajshahi, northeast: Sylhet, central: Dhaka and Mymensingh, southwest: Khulna, south: Barishal, and southeast: Chittagong). According to LULC analysis, agricultural land was the most vulnerable land use class, with notable change.

In **Chap. 4**, the shoreline status was explored via change assessment in 1991, 2006, and 2021 with satellite remote sensing and geographical information system (GIS) approaches. Landsat 5, 7 ETM+, and 8 OLI satellite images were employed for onshore tasseled cap transformation (TCT) and land and sea classification calculation to create shore boundaries and conduct baseline, land accretion, erosion, point distance, and near-feature analysis. This research highlights the changes in shorelines, referring to evidence of active delta formation through accretion and erosion processes, i.e., climate change and sea level rise, in the coastal regions of Bangladesh.

In **Chap. 5**, a novel damage-based crop insurance system was developed to support farmers affected by flash floods. Essential determinants influencing farmers' willingness to adopt WTA insurance were identified, as well as the maximum amount they would be willing to pay (WTP). This determination was achieved through the utilization of binary logistic regression and multiple regression techniques. This study is the first to investigate crop insurance, amalgamating remote sensing and GIS applications with an econometric approach to estimate premium rates based on damage, and holds significant potential in assisting flash flood-affected farmers by subsidizing premiums as part of a standard subsidy program.

In **Chap. 6**, a damaged area index (DAI) was developed for rice crop-based agricultural lands affected by two consecutive tropical cyclones by assessing typical damaged areas in the coastal region of Bangladesh through remote sensing datasets. The developed index could help support farmers with a new adaptive policy for recurrent cyclones in tropical cyclone-prone countries in the global south. In **Chap.**

**7**, land suitability analysis is presented for potential vineyard extensions using fuzzy expert systems and remote sensing approaches. A suitability map was developed using a fuzzy-based multicriteria decision-making system, geographical information system (GIS), and fuzzy membership functions. This study can contribute to decision-making processes for extending vineyards in countries such as Afghanistan. In **Chap. 8**, suitable lands for tea production are determined in the northeastern part of Bangladesh using remote sensing phenological datasets, geospatial datasets of soil–plant biophysical properties, and expert opinions. In **Chap. 9**, land suitability analysis of coffee plantations in Central Aceh, Indonesia, is conducted to support farmers in prioritizing land selection for developing coffee plantations via remote sensing and multicriteria analysis.

In **Chap. 10**, a land suitability model is introduced for rice production based on suitability levels, and insurance premiums were proposed to obtain maximum returns based on the harvest index and subsidy dependence factor for the marginally and moderately suitable areas in the northern part of Bangladesh by applying multicriteria analysis via a geographical information system and analytical hierarchy process. This new crop insurance model, based on land suitability, could provide consistent support for owners of land of different qualities to increase production. **Chapter 11** focuses on an integrated approach of the AHP for monitoring drought and rice yield loss due to drought in the dry zone of Myanmar. This integrated model could facilitate regional drought assessment and support farmers and stakeholders in monitoring crop moisture levels and establishing early warning systems by providing timely information on potential hazards throughout the rice growing season.

In **Chap. 12**, the microclimatic conditions of a coffee-based agroforestry system are characterized to determine the appropriate microclimate and shade level management for coffee plants under the agroforestry system and to determine the resilience and adaptability of coffee plants to climate hazards. In **Chap. 13**, satellite remote sensing techniques are applied to identify renewable energy sources for agricultural production and energy security in Lahendong, North Sulawesi, Indonesia, to reduce global warming. This study describes the use of remote sensing data to determine geothermal areas and identify the potential of land replacement for agricultural production and energy security.

**Chapter 14** provides physio-climatic and socioeconomic suitability assessments of alternative wetting and drying (AWD) irrigation water management in Bangladesh, which is a water-saving, methane emission-reducing and cost-effective technique for rice cultivation. The main objective of this study was to classify the suitability of the AWD technique for irrigated rice areas in Bangladesh.

The application of remote sensing and machine learning approaches to estimate the availability of local food in the urban area of Indonesia is explored in **Chap. 15**. Different vegetation indices are used for yield prediction at the growth stages of plants. The findings of this study suggest that vegetation indices could be used for formulating site-specific management of local crops and for predicting yields. These integrated models could be used for logistics and decision-making regarding local crop production.

In **Chap. 16**, the need for integrated drought monitoring is outlined for supporting evidence-based adaptation in African drought-prone environments. In this study, different satellite-based approaches for characterizing agricultural and hydrological droughts are reviewed, including precipitation anomalies, vegetation indices, evapotranspiration deficits, and surface water declines, which could offer new drought monitoring capabilities to help vulnerable regions anticipate and prepare for droughts exacerbated by climate change.

At the end of this book, Chap. 17 highlights the three main aspects of the climate risk to agriculture, namely, climate risk due to drought and floods, which severely impacts crop production and food security; climatic impacts on land use and land cover changes; microclimatic adaptation through crop selection; and climate risk mitigation through subsidies and crop insurance schemes. A team effort is made in this book to align potential remote sensing, machine learning, and GIS applications from climate change perspectives in certain highly climate risk-prone countries in South and Southeast Asian and African regions.

The overall discussion in this book can help food security planners, policymakers, and researchers in proper planning and managing climate change to mitigate the risk of agricultural production decline to ensure food security.

Tsukuba, Ibaraki, Japan
Tofael Ahamed

# Acknowledgments

I would like to express my heartfelt appreciation to Professor Yoshiro Higano, Editor-in-Chief of the *Asia-Pacific Journal of Regional Science*, for inspiring me to complete this second edition. Over time, our collaboration and the publication of research articles in the *Asia-Pacific Journal of Regional Science* have been the main resources in shaping this book series. In this edited volume, the book focuses solely on the application of remote sensing methods from a climate change perspective. Information on climate-induced impacts of flooding, flash floods, and drought on agricultural crops is provided to address possible solutions for food security in South Asia, Southeast Asia, and some regions of Africa.

I appreciate the *Asia-Pacific Journal of Regional Science*, Remote Sensing, for granting copyright permission to reuse some of their published articles as chapters. Special thanks are due to Professor Tekeshi Mizunoya and Ms. Hatsumi Uchimura of the University of Tsukuba for their support in the copyright process through the Japan Section of the Regional Science Association International (JSRAI).

I extend my gratitude and thanks to the contributors to this book, who helped me to fulfill the aim of crafting all the chapters concurrently. This time, in addition to contributions of our laboratory at the University of Tsukuba, collaborators from the Philippines, Indonesia, and Nigeria dedicated their time to the various stages of reviewing their chapters. I extend my thanks and appreciation to contributors Dr. Janet E. Pablo of Benguet State University, the Philippines; Professor Kudang Boro Seminar of Bogor Agricultural University, Indonesia; Dr. Riska Ayu Purnamasari of the National Research and Innovation Agency (BRIN), Indonesia; and Dr. Emmanuel Owoicho Abah of the Federal Ministry of Agriculture and Food Security, Nigeria.

I would like to express special appreciation to my committed team of PhD students for their tireless efforts in cross-checking, updating, and formatting the book. In this regard, Kazi Faiz Alam and Sara Tokhi Arab have been exceptional in their support, working diligently on content development, literature collection, layout formatting, and reference cross-checking for this book. Munirah Hayati Hamidon also deserves acknowledgment for her dedicated contributions in preparing graphical summaries and figures for several chapters and her tireless invaluable support and assistance until the final stage of the book. This second edition is a testament to the outstanding teamwork and dedication of this remarkable group. Their continuous support, patience, and hard work have been crucial in realizing the goals of this book, and I am sincerely thankful to each team member for their contribution.

Finally, I acknowledge the support I have received from the University of Tsukuba in publishing articles at different times as open-access sources and finally publishing this book in the Series of Frontiers in Regional Science from Springer-Nature.

Tofael Ahamed

# Contents

# Chapter 1
# Application of Remote Sensing in the Analysis of Climate Extremes Due to Global Climate Change

**Kazi Faiz Alam, Md Shamsuzzoha, Sara Tokhi Arab, Ivana Joy Pauline Pangaribuan, and Tofael Ahamed**

**Abstract** In this chapter, several remote sensing applications involving different analytical approaches are illustrated to assess the resulting vulnerability and impacts of climate extremes such as riverbank erosion, floods, cyclones, and droughts. In this regard, at the first stage, an erosion vulnerability analysis is presented where four criteria, namely, the normalized difference water index (NDWI), distance from the river, soil adjusted vegetation index (SAVI), and land use and land cover (LULC), are used in erosion vulnerability assessment. At the second stage, a flood vulnerability extension is highlighted using multicriteria analysis (NDWI, distance from the river, elevation, and rainfall) based on fuzzy expert systems and remote sensing datasets. At the third stage, the damage area attributed to cyclones is assessed using different indices, including the normalized difference vegetation index (NDVI), modified vegetation condition index (mVCI), and damage difference vegetation damage index (DVDI), where the occur-

K. F. Alam
Graduate School of Science and Technology, University of Tsukuba, Tsukuba, Ibaraki, Japan

M. Shamsuzzoha
Graduate School of Science and Technology, University of Tsukuba, Tsukuba, Ibaraki, Japan

Department of Emergency Management, Faculty of Environmental Science and Disaster Management, Patuakhali Science and Technology University, Patuakhali, Bangladesh

S. T. Arab
Division of Intelligent Agricultural Machinery Research, Institute of Agricultural Machinery, National Agriculture and Food Research Organization, Tsukuba, Ibaraki, Japan

I. J. P. Pangaribuan
Graduate School of Science and Technology, University of Tsukuba, Tsukuba, Ibaraki, Japan

Tropical Biodiversity Conservation, Faculty of Forestry and Environment, IPB University, Bogor, West Java, Indonesia

T. Ahamed (✉)
Institute of Life and Environmental Sciences, University of Tsukuba, Tsukuba, Ibaraki, Japan
e-mail: tofael.ahamed.gp@u.tsukuba.ac.jp



T. Ahamed (ed.), *Remote Sensing Application II*, New Frontiers in Regional Science: Asian Perspectives 77, https://doi.org/10.1007/978-981-97-1188-8_1

rence and frequency of cyclones are examined by analyzing global datasets. At the fourth stage, the drought severity is determined using several satellite remote sensing indices, such as the normalized difference vegetation index (NDVI), vegetation condition index (VCI), temperature condition index (TCI), and standard precipitation index (SPI). Finally, an inventory of community perception-based knowledge is established as a source of indigenous knowledge to mitigate climate change impacts. These combined approaches may help further studies related to climate change effects.

**Keywords** Global climate change · Riverbank erosions · Flood · Tropical cyclones · Drought · Indigenous knowledge

## 1.1 Research Background

Climate change has been a significant issue in recent years and refers to the long-term changes in temperature weather patterns both locally and globally (Islam et al. 2022a, b; Karim and Mimura 2008; Rahaman et al. 2020). Low-lying deltaic regions are very vulnerable to natural calamities such as floods, flash floods, tidal surges, cyclones, tsunamis, westerlies, onshore coastal bank erosion, riverbank erosion, droughts, and saline water intrusion (Alam and Ahamed 2023a, b; Al-Maruf et al. 2021; An et al. 2012; Arab et al. 2022; Braun and Aßheuer 2011; Islam et al. 2022a, b; Islam 2016; Kulatunga et al. 2014; Nadiruzzaman and Wrathall 2015; Shamsuzzoha and Ahamed 2023; Talukder et al. 2018; Tinh et al. 2021). All these disasters frequently affect the lives and livelihoods of people living in low-lying developing countries, thereby damaging houses, property, and agricultural crops (Di et al. 2018; Haq et al. 2012; Hongo et al. 2015; Shamsuzzoha et al. 2021). With the change in global climate, the frequency of natural disasters has increased due to irregular rainfall and a rise in the global temperature. On a global level, average economic losses of approximately 170 billion USD/year have been reported due to climatological, hydrological, biological, and geophysical disasters, with peaks in 2011 and 2017 (FAO 2021). Asian and African countries such as Afghanistan, Bangladesh, Bhutan, China, India, Indonesia, Pakistan, Myanmar, Vietnam, Ethiopia, Congo, Ghana, Nigeria, Togo, Tanzania, Uganda, and Zambia are facing the greatest challenges associated with climate change (Ahamed 2023; Baninla et al. 2022). To mitigate the adverse effects of climate change, comprehensive studies are needed that can provide the necessary information for proper planning (Houston et al. 2015; Shaw et al. 2013a, b). Currently, the application of remote sensing technologies has gained increasing research interest in different fields due to the data availability and faster analytical facilities, such as different machine learning techniques and cloud-based computational facilities. Another important factor for the increasing familiarity with remote sensing techniques is their spatial and temporal coverage (Arab et al. 2022). Therefore, this chapter illustrates the application of different analytical approaches for assessing the impacts of floods, riverbank erosion, droughts, and cyclones on regional and global scales using remote sensing approaches (Fig. 1.1).

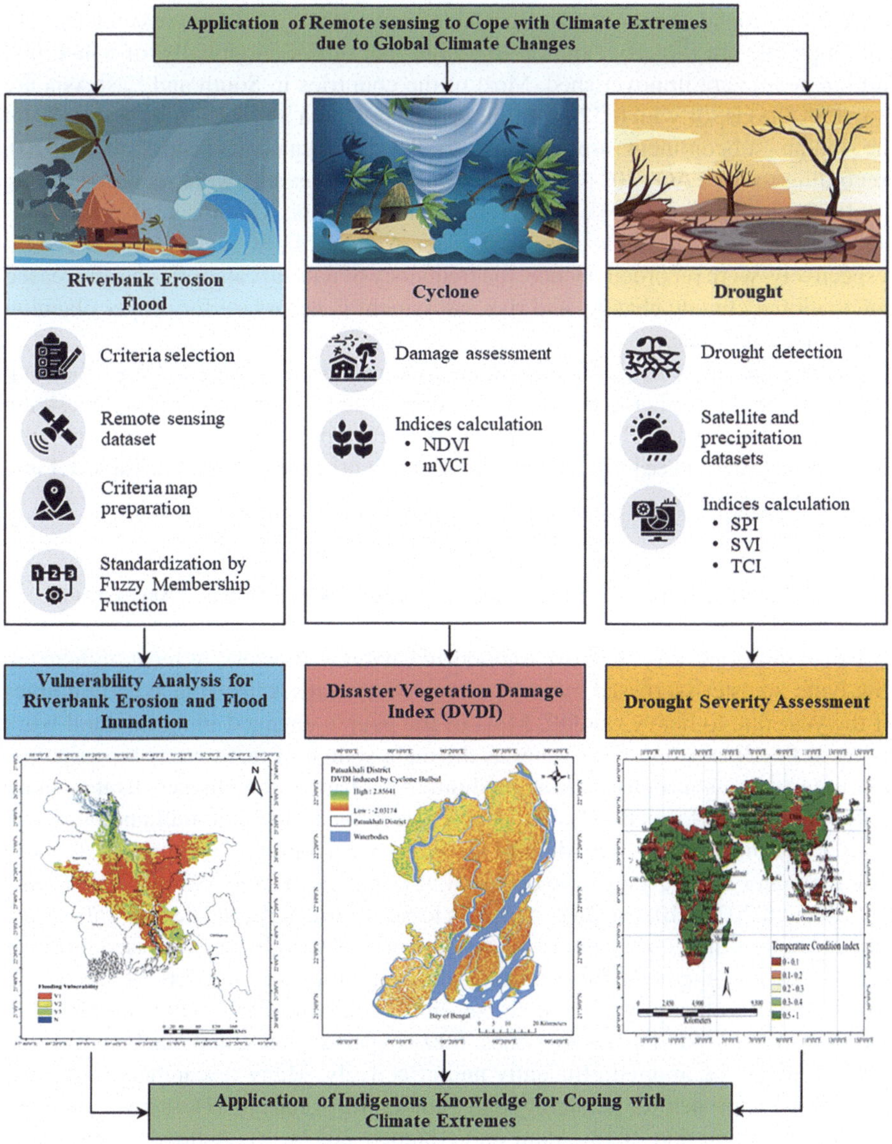


**Fig. 1.1** Application of remote sensing techniques in climate extremes assessment

## 1.2 Application of Remote Sensing in Climate Extremes Assessment

### *1.2.1 Flood*

Floods are the most common and severe hazard that disrupts the lives and livelihood of people in low-income countries, the most where drainage and protection systems are less developed. According to Rentschler and Salha (2020), approximately 1.47 billion people are directly exposed to a high risk of floods globally, of which 600 million people are impoverished. Most of the countries in South and East Asia are at risk of floods, of which 329 million people in China and 225 million people on the Indian subcontinent are exposed to floods (Rentschler and Salha 2020). According to the FAO (2021), an average of 30 flood events were recorded during the 1970s; however, in 1980, there were 60 flood events, which is double the number, while during the 2000s and in 2006, an average of 180 and 246 flood events, respectively, were recorded. Hence, the frequency of natural calamities has increased due to global climate change, and risk assessment is necessary for proper planning and managing floods.

Among the 25 most flood-vulnerable countries worldwide, 58.7% of the total population in the Netherlands and 57.5% of the total population in Bangladesh are exposed to flood risks. The total population and the percentage of the total population exposed to flood risks in 25 countries of the world are provided in Table 1.1 and Fig. 1.2, respectively.

#### 1.2.1.1 Application of Remote Sensing in Flood Vulnerability Assessment

Deltaic regions are very sensitive to flood-like natural disasters, as most of them are low-lying areas crisscrossed by many rivers, which causes flooding at certain times of the year due to heavy rainfall. Most of the people engaged in agricultural work living in floodplains use river water as the main source of fresh water for irrigation and waterways as a transportation medium. As a part of the Ganges–Brahmaputra delta, Bangladesh faces floods almost every year due to local rainfall and rainwater from upstream sources. According to the Water Development of Bangladesh, more than ten major flood events have occurred in Bangladesh from 1954 to 2022 (Alam and Ahamed 2023a, b), causing enormous losses of life and considerable infrastructure and agricultural crop damage. Many researchers worldwide have used different approaches and methods for assessing and mitigating the impacts of floods. To assess flood vulnerability, remote sensing techniques involving multicriteria decision-making (MCDM) approaches, fuzzy expert systems, and machine learning techniques can be applied efficiently and effectively. Many researchers have used the MCDM approach for suitability and vulnerability analysis (Hossain et al. 2009; Ferretti and Pomarico 2013; Alam and Ahamed 2023a, b). A flood vulnerability map is prepared via multicriteria analysis (normalized difference water index

**Table 1.1** Population exposed to flood risks in different countries of the world

| No. | Name of the country | Flood risk by population exposed (%) | Total population exposed (million) |
|---|---|---|---|
| 1 | Netherlands | 58.7 | 10.1 |
| 2 | Bangladesh | 57.5 | 94.42 |
| 3 | Vietnam | 46.0 | 45.5 |
| 4 | Egypt | 40.5 | 38.87 |
| 5 | Myanmar | 39.9 | 19.1 |
| 6 | Laos | 39.7 | 2.99 |
| 7 | Cambodia | 38.1 | 7.43 |
| 8 | Guyana | 37.9 | 0.28 |
| 9 | Suriname | 37.7 | 0.23 |
| 10 | Iraq | 36.8 | 16.35 |
| 11 | Thailand | 33.9 | 25.43 |
| 12 | South Sudan | 32.5 | 5.44 |
| 13 | Pakistan | 31.1 | 71.79 |
| 14 | Nepal | 29.4 | 11.99 |
| 15 | Republic of Congo | 29.3 | 1.17 |
| 16 | Philippines | 29.0 | 30.48 |
| 17 | Japan | 28.7 | 36.06 |
| 18 | Austria | 27.8 | 2.44 |
| 19 | India | 27.7 | 390 |
| 20 | Albania | 27.6 | 0.77 |
| 21 | China | 27.5 | 395 |
| 22 | Chad | 27.4 | 4.55 |
| 23 | Indonesia | 27 | 75.7 |
| 24 | Croatia | 226.9 | 1.09 |
| 25 | Slovakia | 26.7 | 1.4 |

Data source: Rentschler et al. (2022); Conte and Kostandi (2022)

[NDWI], distance from the river, elevation, and rainfall) using fuzzy expert systems and remote sensing datasets (Fig. 1.3). According to flood vulnerability analysis, more than 44% of the areas near the main river systems in Bangladesh are highly vulnerable, and 37% of all areas are moderately vulnerable to flooding (Fig. 1.3).

### 1.2.2 *Riverbank Erosion*

Among natural disasters, riverbank erosion and floods are the most devastating in the delta-like Ganges–Brahmaputra region, Mississippi delta, and Mekong delta. The most dynamic natural process that causes major geomorphological changes in flood plains is riverbank erosion (Hooke 1979). Loss of human settlements, valuable agricultural lands, and urban areas is the most widely recognized phenomenon due to riverbank erosion (Deb and Ferreira 2015; Pahlowan and Hossain 2015).

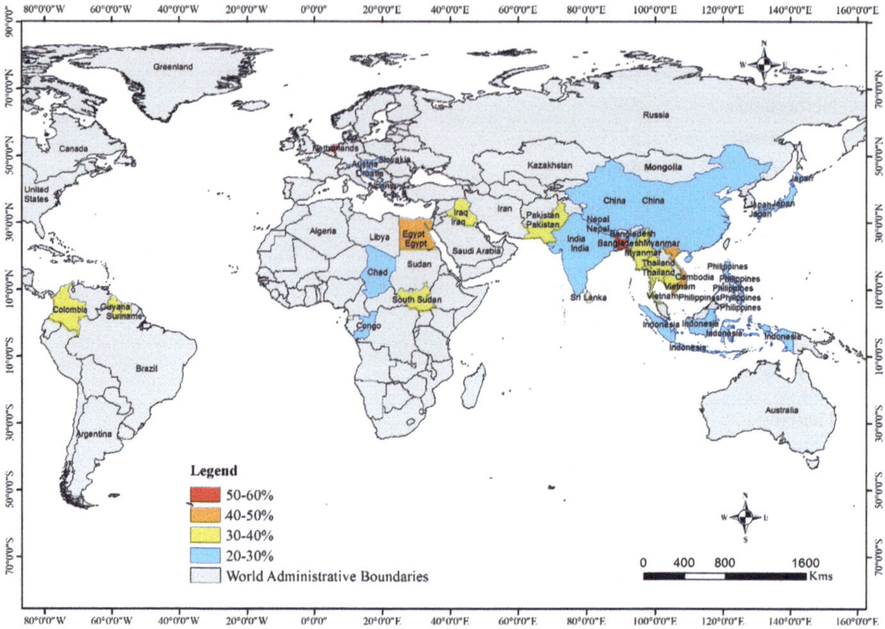


**Fig. 1.2** Percentage of the total population exposed to flood risks in different countries of the world. (Data source: Rentschler et al. 2022; Conte and Kostandi 2022)

According to Mutton and Haque (2004), an estimated average of 60,000 people is displaced every year by riverbank erosion in the Ganges–Brahmaputra–Jamuna River system. The U.S. Army Corps of Engineers spends more than $200 million per year for maintaining channel stability and combatting bank erosion of the Mississippi River (James 2019). Therefore, it is necessary to assess the risk and impact of riverbank erosion in deltaic regions for proper land use and agricultural planning.

#### 1.2.2.1 Application of Remote Sensing in Riverbank Erosion Vulnerability Assessment

Satellite remote sensing offers the opportunity to assess vulnerability and suitability on a microregional scale. For this purpose, the data provided by different satellites, such as Landsat, Moderate Resolution Imaging Spectroradiometer (MODIS), Sentinel, and Advanced Land Observing Satellite (ALOS) images of different spatial resolutions with temporal variations, have been used by researchers. Moreover, faster computational models, including artificial neural networks, fuzzy models, deep learning models, clustering, time series analysis, and Markov chain models based on mathematical and statistical methods, have been used widely in vulnerability and suitability assessment. Recently, cloud computation platforms, such as

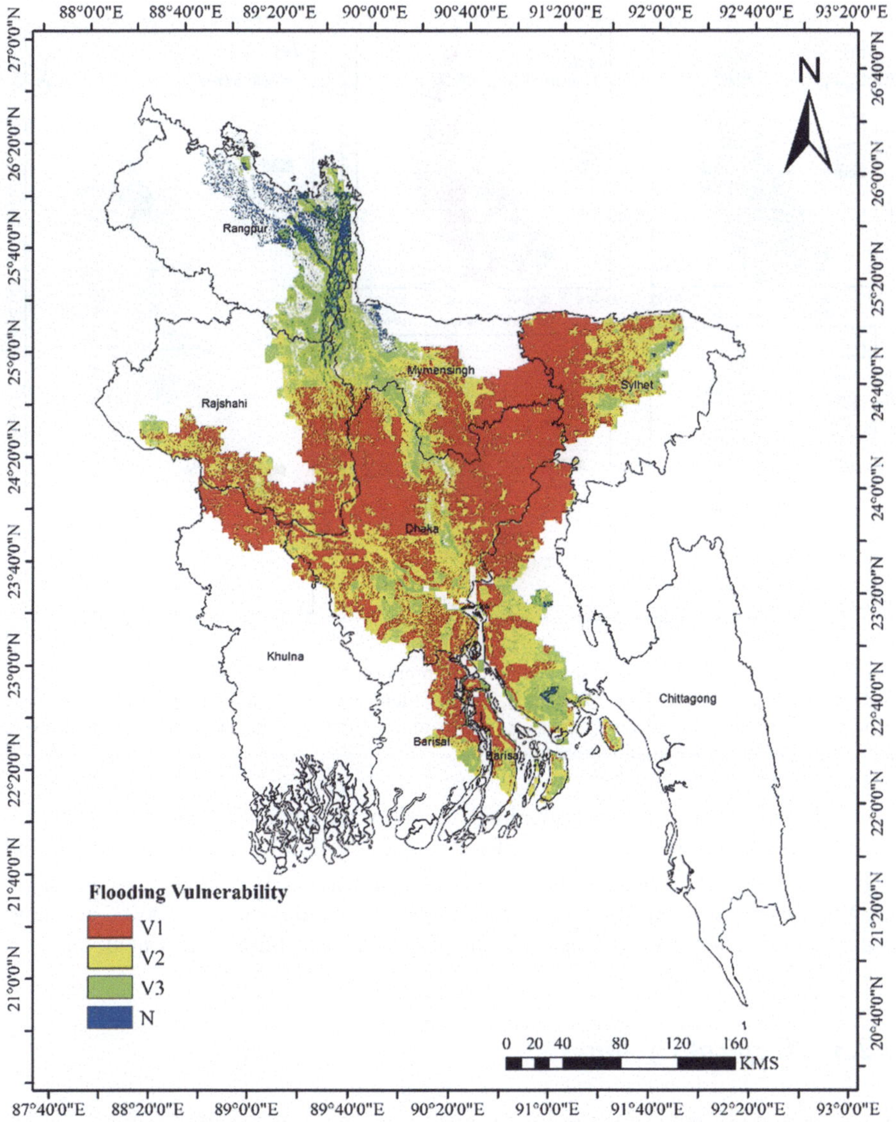


**Fig. 1.3** Flood vulnerability analysis using the MCDM approach and remote sensing datasets. *V1* highly vulnerable, *V2* moderately vulnerable, *V3* marginally vulnerable, and *N* not vulnerable to flooding. (Modified after Alam and Ahamed 2022)

the Google Earth Engine (GEE), have gained popularity among researchers due to their high computational capability and easy accessibility of datasets. According to a recent study on riverbank erosion vulnerability assessment based on multicriteria analysis using remote sensing datasets and fuzzy expert systems, more than 77% of the areas within 3–7 km on both sides of the Jamuna River in Bangladesh are highly

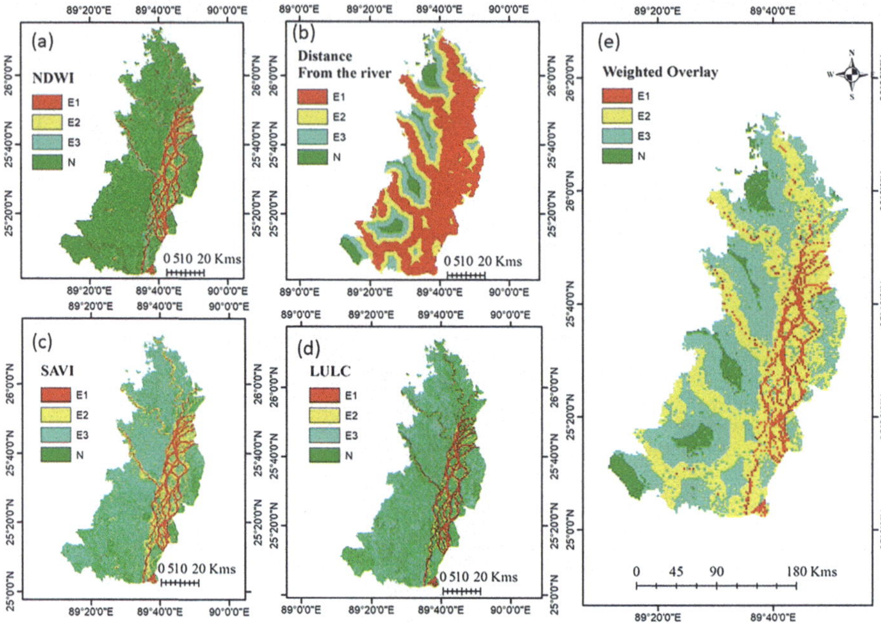


**Fig. 1.4** Erosion vulnerability map based on different criteria: (**a**) NDWI; (**b**) SAVI; (**c**) distance from the river; and (**d**) LULC. Weighted overlay map, where E1 denotes the water body, E2 denotes the highly to moderately erosion-prone area, E3 denotes the marginally erosion-prone area, and N is the area not vulnerable to erosion

to moderately vulnerable to riverbank erosion (Alam and Ahamed 2023a, b). Erosion vulnerability analysis results based on different criteria are shown in Fig. 1.4, in which four criteria, namely, the normalized difference water index (NDWI), distance from the river, soil adjusted vegetation index (SAVI), and land use and land cover (LULC), are used for erosion vulnerability assessment.

### *1.2.3 Tropical Cyclones*

Tropical cyclones are one of the most damaging extreme climatic hazards, causing hundreds of casualties and billions of US dollars in damage globally every year (Klotzbach et al. 2018, 2022; Knutson et al. 2010; Mendelsohn et al. 2012; Shamsuzzoha et al. 2021, 2022). The rate of recurrence and intensity of tropical cyclones have increased in recent times due to global climate change (Chen et al. 2022; Knutson et al. 2019; Kossin et al. 2020).

Tropical cyclones form in the high-sea surface temperature regions of the East Pacific, West Pacific, North Atlantic, North Indian, South Pacific, South Atlantic, and South Indian Ocean basins (Knutson et al. 2010; National Oceanic and Atmospheric Administration 2023). The Tropical Cyclones National Oceanic and

Atmospheric Administration (NOAA), USA, provides open-source data, namely, the International Best Track Archive for Climate Stewardship (IBTrACS), from which a global scenario map of cyclones between 1997 and 2022 is shown in Fig. 1.5. After the analysis of all line- and point-based Geographic Information System( GIS) datasets of cyclones for all ocean basins over 46 years (1977–2022), the cyclone occurrence frequency is shown in Fig. 1.6.

According to the analysis, the Pacific Ocean basin exhibited the highest occurrence frequency of cyclones because of the geospatial conditions. In the West Pacific region, 25.73% of all tropical cyclones were generated, and the frequency was the

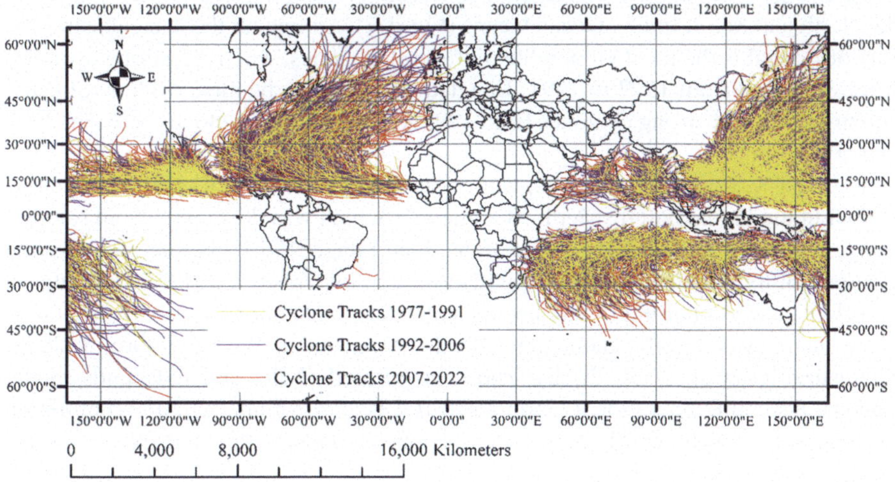


**Fig. 1.5** Global scenario map of tropical cyclones: (1) 1977–1991; (2) 1992–2006; and (3) 2007–2022

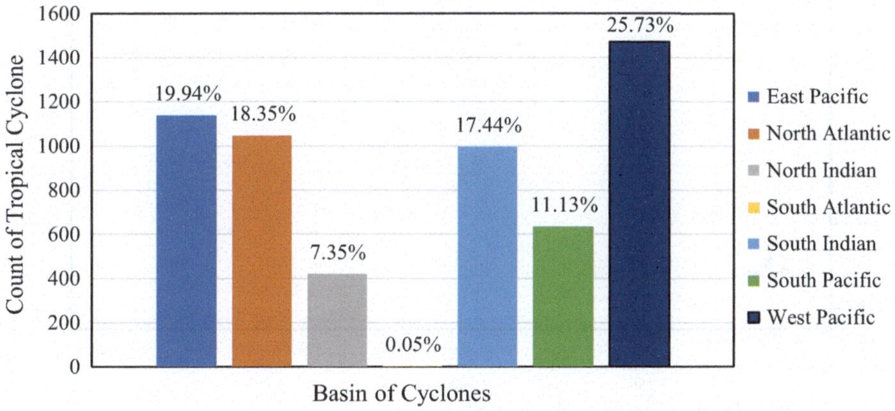


**Fig. 1.6** Total tropical cyclone occurrence from 1977 to 2022 in high-sea surface temperature regions

highest relative to other ocean basins (Fig. 1.6). However, in the East and South Pacific Ocean basins, cyclone formation accounted for 19.94% and 11.13%, respectively, of the total occurrence. In contrast, in the North and South Atlantic Ocean basins, cyclone formation accounted for 18.35% and 0.05%, respectively, of the total occurrence (Fig. 1.6). In the North and South Indian Ocean basins, 7.35% and 17.44% of all tropical cyclones were generated, respectively (Fig. 1.6).

According to Fig. 1.7, an increasing trend of cyclone occurrence was observed in the East Pacific, North Atlantic, and South Atlantic Ocean basins from 1977 to 2022. From a regional perspective, from 1992 to 2006, tropical cyclone formation has significantly increased over the past decades in the Bay of Bengal due to global climate change (Fig. 1.7). However, from 2007 to 2022, decreasing trends were observed in the North and South Indian Ocean basins (Fig. 1.7). Among all the ocean basins, the Bay of Bengal is the most vulnerable to tropical cyclones, where 15 strong cyclones were generated from 1977 to 1991. Bangladesh is one of the most cyclone-affected countries, with 14 strong cyclones from 1992 to 2006 and 17 cyclones between 2007 and 2022 (Fig. 1.7).

Figure 1.8 shows a disaster vegetation damage index (DVDI) map for cyclone Bulbul in the Patuakhali district in Bangladesh, which was developed from satellite images captured by Landsat 8 (operational land imager [OLI]) before and after the cyclone. Bulbul made landfall in Bangladesh on November 9, 2019, as a Category 2 cyclone, and a total of 247,581 hectares of Transplant Aman (T. Aman) rice cultivated land was totally damaged (Needs Assessment Working Group 2019; Shamsuzzoha et al. 2021, 2022). The near-infrared (NIR) and red bands of the Landsat 8 satellite operational land imager (OLI) of November 7, 2019 (before the cyclone) and November 23, 2019 (after cyclone Bulbul), with cloud-free coverage,

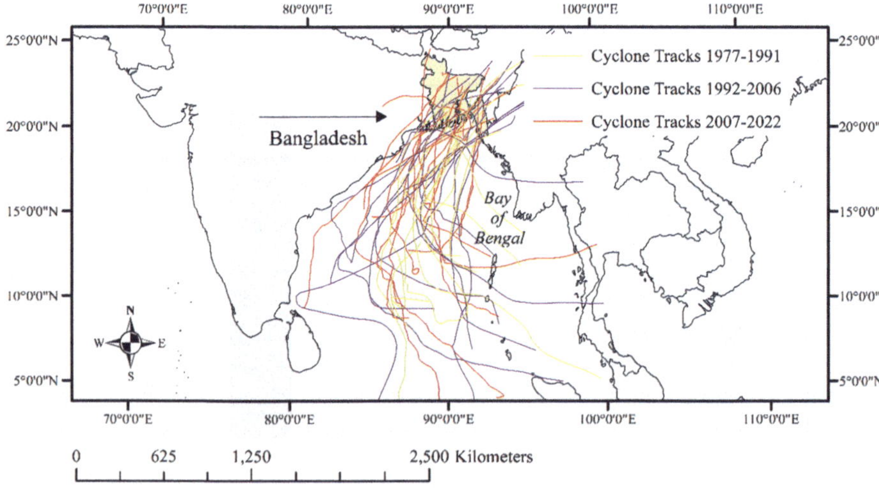


**Fig. 1.7** Regional scenario of Bay of Bengal-induced tropical cyclones toward Bangladesh: (1) 1977–1991; (2) 1992–2006; and (3) 2007–2022

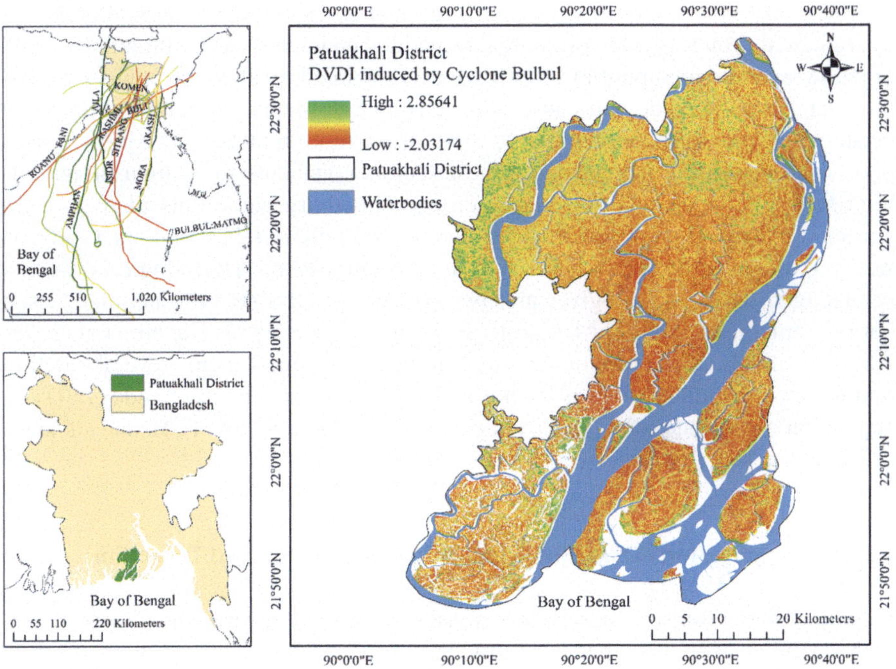


**Fig. 1.8** Disaster vegetation damage index (DVDI) in the aftermath of tropical cyclone Bulbul in Patuakhali district, Bangladesh

retrieved from the United States Geological Survey (USGS) Earth Explorer archive, were employed for the necessary computations in DVDI assessment.

### *1.2.4 Drought*

Drought is one of the most catastrophic natural disasters that occurs every year worldwide, particularly in the arid regions of Asia and Africa. Drought causes crop loss and water scarcity in all parts of the globe, including South and Central Asia and southern, western, eastern, and northwestern Africa (Afghanistan, India, Pakistan, Sri Lanka, Nepal, Bangladesh, Indonesia, China, Morocco, Sahel, Ethiopia, Congo, Ghana, Nigeria, Togo, Tanzania, Uganda, and Zambia) (Arab et al. 2022; Arab and Ahamed 2023; Dai 2011; Rojas et al. 2011; Masih et al. 2014). According to Pei et al. (2019) and Zeng et al. (2019), the estimated drought-related economic losses range from 6 to 8 billion USD per year. Orimoloye (2022) investigated one of the largest commercial land areas in South Africa, and they reported that 30% of the total agricultural sector was severely affected by drought in 2011, 2016, 2018, and 2019. Gichure (2017) and He et al. (2019) reported that 2.5 million people in Kenya faced hunger due to the 2017 drought.

There are typically three drought categories: meteorological, agricultural, and hydrological droughts. Meteorological drought is characterized by months or years of below-average precipitation. It is often accompanied by above-average temperatures, and both precede and cause other types of droughts. Agricultural droughts occur when precipitation decreases, there are intermittent but infrequent downpours, or evaporation increases. All these factors contribute to drought in agricultural settings, which in turn reduces crop productivity and stunts plant growth. During a hydrological drought, water storage in aquifers, lakes, and reservoirs, as well as river streamflow, falls below long-term average levels. Hydrological droughts need longer to develop in a given area because they involve stored water that is used but not replenished (World Meteorological Organization 1992; Hakam et al. 2023; Ali et al. 2023). The drought severity has been measured using several satellite remote sensing indices, such as the normalized difference vegetation index (NDVI), vegetation condition index (VCI), temperature condition index (TCI), and standard precipitation index (SPI).

#### 1.2.4.1 Application of Satellite Remote Sensing in Drought Detection

Satellite remote sensing has been extensively applied in drought detection and assessment because it is a near real-time available and cost-effective method, especially in large areas. The SPI (McKee et al. 1993) is one of the indicators that depends on precipitation data, thus providing a rapid drought index once precipitation data are available. The SPI provides a comparison of precipitation with the climatic mean, estimated as a multiyear average. The SPI can be used to detect the drought severity based on precipitation data. The monthly SPI was computed using daily climate hazards group infrared precipitation with station (CHIRPS) data (since 1981). SPI-1 was used to map precipitation deficits across Africa and Asia. The SPI results indicated that the values in most of South and Western Asia were very low in March 2021, at −3 (Fig. 1.9). The SPI values also indicated extreme drought in the regions of Algeria, Mauritania, Libya, Zambia, Mozambique, Kenya, Uganda, Ethiopia, Saudi Arabia, Iraq, Jordan, Iran, Pakistan, Oman, Nepal, Bangladesh, Cambodia, China, Hong Kong, Afghanistan, etc. In addition, the SPI was used for drought detection.

The standard vegetation index (SVI) was developed by Peters et al. (2002) and indicates the likelihood of deviation from the typical enhanced vegetation index (EVI) across a few years. The SVI results from June 2021 indicated that the northern, western, and southern regions of Asia were all severely affected by drought. However, in Africa, the level of greenness in the central region was higher than that in the northern and southern regions (Fig. 1.10).

Furthermore, the TCI is another key indicator for drought assessment in large areas that reflects the stress conditions of vegetation. The TCI was calculated from MODIS land surface temperature data from 2019 to 2021. The TCI results for

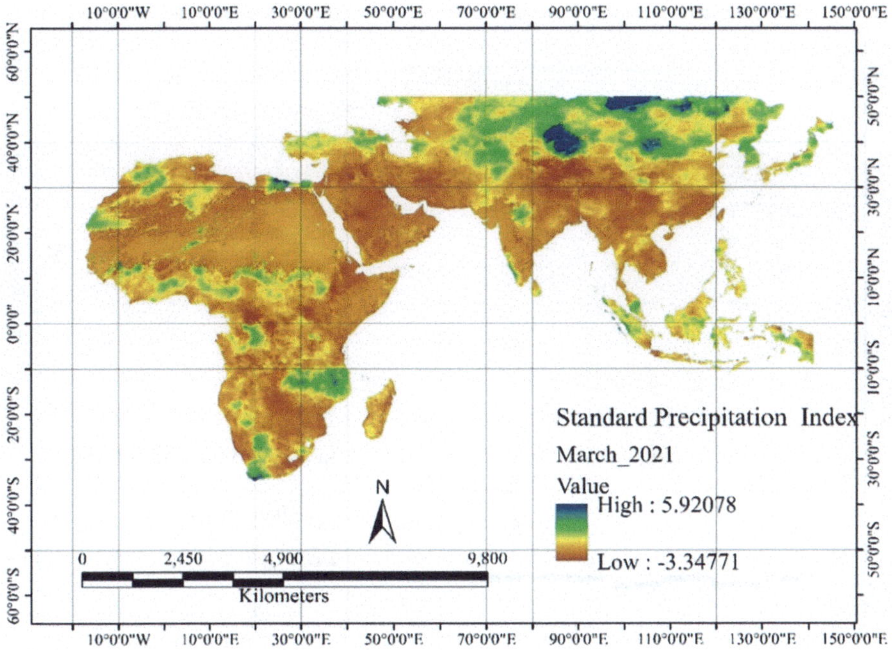


**Fig. 1.9** SPI-1 according to CHIRPS precipitation data (0.05°) across the Asian and African continents in March 2021

August 2021 indicated that Mauritania, Cote d'Ivoire, Ghana, Nigeria, Togo, Chad, Sudan, Somalia, Tanzania, Mozambique, South Africa, Libya, Tunisia, Egypt, Turkey, Syria, Iraq, Saudi Arabia, Iran, Baikonur, Tajikistan, Siachen Glacier, India, Nepal, China, Hong Kong, the Philippines, Malaysia, Indonesia, and Korea were affected by extreme drought (0–10%) (Fig. 1.11).

### *1.2.5 Impact of Climate Change on the Community*

The global climate change problem has become of increasing concern over the last few decades, and it is highly imperative to obtain suitable and sustainable solutions to mitigate the impact on communities. The effects of climate change are notable and common in many places worldwide. According to organizational and research data, the global temperature and rainfall are gradually changing, increasing the frequency of different climatic events, such as floods, droughts, and cyclones, which affect life, livelihood, subsistence, and culture (Barnett and Adger 2003). Developing countries are the most adversely affected by climate change, where most people depend on nature for agriculture. In addition to the main population, a large group of indigenous people living in

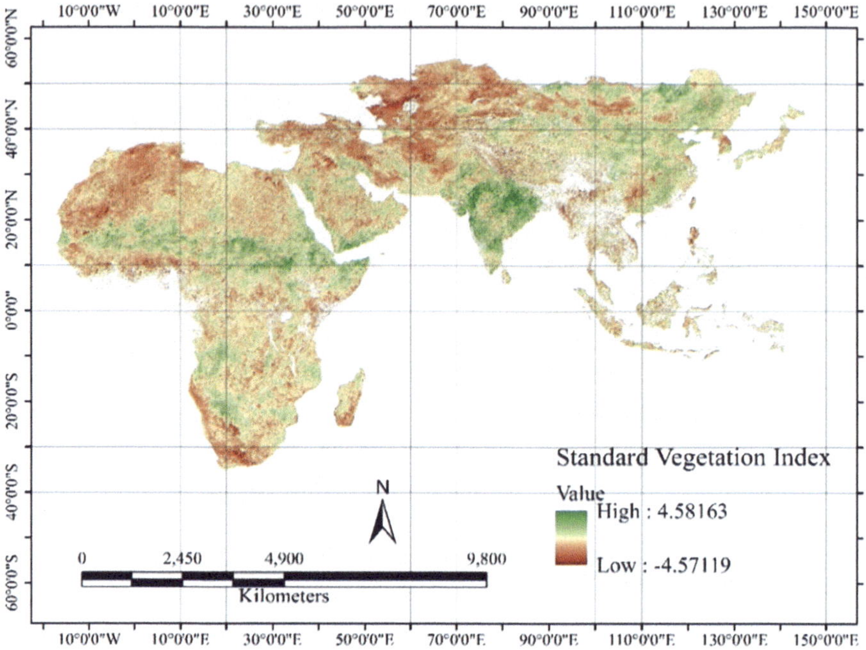


**Fig. 1.10** Standard vegetation index (SVI) for 2021 in June across Asia and Africa

remote areas such as hills and isolated islands in many parts of the world faces severe vulnerability to climate change (Bardsley 2018). The impact of climate change on the lives of indigenous people is considerable, as they live in relatively fragile environments and remote areas with poor communication systems. According to the indigenous peoples, marginalized populations and climate change (PMPCC) (2011) report, due to the continuing reliance of the livelihoods of indigenous people on natural resources, they are directly exposed to continuous climate change extremes.

According to the Asia Indigenous Peoples (2014) report, approximately 370 million indigenous people live in Asia, enriching the enormous cultural and linguistic diversity of the region. Among them, 84.32 million indigenous people of 622–635 ethnic groups live in India, 105 million people of 400 ethnic groups live in China, 1.7–3.7 million people of 45 ethnic groups live in Bangladesh, 14.4–19.2 million people of 135 ethnic groups live in Myanmar, 0.90 million people of 25 ethnic groups live in Thailand, 3.4 million people of 97 ethnic groups live in Malaysia, and 50–70 million people of 700 indigenous groups live in Indonesia. The estimated indigenous population encompassing several groups living in different countries of Asia is provided in Table 1.2.

Indigenous people are not only passive victims of climate change but also observers of environmental change and its impacts based on their accumulated knowledge and experience of nature (Habiba et al. 2013). Indigenous knowledge and experience of nature are integral parts of their way of life (Zhang and Nakagawa 2018).

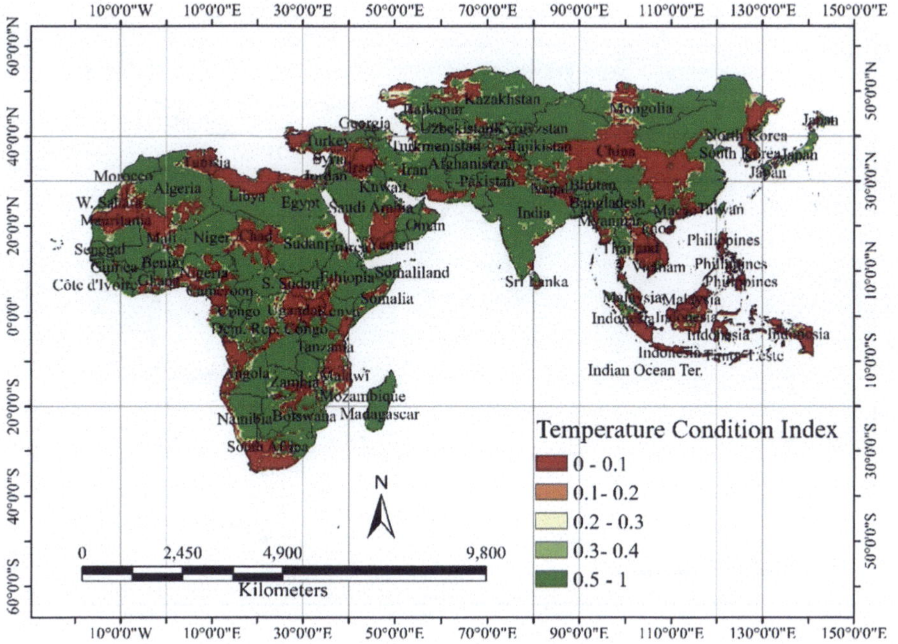


**Fig. 1.11** Vegetation stress generated from temperature condition index (TCI) data for August 2021 in Asia and Africa

**Table 1.2** Indigenous peoples living in Asia

| Country | Estimated indigenous population (million) | Number of indigenous groups |
|---|---|---|
| Pakistan | 18.9 | Unknown |
| Nepal | 10.6 | 80+ |
| India | 84.32 | 622–635 |
| Bangladesh | 1.7–3.7 | 45 |
| Bhutan | 0.10 | Unknown |
| China | 105 | 400+ |
| Myanmar | 14.4–192 | 135 |
| Thailand | 0.90 | 25+ |
| Vietnam | 10 | 90+ |
| Malaysia | 3.4 | 97 |
| Indonesia | 50–70 | 700+ |
| Philippines | 12–15 | 14+ |

Source: Pact, Asia Indigenous Peoples (2014)

Although indigenous people are not highly familiar with the adverse effects of climate change, they follow various coping strategies to overcome these impacts by implementing their vast practical knowledge acquired from nature (Dhungana et al.

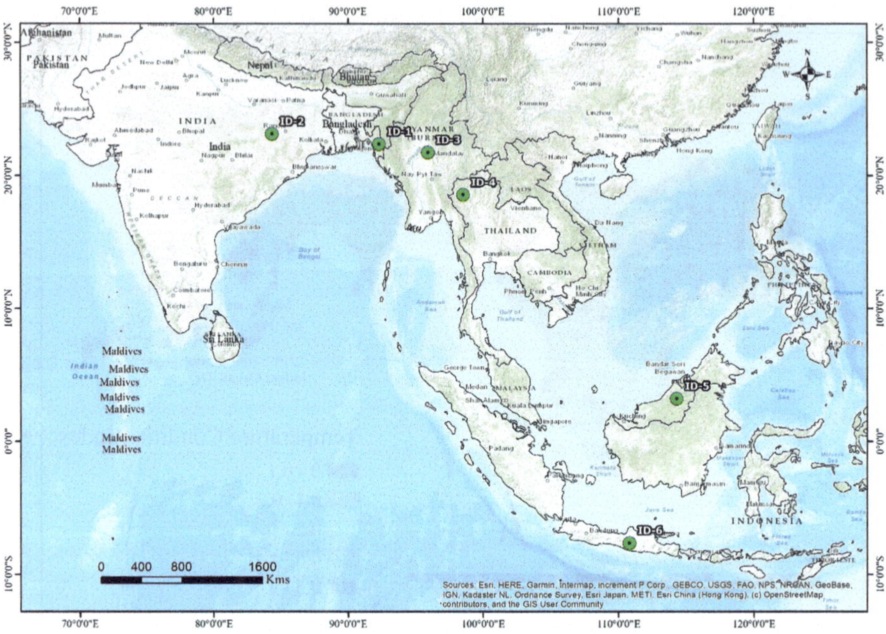


**Fig. 1.12** Living locations of the indigenous communities in India, Bangladesh, Myanmar, Thailand, Malaysia, and Indonesia

2023; Gómez-Baggethun 2022). Recently, many researchers have determined that the bottom-up approach of disaster management can be more effective than the top-down approach. Garai (2022) suggested that the cultural responses of indigenous people to adaptation to climate change could be helpful to the government and policymakers in policy formulation and implementation. Therefore, inventories of the spatial existence and knowledge of indigenous people worldwide over time are important (Fig. 1.12).

## 1.3 Conclusions

A significant long-term change in weather in a certain area is referred to as climate change. The global climate is connected by the Earth's planetary system with factorial components of the Sun, Earth, Moon, oceans, rain, wind, snow, deserts, forests, and everything in nature. Recently, global climate change has become of great concern among scientists and policymakers as the natural disasters associated with climate change (riverbank erosion, floods, cyclones, droughts, landslides, and seawater intrusions) adversely affect the lives and livelihood of people through property and agricultural crop damage. Therefore, occurrence frequency analysis, vulnerability evaluation, and impact assessment of climate extremes are necessary for proper planning to mitigate the adverse effects of global climate change. Remote sensing techniques using multispectral satellite imageries in many GIS environments and cloud computation platforms have been widely adopted by researchers for their easy access and faster computational capabilities. In this regard, in this chapter, a few approaches are elucidated to briefly introduce riverbank erosion vulnerability assessment, flood vulnerability evaluation, assessment of the damage caused by cyclones, and drought detection and intensity assessment covering Asia and Africa. The MCDM approach with a fuzzy expert system in the ArcGIS environment using satellite datasets is used for riverbank erosion and flood vulnerability analysis, and different indices are used in damage assessment and drought detection. Finally, a remote sensing approach involving inventories of the spatial existence and knowledge of indigenous people worldwide over time is proposed for analyzing the coping capabilities under global climate change. The approaches described in this chapter could facilitate further analysis and assessment of global climate change impacts.

## References

Ahamed T (2023) Special issue on the assessment of climate change impacts on regional economics in South Asia. Asia Pac J Reg Sci 7:323–328. https://doi.org/10.1007/s41685-023-00300-y

Alam KF, Ahamed T (2022) Assessment of land use land cover changes for predicting vulnerable agricultural lands in River basins of Bangladesh using remote sensing and a fuzzy expert system. Remote Sens 14(21):5582

Alam KF, Ahamed T (2023a) Climate-adaptive potential crops selection in vulnerable agricultural lands adjacent to the Jamuna River basin of Bangladesh using remote sensing and a fuzzy expert system. Remote Sens 15(8):2201. https://doi.org/10.3390/rs15082201

Alam KF, Ahamed T (2023b) Erosion vulnerable area assessment of Jamuna River system in Bangladesh using a multi-criteria-based geospatial fuzzy expert system and remote sensing. Asia Pac J Reg Sci 7:433. https://doi.org/10.1007/s41685-023-00292-9

Ali S, Khorrami B, Jehanzaib M, Tariq A, Ajmal M, Arshad A et al (2023) Spatial downscaling of GRACE data based on XGBoost model for improved understanding of hydrological droughts in the Indus basin irrigation system (IBIS). Remote Sens 15(4):873. https://doi.org/10.3390/rs15040873

Al-Maruf A, Jenkins JC, Bernzen A, Braun B (2021) Measuring household resilience to cyclone disasters in Coastal Bangladesh. Climate 9(6):97. https://doi.org/10.3390/cli9060097

An Z, Colman SM, Zhou W, Li X, Brown ET, Jull AJT, Cai Y, Huang Y, Lu X, Chang H, Song Y, Sun Y, Xu H, Liu W, Jin Z, Liu X, Cheng P, Liu Y, Ai L et al (2012) Interplay between the Westerlies and Asian monsoon recorded in Lake Qinghai sediments since 32 ka. Sci Rep 2:619. https://doi.org/10.1038/srep00619

Arab ST, Ahamed T (2023) Near-real-time drought monitoring and assessment for vineyard production on a regional scale with standard precipitation and vegetation indices using Landsat and CHIRPS datasets. Asia Pac J Reg Sci 7:591–614. https://doi.org/10.1007/s41685-023-00286-7

Arab ST, Islam MM, Shamsuzzoha M, Alam KF, Muhsin N, Noguchi R, Ahamed T (2022) A review of remote sensing applications in agriculture and forestry to establish big data analytics. In: Remote sensing application: regional perspectives in agriculture and forestry. pp 1–24. https://doi.org/10.1007/978-981-19-0213-0_1

Baninla Y, Sharifi A, Allam Z, Tume SJP, Gangtar NN, George N (2022) An overview of climate change adaptation and mitigation research in Africa. Front Clim 4(976427):1–24. https://doi.org/10.3389/fclim.2022.976427

Bardsley DK (2018) Indigenous knowledge and practice for climate change adaptation. In: Encyclopedia of the anthropocene. Elsevier, pp 359–367. https://doi.org/10.1016/B978-0-12-809665-9.09797-4
Barnett J, Adger WN (2003) Climate dangers and atoll countries. Clim Chang 61(3):321–337
Braun B, Aßheuer T (2011) Floods in megacity environments: vulnerability and coping strategies of slum dwellers in Dhaka/Bangladesh. Nat Hazards 58(2):771–787. https://doi.org/10.1007/s11069-011-9752-5
Chen X, Avtar R, Umarhadi DA, Louw AS, Shrivastava S, Yunus AP, Khedher KM, Takemi T, Shibata H (2022) Post-typhoon forest damage estimation using multiple vegetation indices and machine learning models. Weather Clim Extrem 38:100494. https://doi.org/10.1016/j.wace.2022.100494
Conte N, Kostandi C (2022) Mapped: countries with the highest flood risk. Visual Capitalist
Dai A (2011) Drought under global warming: a review. Wiley Interdiscip Rev Clim Chang 2(1):45–65. https://doi.org/10.1002/wcc.81
Deb M, Ferreira C (2015) Planform channel dynamics and bank migration hazard assessment of a highly sinuous river in the north-eastern zone of Bangladesh. Environ Earth Sci 73:6613–6623
Dhungana G, Pal I, Ghimire R, Dhungana RK, Tuladhar N (2023) Foundation of indigenous knowledge theory for disaster risk reduction. In: Multi-hazard vulnerability and resilience building: cross cutting issues. Elsevier, pp 347–361. https://doi.org/10.1016/B978-0-323-95682-6.00016-4
Di L, Yu E, Shrestha R, Lin L (2018) DVDI: a new remotely sensed index for measuring vegetation damage caused by natural disasters. In: IGARSS 2018—2018 IEEE international geoscience and remote sensing symposium, pp 9067–9069. https://doi.org/10.1109/IGARSS.2018.8518022
FAO (2021) The impact of disasters and crises 2021 on agriculture and food security: 2021, Rome
Ferretti V, Pomarico S (2013) An integrated approach for studying the land suitability for ecological corridors through spatial multicriteria evaluations. Environ Dev Sustain 15:859–885
Gichure RW (2017) Effects of drought on crop production and coping mechanisms undertaken by small scale farmers: a case of Makueni County, Kenya. Doctoral dissertation, University of Nairobi
Garai J, Ku HB, Zhan Y (2022) Climate change and cultural responses of indigenous people: A case from Bangladesh. Current Research in Environmental Sustainability 4:100130
Gómez-Baggethun E (2022) Is there a future for indigenous and local knowledge? J Peasant Stud 49(6):1139–1157. https://doi.org/10.1080/03066150.2021.1926994
Habiba U, Shaw R, Abedin Md A (2013) Community-based disaster risk reduction approaches in Bangladesh. pp 259–279. https://doi.org/10.1007/978-4-431-54252-0_12
Hakam O, Baali A, Azennoud K, Lyazidi A, Bourchachen M (2023) Assessments of drought effects on plant production using satellite remote sensing technology, GIS and observed climate data in Northwest Morocco, case of the Lower Sebou Basin. Int J Plant Prod 17:267–282. https://doi.org/10.1007/s42106-023-00236-5
Haq MZ, Robbani M, Ali M, Mainul Hasan M, Mahmudul Hasan M, Uddin MJ, Begum M, da Silva JAT, Pan XY, Karim MR (2012) Damage and management of cyclone Sidr-affected homestead tree plantations: a case study from Patuakhali, Bangladesh. Nat Hazards 64(2):1305–1322. https://doi.org/10.1007/s11069-012-0299-x
He X, Estes L, Konar M, Tian D, Anghileri D, Baylis K et al (2019) Integrated approaches to understanding and reducing drought impact on food security across scales. Curr Opin Environ Sustain 40:43–54
Hongo C, Tsuzawa T, Tokui K, Tamura E (2015) Development of damage assessment method of rice crop for agricultural insurance using satellite data. J Agric Sci 7(12):59. https://doi.org/10.5539/jas.v7n12p59
Hooke JM (1979) An analysis of the processes of river bank erosion. J Hydrol 42(1–2):39–62
Hossain MS, Chowdhury SR, Das NG, Sharifuzzaman SM, Sultana A (2009) Integration of GIS and multicriteria decision analysis for urban aquaculture development in Bangladesh. Landsc Urban Plan 90(3–4):119–133

Houston JB, Hawthorne J, Perreault MF, Park EH, Goldstein Hode M, Halliwell MR, Turner McGowen SE, Davis R, Vaid S, McElderry JA, Griffith SA (2015) Social media and disasters: a functional framework for social media use in disaster planning, response, and research. Disasters 39(1):1–22. https://doi.org/10.1111/disa.12092

Islam SN (2016) Deltaic floodplains development and wetland ecosystems management in the Ganges–Brahmaputra–Meghna Rivers Delta in Bangladesh. Sustain Water Resour Manag 2(3):237–256. https://doi.org/10.1007/s40899-016-0047-6

Islam MM, Matsushita S, Noguchi R, Ahamed T (2022a) A damage-based crop insurance system for flash flooding: a satellite remote sensing and econometric approach. Asia Pac J Reg Sci 6(1):47–89. https://doi.org/10.1007/s41685-021-00220-9

Islam MS, Samreth S, Islam AHMS, Sato M (2022b) Climate change, climatic extremes, and households' food consumption in Bangladesh: a longitudinal data analysis. Environ Chall 7:100495. https://doi.org/10.1016/j.envc.2022.100495

James RD (2019) Civil works budget of the U.S. Army Corps of Engineers

Karim M, Mimura N (2008) Impacts of climate change and sea-level rise on cyclonic storm surge floods in Bangladesh. Glob Environ Chang 18(3):490–500. https://doi.org/10.1016/j.gloenvcha.2008.05.002

Klotzbach PJ, Bowen SG, Pielke R, Bell M (2018) Continental U.S. Hurricane landfall frequency and associated damage: observations and future risks. Bull Am Meteorol Soc 99(7):1359–1376. https://doi.org/10.1175/BAMS-D-17-0184.1

Klotzbach PJ, Wood KM, Schreck CJ, Bowen SG, Patricola CM, Bell MM (2022) Trends in global tropical cyclone activity: 1990–2021. Geophys Res Lett 49(6):e2021GL095774. https://doi.org/10.1029/2021GL095774

Knutson TR, McBride JL, Chan J, Emanuel K, Holland G, Landsea C, Held I, Kossin JP, Srivastava AK, Sugi M (2010) Tropical cyclones and climate change. Nat Geosci 3(3):157–163. https://doi.org/10.1038/ngeo779

Knutson T, Camargo SJ, Chan JCL, Emanuel K, Ho C-H, Kossin J, Mohapatra M, Satoh M, Sugi M, Walsh K, Wu L (2019) Tropical cyclones and climate change assessment: part I: detection and attribution. Bull Am Meteorol Soc 100(10):1987–2007. https://doi.org/10.1175/BAMS-D-18-0189.1

Knutson T, Camargo SJ, Chan JCL, Emanuel K, Ho C-H, Kossin J, Mohapatra M, Satoh M, Sugi M, Walsh K, Wu L (2020) Tropical cyclones and climate change assessment: part II: projected response to anthropogenic warming. Bull Am Meteorol Soc 101(3):E303–E322. https://doi.org/10.1175/BAMS-D-18-0194.1

Kossin JP, Knapp KR, Olander TL, Velden CS (2020) Global increase in major tropical cyclone exceedance probability over the past four decades. Proc Natl Acad Sci 117(22):11975–11980. https://doi.org/10.1073/pnas.1920849117

Kulatunga U, Wedawatta G, Amaratunga D, Haigh R (2014) Evaluation of vulnerability factors for cyclones: the case of Patuakhali, Bangladesh. Int J Disaster Risk Reduct 9:204–211. https://doi.org/10.1016/j.ijdrr.2014.05.011

Masih I, Maskey S, Mussá FEF, Trambauer P (2014) A review of droughts on the African continent: a geospatial and long-term perspective. Hydrol Earth Syst Sci 18(9):3635–3649. https://doi.org/10.5194/hess-18-3635-2014

McKee TB, Doesken NJ, Kleist J (1993) Analysis of Standardized Precipitation Index (SPI) data for drought assessment. Water (Switzerland) 26(2):1–72

Mendelsohn R, Emanuel K, Chonabayashi S, Bakkensen L (2012) The impact of climate change on global tropical cyclone damage. Nat Clim Chang 2(3):205–209. https://doi.org/10.1038/nclimate1357

Mutton D, Haque CE (2004) Human vulnerability, dislocation and resettlement: adaptation processes of river-bank erosion-induced displaces in Bangladesh. Disasters 28(1):41–62

Nadiruzzaman M, Wrathall D (2015) Participatory exclusion—cyclone Sidr and its aftermath. Geoforum 64:196–204. https://doi.org/10.1016/j.geoforum.2015.06.026

National Oceanic and Atmospheric Administration (2023, April 5) International-best-track-archive-for-climate-stewardship-ibtracs. https://www.ncei.noaa.gov/data/international-best-track-archive-for-climate-stewardship-ibtracs/v04r00/access/shapefile/
Needs Assessment Working Group (2019) Bangladesh Cyclone Bulbul 2019 joint rapid assessment. https://reliefweb.int/report/bangladesh/cyclone-bulbul-2019-joint-rapid-assessment-needs-assessment-working-group-nawg-bangladesh-16-november-2019
Orimoloye IR (2022) Agricultural drought and its potential impacts: enabling decision-support for food security in vulnerable regions. Front Sustain Food Syst 6:15
Pact, Asia Indigenous Peoples (2014) Overview of the state of Indigenous peoples in Asia
Pahlowan EU, Hossain ATMS (2015) Jamuna River erosional hazards, accretion & annual water discharge—a remote sensing & GIS approach. Int Arch Photogramm Remote Sens Spat Inf Sci 40:831–835
Pei W, Fu Q, Liu D, Li T, Cheng K, Cui S (2019) A novel method for agricultural drought risk assessment. Water Resour Manag 33:2033–2047
Peters AJ, Walter-Shea EA, Ji L, Vina A, Hayes M, Svoboda MD (2002) Drought monitoring with NDVI-based standardized vegetation index. Photogramm Eng Remote Sens 68(1):71–75
Rahaman MA, Mursheduzzaman, Ali Reza GAM, Chowdhury AM, Avi AR, Chakraborty TR, Shamsuzzoha M (2020) Nature-based solutions to promote climate change adaptation and disaster risk reduction along the Coastal Belt of Bangladesh. In: The Palgrave handbook of climate resilient societies. https://doi.org/10.1007/978-3-030-32811-5_49-1
Rentschler JE, Salha M (2020) 1.47 billion people face flood risk worldwide: for over a third, it could be devastating. World Bank Blogs
Rentschler J, Salhab M, Jafino BA (2022) Flood exposure and poverty in 188 countries. Nat Commun 13(1):3527
Rojas O, Vrieling A, Rembold F (2011) Assessing drought probability for agricultural areas in Africa with coarse resolution remote sensing imagery. Remote Sens Environ 115(2):343–352. https://doi.org/10.1016/j.rse.2010.09.006
Shamsuzzoha M, Ahamed T (2023) Shoreline change assessment in the Coastal Region of Bangladesh delta using tasseled cap transformation from satellite remote sensing dataset. Remote Sens 15(2):295. https://doi.org/10.3390/rs15020295
Shamsuzzoha M, Noguchi R, Ahamed T (2021) Damaged area assessment of cultivated agricultural lands affected by cyclone bulbul in coastal region of Bangladesh using Landsat 8 OLI and TIRS datasets. Remote Sens Appl Soc Environ 23:100523. https://doi.org/10.1016/j.rsase.2021.100523
Shamsuzzoha M, Noguchi R, Ahamed T (2022) Rice yield loss area assessment from satellite-derived NDVI after extreme climatic events using a fuzzy approach. Agric Inf Res 31(1):32–46. https://doi.org/10.3173/air.31.32
Shaw R, Islam A, Mallick F (2013a) Disaster, risk and evolution of the concept. pp 3–21. https://doi.org/10.1007/978-4-431-54252-0_1
Shaw R, Islam A, Mallick F (2013b) National perspectives of disaster risk reduction in Bangladesh. pp 45–62. https://doi.org/10.1007/978-4-431-54252-0_3
Talukder MF, Shamsuzzoha M, Hasan I (2018) Damage and agricultural rehabilitation scenario of post cyclone Mahasen in Coastal Zone of Bangladesh. J Soc Anthropol 2(1):36–43. https://doi.org/10.12691/jsa-2-1-7
Tinh NX, Tanaka H, Abe G, Okamoto Y, Pakoksung K (2021) Mechanisms of flood-induced levee breaching in Marumori town during the 2019 Hagibis typhoon. Water (Switzerland) 13(2):244. https://doi.org/10.3390/w13020244
World Meteorological Organization (1992) International meteorological vocabulary. 2nd ed. WMO Rep 182, 784 p
Zeng Z, Wu W, Li Z, Zhou Y, Guo Y, Huang H (2019) Agricultural drought risk assessment in Southwest China. Water 11(5):1064

Zhang H, Nakagawa H (2018) Validation of indigenous knowledge for disaster resilience against river flooding and bank erosion. In: Science and technology in disaster risk reduction in Asia. Elsevier, pp 57–76. https://doi.org/10.1016/B978-0-12-812711-7.00005-5

# Chapter 2
# An Assessment of Drought Stress in Tea Plantation Areas in Bangladesh Using Optical and Thermal Remote Sensing: A Climate Change Perspective

**Animesh Chandra Das, Ryozo Noguchi, and Tofael Ahamed**

**Abstract** Drought is one of the deleterious climatic events that affects the productivity and quality of tea by limiting the growth and development of the plants. The aim of this research was to determine drought severity in tea plantation areas using a remote sensing technique with the standardized precipitation index (SPI). Landsat 8 OLI/TIRS (Operational Land Imager/Thermal Infrared Sensor) image data were processed to measure the land surface temperature (LST) and soil moisture index (SMI). Maps for the normalized difference moisture index (NDMI), normalized difference vegetation index (NDVI), leaf area index (LAI), and yield maps were developed from Sentinel-2 satellite image data. The drought frequency was measured from the classification of droughts employing the SPI. The results of this study demonstrate that the drought frequency for the Sylhet meteorological station was 38.46% for near-normal, 35.90% for normal, and 25.64% for moderately dry months. In contrast, the Sreemangal meteorological station showed frequencies of 28.21%, 41.02%, and 30.77% for near-normal, normal, and moderately dry months, respectively. The correlation coefficients between the SMI and NDMI were 0.84, 0.77, and 0.79 for the drought periods of 2018–2019, 2019–2020, and 2020–2021, respectively, indicating a strong relationship between soil and plant canopy moisture. The yield prediction results with respect to drought stress in tea plantation areas show that 61%, 60%, and 60% of tea estates in the study area had lower predicted yields than the actual yield during the drought period, which accounted for 7.72%, 11.92%, and 12.52% yield losses in 2018, 2019, and 2020, respectively. The outcome of this research suggests that satellite remote sensing with the SPI could be

A. C. Das
Faculty of Agriculture, Sylhet Agricultural University, Sylhet, Bangladesh

R. Noguchi
Graduate School of Agriculture, Kyoto University, Kyoto, Japan

T. Ahamed (✉)
Institute of Life and Environmental Sciences, University of Tsukuba, Tsukuba, Ibaraki, Japan
e-mail: tofael.ahamed.gp@u.tsukuba.ac.jp



T. Ahamed (ed.), *Remote Sensing Application II*, New Frontiers in Regional Science: Asian Perspectives 77, https://doi.org/10.1007/978-981-97-1188-8_2

a valuable tool for land policy makers, land use planners, and scientists to measure drought stress in tea estates.

**Keywords** Drought · Tea plantation areas · SPI · Remote sensing · Yield losses

## 2.1 Introduction

Drought is one of the most detrimental climatic events with regard to the changing climate that influences the growth and development of plants and consequently affects sustainable tea production (Nalina et al. 2018; Ahmed et al. 2016). Drought influences plant physiological activities more than other environmental factors, changing cellular mechanisms and ultimately affecting productivity of tea (Anjum et al. 2011; Shao et al. 2009). The duration and severity of drought affect the production and quality of tea, and the crop yield losses caused by drought stress are becoming unpredictable because of climate change and increasing water stress in tea plantations (Zhou et al. 2014). Tea (*Camellia sinensis* (L) O. Kuntze) is a perennial evergreen shrub cultivated in a variety of edaphic and climatic conditions across the tropics, subtropics, and Mediterranean regions (Nalina et al. 2021; Hajiboland 2017). Both the yield and quality of tea depend on various factors, such as cultivar, cultural practices, and plant growth stages, along with season and climate. Tea is a rainfed crop, and its productivity depends mainly on climatic conditions for its optimal growth. Therefore, changes in climate, especially a reduction in rainfall, cause water stress that affects the yield and quality of tea (Wijeratne 1996). Bangladesh is one of the leading tea-producing countries, ranking ninth and producing 83 million kg of tea (International Tea Committee 2017). The tea industry of Bangladesh annually earns 20.90 million USD (0.81% of GDP), exporting approximately 18 million kilograms of tea, which is accounted for 1.37% of the export of the global tea trade (Kamruzzaman et al. 2015). One of the important reasons for the lower yield of tea is adverse climatic conditions such as drought. The tea estates of Bangladesh mainly experience drought stress during the winter months from December to February (Das et al. 2020). It has been reported that drought stress causes yield losses in tea of 14–33%, and the rate of mortality, especially in young plants, ranges between 6 and 19% (Nalina et al. 2021; Guo et al. 2017). Therefore, it is necessary to measure the drought severity in tea plantation areas to take further initiatives to mitigate its adverse effects.

Droughts are predominantly classified into four types: meteorological drought denotes a deficit in precipitation; agricultural drought indicates a deficit in soil moisture; hydrological drought implies a reduction in surface runoff, groundwater, and water storage; and socioeconomic drought involves supply and demand of economic goods with the elements of the other types of droughts and social responses as well. However, all types of droughts are associated with sustained precipitation deficits (AghaKouchak et al. 2015; Wilhite 2005). Several ground-based and remote sensing indices with their advantages and limitations for drought monitoring and assessment are being used. Ground-based indices incorporate information on precipitation, soil moisture status, and water supply rather than emphasizing many spatial details. Furthermore, these drought indices are measured for dispersed locations, and thus the limitation of these indices is that they are dependent on data collected from weather stations, resulting in sparsely distributed drought indicators, which influences the authenticity of the drought indicators (Brown et al. 2002). In contrast, remote sensing drought indices are measured using the surface reflectance of different bands derived from satellite data over the entire area and have been widely used for drought stress detection (Murad and Islam 2011).

Historical drought indicators are used to represent droughts in terms of duration, magnitude, and areal extent (Mishra and Singh 2010). Several drought indicators, such as the standardized precipitation index (SPI), Palmer drought severity index (PDSI), standardized precipitation evapotranspiration index (SPEI), and effective drought index (EDI), have been derived and proposed to quantify the pattern and magnitude of droughts. Among these indicators, the SPI is the most widely used index with the advantages of its simplicity in use, and spatial consistency in interpretation that can be used in drought risk assessment and decision-making. Thus, the SPI can be used efficiently for different time periods, from 1 month to several years according to the user's interest (Kamruzzaman et al. 2019; Guttman 1998).

With advances in remote sensing technology, the use of historical drought indicators has been strengthened by newly developed remote sensing indices with the advantage of real-time monitoring. Thus, the remote sensing technique is increasingly being considered a robust tool for drought monitoring. From a climatological perspective, monitoring precipitation, soil moisture, ground and terrestrial water storage, evapotranspiration, and snow are considered for both ground- and satellite-based drought monitoring; while from an ecological perspective, magnitude of drought can be assessed by monitoring vegetation health condition and land cover changes using remote sensing data (AghaKouchak et al. 2015; Nemani et al. 2009). Different satellite-based drought indices, such as the normalized difference vegetation index (NDVI), soil-adjusted vegetation index (SAVI), enhanced vegetation index (EVI), vegetation condition index (VCI), temperature condition index (TCI), perpendicular drought index (PDI), normalized difference moisture index (NDMI), modified perpendicular drought index (MPDI), distance drought index (DDI), and soil moisture index (SMI), are widely used indicators for drought monitoring (AghaKouchak et al. 2015). According to recent studies on drought monitoring, it is evident that vegetation moisture indicators such as NDMI are more powerful than other vegetation indices, as they quantify the water content in the spongy mesophyll tissues of the plant canopies in higher biomass ecosystems, which is directly related to the soil moisture status in the vegetated lands (Caccamo et al. 2011; Van Niel et al. 2003). On the other hand, drought stress can be measured using surface brightness temperature obtained from thermal bands of different satellite instruments such as MODerate resolution Imaging Spectroradiometer (MODIS), Advanced Very-High-Resolution Radiometer (AVHRR), Visible Infrared Imaging Radiometer Suite (VIIRS), Thematic Mapper (TM), Enhanced Thematic Mapper Plus (ETM+), and Thermal Infrared Sensor (TIRS). Land surface temperature (LST) measured from thermal infrared (TIR) channels has been observed to be a powerful indicator for providing valuable information on land surface moisture, which is an inevitable tool to measure the magnitude of drought (Gutman 1990).

Several studies have been performed for assessing and monitoring drought using remote sensing tools. A study on NDVI-based agricultural drought assessment was conducted to measure drought severity in the Palar Basin in Tamil Nadu, India, using IRS (Indian Remote Sensing Satellites)-1C and 1D datasets (Krishna et al. 2009). Another study was performed in the western tracts of Tamil Nadu using satellite datasets of the National Oceanic and Atmospheric Administration (NOAA) (Muthumanickam et al. 2011). Drought stress assessment was performed for hickory plantations in western Zhejiang, China, using a change detection approach from multitemporal Landsat 8 imagery (Xi et al. 2016). A study was performed to investigate vegetation droughts in natural forests, rubber, and oil palm plantations in peninsular Malaysia using MODIS datasets (Razali et al. 2016). Another study was carried out in South Africa to measure the effects of drought on forest plantations using MODIS time-series and meteorological data, suggesting that the normalized difference moisture index (NDMI) is a robust indicator to understand the interactions among precipitation, soil moisture, and moisture content in the plant biomass (Xulu et al. 2018). A comprehensive study on the effects of drought and other disturbances on forest landscapes in North Carolina, USA, was performed using MODIS and Landsat ET (Evapotranspiration) product time series datasets (Yang et al. 2020). In addition, several studies have been conducted for small-scale drought monitoring in tea plantations using wireless sensor networks (WSNs) (Gupta et al. 2014; Jiang et al. 2011; Sun et al. 2010). The WSN-based drought stress assessment system is not suitable for large and remote areas in many developing countries, such as Bangladesh, due to the disruption in power supply and Internet networks, which are the major impediments in the sensor network system. Those studies provide information regarding the research gap in measuring drought severity for tea estates to a greater extent using satellite remote sensing datasets.

To the best of our knowledge, no studies regarding drought stress assessment for tea plantation areas on a large scale have been performed to identify drought severity in tea estates using satellite remote sensing datasets. Furthermore, no standard drought index method has been developed for assessing drought severity. Therefore, a comprehensive study utilizing weather station-based precipitation datasets and multisatellite remote sensing datasets to develop a drought classification system was undertaken. The aim of the present study was to characterize the spatiotemporal pattern of drought in tea plantations and to develop a drought classification system for tea plantation areas to ensure early warning for risk mitigation and sustainable tea production in Bangladesh.

## 2.2 Materials and Methods

Precipitation data collected from the Bangladesh Meteorological Department (BMD) were utilized to calculate the SPI and drought frequency. Both Landsat 8 OLI/TIRS (Operational Land Imager/Thermal Infrared Sensor) and Sentinel-2 multispectral instrument (MSI) satellite imagery were used as remote sensing datasets. Ground reference geographic positioning system (GPS) information was gathered during the field survey to extract the points over tea plantation areas to quantify the LST, NDMI, NDVI, and leaf area index (LAI) values.

### 2.2.1 Study Area

The study area is considered to lie within the tea-growing zone between latitudes 23°58′40″ and 25°11′25″ North and longitudes 91°16′09″ and 92°30′17″ East and is located in the northeastern part of Bangladesh and consists of 21 subdistricts in the Sylhet Division (Fig. 2.1).

The population size of this locality is approximately ten million, which is less than 7% of the total population of Bangladesh. The study area lies between the latitudes of 23°58′ and 25°12′ North and the longitudes of 90°56′ and 92°30′ East. The area is surrounded by the Indian states of Meghalaya, Assam, and Tripura to the north, east and south, respectively, and divisions of Chattagram to the southwest and Dhaka and Mymensingh to the west. The land within the study area is 1,229,840 hectares, and the elevation is less than 335 m. The study area receives an adequate amount of rainfall in the monsoon season that is favorable for tea cultivation.

The subdistricts include Moulvibazar, Kamalganj, Kulaura, Rajnagar, Sreemangal, Barlekha, and Juri in the Moulvibazar district; Habiganj Bahubal, Chunarughat, Madhabpur, and Nabiganj in the Habiganj district; and Sylhet, Balaganj, Beanibazar, Fenchuganj, Golapganj, Gowainghat, Jaintiapur, Kanaighat, and Zakiganj in the Sylhet district. The area is bordered by the Indian provinces of Meghalaya, Assam, and Tripura to the north, east, and south, respectively, and subdistricts of Companiganj, Chatak, Jagannathpur, Baniachung, and Lakhai to the west and the Division of Chattogram to the southwest. There are 135 tea estates in the study area with 98,413 hectares of land, contributing approximately 85% of the total tea production of the country (63.42 million kg, 2018) (Bangladesh Tea Board 2020). This area is characterized by a sufficient amount of rainfall during the monsoon season that is favorable for tea production but experiences drought stress during the winter season.

### 2.2.2 Precipitation Data and the SPI Calculation

The daily precipitation data for the local weather stations, Sylhet and Sreemangal, were collected from January 2018 to March 2021 to prepare the monthly precipitation sums, which represent the rainfall scenario of the study area. Drought monitoring was performed employing the standardized precipitation index (SPI). Positive SPI values indicate wet conditions, and negative values denote drought periods with less precipitation than normal (Habibie et al. 2020). The precipitation data gathered

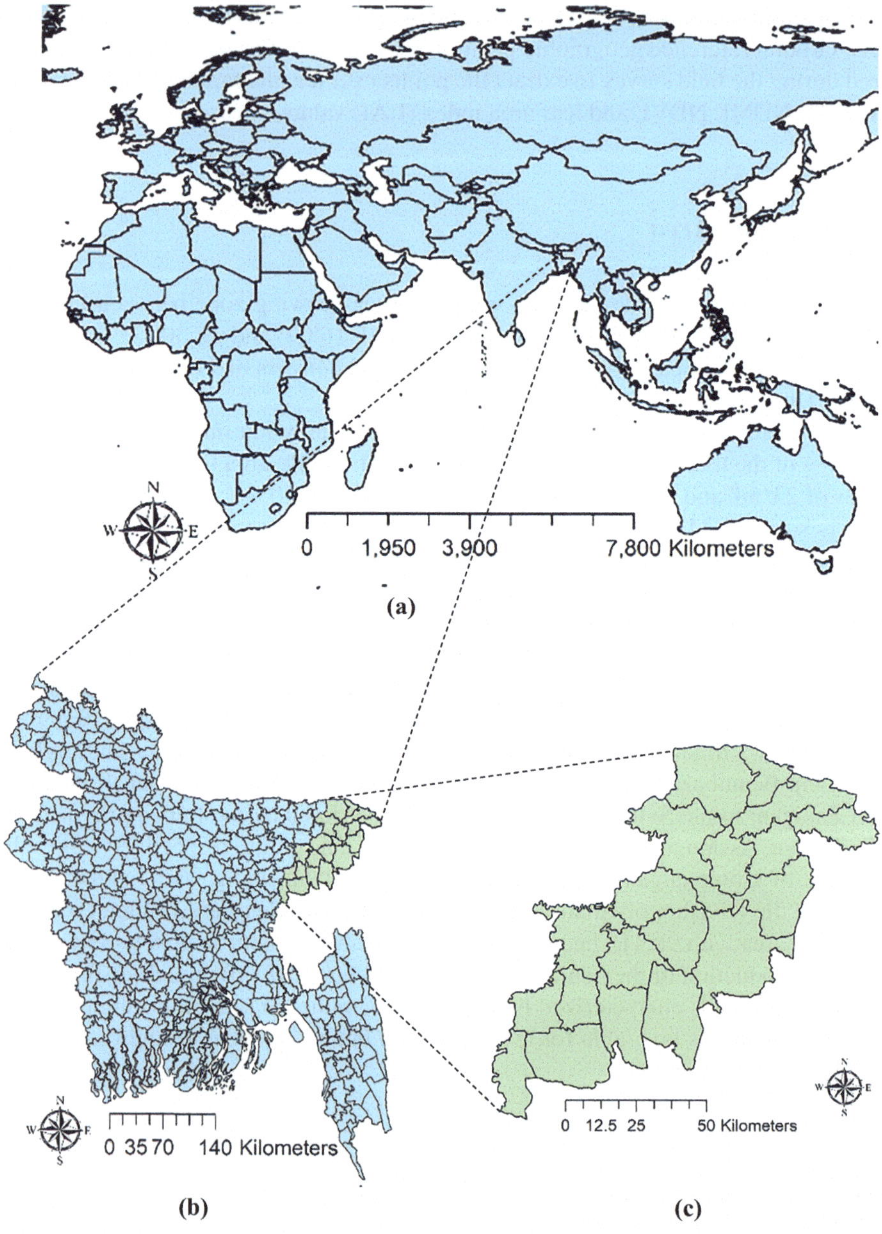


**Fig. 2.1** Geographic location of the study area: (**a**) Bangladesh on the world map, (**b**) Bangladesh, and (**c**) tea cultivation area in the Sylhet Division

from the local weather stations were utilized to classify wet and dry conditions (Paulo et al. 2012). Drought characterization was performed using the SPI to measure the severity, duration, and monthly coverage of drought of each category (Paulo and Pereira 2006). The accumulated SPI values were utilized to classify the drought category. The SPI value, $\leq -1$, refers to SPI-1, SPI-3, SPI-6, and SPI-12. Thus, the negative SPI values represent the corresponding drought severity for quantitative assessment of drought (World Meteorological Organization 2012; Peterson and Vose 1997; McKee et al. 1993). The SPI-1 computation was performed for 2018–2019, 2019–2020, and 2020–2021 using monthly precipitation data to investigate future drought events. Drought classification was performed according to Razali et al. (2016).

The SPI in a cumulative distribution is expressed using the gamma function. The gamma distribution is used to define the probability density or the function of frequency (McKee et al. 1993). The gamma distribution can be represented as follows:

$$G(x_i) = \int_{x_i}^{0} g(x_i)\,dx_i = \frac{1}{\beta^{\alpha}\Gamma(\alpha)} \int_{x_i}^{0} t^{\alpha-1} e^{-x_i/\beta} dx_i \tag{2.1}$$

Here, $\alpha > 0$ and $\beta > 0$ refer to the shape parameter and scale parameter, respectively (Alam et al. 2013).

$x$ denotes the amount of precipitation in millimeters in consecutive months $i$, and $\Gamma(\alpha)$ is the gamma function. When $x_i = 0$, the cumulative gamma distribution is unknown, and the cumulative probability is encountered. The SPI estimation also involves a matching density function of the gamma likelihood to the frequency distribution of precipitation:

$$\alpha = \frac{1}{4A}\left(1+\sqrt{1+\frac{4A}{3}}\right) \tag{2.2}$$

$$\beta = \frac{x}{\alpha} \tag{2.3}$$

$$A = \ln(x) - \frac{\sum \ln(x)}{n} \tag{2.4}$$

For an undefined gamma function, when $x_i = 0$, the value of G $(x_i)$ is expressed as follows:

$$H(x_i) = q + (1-q).G(x_i) \tag{2.5}$$

where $q$ is the probability of zero. Then, the cumulative probability $H(x_i)$ is transformed to the standard normal distribution for the SPI value (McKee et al. 1993). To calculate the SPI, the cumulative probability distribution is transformed into a normal distribution that can be expressed as follows:

$$z = \text{SPI} = -\left(t - \frac{c_0 + c_1 t + c_2 t^2}{1 + d_1 t + d_2 t^2 + d_3 t^3}\right), t = \sqrt{\ln\left(\frac{1}{(H(x))^2}\right)} \tag{2.6}$$

When $0 < H(x) < 0.5$, the following expression is true:

$$z = \text{SPI} = +\left(t - \frac{c_0 + c_1 t + c_2 t^2}{1 + d_1 t + d_2 t^2 + d_3 t^3}\right), t = \sqrt{\ln\left(\frac{1}{1-(H(x))^2}\right)} \tag{2.7}$$

The drought frequency was calculated according to the following formula:

$$\text{Drought frequency}(\%) = \frac{\text{Number of months affected by drought}}{\text{Total number of months}} \times 100 \tag{2.8}$$

### 2.2.3 Field Survey and Ground Reference Information

Ground reference information was collected during the 2019 field survey. GPS locations of tea estates were recorded around the study area using a handheld GPS locator (eTrex, Garmin, USA). Furthermore, this GPS information was used to extract point values from the remote sensing data employing the Spatial Analyst Tools in ArcGIS (Fig. 2.2). According to the field survey, among the 135 tea estates in the study area, 91 were located in Moulvibazar, 25 in Habiganj, and 19 in Sylhet District (Fig. 2.3).

### 2.2.4 Remote Sensing Data

Multisatellite image data from the Landsat 8 OLI/TIRS and Sentinel-2 MSI were collected for drought estimation from December to February for the years 2018–2019, 2019–2020, and 2020–2021. Landsat 8 is equipped with the OLI (Operational Land Imager) and TIRS (Thermal Infrared Sensor) instruments. The OLI sensor provides image data in the coastal aerosol, visible (blue, green and red), near-infrared (NIR), and shortwave infrared (SWIR) spectra with a spatial resolution of 30 m and in a panchromatic band with a spatial resolution of 15 m. The TIRS sensor provides image data in two thermal infrared bands with a spatial resolution of 100 m. The TIRS dataset, registered to the OLI data, is geometrically, radiometrically, and terrain-corrected 12-bit product. Landsat 8 C1 level-1 images with path 136 and row 43 were processed to develop LST and SMI maps. The collected Landsat 8 image data were processed at a resolution of 30 m using the resampling technique in ArcGIS followed by subsequent mosaicking and masking. All selected

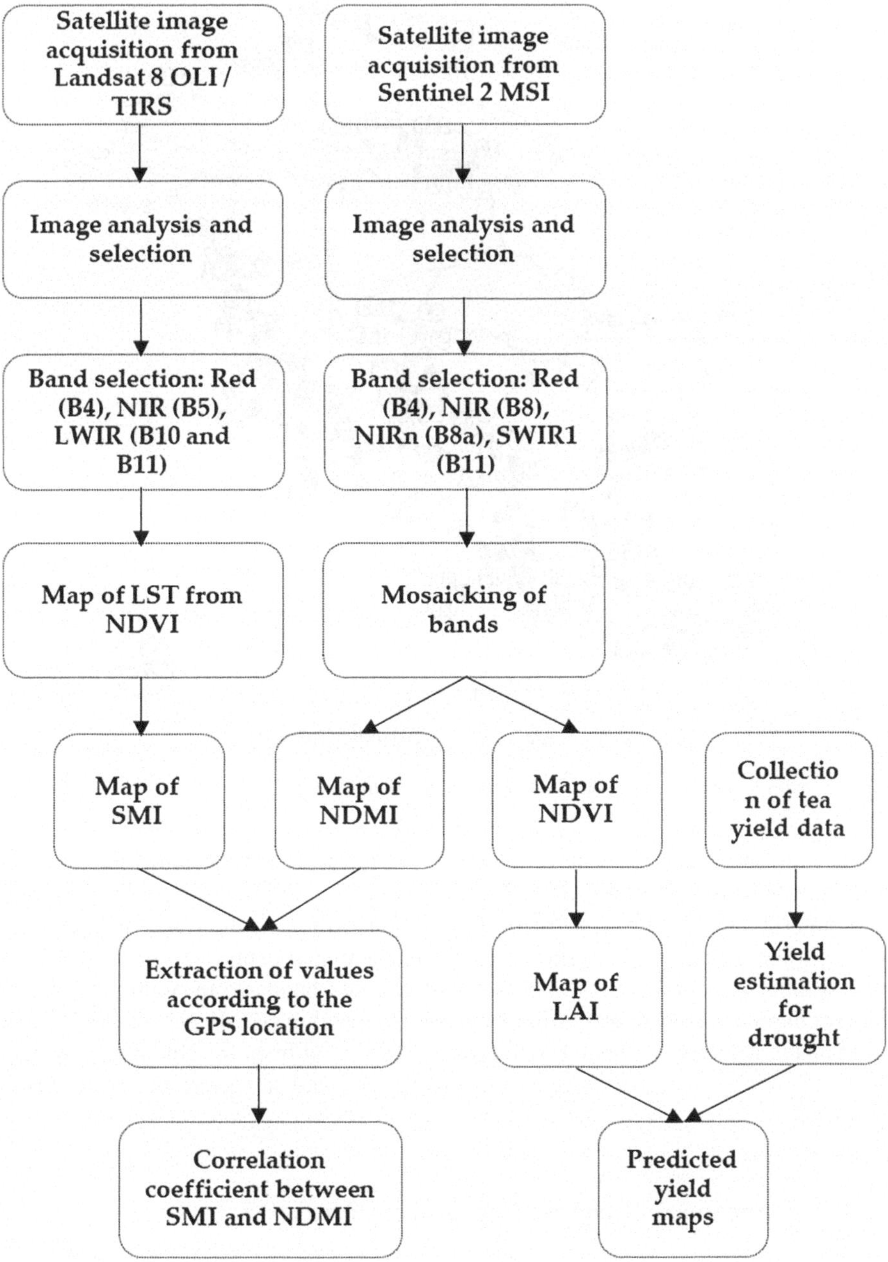


**Fig. 2.2** Flowchart for drought stress assessment in tea estates from remote sensing data

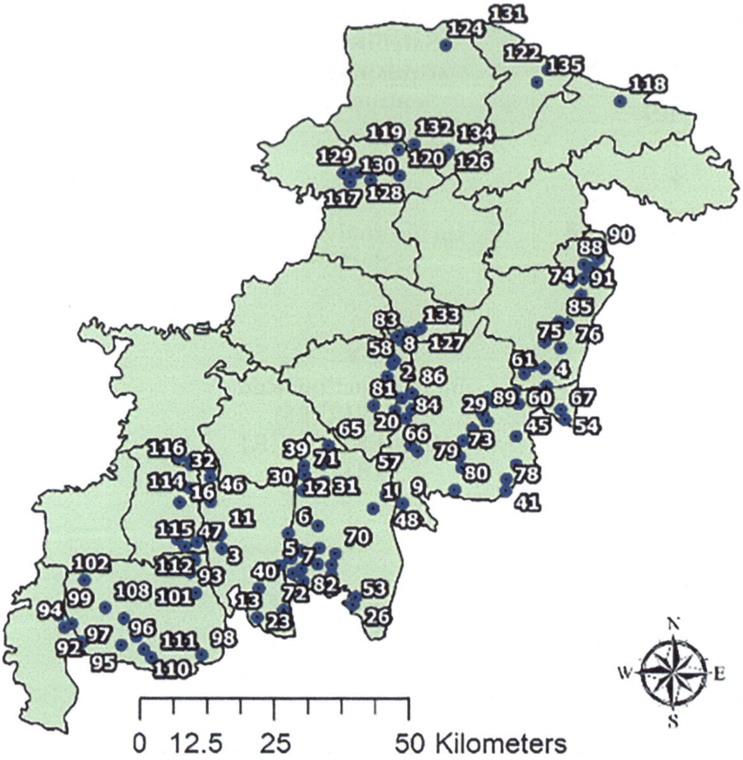


**Fig. 2.3** GPS location of tea estates according to the field survey

bands from Landsat 8 were resampled to a 30 m resolution for uniformity in cell size and data. Furthermore, an algebraic operation, radiometric calibration, and atmospheric and geometric corrections were performed on the raster images. The average reflectance for the raster images was calculated using the raster calculator tool to minimize the spatial variability. All selected bands from Sentinel-2 image data resampled to a 20 m resolution were used to develop NDMI, NDVI, NDVI, and yield maps. All remote sensing indices were calculated from satellite data with less than 2% cloud cover for three consecutive years and analyzed in the ArcGIS® environment.

#### 2.2.4.1 Estimation of Land Surface Temperature (LST)

Land surface temperature is a controlling parameter of canopy temperature and water content in vegetation that can be used effectively to monitor soil moisture in vegetated land (Zhang and Zhou 2016).

LST was calculated for the tea estates using temporal information from the Landsat 8 OLI/TIRS dataset with less cloud coverage (Binte Mostafiz et al. 2021; Habibie et al. 2019). There were two steps involved in the estimation of the

LST. First, the NDVI data were retrieved using two wave bands, near-infrared ($B5$) and red ($B4$), according to the following formula:

$$\mathrm{NDVI} = \frac{R_{\mathrm{NIR}} - R_{\mathrm{RED}}}{R_{\mathrm{NIR}} + R_{\mathrm{RED}}} \tag{2.9}$$

Proportion vegetation (PV) was calculated using the NDVI, which can be expressed as follows:

$$\mathrm{PV} = \left( \frac{\mathrm{NDVI} - \mathrm{NDVI}_{\mathrm{min}}}{\mathrm{NDVI}_{\mathrm{max}} - \mathrm{NDVI}_{\mathrm{min}}} \right)^2 \tag{2.10}$$

Here, $\mathrm{NDVI}_{\mathrm{min}}$ and $\mathrm{NDVI}_{\mathrm{max}}$ were obtained from the respective NDVI images. The land surface emissivity (LSE) was retrieved from the PV and calculated according to the following expression (Jesus and Santana 2017):

$$\mathrm{LSE} = 0.004 * \mathrm{PV} + 0.986 \tag{2.11}$$

In the second step, the thermal bands were transformed to digital numbers to calculate the radiance. The spectral radiance was calculated from the thermal bands ($B10$ and $B11$), which can be expressed as follows:

$$L\lambda = \mathrm{ML} + Q_{\mathrm{CAL}} + \mathrm{AL} \tag{2.12}$$

$$L\lambda = 0.0003342 * \mathrm{Band10} + 0.1 \,\mathrm{and}\, L\lambda = 0.0003342 * \mathrm{Band11} + 0.1$$

where $L\lambda$ denotes the spectral radiance of the TOA (top of atmosphere) at the sensor aperture, ML is the multiplicative rescaling factor of the corresponding bands that were derived from the metadata, $Q_{\mathrm{CAL}}$ refers to the pixel values (DN) of the quantized and calibrated standard products, and AL is the additive rescaling factor of the specific band. The brightness temperature (BT) was calculated using the following expression:

$$\mathrm{BT} = \frac{K2}{\ln\left[\left(K1/L\lambda\right) + 1\right]} - 273.15 \tag{2.13}$$

where BT denotes the satellite brightness temperature in Celsius. $K1$ and $K2$ refer to the calibration constants in Kelvin derived from the metadata of the thermal conversion constants of the respective bands. Finally, the LST was calculated using the following formula:

$$\mathrm{LST} = \frac{\mathrm{BT}}{1 + \left( \frac{\lambda * \mathrm{BT}}{\mathrm{PV}} \right) * \ln \mathrm{LSE}} \tag{2.14}$$

#### 2.2.4.2 Estimation of Soil Moisture Index (SMI)

Soil moisture is a potential determinant to understand drought severity and is primarily driven by precipitation through the process of infiltration as well as evapotranspiration (Fang et al. 2021; Yin et al. 2019). Soil moisture is variable both spatially and temporally due to differences in soil properties and drainage (Wang and Qu 2009). Among various indices, the LST-based SMI is a widely used index to monitor soil moisture in vegetated land. The SMI is the proportion of the difference between the current soil moisture and the permanent wilting point to the field capacity as well as the residual soil moisture (Saha et al. 2019). The value of the SMI ranges between 0 and 1, where 0 refers to very dry conditions and 1 denotes extremely wet conditions. The SMI is associated with the derived LST of a particular area that can be calculated using the following expression:

$$\mathrm{SMI} = \frac{\mathrm{LST}_{\max} - \mathrm{LST}}{\mathrm{LST}_{\max} - \mathrm{LST}_{\min}} \tag{2.15}$$

where $\mathrm{LST}_{\max}$ and $\mathrm{LST}_{\min}$ refer to the maximum and minimum surface temperatures of the study area, respectively.

#### 2.2.4.3 Estimation of Normalized Difference Moisture Index (NDMI)

The normalized difference moisture index determines the liquid water molecules in plant canopies that interact with solar radiation, and, thus, water deficit results in a decrease in photosynthesis affecting the growth and development of plants (Enquist and Ebersole 1994). The NDMI is a measure of water content in the spongy mesophyll tissues of vegetation canopies, which acts as a functioning indicator of climate and soil properties regulating water availability. The NDMI responds to variations in both water content (absorption of *B8A*) and spongy mesophyll tissues in plant canopies (reflectance of *B*11) (Serrano et al. 2019; Gao 1996). The NDMI was calculated from Sentinel-2 satellite images using the following formula (Korhonen et al. 2017):

$$\mathrm{NDMI} = \frac{\mathrm{NIR}_n - \mathrm{SWIR}_1}{\mathrm{NIR}_n + \mathrm{SWIR}_1} = \frac{B8a - B11}{B8a + B11} \tag{2.16}$$

where $\mathrm{NIR}_n$ is the near-infrared (narrow) reflectance of band 8A with wavelengths ranging from 855 to 875 nm and $\mathrm{SWIR}_1$ is the shortwave infrared reflectance of band 11 with wavelengths of 1565–1655 nm.

### 2.2.5 *Statistical Analysis*

Mean values per pixel of individual estates were used for drought assessment in tea estates. Correlation analyses were performed to assess the relationship between the soil moisture and vegetation canopy moisture. Regression analyses were also performed to show the association of the LAI and time-series yield information for yield prediction in the drought season.

### 2.2.6 *Validation of Yield in the Drought Period*

Yield data for 2018, 2019, and 2020 were collected from the Bangladesh Tea Board (BTB). The normalized difference vegetation index (NDVI) for the drought period was calculated from the reflectance of band 4 *(red)* and band 8 *(near-infrared)* of Sentinel-2 satellite imagery using the following expression:

$$\mathrm{NDVI} = \frac{\mathrm{NIR} - R}{\mathrm{NIR} + R} = \frac{B8 - B4}{B8 + B4} \tag{2.17}$$

The leaf area index (LAI) is an important determinant of tea yield and was also calculated from the NDVI using geospatial techniques. The LAI was calculated according to the equation used in previous studies based on the correlation between the LAI and NDVI utilizing the least square method (Tewari et al. 2003; Tuvshinbayar et al. 2017; Arab et al. 2021):

$$\mathrm{LAI} = 0.57 \times \exp\left(2.33 \times \mathrm{NDVI}\right) \tag{2.18}$$

Yield data were compared with the corresponding extracted LAI values in the scatter plot using regression analysis. The predicted yield maps in the drought period were developed using the linear regression equation in the ArcGIS environment. Finally, the extracted values from the yield maps were compared with the actual yield in the drought period to measure the yield loss in tea.

## 2.3 Results

### 2.3.1 *Standardized Precipitation Index (SPI) and Drought Classification*

Drought characterization based on the SPI was performed using the software provided by the National Drought Mitigation Center (NMDC), University of Nebraska, USA (Fig. 2.4). The results of drought frequency observed in the 39-month period for the Sylhet station demonstrated that 15 months belonged to the near-normal,

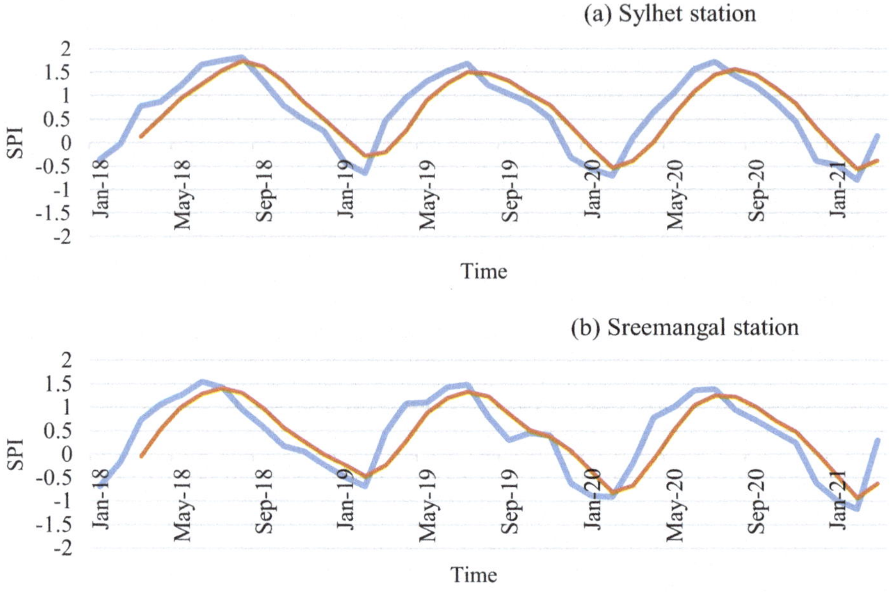


**Fig. 2.4** SPI trajectories for drought monitoring

**Table 2.1** Drought frequency in different categories

| Drought classes | SPI level | No. of months affected | | Frequency (%) | |
|---|---|---|---|---|---|
| | | Sylhet | Sreemangal | Sylhet | Sreemangal |
| Extremely wet | >3.0 to 3.0 | – | – | – | – |
| Moderately wet | 3.0 to 2.0 | – | – | – | – |
| Near-normal | 2.0 to 1.0 | 15 | 11 | 38.46 | 28.21 |
| Normal | 1.0 to 0 | 14 | 16 | 35.90 | 41.02 |
| Moderately dry | 0 to −1.0 | 10 | 12 | 25.64 | 30.77 |
| Severely dry | −1.0 to −2.0 | – | – | – | – |
| Extremely dry | −2.0 to −3.0 | – | – | – | – |

14 months normal, and 10 months moderately dry categories, which accounted for 38.46%, 35.90%, and 25.64%, respectively. In contrast, the Sreemangal station showed that the frequencies for near-normal, normal, and moderately dry conditions were 28.21%, 41.02%, and 30.77%, respectively (Table 2.1).

### 2.3.2 *Land Surface Temperature (LST) and Soil Moisture Index (SMI)*

The average LST in the drought months (December to February) for the three consecutive years derived from the Landsat 8 remote sensing data showed that the

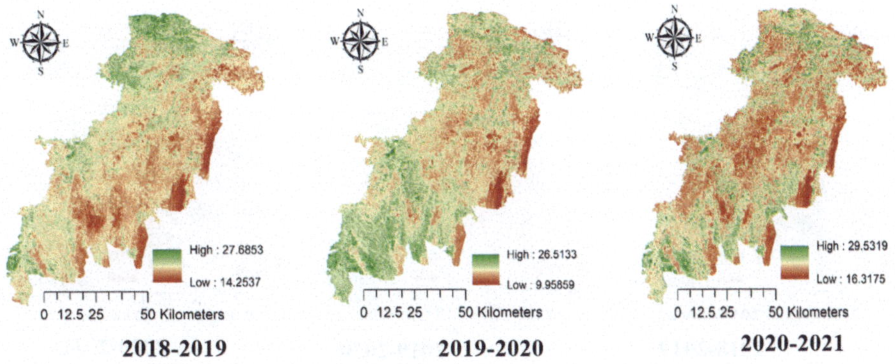


**Fig. 2.5** Land surface temperature (LST) in the drought period

surface temperature ranged between 14.25 and 27.68, 9.95 and 26.51, and 16.31 and 29.53, respectively (Fig. 2.5).

The SMI in the drought period demonstrated that the status of soil moisture ranged between 0.02 and 1, 0.01 and 0.99, and 0.04 and 0.99 for 2018–2019, 2019–2020, and 2020–2021, respectively (Fig. 2.6).

### *2.3.3 Normalized Difference Moisture Index (NDMI)*

The average NDMI in plant canopies observed in the drought period ranged between −0.522 and 0.811, −0.365 and 0.824, and −0.360 and 0.853 in 2018–2019, 2019–2020, and 2020–2021, respectively (Fig. 2.7). The red color in the NDMI maps indicates the plant biomass with a lower moisture level, and the green color denotes the high moisture content in canopies.

### *2.3.4 Spatial Correlation Between the SMI and NDMI*

The correlation coefficients between the SMI and NDMI were 0.84, 0.77, and 0.79 in the three consecutive years 2018–2019, 2019–2020, and 2020–2021, respectively. This spatial correlation provides evidence that there is a good inter-relationship between moisture in soil and plant biomass during the drought period (Fig. 2.8).

### *2.3.5 Validation of Tea Yield in the Drought Period*

The validation of tea yield was performed using the NDVI and LAI-based yield prediction technique. The trendline obtained from the regression analysis between

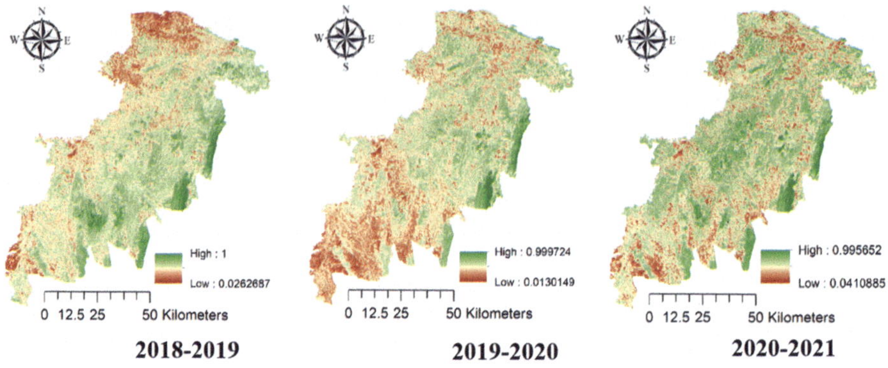


**Fig. 2.6** Soil moisture index (SMI) in the drought period

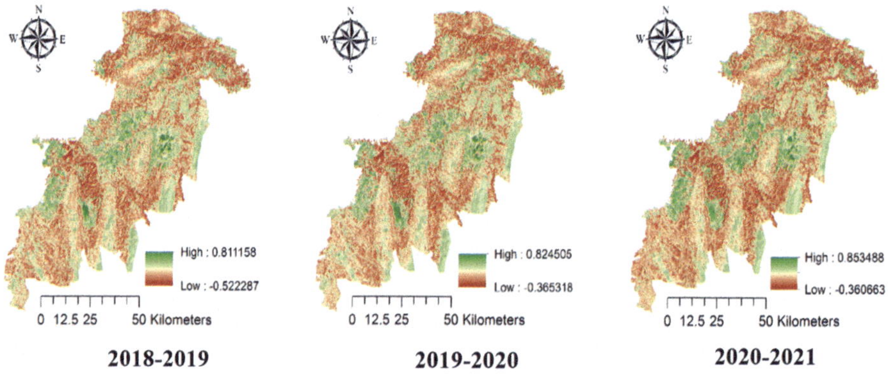


**Fig. 2.7** Normalized difference moisture index (NDMI) in the drought period

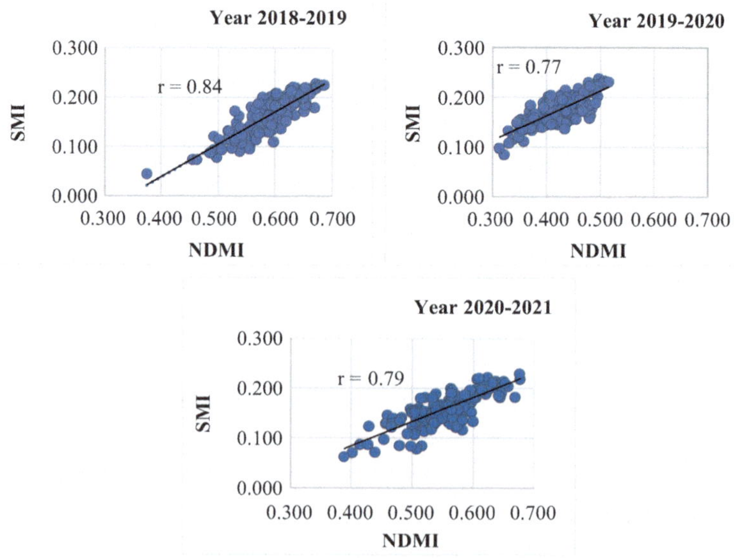


**Fig. 2.8** Correlation coefficient between the SMI and NDMI in the three consecutive years

the observed LAI and time-series yield information during the drought period in tea estates showed a strong relationship with $R^2$ values, 0.66, 0.71, and 0.68, for the three consecutive years 2018–2019, 2019–2020, and 2020–2021, respectively. The regression analysis between LAI and the corresponding yield in the drought period of the three consecutive years shows a strong relationship between LAI and tea yield (Figs. 2.9, 2.10, 2.11, 2.12, and 2.13). According to the trajectories of predicted yield during the drought period, a lower yield was observed than the actual yield for 83 estates in 2018, whereas the number of responsible estates was 81 each in 2019 and 2020. According to the predicted yield, it was also evident that the tea estates in the study area demonstrated yield losses of 7.72%, 11.92%, and 12.52% in the three consecutive years 2018–2019, 2019–2020, and 2020–2021, respectively.

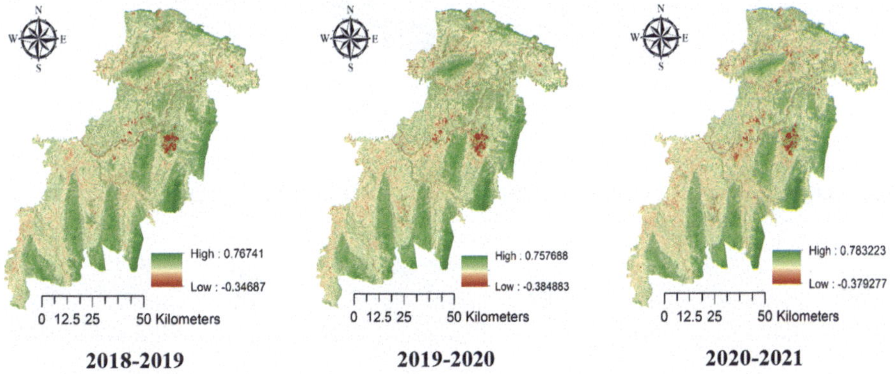


**Fig. 2.9** Normalized difference vegetation index (NDVI) in the drought period

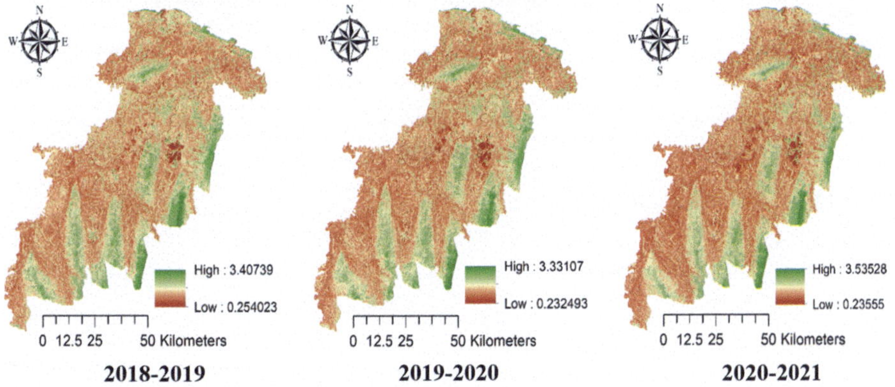


**Fig. 2.10** Leaf area index (LAI) in the drought period

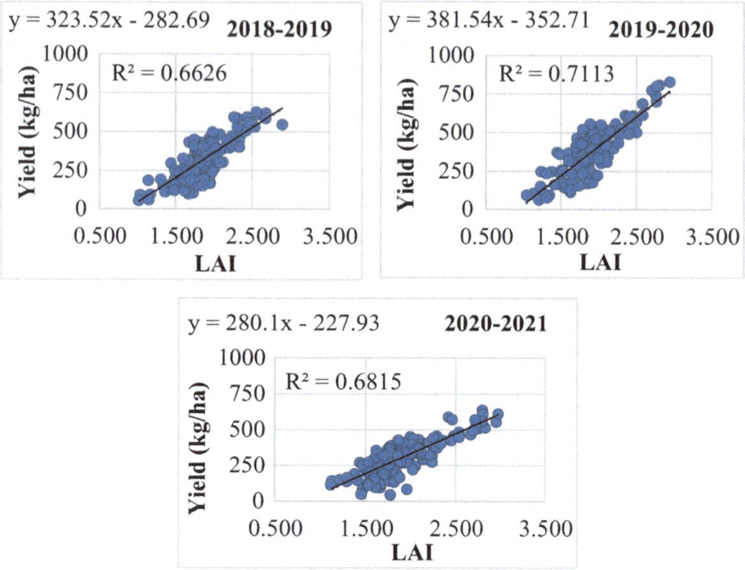


**Fig. 2.11** Regression analysis for yield prediction of tea during the drought period

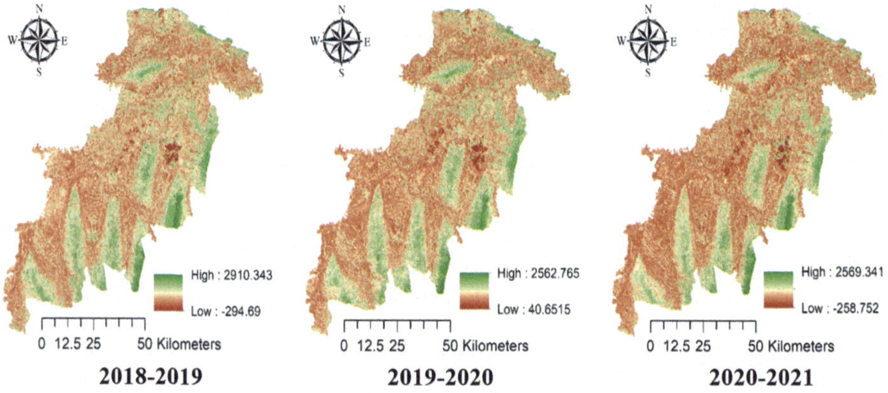


**Fig. 2.12** Yield map in the drought period

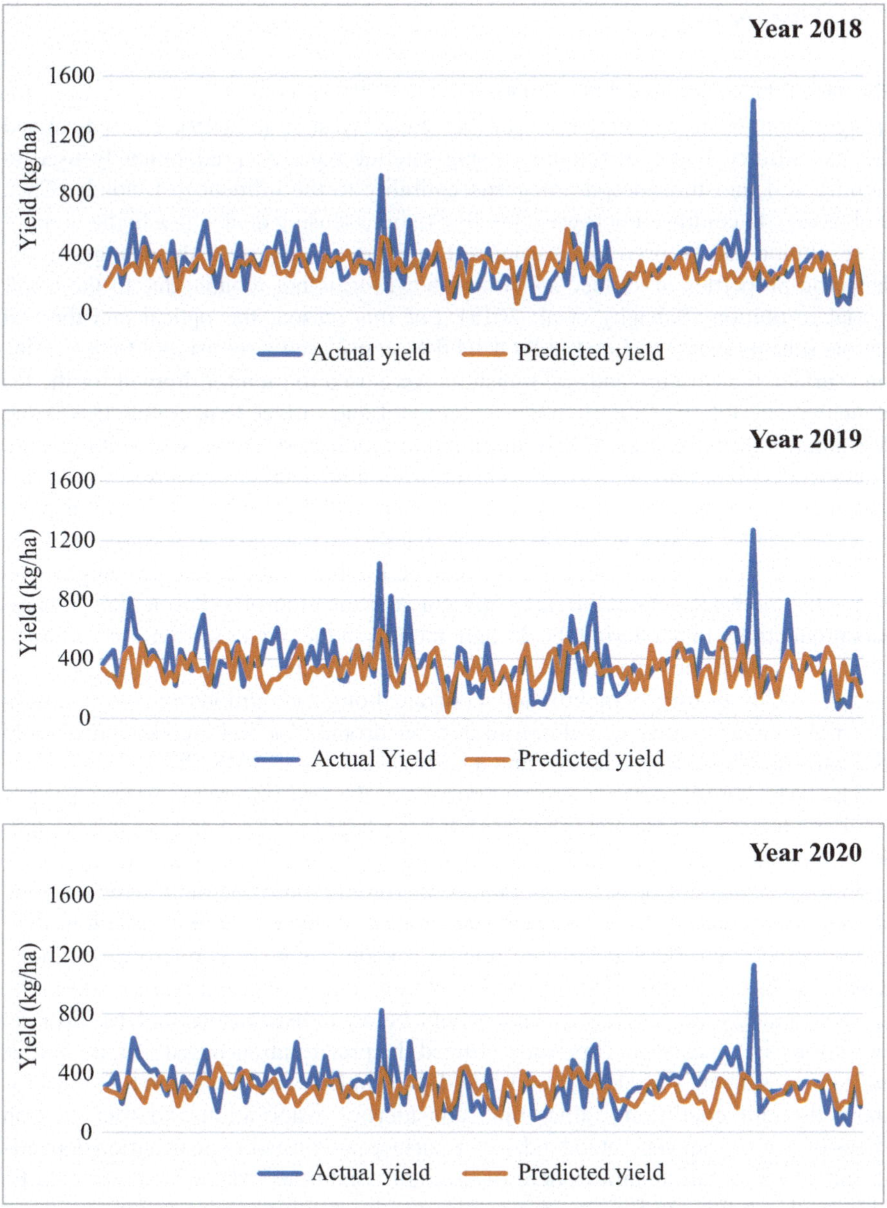


**Fig. 2.13** Predicted yield trajectories in tea estates during the drought period

## 2.4 Discussion

The use of the standardized precipitation index (SPI) with remote sensing datasets for drought monitoring was important for this study (West et al. 2019). The optical and thermal infrared bands of remote sensing satellite data are predominantly used to monitor soil moisture, despite their susceptibility to the influence of clouds (Chen et al. 2020). In contrast, microwave satellite remote sensing data with a higher temporal resolution can be used to measure soil moisture; recognizing the changes in the dielectric properties of surface soil water, however, is not reliable due to the lower spatial resolution (Sabaghy et al. 2018). For this reason, the optical and thermal remote sensing bands of Landsat 8 OLI/TIRS, considered a robust tool to determine land surface temperature and soil moisture, were used to monitor drought for the tea plantations during the cloud-free winter season. Land surface temperature (LST) and vegetation indices such as NDMI are important indicators to measure soil moisture (Rahimzadeh-Bajgiran et al. 2012). The utilization of Sentinel-2 datasets for NDMI measurements was useful because of its higher spatial resolution to monitor plant water content by measuring the reflectance that served the aim of this study.

It is evident that this area receives adequate rainfall in the active growing period during the monsoon season but faces drought after the monsoon season; thus, it hampers productivity in tea during the drought months in the winter season. According to the drought frequency classification, this area is characterized by normal and near-normal conditions for 8–9 months and moderate drought conditions for 3–4 months in a year. As tea is sensitive to soil moisture stress, drought for 3–4 months causes yield loss, affecting both tea production and quality (Nalina et al. 2018, 2021).

The SMI–NDMI model based on optical and thermal infrared remote sensing is efficient and easy to calculate; however, it is susceptible to adverse weather conditions and cannot provide valid data in the presence of clouds (Wang et al. 2020). Therefore, it was not possible to obtain year-round cloud-free data, which was an obstacle to preparing the year-round trajectories. To overcome this situation, integrated optical and thermal infrared remote sensing, as well as microwave remote sensing techniques, should be adopted to obtain year-round land surface soil moisture and plant water content, which would be an indispensable tool for drought monitoring in tea estates. This study utilized the precipitation-based SPI, as well as the remote sensing-based SMI and NDMI, to understand the water relationships under drought conditions rather than considering evaporation and transpiration losses of water from vegetated land, as the surface soil moisture distribution depends on the soil–vegetation–atmosphere of drought conditions. These issues should be addressed in future studies for precise monitoring of drought in tea estates.

## 2.5 Conclusions

This study suggests a drought stress assessment and classification system for tea plantations in Bangladesh using a standardized precipitation index and optical and thermal infrared remote sensing techniques. According to the results of drought frequency, the ratios of near-normal, normal, and moderately dry months for the Sylhet station were 38.46%, 35.90%, and 25.64%, respectively. In contrast, the Sreemangal station showed frequencies of 28.21%, 41.02%, and 30.77% for near-normal, normal, and moderately dry conditions, respectively. The spatial correlations between the SMI and NDMI were 0.84, 0.77, and 0.79 in the three consecutive years 2018–2019, 2019–2020, and 2020–2021, respectively, indicating that there was a strong interrelationship between soil and vegetation canopy moisture during the drought period. This study also predicts that 83, 81, and 81 tea estates among the 135 estates showed lower yields than the actual yield during the drought period; thus, they incurred yield losses, which accounted for 7.72%, 11.92%, and 12.52% in 2018, 2019, and 2020, respectively. The results of the drought stress assessment, as well as drought classification for tea plantations in Bangladesh, could be significant for policy makers, land use planners, and scientists in the decision-making process in ensuring the early warning of drought conditions and thus would help tea growers to understand future drought events and to take the necessary initiatives for increasing production.

**Acknowledgments** The authors express their thanks to the open-access journal *Remote Sensing* published by Multidisciplinary Digital Publishing Institute (MDPI) for their policy to support the authors for reusing the published article. In this regard, we would like to extend our gratitude to *Remote Sensing* journal in publishing this article (Animesh Chandra Das, Ryozo Noguchi, Tofael Ahamed. An Assessment of Drought Stress in Tea Estates Using Optical and Thermal Remote Sensing. *Remote Sensing*. *13*(14): 2730. https://doi.org/10.3390/rs13142730, 2021). Some minor modifications have been made to this book chapter. Furthermore, we would like to thank the University of Tsukuba for supporting this research to develop a drought classification system for tea plantations in Bangladesh. We express our sincere thanks to the National Aeronautics and Space Administration (NASA), the European Space Agency (ESA), the Bangladesh Agricultural Research Council (BARC), the Bangladesh Meteorological Department (BMD), and the Bangladesh Tea Board (BTB) for remote sensing, geospatial, geographic, meteorological, and statistical data. We also express our sincere gratitude to the field enumerators involved in this research.

**Author Contributions** Research Investigation, Methodology, Data Curation, Analysis, Interpretation of Results, and Writing—Original Draft, A.C.D.; Resources and Editing, R.N.; Research Conceptualization, Editing and Supervision T.A. All authors have read and agreed to the published version of the manuscript.

**Funding** This research received no external funding.

**Conflicts of Interest** The authors declare no conflicts of interest.

## References

AghaKouchak A, Farahmand A, Melton FS, Teixeira J, Anderson MC, Wardlow BD, Hain CR (2015) Remote sensing of drought: progress, challenges, and opportunities. Rev Geophys 53:452–480. https://doi.org/10.1002/2014RG000456

Ahmed S, Stepp JR, Kapuscinski AR, Méndez E (2016) Beyond yields: climate change effects on specialty crop quality and agroecological management. Elem Sci Anth 4:000092. https://doi.org/10.12952/journal.elementa.000092

Alam ATMJ, Rahman MS, Saadat AHM, Huq MM (2013) Gamma distribution and its application of spatially monitoring meteorological drought in Barind, Bangladesh. J Environ Sci Nat Resour 5:287–293. https://doi.org/10.3329/jesnr.v5i2.14832

Anjum SA, Xie XY, Wang LC, Saleem MF, Man C, Lei W (2011) Morphological, physiological, and biochemical responses of plants to drought stress. Afr J Agric Res 6:2026–2032. https://doi.org/10.5897/AJAR10.027

Arab ST, Noguchi R, Matsushita S, Ahamed T (2021) Prediction of grape yields from time-series vegetation indices using satellite remote sensing and a machine-learning approach. Remote Sens Appl Soc Environ 22:100485. https://doi.org/10.1016/j.rsase.2021.100485

Bangladesh Tea Board (2020) Statistical handbook on Bangladesh tea industry 2019, vol 8. Project Development Unit; Bangladesh Tea Board, Sreemangal, pp 1–146

Binte Mostafiz R, Noguchi R, Ahamed T (2021) Agricultural land suitability assessment using satellite remote sensing-derived soil-vegetation indices. Land 10:223. https://doi.org/10.3390/land10020223

Brown JF, Reed BC, Hyes MJ, Wilhite AD, Hubbard K (2002) A prototype drought monitoring system integrating climate and satellite data, Pecora 15/Land Satellite Information IV/ISPRS Commission I/FIEOS. http://www.isprs.org/commission1/proceedings02/paper/00074

Caccamo G, Chisholm LA, Bradstock RA, Puotinen ML (2011) Assessing the sensitivity of MODIS to monitor drought in high biomass ecosystems. Remote Sens Environ 115:2626–2639. https://doi.org/10.1016/j.rse.2011.05.018

Chen M, Zhang Y, Yao Y, Lu J, Pu X, Hu T, Wang P (2020) Evaluation of the OPTRAM model to retrieve soil moisture in the Sanjiang Plain of Northeast China. Earth Space Sci 7:e2020EA001108. https://doi.org/10.1029/2020EA001108

Das AC, Noguchi R, Ahamed T (2020) Integrating an expert system, GIS, and satellite remote sensing to evaluate land suitability for sustainable tea production in Bangladesh. Remote Sens 12:4136. https://doi.org/10.3390/rs12244136

Enquist BJ, Ebersole JJ (1994) Effects of added water on photosynthesis of *Bistorta vivipara*: the importance of water relations and leaf nitrogen in two alpine communities, Pikes peak, Colorado, U.S.A. Arct Alp Res 26:29–34. https://doi.org/10.1080/00040851.1994.12003035

Fang B, Kansara P, Dandridge C, Lakshmi V (2021) Drought monitoring using high spatial resolution soil moisture data over Australia in 2015–2019. J Hydrol 594:125960. https://doi.org/10.1016/j.jhydrol.2021.125960

Gao BC (1996) NDWI—a normalized difference water index for remote sensing of vegetation liquid water from space. Remote Sens Environ 58:257–266. https://doi.org/10.1016/S0034-4257(96)00067-3

Guo Y, Zhao S, Zhu C, Chang X, Yue C, Wang Z, Lin Y, Lai Z (2017) Identification of drought-responsive miRNAs and physiological characterization of tea plant (Camellia sinensis L.) under drought stress. BMC Plant Biol 17:211. https://doi.org/10.1186/s12870-017-1172-6

Gupta N, Gupta PP, Pramanik P, Saikia A, Sengupta L, Bhagat RM, Bhattacharya N (2014) Integration of geoinformatics and wireless sensors for smart agriculture in tea. In: Second international conference on remote sensing and geoinformation of the environment (RSCy2014), Paphos, Cyprus, 12 August 2014; SPIE: WA, USA, vol 9229, p 92290W. https://doi.org/10.1117/12.2066366

Gutman GG (1990) Towards monitoring droughts from space. J Clim 3:282–295. https://doi.org/10.1175/1520-0442(1990)003<0282:TMDFS>2.0.CO;2

Guttman NB (1998) Comparing the palmer drought index and the standardized precipitation index. J Am Water Resour Assoc 34:113–121. https://doi.org/10.1111/j.1752-1688.1998.tb05964.x

Habibie MI, Noguchi R, Shusuke M, Ahamed T (2019) Land suitability analysis for maize production in Indonesia using satellite remote sensing and GIS-based multicriteria decision support system. GeoJournal 86:777–807. https://doi.org/10.1007/s10708-019-10091-5

Habibie MI, Noguchi R, Matsushita S, Ahamed T (2020) Development of micro-level classifiers from land suitability analysis for drought-prone areas in Indonesia. Remote Sens Appl Soc Environ 20:100421. https://doi.org/10.1016/j.rsase.2020.100421

Hajiboland R (2017) Environmental and nutritional requirements for tea cultivation. Folia Hortic 29:199–220. https://doi.org/10.1515/fhort-2017-0019

International Tea Committee (2017) Annual bulletin of statistics. International Tea Committee Ltd, London

Jesus JBD, Santana IDM (2017) Estimation of land surface temperature in caatinga area using Landsat 8 data. J Hyperspectr Remote Sens 7:150–157. https://periodicos.ufpe.br/revistas/jhrs/article/viewFile/22766/pdf. Accessed 15 Apr 2021

Jiang S, Wang W, Xiao G, Li Z (2011) Design of wireless sensor node for drought monitoring in tea plantation. In: 2011 international conference on electric information and control engineering, Wuhan, China, 15–17 April 2011; IEEE: NY, USA, pp 1328–1332. https://doi.org/10.1109/ICEICE.2011.5777293

Kamruzzaman M, Parveen S, Das AC (2015) Livelihood improvement of tea garden workers: a scenario of marginalized women group in Bangladesh. Asian J Agric Ext Econ Soc 7:1–7. https://doi.org/10.9734/AJAEES/2015/18690

Kamruzzaman M, Hwang S, Cho J, Jang MW, Jeong H (2019) Evaluating the spatiotemporal characteristics of agricultural drought in Bangladesh using effective drought index. Water 11:2437. https://doi.org/10.3390/w11122437

Korhonen L, Packalen P, Rautiainen M (2017) Comparison of Sentinel-2 and Landsat 8 in the estimation of boreal forest canopy cover and leaf area index. Remote Sens Environ 195:259–274. https://doi.org/10.1016/j.rse.2017.03.021

Krishna TM, Ravikumar G, Krishnaveni M (2009) Remote sensing based agricultural drought assessment in Palar basin of Tamil Nadu state, India. J Indian Soc Remote Sens 37:9–20. https://doi.org/10.1007/s12524-009-0008-8

McKee TB, Doesken NJ, Kleist J (1993) The relationship of drought frequency and duration to time scale. In: Proceedings of the 8th conference on applied climatology, Anaheim, California, 17–22 Jan 1993, vol 17, pp 179–183. https://climate.colostate.edu/pdfs/relationshipofdrought-frequency.pdf

Mishra AK, Singh VP (2010) A review of drought concepts. J Hydrol 354:202–216. https://doi.org/10.1016/j.jhydrol.2010.07.012

Murad H, Islam AKMS (2011) Drought assessment using remote sensing and GIS in north-west region of Bangladesh. In: Proceedings of the 3rd international conference on water & flood management, pp 797–804

Muthumanickam D, Kannan P, Kumaraperumal R, Natarajan S, Sivasamy R, Poongodi C (2011) Drought assessment and monitoring through remote sensing and GIS in western tracts of Tamil Nadu, India. Int J Remote Sens 32:5157–5176. https://doi.org/10.1080/01431161.2010.494642

Nalina M, Saroja S, Rajkumar R, Radhakrishnan B, Chandrashekara KN (2018) Variations in quality constituents of green tea leaves in response to drought stress under south Indian condition. Sci Hortic 233:359–369. https://doi.org/10.1016/j.scienta.2018.02.009

Nalina M, Saroja S, Chakravarthi M, Rajkumar R, Radhakrishnan B, Chandrashekara KN (2021) Water deficit-induced oxidative stress and differential response in antioxidant enzymes of tolerant and susceptible tea cultivars under field condition. Acta Physiol Plant 43:1–17. https://doi.org/10.1007/s11738-020-03174-1

Nemani R, Hashimoto H, Votava P, Melton F, Wang W, Michaelis A, Mutch L, Milesi C, Hiatt S, White M (2009) Monitoring and forecasting ecosystem dynamics using the terrestrial observation and prediction system (TOPS). Remote Sens Environ 113:1497–1509. https://doi.org/10.1016/j.rse.2008.06.017

Paulo AA, Pereira LS (2006) Drought concepts and characterization: comparing drought indices applied at local and regional scales. Water Int 31:37–49. https://doi.org/10.1080/02508060608691913

Paulo AA, Rosa RD, Pereira LS (2012) Climate trends and behaviour of drought indices based on precipitation and evapotranspiration in Portugal. Nat Hazards Earth Syst Sci 12:1481–1491. https://doi.org/10.5194/nhess-12-1481-2012

Peterson TC, Vose RS (1997) An overview of the global historical climatology network temperature database. Bull Am Meteorol Soc 78:2837–2849. https://doi.org/10.1175/1520-0477(1997)078<2837:AOOTGH>2.0.CO;2

Rahimzadeh-Bajgiran P, Omasa K, Shimizu Y (2012) Comparative evaluation of the vegetation dryness index (VDI), the temperature vegetation dryness index (TVDI) and the improved TVDI (iTVDI) for water stress detection in semi-arid regions of Iran. ISPRS J Photogramm Remote Sens 68:1–12. https://doi.org/10.1016/j.isprsjprs.2011.10.009

Razali SM, Atucha AAM, Nuruddin AA, Hamid HA, Shafri HZM (2016) Monitoring vegetation drought using MODIS remote sensing indices for natural forest and plantation areas. J Spat Sci 61:157–172. https://doi.org/10.1080/14498596.2015.1084247

Sabaghy S, Walker JP, Renzullo LJ, Jackson TJ (2018) Spatially enhanced passive microwave derived soil moisture: capabilities and opportunities. Remote Sens Environ 209:551–580. https://doi.org/10.1016/j.rse.2018.02.065

Saha A, Patil M, Goyal VC, Rathore DS (2019) Assessment and impact of soil moisture index in agricultural drought estimation using remote sensing and GIS techniques. Proceedings 7:2. https://doi.org/10.3390/ECWS-3-05802

Serrano J, Shahidian S, Marques da Silva J (2019) Evaluation of Normalized Difference Water Index as a tool for monitoring pasture seasonal and inter-annual variability in a Mediterranean agro-silvo-pastoral system. Water 11:62. https://doi.org/10.3390/w11010062

Shao HB, Chu LY, Jaleel CA, Manivannan P, Panneerselvam R, Shao MA (2009) Understanding water deficit stress-induced changes in the basic metabolism of higher plants–biotechnologically and sustainably improving agriculture and the ecoenvironment in arid regions of the globe. Crit Rev Biotechnol 29:131–151. https://doi.org/10.1080/07388550902869792

Sun D, Jiang S, Wang W, Tang J (2010) WSN design and implementation in a tea plantation for drought monitoring. In: 2010 international conference on cyber-enabled distributed computing and knowledge discovery, Huangshan, China, 10–12 October 2010; IEEE: New York, pp 156–159. https://doi.org/10.1109/CyberC.2010.36

Tewari S, Kulhavy J, Rock BN, Hadas P (2003) Remote monitoring of forest response to changed soil moisture regime due to river regulation. J For Sci 49:429–438. https://doi.org/10.17221/4716-jfs

Tuvshinbayar D, Erdenetuya B, Erkhembayar E, Batbileg B, Sarangerel J (2017) Some results of crop stress monitoring by remote sensing in northern Mongolia. Mong J Agric Sci 21:59–63. https://doi.org/10.5564/mjas.v21i02.906

Van Niel TG, McVicar TR, Fang H, Liang S (2003) Calculating environmental moisture for per-field discrimination of rice crops. Int J Remote Sens 24:885–890. https://doi.org/10.1080/0143116021000009921

Wang L, Qu JJ (2009) Satellite remote sensing applications for surface soil moisture monitoring: a review. Front Earth Sci China 3:237–247. https://doi.org/10.1007/s11707-009-0023-7

Wang Z, Guo P, Wan H, Tian F, Wang L (2020) Integration of microwave and optical/infrared derived datasets from multi-satellite products for drought monitoring. Water 12:1504. https://doi.org/10.3390/w12051504

West H, Quinn N, Horswell M (2019) Remote sensing for drought monitoring & impact assessment: progress, past challenges and future opportunities. Remote Sens Environ 232:111291. https://doi.org/10.1016/j.rse.2019.111291

Wijeratne MA (1996) Vulnerability of Sri Lanka tea production to global climate change. Water Air Soil Pollut 92:87–94. https://doi.org/10.1007/BF00175555

Wilhite DA (2005) Drought and water crises: science, technology, and management issues. CRC Press: Taylor and Francis Group, NW Suite 3000, FL, USA, pp 1–432. ISBN 13: 978-1-4200-2838-6

World Meteorological Organization (2012) Standardized precipitation index user guide (WMO-No.1090). WMO, Geneva. https://wamis.org/agm/pubs/SPI/WMO_1090_EN.pdf

Xi Z, Lu D, Liu L, Ge H (2016) Detection of drought-induced Hickory disturbances in western Lin An county, China, using multitemporal Landsat imagery. Remote Sens 8:345. https://doi.org/10.3390/rs8040345

Xulu S, Peerbhay K, Gebreslasie M, Ismail R (2018) Drought influence on forest plantations in Zululand, South Africa, using MODIS time-series and climate data. Forests 9:528. https://doi.org/10.3390/f9090528

Yang Y, Anderson M, Gao F, Hain C, Noormets A, Sun G, Wynne R, Thomas V, Sun L (2020) Investigating impacts of drought and disturbance on evapotranspiration over a forested landscape in North Carolina, USA using high spatiotemporal resolution remotely sensed data. Remote Sens Environ 238:111018. https://doi.org/10.1016/j.rse.2018.12.017

Yin J, D'Odorico P, Porporato A (2019) Soil moisture dynamics in water-limited ecosystems. In: D'Odorico P, Porporato A, Runyan CW (eds) Dryland ecohydrology, 2nd edn. Springer, Cham, pp 31–48. ISBN: 978-3-030-23269-6

Zhang D, Zhou G (2016) Estimation of soil moisture from optical and thermal remote sensing: a review. Sensors 16:1308. https://doi.org/10.3390/s16081308

Zhou L, Xu H, Mischke S, Meinhardt LW, Zhang D, Zhu X, Li X, Fang W (2014) Exogenous abscisic acid significantly affects proteome in tea plant (Camellia sinensis) exposed to drought stress. Hortic Res 1:14029. https://doi.org/10.1038/hortres.2014.29

# Chapter 3
# Assessment of Land Use Land Cover Changes and Prediction of Vulnerable Agricultural Land in River Basins of Bangladesh Using Remote Sensing and a Fuzzy Expert System

**Kazi Faiz Alam and Tofael Ahamed**

**Abstract** Land use land cover (LULC) assessment is a key process for the successful planning and proper management of land resources in riverine countries. Bangladesh is a riverine country with a large river network and Delta, where land use changes are significant due to water inundation at adjacent distances. The LULC describes how the urban area, natural occupancy by water bodies, vegetation, and agricultural land have become a challenging issue for the sustainability of agricultural lands over time. Therefore, the aim of this study was to assess the LULC changes over 26 years from 1995 to 2021 to find the most changed land use conditions within the 25 km territory of the main river systems of Bangladesh. In addition, the prediction of vulnerable areas for agricultural land use in terms of inundation by river water was also analyzed. The study area includes river networks distributed through eight administrative divisions (Rangpur, Rajshahi, Mymensingh, Sylhet, Dhaka, Khulna, Barishal, and Chittagong) of Bangladesh, covering an area of 64,556 $km^2$. The study was conducted by identifying permanent water bodies from NDWI indices and preparing LULC maps that include the five main land use classes (water body, bare land, vegetation, agricultural land, and urban area) in the Google Earth Engine platform using supervised classification. The LULC maps were then analyzed in the ArcGIS® environment. A vulnerability map for agricultural land use was also prepared using a fuzzy expert-based system applying multi-criteria analysis. From the land use land cover map of the study area, it was found that among the five land use classes, water bodies, bare land, vegetation, and urban areas increased in size by 3.65%, 2.18%, 3.31%, and 2.55%, respectively, whereas

K. F. Alam
Graduate School of Science and Technology, University of Tsukuba, Tsukuba, Ibaraki, Japan

T. Ahamed (✉)
Institute of Life and Environmental Sciences, University of Tsukuba, Tsukuba, Ibaraki, Japan
e-mail: tofael.ahamed.gp@u.tsukuba.ac.jp



T. Ahamed (ed.), *Remote Sensing Application II*, New Frontiers in Regional Science: Asian Perspectives 77, https://doi.org/10.1007/978-981-97-1188-8_3

agricultural land use significantly decreased by 11.68%. This decrease in agricultural land use was common for the analyzed area of all administrative divisions. According to the vulnerable area map of the eight divisions, more than 50% of the analyzed area of the Khulna and Dhaka divisions and more than 40% of the analyzed area of the Rajshahi, Mymensingh, Sylhet, Barishal, and Chittagong divisions were highly vulnerable to agricultural land use due to the possibility of inundation by water. However, approximately 44% of the analyzed area of the Rangpur division was not vulnerable for agricultural land use. The prepared LULC and vulnerability maps can be helpful for the future land use planning of Bangladesh to meet the increasing demand for food production and livelihoods for increasing populations.

**Keywords** Land use land cover (LULC) · River basin · Agricultural land use · Vulnerability · Fuzzy expert system · Remote sensing · ArcGIS® · Google Earth Engine · Bangladesh

## 3.1 Introduction

Low-lying deltaic regions are very vulnerable to the effects of global climate change. These regions are continuously facing natural calamities like floods, flash floods, cyclones, saline water intrusions, and riverbank erosion. According to FAO (2021), an averaged 170 billion USD economic losses occurred due to climatological, hydrological, biological, and geophysical associated disasters. Among the natural calamities, floods are the most common in deltaic regions and flood events are increasing due to global climate change. An average of 30 flood events were recorded during the 1970s but in 1980 flood events were 60, which was double; in 2000s and 2006, an average 180 and 246 flood events were recorded (Fig. 3.1). Bangladesh is a riverine country and part of Ganges–Brahmaputra (GBM) delta where many small to large rivers flow toward the Bay of Bengal. Padma, Meghna, Jamuna, and Brahmaputra are the main large rivers of Bangladesh. All rivers flow from the north to the south over the country and contribute to the Bay of Bengal in the south. In addition to these main rivers, there are many tributaries and distributaries of these rivers. These rivers experience severe inundation during the monsoon periods of the year. In the monsoon season, most rainfall occurs, and low-lying areas become flooded with rainwater. The rivers of Bangladesh also become overloaded by rainwater, and water comes from upstream, which causes seasonal floods on both sides of its banks. On the other hand, the land area is continuously changing from one land use to another, such as agricultural to urban land, agricultural land to man-made vegetation, and bare land to natural vegetation. This land use change mostly affects agricultural land (Hossain et al. 2020), which is decreasing to meet the increasing demand of higher populations. Land use land cover (LULC) assessment is an important process for meaningful planning and management of land resources. The land use changes over the

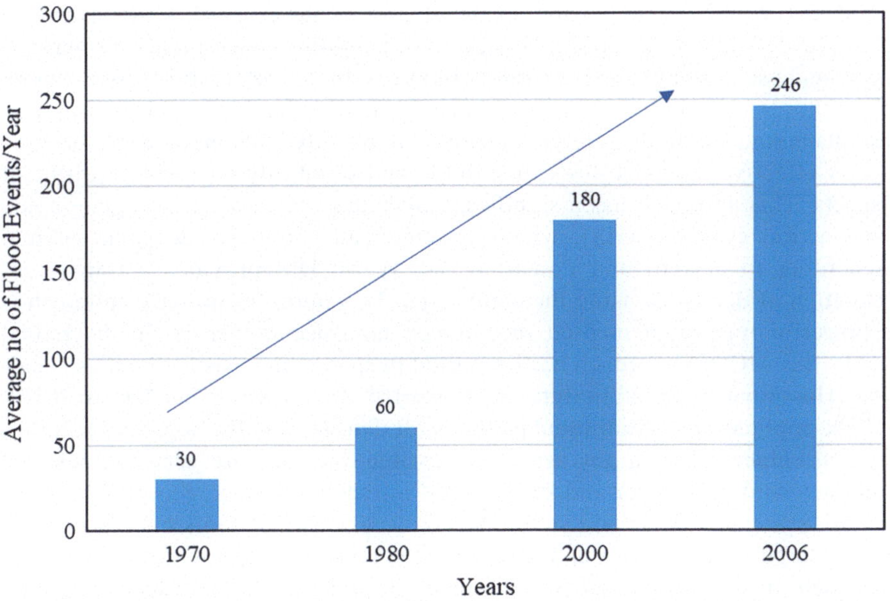


**Fig. 3.1** Average flood events per year from 1970 to 2006 (FAO 2021)

years can be recognized through changing satellite imagery; this is a potential source of high-quality information at the local, regional, and global scales (Csaplovics 1998; Foody 2002).

Remote sensing (RS) technologies have become the best way to obtain and update information about current conditions with dynamic changes in LULC from the Earth's surface over long periods of time. Therefore, in LULC classification, information about cultural features, urban areas, natural vegetative surfaces, and agricultural land are described (Akıncı et al. 2013). To achieve environmental security and sustainable development, pattern and change detection assessments of LULC have become a great concern of scientists worldwide (Li 1996). Land use inventories are becoming increasingly important in various sectors, such as urban planning, environmental studies, operational planning, and agricultural planning. LULC data also play an important role as an input for modeling, and they are related to the development of policy as well as to face the adverse effects of climate change (Disperati and Virdis 2015).

As a deltaic area, Bangladesh faces climate change risks and outbreaks such as cyclones, riverbank erosion, salinity intrusion, floods, and flash floods. All these natural calamities affect the livelihoods of farmers and related stakeholders through their adverse effects by destroying land and property as well as damaging crops. Many low-lying agricultural areas become inundated by floods and flash floods during the rainy season in Bangladesh, which causes an enormous deficit in agricultural production. Several studies have been conducted to understand the adverse effects of cyclones, riverbank erosion, floods, and flash floods covering different portions of Bangladesh (Khan et al. 2014; Shamsuzzoha et al. 2021; Islam et al. 2022). However, vulnerability assessments for agricultural land use in Bangladesh are rarely reported. In the vulnerability assessment, the multiple criteria-based decision-making method has potential uses in land use planning. The multicriteria decision-making (MCDM) method has become a very effective approach for vulnerability and suitability analysis for different aspects. This approach has a significant advantage when used with geographic information systems (GIS), which systematically supports decision-making. According to Ferretti and Pomarico (2013), MCDM provides a transparent reflection of decisions using thematic maps. This multidisciplinary approach is a powerful integrated method in complex land use scenarios. In diversified fields, this MCDM approach has been used in spatial analysis for land use analysis (Hossain et al. 2009; Ferretti and Pomarico 2012; Hassan and Nazem 2016).

The application of traditional methods for determining the degree of suitability and vulnerability in any aspect is possible, but they are very difficult and time-consuming (Frazier and Page 2000). In recent decades, geographic information systems (GIS) and satellite remote sensing (RS) tools have been suggested to be very effective and have attracted considerable attention in the field of suitability and vulnerability assessment (Tehrany et al. 2014; Khosravi et al. 2016). MCDM contributes significantly to land suitability assessment in industrial development and the production of different crops, such as cassava, tea, and maize (Muhsin et al. 2018; Purnamasari et al. 2019; Das et al. 2020; Habibie et al. 2021). However, LULC in the large riverine networks was not exclusively projected in any of the studies, particularly agricultural land use change to ensure the sustainability of agricultural lands over time. It is necessary to predict the land use changes for deltaic countries for sustainable planning of agricultural lands, settlements, and livelihoods. Additionally, LULC change detection is important to identify water hazards or lands vulnerable to inundation in riverine-adjacent territories. Thus, the first research question focuses on assessing the LULC in the large river basin for long time periods, especially for agricultural land use change to ensure land use planning for food production. Another important research question is how to predict the inundation vulnerability for agricultural lands in spatial and time scales. In this regard, risk-prone flood inundation mapping for policy implications in agriculture is important for agricultural land use planning. Satellite remote sensing can predict the inundation vulnerability from multispectral and multitemporal perspectives in the distributed river basin. In this regard, multicriteria considering climate change, land use planning, distance, and geographical extent can help to estimate the inundation hazard.

Therefore, the objectives of the present study were to assess the land use land cover changes (LULC) in the river basin of Bangladesh from 1995 to 2021 to find the most changed land use classes and to predict the lands vulnerable to agricultural land use by preparing a vulnerability map using a fuzzy expert system and satellite remote sensing datasets.

## 3.2 Methodology

The study was conducted using Landsat 5 and Landsat 8 Operational Land Imager (OLI) multispectral images with 30 m spatial resolutions retrieved between 1995 and 1996 and between 2021 and 2022. Satellite images were used to prepare LULC maps and different criteria maps for assessing change detection. Shuttle Radar Topography Mission ( SRTM) data were used to make the elevation map (Table 3.1). The distance from the river map was calculated using the Euclidian distance utilizing the river shape file. Annual average precipitation data from 2021 were collected from CHIRPS (1995–2022). A rainfall map of the study area was prepared from the Climate Hazards Group InfraRed Precipitation with Station data (CHIRPS) precipitation data. The administrative and river shape files were retrieved from the Bangladesh Bureau of Statistics (BBS), which were updated in 2020 (Table 3.1). Then, the study area was masked according to riverine networks and adjacent territories. All LULC maps and criteria maps were prepared in the Google Earth Engine (GEE) platform. Satellite images of less than 1% cloud coverage taken from the Landsat collection were utilized for the analysis. Finally, all other analyses were carried out in the ArcGIS® environment.

### *3.2.1 Study Area*

Bangladesh is a deltaic area with many small to large rivers flowing through the country. These rivers are directly related to the human lives and livelihoods of farmers. The study area covers eight administrative divisions of Bangladesh (Rangpur, Rajshahi, Mymensingh, Sylhet, Dhaka, Khulna, Barishal, and Chittagong) from north to south. The analysis was conducted within the 25 km territory of both banks of the main rivers to understand the effect of rivers on LULC (Jamuna, Padma, Brahmaputra, and Meghna, including their tributaries and distributaries) covering an area of 64,556 km$^2$. The analyzed areas for the eight administrative divisions

**Table 3.1** Data used for LULC classification and predicting lands vulnerable for agricultural activities

| Source | Type of data |
|---|---|
| Bangladesh Bureau of Statistics (BBS) | Administrative area and river shape files |
| United States Geological Survey (USGS) | Landsat 8 OLI satellite images with bands (1–7) of 2021, 30 m spatial resolution |
| USGS | Landsat 5 satellite images of 1990, including all bands (b1–b7) with 30 m spatial resolution |
| SRTM (USGS) | Digital Elevation Model (DEM) data |
| www.chrsdata.eng.uci.edu (accessed on 31 July 2022) | Precipitation data for the year 2021 |

were 7368 $km^2$, 8132 $km^2$, 18,112 $km^2$, 1735 $km^2$, 7534 $km^2$, 6444 $km^2$, 4828 $km^2$, and 10,235 $km^2$ for Rangpur, Rajshahi, Dhaka, Khulna, Mymensingh, Sylhet, Barishal, and Chittagong, respectively (Fig. 3.2). The areas vulnerable to agricultural land use were also assessed for riverine network cover in all divisions.

Geologically, two major tectonic units of Bangladesh are: (1) In the Northwest, the Stable Precambrian Platform and (2) In the Southeast, Geosynclinal basin. Another Northeast–Southwest trending narrow unit that separates the above two units is Hinge Zone (Johnson and Nur Alam 1991; Uddin and Lundberg 1999). Among the eight administrative divisions of Bangladesh, Rangpur is the most northern division and geologically falls in the North–Northwestern part of the Bengal basin. The surface area is a recent floodplain deposit (Khan 1991a, b) composed of clay, silt, and fine- to medium-grained sand (BWDB 1974). This division's most common land use classes are water bodies, urban areas, agricultural land, and natural vegetation. The coastal part of the Khulna division is highly vulnerable to cyclones (Shamsuzzoha et al. 2021). Barishal is in the south of Bangladesh, and the division contains many small to large rivers. The Bay of Bengal is in the south of the division. The lands are mainly covered by water bodies, agricultural land, natural vegetation, and urban areas. The coastal portions are highly vulnerable to cyclones. Chittagong is the southernmost division of Bangladesh, called the port city. The eastern portion of this division is a hilly region that mostly contains natural vegetation, and the southern part contains the longest coast of the Bay of Bengal. The most

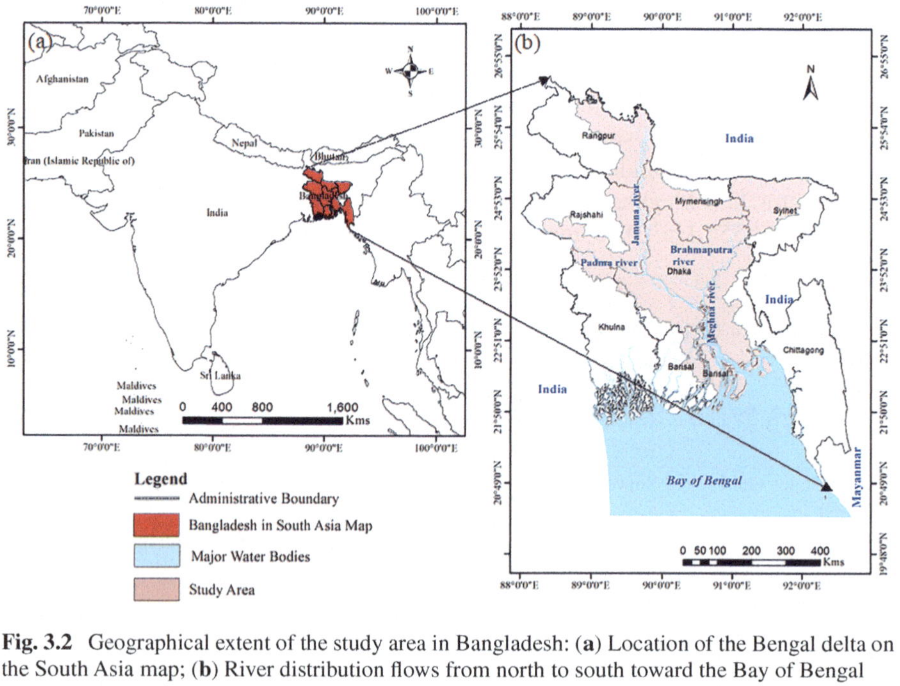


**Fig. 3.2** Geographical extent of the study area in Bangladesh: (**a**) Location of the Bengal delta on the South Asia map; (**b**) River distribution flows from north to south toward the Bay of Bengal

common land use classes are water bodies, natural vegetation, urban area, and agricultural land.

## 3.2.2 Research Framework

The study was conducted in two stages. In the first stage, LULC maps for 1995 and 2021 were prepared for the study area to assess the LULC change over 26 years. Then, permanent water bodies between 1995 and 2021 were detected from the NDWI (Normalized Difference Water Index) map to understand the changes in the river basin area. In the second stage, four criteria related to inundation by water were selected to prepare the vulnerability map to predict the areas vulnerable to agricultural land use. A fuzzy expert base multicriteria decision-making (MCDM) method was used for the vulnerability analysis. A stepwise workflow has been described in Fig. 3.3.

### 3.2.2.1 Land Use Land Cover Change Assessment

The LULC maps of the study area for 1995 and 2021 were prepared under five major land use classes: water bodies, bare land, vegetation, agricultural land, and urban areas. Here, water bodies include main rivers as well as other permanent water bodies; bare land includes the sandy land (chor land) near the rivers; vegetation includes both natural forests and other vegetation such as fruit orchards; agricultural land includes land with crops and land with no crops but still used for agricultural activity, and urban areas were lands with built-up features. The maps were prepared in the Google Earth Engine platform, which can easily handle a large study area. Working on this platform can also help to avoid hardware limitations. Several studies have already been conducted on the Google Earth Engine platform (Kumar and Mutanga 2018; Mugiraneza et al. 2020; Zhao et al. 2021; Shafizadeh-Moghadam et al. 2021).

Landsat 5 image collections with less than 1% cloud coverage captured between January 1, 1995, and March 30, 1996, were used for preparing the LULC map for 1995. A total of 2564 reference points were collected for 5 land use classes, where 713 sample points for water bodies, 148 points for bare land, 302 points for vegetation, 1116 points for agricultural land, and 285 points for urban areas were collected for the training dataset. The reference points of every pixel value were collected and compared with the Google Earth Pro® map. Of the datasets, 70% of the data were used for training, and 30% were used for testing.

To prepare the LULC map for 2021, Landsat 8 OLI images with less than 1% cloud coverage were captured between January 2021 and March 2022. A total of 1755 reference points were collected for 5 land use classes, where 726 reference points were for water bodies, 130 for bare land, 237 for vegetation, 415 for agricultural land, and 247 points were for urban areas. The reference points of every pixel

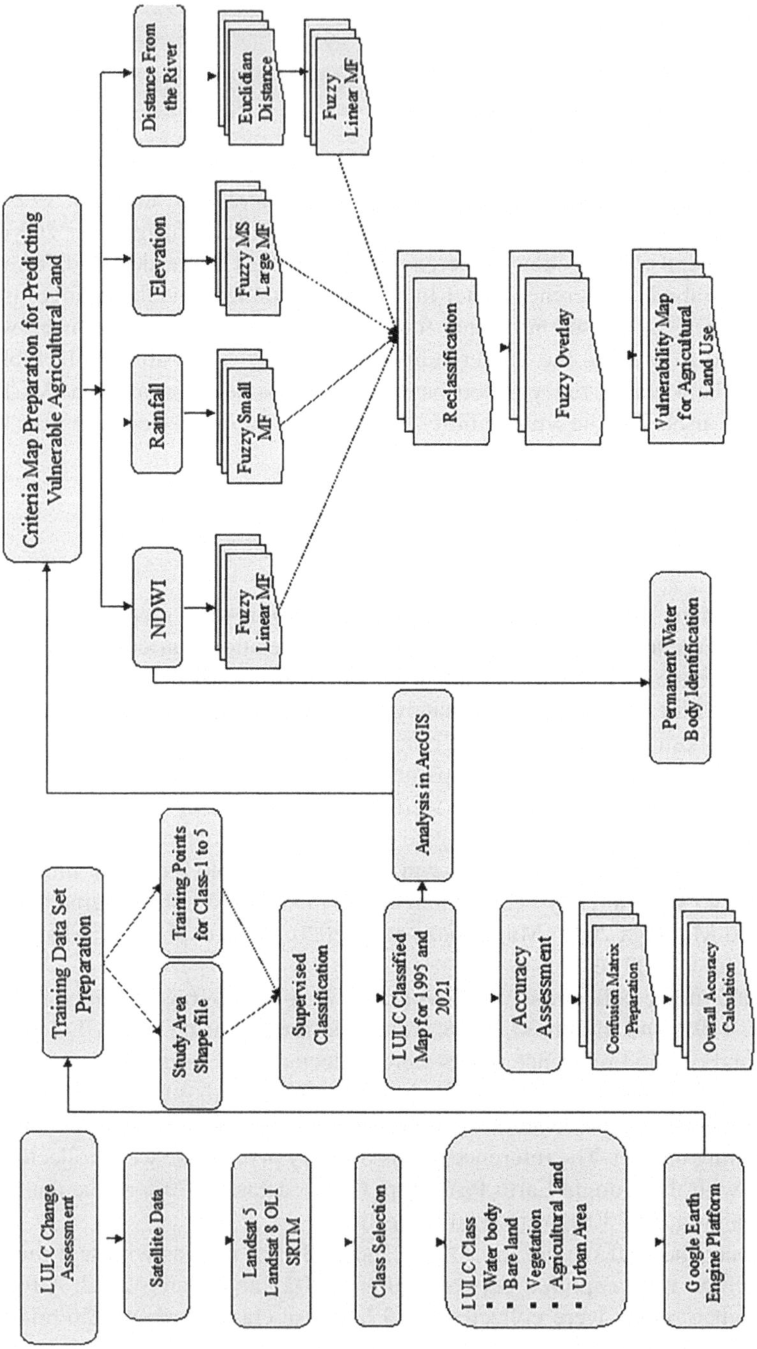


**Fig. 3.3** Research framework for LULC change assessment for predicting vulnerable agricultural lands

value were collected and compared with the Google Earth Pro® map. Of the datasets, 70% of the data were used for training, and 30% were used for testing.

Eight Landsat image tiles from three paths were analyzed for preparing both LULC maps. They were retrieved from path-136, row-42, 43, 44; path-137, row-43, 44; and path-138, row 42, 43, 44. It is very hard to find a cloud-free image at the same time; therefore, a long time-span was taken for image collection. Most of the agricultural lands in Bangladesh are used for multicrop cultivation. When few lands have crops, others may have none or may be preparing crops in the same area. To classify the agricultural land, we compared the training dataset for both lands with crops and with no crop to the images from Google Earth Pro®. We have also taken sample points for different growth stages of crops as their spectral signature also varied in different stages.

The following scripts were used for preparing the LULC map. Three algorithms were used for preparing the LULC maps for 1995 and 2021, Algorithms 1 and 2 were used for LULC map preparation and Algorithm 3 was used for accuracy assessment of the prepared land use land class maps. Original GEE codes are given in Appendices 1 and 2.

Algorithm 1: "Image retrieval of Landsat 5 or 8, False colour composite image preparation and LULC map preparation for 1995"

| | Script |
|---|---|
| 1 | var L5 or L8 = ee.imagecollection("Collection Snippet for landsat 5 or 8") |
| 2 | .filterBounds (ROI) |
| 3 | .filterDate ("start date", "end date") |
| 4 | . filterMetadata('CLOUD_COVER','less_than', 1) |
| 5 | .mean() |
| 6 | .clip(ROI) |
| 7 | Map.addLayer(L5 or 8, {bands:["B4" , "B3" , "B2"]}); |
| 8 | var training_points =landuse class1.merge(landuse class2).merge(..); |
| 9 | var training_data = L5.sampleRegions({collection:training_points,properties: ['LC'],scale:30}) |
| 10 | var classifier = ee.Classifier.smileCart() |
| 11 | var classifier = classifier.train({features:training_data,classProperty: 'LC',inputProperties:["B1", "B2", "B3", "B4", "B5", "B6", "B7"]}); |
| 12 | var classified_image = L5 or L8.classify(classifier); |
| 13 | Map.addLayer(classified_image,{min:0, max:4, palette:[ 'colour1', 'colour2', 'colour3', 'colour4', 'colour']},'classified image'); |
| | Run |

Algorithm 2: "Image retrieval of Landsat 5 or 8, False colour composite image preparation and LULC map preparation for 2021"

| | Script |
|---|---|
| 1 | var L5 or L8 = ee.imagecollection("Collection Snippet for landsat 5 or 8") |
| 2 | .filterBounds (ROI) |

| Algorithm 2: "Image retrieval of Landsat 5 or 8, False colour composite image preparation and LULC map preparation for 2021" | |
|---|---|
| | Script |
| 3 | .filterDate ("start date", "end date") |
| 4 | . filterMetadata('CLOUD_COVER','less_than', 1) |
| 5 | .mean() |
| 6 | .clip(ROI) |
| 7 | Map.addLayer(L5 or 8, {bands:["B4" , "B3" , "B2"]}); |
| 8 | var training_points =landuse class1.merge(landuse class2).merge(..); |
| 9 | var training_data = L5.sampleRegions({collection:training_points,properties: ['LC'],scale:30}) |
| 10 | var classifier = ee.Classifier.smileCart() |
| 11 | var classifier = classifier.train({features:training_data,classProperty: 'LC',inputProperties:["B1", "B2", "B3", "B4", "B5", "B6", "B7"]}); |
| 12 | var classified_image = L5 or L8.classify(classifier); |
| 13 | Map.addLayer(classified_image,{min:0, max:4, palette:[ 'colour1', 'colour2', 'colour3', 'colour4', 'colour']},'classified image'); |
| | Run |

| Algorithm 3: "Accuracy assessment" | |
|---|---|
| | Script |
| 1 | var trainingData = training_data.randomColumn(); |
| 2 | var trainSet = trainingData.filter(ee.Filter.lessThan('random', 0.7)); |
| 3 | var testSet = trainingData.filter(ee.Filter.greaterThanOrEquals('random', 0.7)); |
| 4 | var confusionMatrix = ee.ConfusionMatrix(testSet.classify(classifier) |
| 5 | .errorMatrix({ |
| 6 | actual: 'LC', |
| 7 | predicted: 'classification' |
| 8 | })) |
| 9 | print('Confusion Matrix', confusionMatrix); |
| 10 | print("Overall Accuracy:", confusionMatrix.accuracy()); |

The prepared LULC images were resampled by 30 × 30 m cell sizes as the GEE platform produces an image in pixels, and final maps were prepared in the ArcGIS® environment using resampling tools for 1995 and 2021 (Fig. 3.4a, b).

NDWI maps were also prepared in the Google Earth Engine platform for 1995 and 2021 to identify permanent water bodies. Landsat 5 and Landsat 8 OLI image collections were used to prepare the NDWI map. Algorithm 4 was used to calculate the NDWI indices and map preparation. Original GEE codes are given in Appendix 3.

| Algorithm 4: "NDWI map preparation" | |
|---|---|
| | Script |
| Export | var ROI: Table shapefile |
| 1 | var L5 or L8 = ee.ImageCollection("Collection Snippet for Landsat 5 or 8") |

Algorithm 4: "NDWI map preparation"

Script

```
.filterBounds(ROI)
.filterDate("start date", "end date")
.filterMetadata('CLOUD_COVER','less_than', 1)
.mean()
.clip(ROI);
var green = L5.select('B2'); or = L8.select('B3');
var nir = L5.select('B4'); or = L8.select('B5');
var ndwi = green.subtract(nir).divide(green.add(nir)).rename('NDWI');
var ndwiParams = {min: −1, max: 1, palette: ['black', 'white', 'blue']};
Map.addLayer(ndwi, ndwiParams, 'NDWI image');
```

The prepared NDWI images were downloaded from the Google Earth Engine cloud server to Google Drive and analyzed in the ArcGIS® environment (Environmental Systems Research Institute, Inc. (ESRI), Redlands, California, USA) after resampling 30 × 30 m cells using resampling tools. Both NDWI images were classified into two classes (water body and non-water body) using the index value for water 0 (Özelkan 2020). Finally, maps representing only permanent water bodies were prepared (Fig. 3.4c, d).

To represent the LULC changes, maps of individual land use classes against water bodies were prepared (Figs. 3.5 and 3.6).

At the end, both LULC maps (1995 and 2021) were masked according to the administrative divisions (Figs. 3.7 and 3.8) and analyzed to determine the division of land use changes.

#### 3.2.2.2 Vulnerability Analysis for Agricultural Land Use

Considering the inundation risk of agricultural land, the vulnerability analysis for agricultural land use in terms of inundation by water was performed on the same area taken for land use/land cover change analysis. In this regard, the second stage of this research was conducted to prepare a vulnerability map of the area for agricultural land use. Vulnerability analysis was performed using reclassification of multicriteria analysis based on the fuzzy membership functions.

#### 3.2.2.3 Reclassification by Fuzzy Membership Function

Fuzzy set theories allow the modeling of both suitability and vulnerability assessment within a GIS of a specific domain. In the fuzzy standard approach, it is clearly and crisply defined within a membership class or set, and they are either in the class or not (Bellman and Zadeh 1970). In this study, to assign vulnerability classes for agricultural land use, fuzzy membership classification was used to accommodate the high uncertainty scoring method. For standardization, fuzzy membership functions were used. According to the literature review and based on references, several

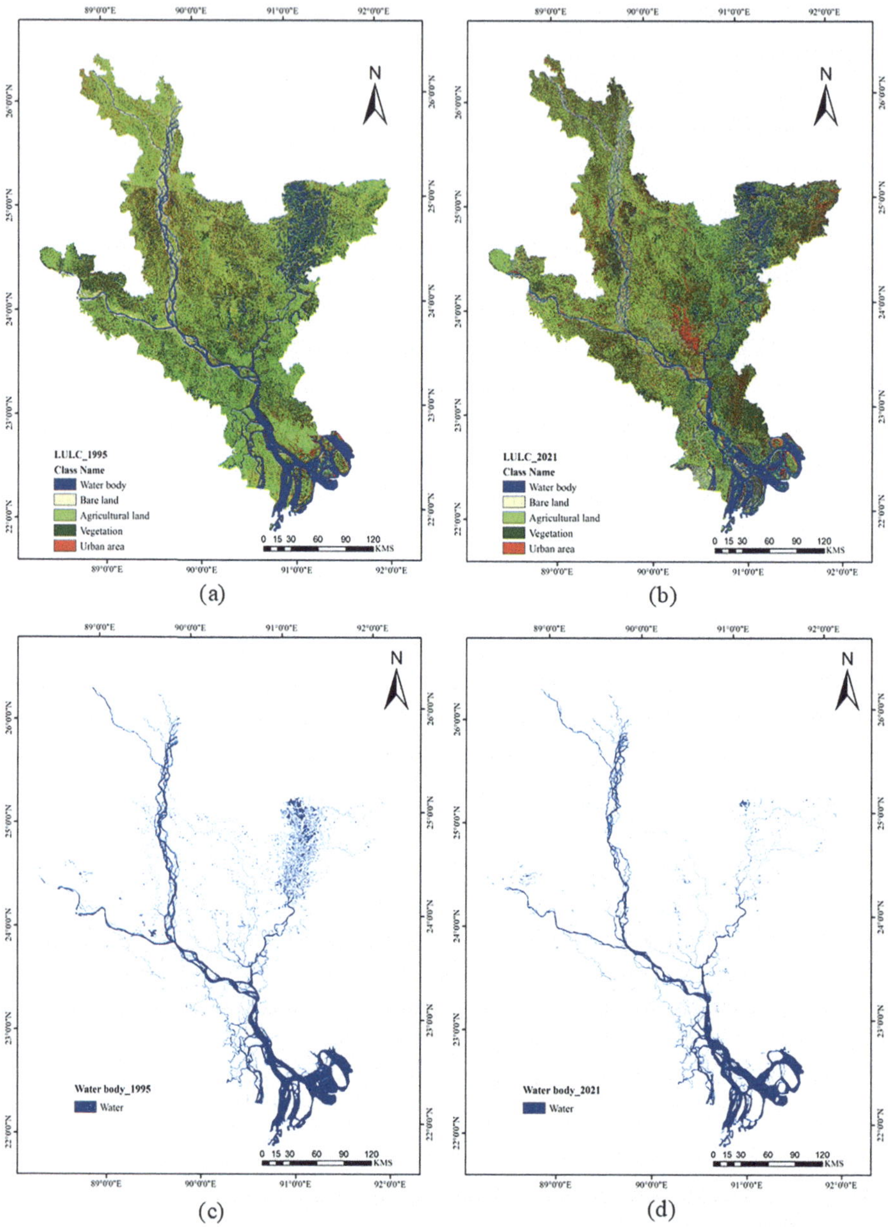


**Fig. 3.4** LULC and river distribution as permanent water bodies: (**a**) LULC map of 1995; (**b**) LULC map of 2021; (**c**) permanent water bodies in 1995; and (**d**) permanent water bodies in 2021

membership functions were selected. Seven fuzzy membership functions were built, and three were selected based on ecological, climatic, geological, and geographical criteria. The membership functions are Linear, Small, and Member Ship (MS) Large, which produce continuous fuzzy classifications of standardized criteria (Eqs.

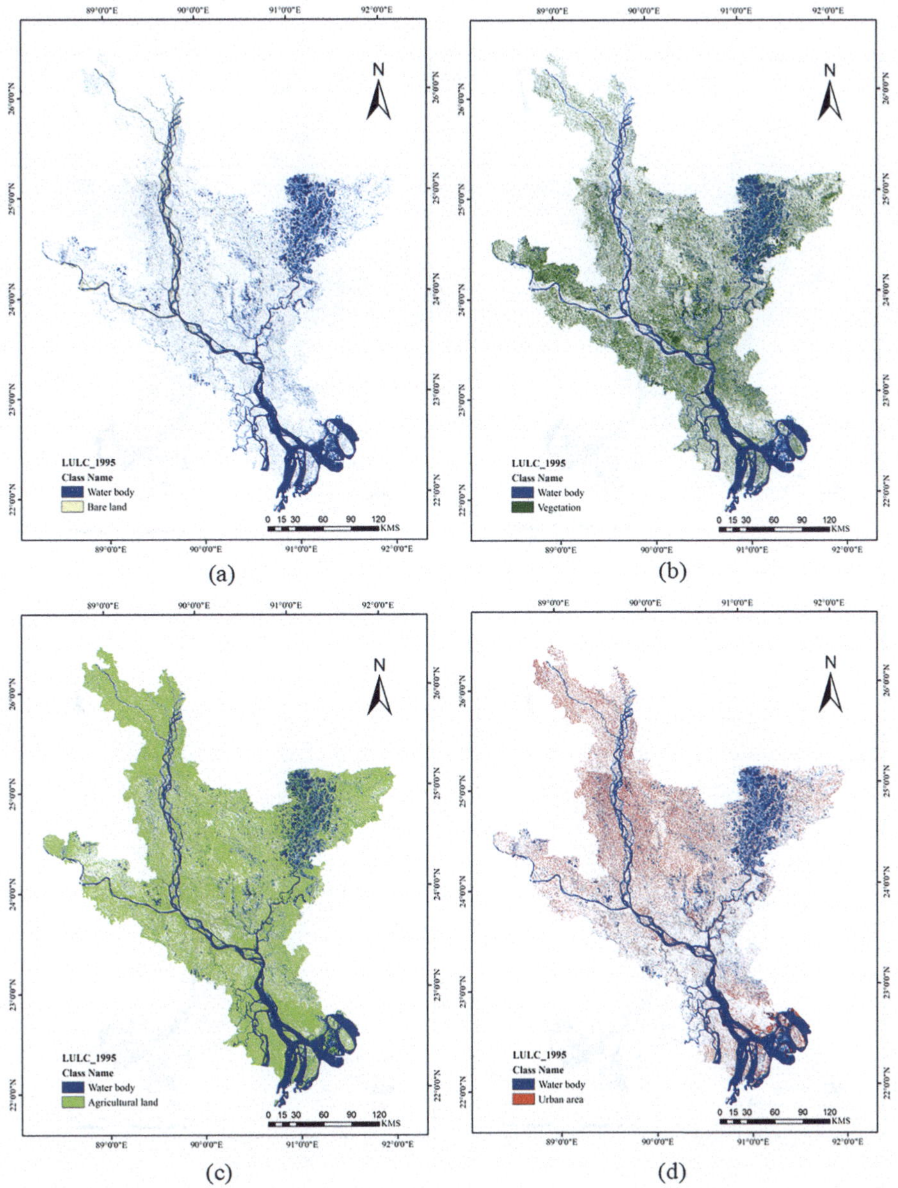


**Fig. 3.5** LULC map for different classes with permanent water bodies: (**a**) bare land; (**b**) vegetation; (**c**) agricultural land; and (**d**) urban area in 1995

(3.1)–(3.3)). A fuzzy small membership function was used when smaller input values were more likely to be a member of the set, a fuzzy MS large membership function was used when most of the input values were large members of the set, and a fuzzy linear membership function applied a linear function between the minimum and maximum values. The same equation was used for MS large membership and

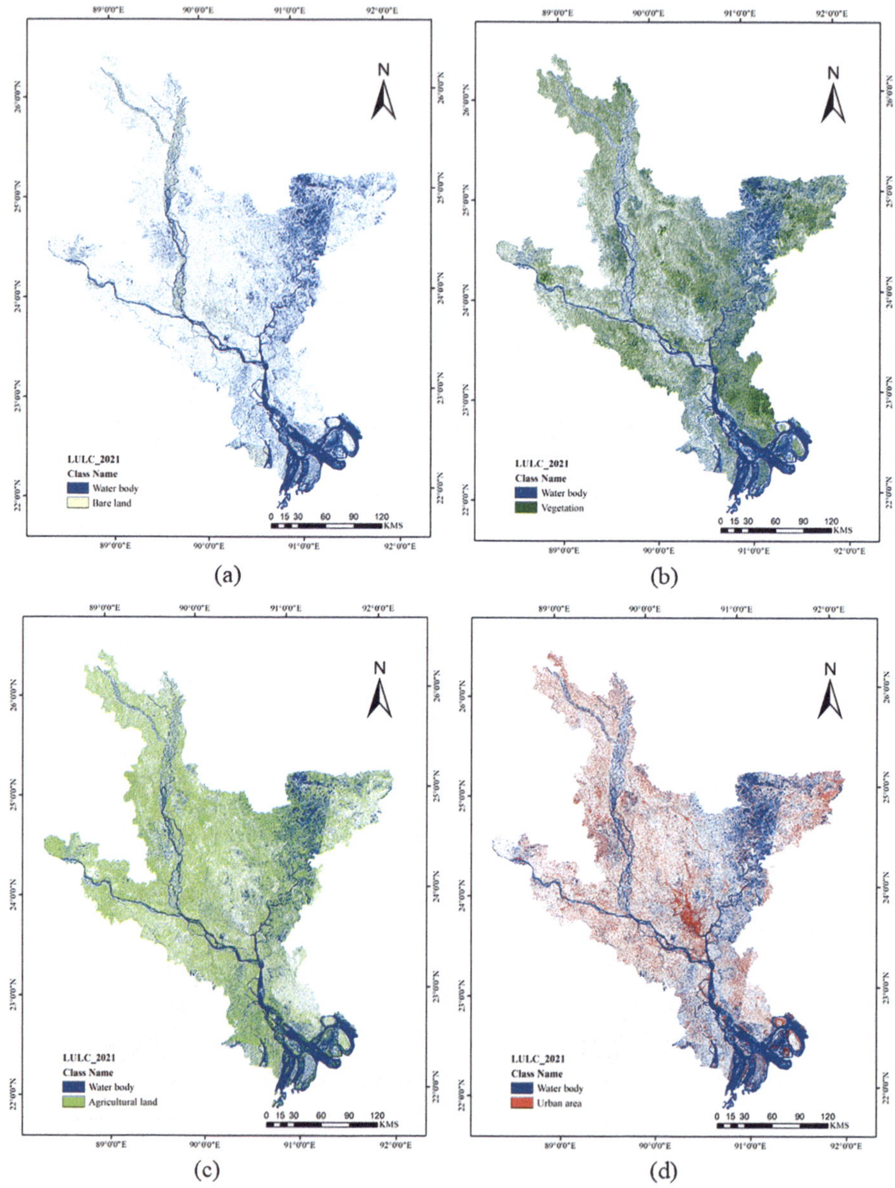


**Fig. 3.6** LULC map for different classes with permanent water bodies: (**a**) bare land; (**b**) vegetation; (**c**) agricultural land; and (**d**) urban area of 2021

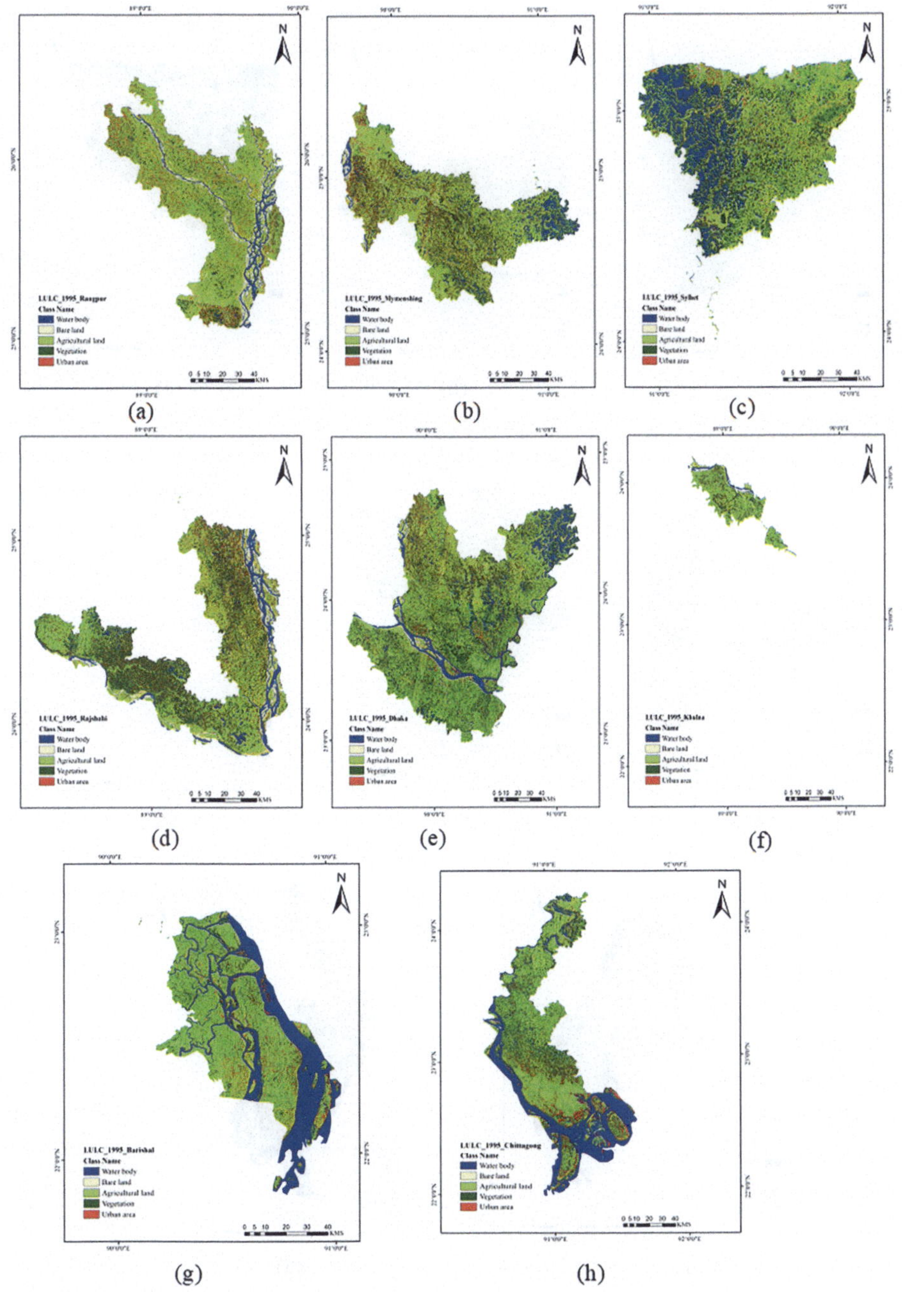


**Fig. 3.7** LULC maps for all administrative divisions of Bangladesh for 1995: (**a**) Rangpur; (**b**) Mymensingh; (**c**) Sylhet; (**d**) Rajshahi; (**e**) Dhaka; (**f**) Khulna; (**g**) Barishal; and (**h**) Chittagong

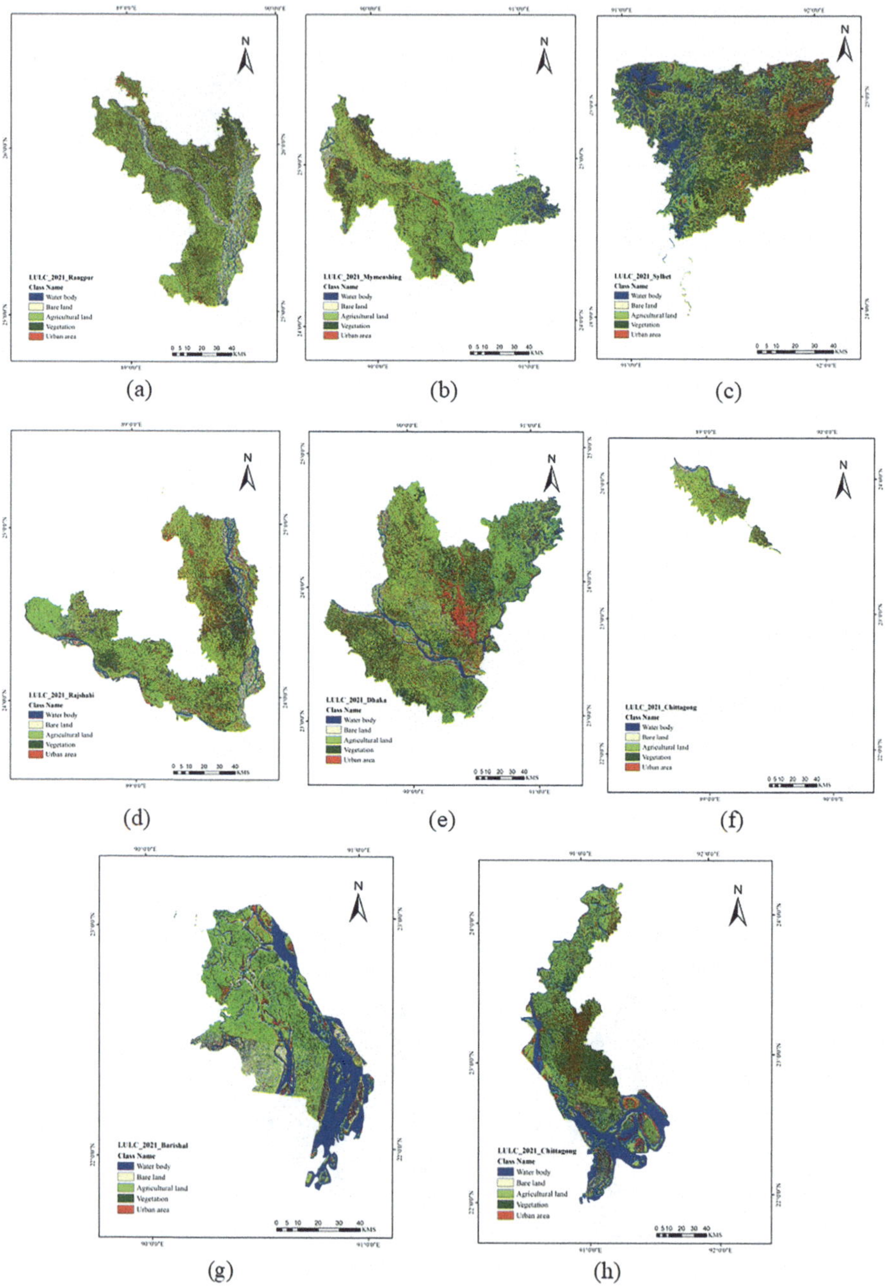


**Fig. 3.8** LULC map for all administrative divisions of Bangladesh for 2021: (**a**) Rangpur; (**b**) Mymensingh; (**c**) Sylhet; (**d**) Rajshahi; (**e**) Dhaka; (**f**) Khulna; (**g**) Barishal; and (**h**) Chittagong

large membership functions, according to the ArcGIS® environment provided by ESRI. For reclassification, the natural breaks (Jenks) method was used, as there were limited references regarding the vulnerability assessment for agricultural land use in terms of water inundation in the river basin area.

$$? = f(x) = \begin{cases} 0 & x ? a \\ \dfrac{x ? a}{b ? a} & a < x < b \\ 1 & x ? b \end{cases} \tag{3.1}$$

$$?(x) = \frac{1}{1 + \left(\dfrac{x}{f_2}\right)^{f_1}} \tag{3.2}$$

$$?(x) = \frac{1}{1 + \left(\dfrac{x}{f_2}\right)^{?f_1}} \tag{3.3}$$

In Eq. (3.1), $a$ and $b$ represent the value for the $x$ coordinate, where $x$ represents the real value (Crisp value). In Eqs. (3.2) and (3.3), $x$ is the crisp value, $f_1$ is the spread, and $f_2$ is the midpoint for the fuzzy large and MS large membership functions, which vary for different criteria.

#### 3.2.2.4 Criteria Selection for Vulnerability Assessment for Agricultural Land Use

For this study, four main criteria related to inundation by the water were selected. These criteria were NDWI, rainfall, elevation, and distance from the river, which are geological and geographical criteria. All the criteria were directly related to water inundation in the river basin area of Bangladesh. After calculating all the criteria, the vulnerability maps were reclassified into four classes: V1 (highly vulnerable area for agricultural land use), V2 (moderately vulnerable area for agricultural land use), V3 (marginally vulnerable area for agricultural land use), and N (nonvulnerable area for agricultural land use).

#### 3.2.2.5 Normalized Difference Water Index (NDWI)

The NDWI was taken as the criterion for the vulnerability analysis and was used to predict vulnerable areas for agricultural land use in terms of water inundation. Furthermore, NDWI was used to identify permanent water bodies, such as rivers, and permanently waterlogged areas were very important for vulnerability assessment. The areas near the water source were much more suitable for agriculture, but those areas were also vulnerable to inundation during the rainy season by flood water. As our study area was within the 25 km territory of the main river systems of Bangladesh, NDWI could be an important criterion for the prediction of vulnerable agricultural lands. In this study, Landsat 8 OLI and Landsat 5 images were used to calculate the NDWI, and two spectral bands, green and near-infrared (NIR), were utilized for the calculation. The following equation was used to calculate the NDWI (McFeeters 1996):

$$\text{NDWI} = \frac{\text{Green} ? \text{NIR}}{\text{Green} + \text{NIR}} \tag{3.4}$$

Finally, the NDWI was standardized with the fuzzy linear membership function (Fig. 3.9a), and the study area was reclassified into four vulnerability classes for agricultural land use (Fig. 3.10a).

#### 3.2.2.6 Rainfall

In this study, the average rainfall map for 2021 was used from the Center for Hydrometeorology and Remote Sensing (CHRS) Data Portal. According to the rainfall map for 2021 in Bangladesh, the highest and lowest rainfall was recorded as 2455 mm and 1117 mm, respectively. The rainfall map was resampled with a 30 × 30 m cell size using the resampling tool and then standardized by applying the fuzzy small membership function in ArcGIS® (Fig. 3.9b). Finally, the rainfall map was reclassified according to four vulnerability classes (Fig. 3.10d).

#### 3.2.2.7 Elevation

There is a direct relationship between the elevation of land and inundation by water. Lands of higher elevation are less likely to be affected by water than lands of low elevation. Our analyzed area in this study was the 25 km territory of the main river systems of Bangladesh. Landsat SRTM elevation data were analyzed for this study to prepare the elevation map. From the analysis, the highest and lowest elevations were 225 m and −25 m, respectively. As most areas of Bangladesh are plain lands and are suitable for agriculture, they are at the same time vulnerable to inundation by flood water due to lower elevations relative to the mean sea level. During floods and monsoons, most areas near the river go underwater, which causes major damage to crops. The prepared elevation map from Landsat SRTM data was standardized using the fuzzy MS large membership function (Fig. 3.9b), and the area was reclassified into the four vulnerability classes (Fig. 3.10c).

#### 3.2.2.8 Distance from the River

In this study, distance from the river was considered the most important criterion, as the areas nearest to the river have the highest inundation risk. To calculate the distance from the river, a river shape file was collected from the Bangladesh Bureau of

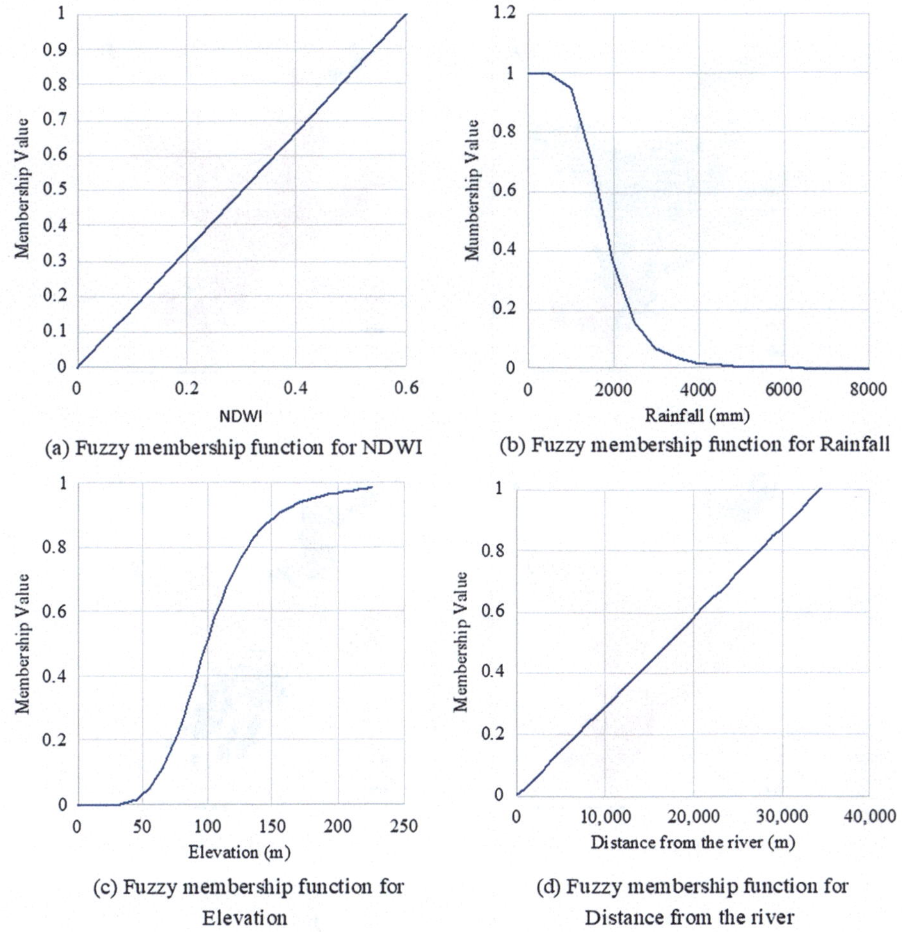


**Fig. 3.9** Fuzzy membership functions used for vulnerability analysis of agricultural land use: (**a**) fuzzy linear; (**b**) fuzzy small; (**c**) fuzzy MS large; and (**d**) fuzzy linear

Statistics (BBS), and the Euclidean distance tool was used to calculate the distance in the ArcGIS® spatial environment. Finally, a fuzzy linear membership function was used to standardize the value (Fig. 3.9d), and then the map was classified based on the four vulnerability classes (Fig. 3.10b).

#### 3.2.2.9 Fuzzy Overlay for Vulnerability Analysis of Agricultural Land Use

In a multicriteria overlay analysis, the possibility of a phenomenon belonging to multiple sets can be analyzed using fuzzy overlay tools. Fuzzy overlays analyze the relationships between the membership of the multiple sets and determine in which

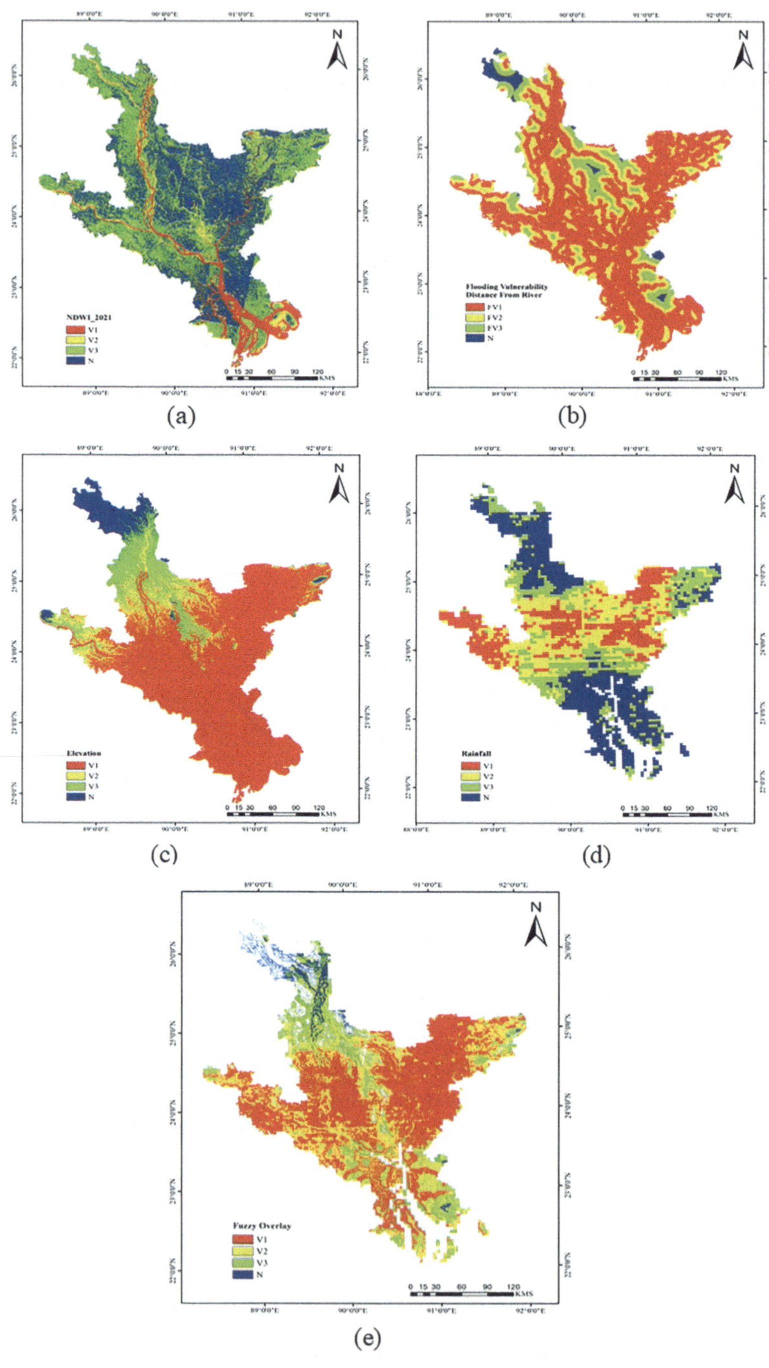


**Fig. 3.10** Reclassified vulnerability map of different indices for agricultural land use: (**a**) NDWI; (**b**) distance from the river; (**c**) elevation; (**d**) rainfall; and (**e**) fuzzy gamma overlay, where V1, V2, V3, and N refer to high, moderate, marginal, and nonvulnerable areas, respectively

sets the phenomenon is possibly present. Five overlay methods are available to combine the data based on set theory analysis. They are Fuzzy and, Fuzzy or, Fuzzy product, Fuzzy sum, and Fuzzy gamma. In this study, a fuzzy gamma overlay was conducted to prepare a vulnerability map from the criteria map produced from the fuzzy linear, fuzzy small, and fuzzy MS large membership functions (Fig. 3.10e). Fuzzy gamma is an algebraic product of the fuzzy sum and fuzzy product, which can be expressed by the following expression (ESRI 2016):

$$?(x) = (\text{FuzzySum})^{y}\ ?(\text{Fuzzy Product})^{1?y} \tag{3.5}$$

The default gamma value of 0.9 was used to generalize the fuzzy gamma overlay function. Then, the prepared vulnerability map was reclassified into four classes using the Jenks Natural Breaks algorithm (Jenks 1967). In this classification, the classes were based on inherent natural groupings in the database (Jenks and Caspall 1971). Finally, the prepared vulnerability map was masked by the eight administrative division shape files of Bangladesh for further analysis (Fig. 3.11).

### 3.2.3 *Accuracy Assessment*

A total of 2564 sample points were taken for the training dataset as feature collections of 5 land use classes for preparing the LULC map of 1995. Of the sample points, 713 were selected for water bodies, 148 points for bare land, 302 points for vegetation, 1116 points for agricultural land, and 285 points for urban areas. To prepare the LULC map for 2021, a total of 1755 sample points were taken as feature collections of 5 land use classes, where 726 sample points were for water bodies, 130 points were for bare land, 237 points were for vegetation, 415 points were for agricultural land, and 247 points were for urban areas. In the accuracy assessment, the training datasets consisted of 70% of the total samples, and 30% were used for testing in both 1995 and 2021. A confusion matrix was generated from 30% of the training dataset points to identify the degree of misclassification in this classification. Higher accuracy was observed in the Google Earth Engine platform, and the script has been illustrated in the Materials and Methods section in Algorithm 4 (Appendices 1 and 2).

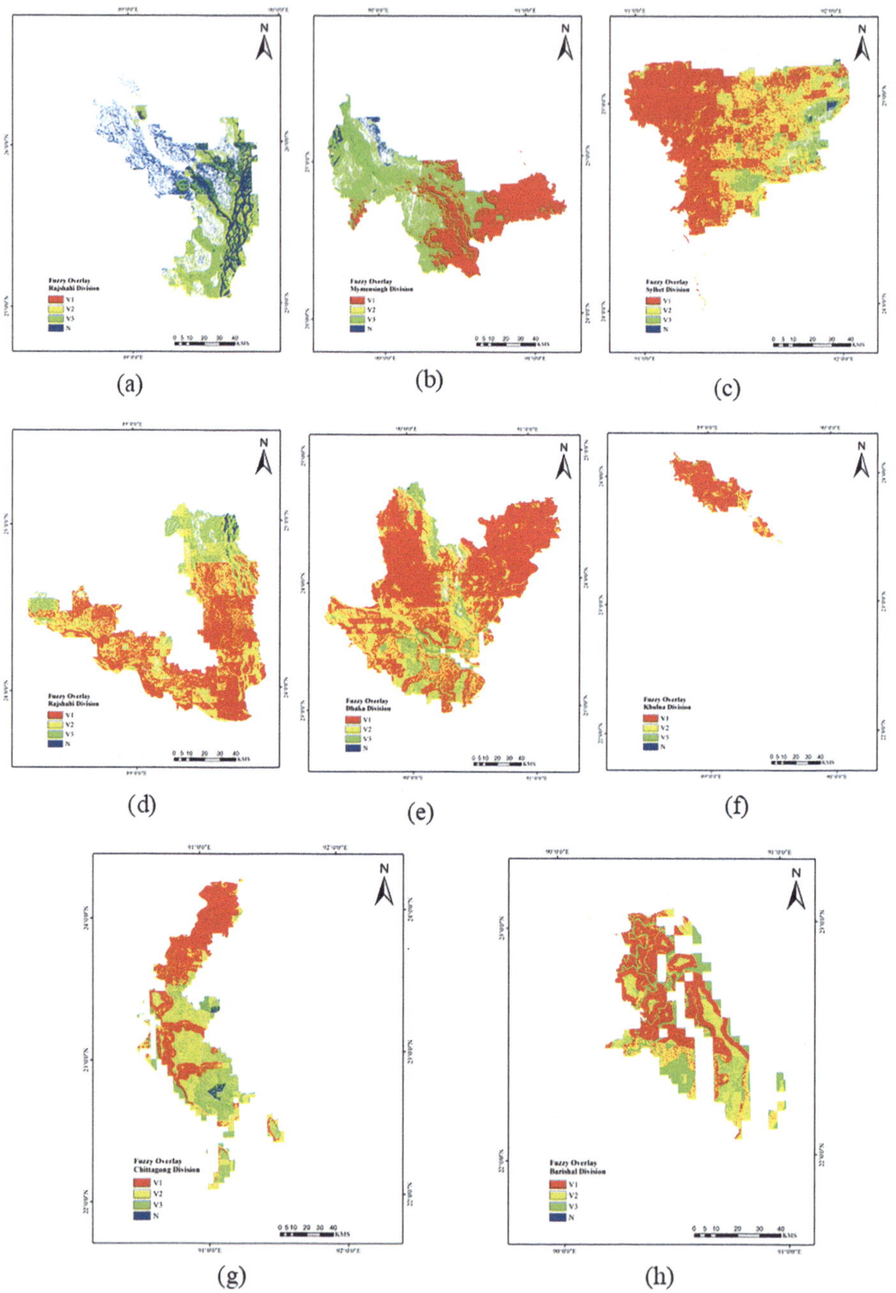


**Fig. 3.11** Reclassified fuzzy overlay vulnerability map, where V1, V2, V3, and N refer to high, moderate, marginal, and nonvulnerable areas, respectively, of different administrative divisions of Bangladesh: (**a**) Rangpur; (**b**) Mymensingh; (**c**) Sylhet; (**d**) Rajshahi; (**e**) Dhaka; (**f**) Khulna; (**g**) Barishal; and (**h**) Chittagong

## 3.3 Results and Discussion

### *3.3.1 Land Use/Land Cover Assessment*

The land use land cover (LULC) maps for 1995 and 2021 were prepared using Landsat-5 and Landsat 8 (OLI) images, respectively, for a total area of 64,555 km$^2$. The analysis area was distributed within a 25-km territory of the main river systems of Bangladesh. From the LULC analysis for 1995, it was found that among the five land use classes, agricultural land (44.41%) and bare land (2.96%) occupied the highest and the lowest amount of analyzed area.

From the LULC map for 2021, it was also found that both agricultural land (32.73%) and bare land (5.13%) possessed the highest and the lowest amount of studied area. According to our analysis, water bodies, bare land, vegetation, and urban areas increased in 26-year span from 1995 to 2021, whereas a significant reduction of 11.68% was observed in agricultural land. The above results suggested that all other LULC classes occupied the former agricultural land (Fig. 3.12; Table 3.2). Agricultural lands were reduced to meet the accommodation places, infrastructure, as well as industrial facilities development for the increasing population of Bangladesh. They have also been converted into orchards for fruit production as fruit cultivations are less vulnerable to water inundation.

After analyzing the LULC map of eight administrative divisions, varied changes in land use classes were observed. Among the five LULC classes, water bodies, bare land, and urban areas increased in all the administrative divisions except Rangpur and Mymensingh. The areas of vegetation increased significantly in the Rangpur, Mymensingh, Sylhet, and Chittagong divisions. In the Rangpur division, vegetation increased by 19.44%, whereas agricultural land decreased by 23.3%, which suggested that the major agricultural land in this division was converted into areas of vegetation. Most of the areas of vegetation were used for fruit cultivation. In the Dhaka, Khulna, Sylhet, and Chittagong divisions, agricultural land decreased significantly from 1995 to 2021. In the Dhaka division, the urban area increased by 4.2%, whereas agricultural land decreased by 7%, and this agricultural land was converted into urban areas. The same scenario was observed for the Khulna, Sylhet, and Chittagong divisions; however, vegetation also increased. The one change that

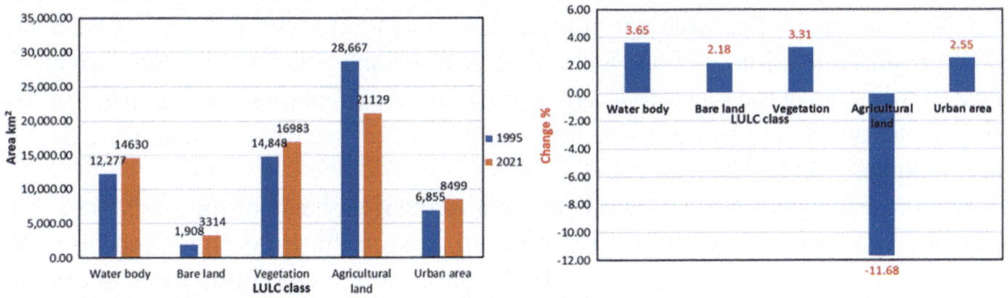


**Fig. 3.12** LULC of five classes by area and changes from 1995 to 2021

**Table 3.2** LULC area of 1995 and 2021 for the study area

| | For the year 1995 | | For the year 2021 | | Changes |
|---|---|---|---|---|---|
| LULC classes | Area (km$^2$) | Area (%) | Area (km$^2$) | Area (%) | Area (%) |
| Water body | 12,277 | 19.02 | 14,630 | 22.66 | 3.65 |
| Bare land | 1908 | 2.96 | 3314 | 5.13 | 2.18 |
| Vegetation | 14,848 | 23.00 | 16,983 | 26.31 | 3.31 |
| Agricultural land | 28,667 | 44.41 | 21,129 | 32.73 | −11.68 |
| Urban area | 6855 | 10.62 | 8499 | 13.17 | 2.55 |

was common for all the divisions was a reduction in agricultural land area, which is alarming for the food security of Bangladesh (Table 3.3; Fig. 3.13).

### 3.3.2 *Vulnerability Analysis for Agricultural Land Use*

According to our analysis of the 53,037 km$^2$ area (total land area), 44.34% was found to be highly vulnerable for agricultural land use, 37.75% was moderately vulnerable, 13.92% was marginally vulnerable, and 3.98% was not vulnerable for agricultural use in terms of inundation by water (Table 3.4). Similar research was also reported for the northeastern region of Bangladesh, where every year, thousands of hectares of crops become damaged in the haor area by flash floods (Islam et al. 2022).

The vulnerability map was finally divided into the eight administrative divisions of Bangladesh according to their coverage in the analyzed area. Of all the divisions, Dhaka, Khulna, Mymensingh, and Barishal have the highest coverage of lands vulnerable for agricultural land use, whereas Rangpur, Rajshahi, and Chittagong have less vulnerable areas, and those divisions are relatively higher in elevation than the mean sea level. According to the land use land cover change assessment, it was found that water bodies and urban areas were increased in all administrative divisions, which made the agricultural land more vulnerable in these divisions. Rangpur had the largest number of nonvulnerable areas for agricultural land use. However, most of the agricultural lands were transformed into fruit cultivation, which was reflected by the increment of vegetated land in the LULC change assessment. In contrast, Khulna had the largest number of vulnerable areas for agricultural land use. However, nonvulnerable agricultural land was found within the analyzed zone of the Khulna division, as only a small area was analyzed for this study. Overall, most of the areas of the eight divisions were in high to moderately vulnerable classes for inundation by floods, but those areas were also reported as suitable for agricultural land use (Table 3.5; Fig. 3.14).

Land use conversion is a dynamic process in the river basin of the study area, and it makes agricultural land vulnerable through erosion, flash flooding, and flooding during the rainy season (Islam et al. 2022). There were some limitations in obtaining permanent bench points and a lack of yearly observations in the large deltaic

**Table 3.3** LULC changes from 1995 to 2021 in eight administrative divisions in Bangladesh

| LULC classes | Administrative division | For the year 1995 Area (%) | For the year 1995 Area (km²) | For the year 2021 Area (%) | For the year 2021 Area (km²) | Changes Area (%) |
|---|---|---|---|---|---|---|
| Water body | Rangpur | 9.07 | 668 | 11.28 | 831 | 2.21 |
| | Rajshahi | 13.52 | 1099 | 14.61 | 1188 | 1.09 |
| | Dhaka | 15.03 | 2723 | 15.91 | 2881 | 0.87 |
| | Khulna | 7.45 | 129 | 10.19 | 177 | 2.74 |
| | Mymensingh | 13.15 | 991 | 14.41 | 1086 | 1.26 |
| | Sylhet | 25.20 | 1624 | 27.72 | 1786 | 2.53 |
| | Barishal | 35.37 | 1708 | 41.83 | 2020 | 6.46 |
| | Chittagong | 32.62 | 3368 | 38.04 | 3893 | 5.42 |
| Bare land | Rangpur | 8.21 | 605 | 9.98 | 736 | 1.77 |
| | Rajshahi | 5.80 | 471 | 7.21 | 586 | 1.41 |
| | Dhaka | 2.28 | 413 | 4.35 | 788 | 2.07 |
| | Khulna | 4.62 | 80 | 7.99 | 139 | 3.37 |
| | Mymensingh | 3.32 | 250 | 3.61 | 272 | 0.29 |
| | Sylhet | 0.40 | 26 | 0.98 | 63 | 0.58 |
| | Barishal | 0.16 | 8 | 7.50 | 362 | 7.33 |
| | Chittagong | 0.38 | 39 | 2.20 | 225 | 1.82 |
| Vegetation | Rangpur | 11.96 | 881 | 31.39 | 2313 | 19.44 |
| | Rajshahi | 27.56 | 2241 | 23.87 | 1941 | −3.69 |
| | Dhaka | 28.82 | 5219 | 28.64 | 5188 | −0.18 |
| | Khulna | 29.65 | 514 | 27.43 | 476 | −2.22 |
| | Mymensingh | 19.80 | 1492 | 26.61 | 2005 | 6.81 |
| | Sylhet | 23.18 | 1493 | 29.34 | 1891 | 6.17 |
| | Barishal | 16.42 | 793 | 11.33 | 547 | −5.10 |
| | Chittagong | 21.35 | 2205 | 25.17 | 2576 | 3.82 |
| Agricultural land | Rangpur | 56.99 | 4199 | 33.69 | 2483 | −23.30 |
| | Rajshahi | 39.62 | 3222 | 39.91 | 3246 | 0.29 |
| | Dhaka | 44.13 | 7993 | 37.16 | 6731 | −6.97 |
| | Khulna | 50.39 | 874 | 41.77 | 724 | −8.62 |
| | Mymensingh | 48.24 | 3635 | 43.82 | 3302 | −4.42 |
| | Sylhet | 42.76 | 2756 | 27.76 | 1789 | −15.01 |
| | Barishal | 42.18 | 2037 | 33.36 | 1611 | −8.82 |
| | Chittagong | 37.65 | 3888 | 24.41 | 2499 | −13.24 |
| Urban area | Rangpur | 13.77 | 1015 | 13.65 | 1006 | −0.12 |
| | Rajshahi | 13.51 | 1099 | 14.40 | 1171 | 0.89 |
| | Dhaka | 9.73 | 1763 | 13.94 | 2524 | 4.20 |
| | Khulna | 7.89 | 137 | 12.62 | 219 | 4.73 |
| | Mymensingh | 15.48 | 1167 | 11.55 | 870 | −3.94 |
| | Sylhet | 8.47 | 546 | 14.20 | 915 | 5.73 |
| | Barishal | 5.86 | 283 | 5.99 | 289 | 0.13 |
| | Chittagong | 8.00 | 826 | 10.19 | 1043 | 2.19 |

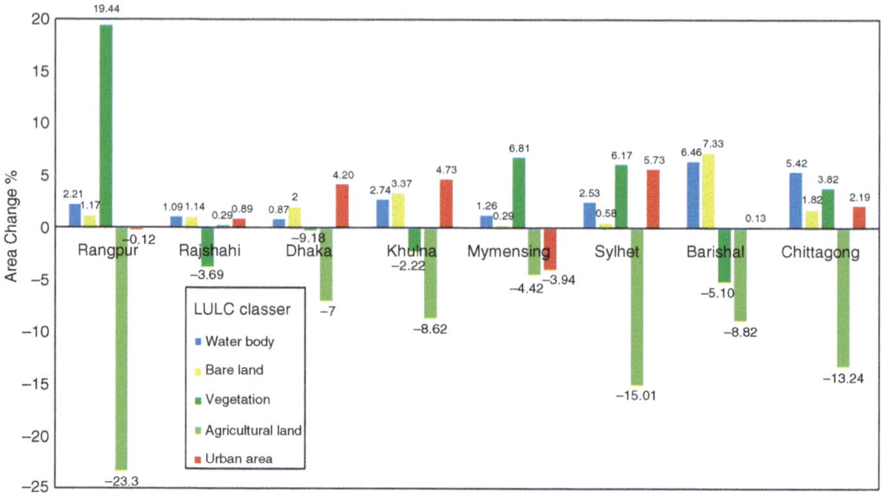


**Fig. 3.13** Administrative division-wise comparison between changes of five land use land cover classes from 1995 to 2021 in the river basin of Bangladesh

**Table 3.4** Vulnerable areas for agricultural land use according to their vulnerability classes

| Vulnerability classes | Area ($km^2$) | Area (%) |
|---|---|---|
| $V_1$ (highly vulnerable area for agricultural land use) | 23,518 | 44.34 |
| $V_2$ (moderately vulnerable area for agricultural land use) | 20,023 | 37.75 |
| $V_3$ (marginally vulnerable area for agricultural land use) | 7385 | 13.92 |
| N (nonvulnerable area for agricultural land use) | 2111 | 3.98 |

region to compare the study results with ground truth data. The vulnerability map for agricultural land use was compared with the flood inundation map of Bangladesh developed from the 120-h forecast dataset by the Bangladesh Water Development Board (BWDB) on August 16, 2017 (Importance of making more water dams and construction of concretes river embankment in Bangladesh to protect flood 2022).

**Table 3.5** Vulnerable areas for agricultural land use according to the vulnerability classes in eight administrative divisions in Bangladesh

| Vulnerability classes | Administrative divisions | Area (%) | Area (km²) |
|---|---|---|---|
| V1 | Rangpur | 0.05 | 2 |
| | Rajshahi | 42.42 | 3190 |
| | Dhaka | 54.26 | 9497 |
| | Khulna | 73.31 | 1217 |
| | Mymensingh | 40.81 | 2673 |
| | Sylhet | 49.16 | 3144 |
| | Barishal | 40.04 | 1149 |
| | Chittagong | 37.52 | 2649 |
| V2 | Rangpur | 13.87 | 533 |
| | Rajshahi | 46.60 | 3504 |
| | Dhaka | 38.66 | 6766 |
| | Khulna | 26.17 | 434 |
| | Mymensingh | 36.90 | 2418 |
| | Sylhet | 37.31 | 2385 |
| | Barishal | 39.64 | 1137 |
| | Chittagong | 40.69 | 2873 |
| V3 | Rangpur | 41.21 | 1584 |
| | Rajshahi | 10.23 | 769 |
| | Dhaka | 7.03 | 1231 |
| | Khulna | 0.52 | 9 |
| | Mymensingh | 19.52 | 1279 |
| | Sylhet | 8.04 | 5134 |
| | Barishal | 20.30 | 582 |
| | Chittagong | 20.31 | 1434 |
| N | Rangpur | 44.87 | 1724 |
| | Rajshahi | 0.74 | 56 |
| | Dhaka | 0.05 | 8 |
| | Khulna | 0.00 | 0 |
| | Mymensingh | 2.77 | 181 |
| | Sylhet | 5.49 | 351 |
| | Barishal | 0.02 | 1 |
| | Chittagong | 1.48 | 104 |

$V_1$ highly vulnerable area, $V_2$ moderately vulnerable area, $V_3$ marginally vulnerable area, and *N* nonvulnerable area for agricultural land use

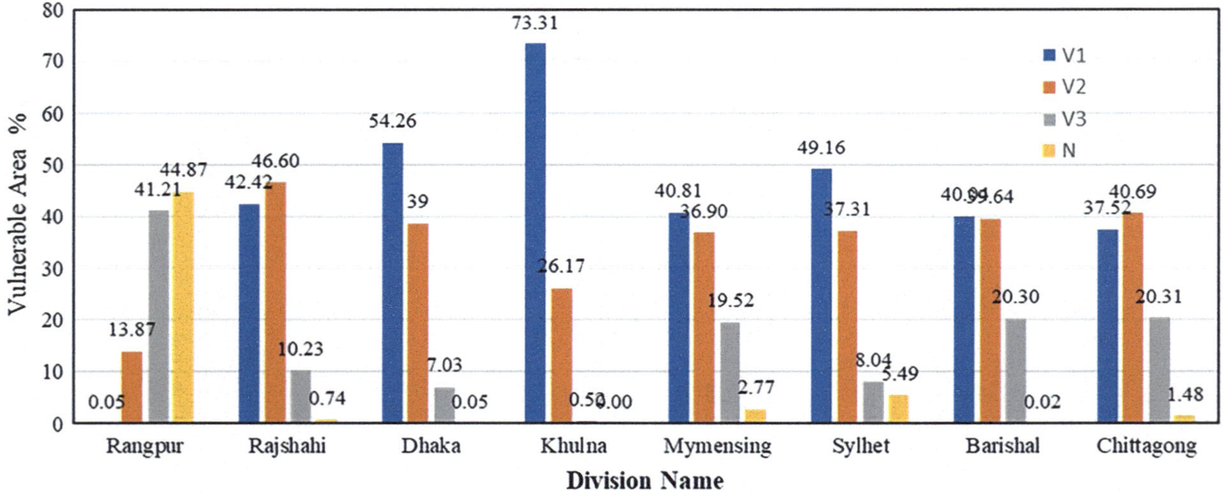


**Fig. 3.14** Vulnerable areas for agricultural land use by division according to the vulnerability classes in Bangladesh (*V1* highly vulnerable area, *V2* moderately vulnerable area, *V3* marginally vulnerable area, and *N* nonvulnerable area)

The vulnerability map of agricultural land use looks like the flood inundation map of 2017.

## 3.4 Conclusion

The present study was conducted to assess the LULC changes from 1995 to 2021 over 26 years within the 25 km territory of the main river systems in Bangladesh. Furthermore, the prediction of vulnerable areas for agricultural land use in terms of inundation by river water was also addressed. All the river and administrative shape files were collected from the Bangladesh Bureau of Statistics. The study was performed in the Google Earth Engine platform and analyzed in the ArcGIS® environment. According to our LULC maps for 1995 and 2021, of the five land use classes, water bodies, bare land, vegetation, and urban areas increased by 3.65%, 2.18%, 3.31%, and 2.55%, respectively, from 1995 to 2021 in the analyzed area, whereas a significant reduction of 11.68% was observed in agricultural land. The total reduction in agricultural land was approximately equal to the sum of the increases of the other land use classes. The main causes of the reduction of agricultural lands are conversion to urban and industrial settlements and the risk of crop loss due to water inundation threatening the livelihoods of the inhabitants of the large river tracks. That is why long-term land use planning, especially vulnerable lands in the river course, is very much required throughout the country. This research is one of the first attempts to identify the vulnerable land uses in the country's river basins for assessments. According to our vulnerability map of the study area for agricultural land use, which was prepared by multicriteria analysis using a fuzzy expert system, the highly vulnerable area coverage was 44.34%, the moderately vulnerable area was 37.75%, the marginally vulnerable area was 13.98%, and the nonvulnerable area was 3.98% for agricultural land use. From the analysis, it was observed that the most changed land use class was agricultural land, and approximately 50% of the studied area was highly vulnerable to agricultural land use in terms of water inundation.

The developed method for land use and land cover based on the GEE had higher accuracy in interpreting changes in the country scale from the Ganges–Brahmaputra (GBM) track to the deltaic transformation. Therefore, the presented LULC and vulnerability maps can be helpful for the future land use planning of Bangladesh to meet the increasing demand for agricultural land use for food production. Furthermore, the methodological process and outcomes of this research can be used for making an effective agricultural policy mitigating the damage of agricultural lands during a specific time of the year.

**Acknowledgments** The authors express their gratitude to the open-access journal *Remote Sensing* published by Multidisciplinary Digital Publishing Institute (MDPI) for their policy to support the authors for reusing the published article. We would like to thank *Remote Sensing* journal in publishing this article (Kazi Faiz Alam, Tofael Ahamed. Assessment of Land Use Land Cover Changes for Predicting Vulnerable Agricultural Lands in River Basins of Bangladesh Using Remote Sensing and a Fuzzy Expert System. *Remote Sensing*. *14*(21), 5582; https://doi.org/10.3390/rs14215582. 2022). Furthermore, the authors would like to acknowledge the University of Tsukuba for its support by providing software and laboratory facilities. We also express our sincere gratitude to the Bangladesh Bureau of Statistics (BBS) for providing the administrative and river shape files and the United States Geological Survey (USGS) for allowing free access to the Landsat satellite images and Digital Elevation Model (DEM) data. We are also grateful to the Google Earth Engine Platform for free access for students and researchers for their studies. The author is grateful to the Rotary Yoneyama Foundation for financial support through a monthly scholarship.

**Funding** This research received no external funding.

**Conflicts of Interest** The authors declare no conflict of interest.

# Appendix 1. LULC-1995

```
var L5 = ee.ImageCollection("LANDSAT/LT05/C01/T1_TOA")//Image retrieval of Landsat 5
        .filterBounds(ROI)
        .filterDate("1995-01-01", "1996-3-30")
        .filterMetadata('CLOUD_COVER','less_than', 1)
        .mean()
        .clip(ROI);

Map.addLayer(L5, {bands:["B4" , "B3" , "B2"]}); // False color Composite image preparation
var training_points = Water.merge(Chorland).merge(Vegetation).merge(Agriculture).merge(Urban);

var training_data = L5.sampleRegions({collection:training_points,properties:['LC'],scale:30})
var classifier = ee.Classifier.smileCart()
var classifier = classifier.train({features:training_data,classProperty: 'LC',inputProperties:["B1",
"B2", "B3", "B4", "B5", "B6", "B7"]});

var classified_image = L5.classify(classifier);
var trainingData = training_data.randomColumn();
var trainSet = trainingData.filter(ee.Filter.lessThan('random', 0.7));
var testSet = trainingData.filter(ee.Filter.greaterThanOrEquals('random', 0.7));
Map.addLayer(classified_image,{min:0, max:4, palette:[ 'pink', 'yellow', 'green', 'blue',
'red']},'classified image');

//Accuracy Assessment
//Classify the dataSet and get confusion matrix
var confusionMatrix = ee.ConfusionMatrix(testSet.classify(classifier)
        .errorMatrix({
  actual: 'LC',
  predicted: 'classification'
        }))
        print('Confusion Matrix', confusionMatrix);
        print("Overall Accuracy:", confusionMatrix.accuracy());

//Map export to google drive
Export.image.toDrive({
        image: classified_image,
        description:'ExportedData',
        folder:"GEE",
        region: ROI,
        scale: 30,
        maxPixels: 1e13,
        });
```

# Appendix 2. LULC-2021

```
//LULC-2021

var L8 = ee.ImageCollection("LANDSAT/LC08/C01/T1_TOA")//Image retrieval of Landsat 8(OLI)

        .filterBounds(ROI)

        .filterDate("2021-01-01", "2022-3-30")

        .filterMetadata('CLOUD_COVER','less_than', 1)

        .mean()

        .clip(ROI);


Map.addLayer(L8, {bands:["B5" , "B4" , "B3"]});//False color Composite image preparation

var training_points =  Water.merge(Chorland).merge(Vegetation).merge(Agriculture).merge(Urban);

var training_data = L8.sampleRegions({collection:training_points,properties:['LC'],scale:30})

var classifier = ee.Classifier.smileCart()

var classifier = classifier.train({features:training_data,classProperty: 'LC',inputProperties:["B1",
"B2", "B3", "B4", "B5", "B6", "B7"]});

var classified_image = L8.classify(classifier);

var trainingData = training_data.randomColumn();

var trainSet = trainingData.filter(ee.Filter.lessThan('random', 0.8));

var testSet = trainingData.filter(ee.Filter.greaterThanOrEquals('random', 0.8));

Map.addLayer(classified_image,{min:0, max:4, palette:[ 'pink', 'yellow', 'green', 'blue',
'red']},'classified image');


//Accuracy Assessment

//Classify the testSet and get a confusion matrix

var confusionMatrix = ee.ConfusionMatrix(testSet.classify(classifier)

        .errorMatrix({

        actual: 'LC',

        predicted: 'classification'

        }));

        print('Confusion Matrix', confusionMatrix);

        print("Overall Accuracy:", confusionMatrix.accuracy());


//Map export to google drive

Export.image.toDrive({
image: classified_image,

description:'ExportedData',

folder:"GEE",

region: ROI,

scale: 30,

maxPixels: 1e13,

});
```

# Appendix 3. NDWI Map for 1995 and 2021

```
//NDWI-1995
var L5 = ee.ImageCollection("LANDSAT/LT05/C01/T1_TOA")//Image retrieval of Landsat 5
        .filterBounds(ROI)
        .filterDate("1995-01-01", "1996-3-30")
        .filterMetadata('CLOUD_COVER','less_than', 1)
        .mean()
        .clip(ROI);
        var green = L5.select('B2');
        var nir = L5.select('B4');
var ndwi = green.subtract(nir).divide(green.add(nir)).rename('NDWI');
var ndwiParams = {min: -1, max: 1, palette: ['black', 'white', 'blue']};
        Map.addLayer(ndwi, ndwiParams, 'NDWI image');

//Map export to google drive
Export.image.toDrive({
        image: ndwi,
        description:'ExportedData',
        folder:"GEE",
        region: ROI,
        scale: 30,
        maxPixels: 1e13,
        });
NDWI-2021
var L8 = ee.ImageCollection("LANDSAT/LC08/C01/T1_TOA") //Image retrieval of Landsat 8(OLI)
        .filterBounds(ROI)
        .filterDate("2021-01-01", "2022-3-30")
        .filterMetadata('CLOUD_COVER','less_than', 1)
        .mean()
        .clip(ROI);
var green = L8.select('B3');
var nir = L8.select('B5');
var ndwi = green.subtract(nir).divide(green.add(nir)).rename('NDWI');
var ndwiParams = {min: -1, max: 1, palette: ['black', 'white', 'blue']};
Map.addLayer(ndwi, ndwiParams, 'NDWI image');

//Map export to google drive
Export.image.toDrive({
        image: ndwi,
        description:'ExportedData',
        folder:"GEE",
        region: ROI,
        scale: 30,
        maxPixels: 1e13,
        });
```

## References

Akıncı H, Özalp AY, Turgut B (2013) Agricultural land use suitability analysis using GIS and AHP technique. Comput Electron Agric 97:71–82

Bangladesh Water Development Board (BWDB) (1974) Land reclamation project report. Government of Bangladesh, Dhaka

Bellman RE, Zadeh LA (1970) Decision-making in a fuzzy environment. Manag Sci 17(4):B-141

Csaplovics E (1998) High-resolution space imagery for regional environmental monitoring-status quo and future trends. Int Arch Photogramm Remote Sens 32:211–216

Das AC, Noguchi R, Ahamed T (2020) Integrating an expert system, gis, and satellite remote sensing to evaluate land suitability for sustainable tea production in Bangladesh. Remote Sens 12(24):4136

Disperati L, Virdis SGP (2015) Assessment of land-use and land-cover changes from 1965 to 2014 in Tam Giang-Cau Hai Lagoon, central Vietnam. Appl Geogr 58:48–64

ESRI (2016) Arithmetic function. https://desktop.arcgis.com/en/arcmap/10.3/manage-data/raster-and-images/arithmetic-function.htm. Accessed 10 Feb 2021

FAO (2021) The impact of disasters and crises 2021 on agriculture and food security: 2021. Rome

Ferretti V, Pomarico S (2012) Integrated sustainability assessments: a spatial multicriteria evaluation for siting a waste incinerator plant in the Province of Torino (Italy). Environ Dev Sustain 14:843–867

Ferretti V, Pomarico S (2013) An integrated approach for studying the land suitability for ecological corridors through spatial multicriteria evaluations. Environ Dev Sustain 15:859–885

Foody GM (2002) Status of land cover classification accuracy assessment. Remote Sens Environ 80(1):185–201

Frazier PS, Page KJ (2000) Water body detection and delineation with Landsat TM data. Photogramm Eng Remote Sens 66(12):1461–1468

Habibie MI, Noguchi R, Shusuke M, Ahamed T (2021) Land suitability analysis for maize production in Indonesia using satellite remote sensing and GIS-based multicriteria decision support system. GeoJournal 86:777–807

Hassan MM, Nazem MNI (2016) Examination of land use/land cover changes, urban growth dynamics, and environmental sustainability in Chittagong city, Bangladesh. Environ Dev Sustain 18:697–716

Hossain MS, Chowdhury SR, Das NG, Sharifuzzaman SM, Sultana A (2009) Integration of GIS and multicriteria decision analysis for urban aquaculture development in Bangladesh. Landsc Urban Plan 90(3–4):119–133

Hossain MS, Arshad M, Qian L, Kächele H, Khan I, Islam MDI, Mahboob MG (2020) Climate change impacts on farmland value in Bangladesh. Ecol Indic 112:106181

Importance of Making More Water Dams and Construction of Concretes River Embankment in Bangladesh to Protect Flood (2022). Sci[1]entific Figure on ResearchGate.. Available online: https://www.researchgate.net/figure/Flood-Inundation-Map-of-Bangladesh[1]120hr-Forecast-Based-on-16-August-2017-BWDB_fig4_346082375 (accessed on 18 July 2022)

Islam MM, Ujiie K, Noguchi R, Ahamed T (2022) Flash flood-induced vulnerability and need assessment of wetlands using remote sensing, GIS, and econometric models. Remote Sens Appl Soc Environ 25:100692

Jenks GF (1967) The data model concept in statistical mapping. Int Yearb Cartogr 7:186–190

Jenks GF, Caspall FC (1971) Error on choroplethic maps: definition, measurement, reduction. Ann Assoc Am Geogr 61(2):217–244

Johnson SY, Nur Alam AM (1991) Sedimentation and tectonics of the Sylhet trough, Bangladesh. Geol Soc Am Bull 103(11):1513–1527

Khan FH (1991a) Geology of Bangladesh. University Press Limited, Dhaka, p 207

Khan LR (1991b) Impacts of recent floods on the rural environment of Bangladesh: a case study. Int J Water Resour Dev 7(1):45–52

Khan I, Ahammad M, Sarker S (2014) A study on River Bank erosion of Jamuna River using GIS and remote sensing technology. Int J Eng Dev Res 2(4):3365–3371

Khosravi K, Nohani E, Maroufinia E, Pourghasemi HR (2016) A GIS-based flood susceptibility assessment and its mapping in Iran: a comparison between frequency ratio and weights-of-evidence bivariate statistical models with multi-criteria decision-making technique. Nat Hazards 83:947–987

Kumar L, Mutanga O (2018) Google Earth Engine applications since inception: usage, trends, and potential. Remote Sens 10(10):1509

Li X (1996) A review of the international researches on land use/land cover change. Acta Geogr Sin 51:558–565. Chinese Edition

McFeeters SK (1996) The use of the Normalized Difference Water Index (NDWI) in the delineation of open water features. Int J Remote Sens 17(7):1425–1432

Mugiraneza T, Nascetti A, Ban Y (2020) Continuous monitoring of urban land cover change trajectories with Landsat time series and LandTrendr-Google Earth Engine cloud computing. Remote Sens 12(18):2883

Muhsin N, Ahamed T, Noguchi R (2018) GIS-based multi-criteria analysis modeling used to locate suitable sites for industries in suburban areas in Bangladesh to ensure the sustainability of agricultural lands. Asia Pac J Reg Sci 2:35–64

Özelkan E (2020) Water body detection analysis using NDWI indices derived from Landsat-8 OLI. Pol J Environ Stud 29(2):1759–1769

Purnamasari RA, Noguchi R, Ahamed T (2019) Land suitability assessments for yield prediction of cassava using geospatial fuzzy expert systems and remote sensing. Comput Electron Agric 166:105018

Shafizadeh-Moghadam H, Khazaei M, Alavipanah SK, Weng Q (2021) Google Earth Engine for large-scale land use and land cover mapping: an object-based classification approach using spectral, textural and topographical factors. GIScі Remote Sens 58(6):914–928

Shamsuzzoha M, Noguchi R, Ahamed T (2021) Damaged area assessment of cultivated agricultural lands affected by cyclone bulbul in coastal region of Bangladesh using Landsat 8 OLI and TIRS datasets. Remote Sens Appl Soc Environ 23:100523

Tehrany MS, Lee MJ, Pradhan B, Jebur MN, Lee S (2014) Flood susceptibility mapping using integrated bivariate and multivariate statistical models. Environ Earth Sci 72:4001–4015

Uddin A, Lundberg N (1999) A paleo-Brahmaputra? Subsurface lithofacies analysis of Miocene deltaic sediments in the Himalayan–Bengal system. Bangladesh Sediment Geol 123(3–4):239–254

Zhao Q, Yu L, Li X, Peng D, Zhang Y, Gong P (2021) Progress and trends in the application of Google Earth and Google Earth Engine. Remote Sens 13(18):3778

# Chapter 4
# Assessment of Shoreline and Agricultural Land Use Changes in the Onshore Coastal Region of Bangladesh Delta Using Satellite Remote Sensing and GIS

**Md Shamsuzzoha and Tofael Ahamed**

**Abstract** Bangladesh is a global south hotspot due to climate change and sea level rise concerns. It is a highly disaster-prone country in the world with active deltaic shorelines. The shorelines are quickly changing to coastal accretion and erosion. Erosion is one of the water hazards to landmass sinking, and accretion related to land level rises due to sediment load deposition on the Bay of Bengal continental shelf. Therefore, this study aimed to explore shoreline status with change assessment for the three study years 1991, 2006, and 2021 using satellite remote sensing and geographical information system (GIS) approaches. Landsat 5, 7 (enhanced thematic mapper plus, ETM+), and 8 (operational land imager, OLI) satellite imageries were employed for onshore tasseled cap transformation (TCT) and land and sea classification calculations to create shore boundaries, baseline assessment, land accretion, erosion, point distance, and near-feature analysis. We converted 16,550 baseline vertices to points as the study ground reference points (GRPs) and validated those points using the country data sheet collected from the Survey of Bangladesh (SoB). We observed that the delta's shorelines were changed, and the overall lands were accredited for the land-increasing characteristics analysis. The total accredited lands in the coastal areas observed during the time periods from 1991 to 2006 were 825.15 km$^2$, from 2006 to 2021 were 756.69 km$^2$, and from 1991 to 2021 were 1223.94 km$^2$ for the 30-year period. Similarly, coastal erosion assessment analysis indicated that the results gained for the period 1991 to 2006 and 2006

M. Shamsuzzoha
Graduate School of Life and Environmental Sciences, University of Tsukuba, Tsukuba, Ibaraki, Japan

Department of Emergency Management, Faculty of Environmental Science and Disaster Management, Patuakhali Science and Technology University, Patuakhali, Bangladesh
e-mail: zohageo@pstu.ac.bd

T. Ahamed (✉)
Institute of Life and Environmental Sciences, University of Tsukuba, Tsukuba, Ibaraki, Japan
e-mail: tofael.ahamed.gp@u.tsukuba.ac.jp



T. Ahamed (ed.), *Remote Sensing Application II*, New Frontiers in Regional Science: Asian Perspectives 77, https://doi.org/10.1007/978-981-97-1188-8_4

to 2021 were 475.87 km$^2$ and 682.75 km$^2$, respectively. Therefore, the total coastal erosion was 800.72 km$^2$ from 1991 to 2021. Neat accretion was 73.94 km$^2$ for the 30-year period from 1991 to 2021. This research indicates the changes in shorelines, referring to the evidence for the delta's active formation through accretion and erosion processes of "climate change" and "sea level rise." This research projects the erosion process and threatens land use changes toward agriculture and settlements in the coastal regions of Bangladesh. The agricultural land use class was ranked one among other major land use and land cover (LULC) classes. In the study area, 34.03%, 34.47%, and 33.52% were occupied by agricultural land use for the study years 1991, 2006, and 2021, respectively, within the studied area of interest (AoI). Among all LULC patterns in the study, the agricultural lands increased in 2006 from the year 1991. On the other hand, the cultivated agricultural land had decreased in the year 2021 from the study year 2006. Hence, the agricultural lands were under great threat of loss of lands due to the coastal erosional processes in the Bangladesh delta. Therefore, it is urgent to develop new croplands area in the accredited new mud, soil, and intertidal (MSI) land-covered area for mitigating the loss of agricultural lands in onshore coastal areas of Bangladesh to strengthen the food security of the coastal region in Bangladesh.

**Keywords** Agricultural land use · Bangladesh delta · Coastal baseline assessment · Onshore environment · Satellite remote sensing and GIS · Shoreline assessment · Shoreline changes assessment · Tasseled cap transformation

## 4.1 Introduction

The shoreline is a time-determined changing boundary between land and sea (Boak and Turner 2005) where coastal accretion and erosion processes have arisen with the worldwide evidence of sea level rise and climate change worries (Matin and Jahid Hasan 2021). Shoreline information analysis is also essential for coastal management and zone protection to assess sea level prediction (Leatherman 2001; Rahman et al. 2011), geospatial research (Ford 2013), and shoreline mapping to be projected to our future Earth (Pardo-Pascual et al. 2012). Due to climate change in the global south, countries face saltwater intrusion due to seawater levels causing significant damage to agricultural crops, changing livelihoods, and settlements affecting coastal areas regionally (Al-Maruf et al. 2021). Land accretion and erosion are significantly affected by shoreline changes. Therefore, shoreline change assessment over decades with the world's enormous number of people through geomorphological and anthropogenic activities has emerged as a wide-reaching concern about coastal stretch cover, land use, and shoreline changes (Wang et al. 2013). In Southeast Asia, a disaster-prone region suffers significantly every year from tidal surges, cyclones, and saltwater intrusions to agricultural lands. In this regard, the shoreline expands, covering India, Bangladesh, and Myanmar adjacent to the Bay of Bengal, and is repeatedly impacted by hydrometeorological extreme calamities, causing massive life and property that loosen every year (Dube et al. 2004; Shamsuzzoha et al. 2022). Among the South Asian countries, the most affected coastal region of the Bangladesh delta encompasses an area of 47,201 $km^2$ that provides shelter and livelihood to 46 million inhabitants, with approximately 2.85 million hectares of agricultural lands (Bala and Hossain 2010). The diverse and dynamic Ganges–Brahmaputra delta, merged with the low-altitude shorelines toward the Meghna estuary, is the most disaster-prone region among the three neighboring countries (Dube et al. 2004; Rahman et al. 2011; Hussain et al. 2014; Rahaman et al. 2020).

Frequent natural disasters such as tropical cyclones, tidal surges, coastal floods, and salinity intrusions are causing recurrent changes to the deltaic shoreline of Bangladesh through live erosion and accretion within the continental shelf of the Bay of Bengal (Shamsuzzoha and Al-Maruf 2012; Mahamud and Takewaka 2018; Talukder et al. 2018; Kabir et al. 2020; Matin and Jahid Hasan 2021). Therefore, to explore the geomorphological activities of the nearshore coast, comprehensive research on shoreline change assessment in the Bay of Bengal is highly needed. Various remote sensing methods have been applied for shoreline extraction and change assessment in several studies using multispectral datasets at-satellite reflectance (Sarwar and Woodroffe 2013; Islam et al. 2016; Mahamud and Takewaka 2016; Dominici et al. 2019; Costantino et al. 2020; Crawford et al. 2021; Natarajan et al. 2021; Ataol and Murat Kale 2022). By using at-satellite reflectance, three uncorrelated indices were mainly used: (1) the brightness index (BI) as a measurement value for the ground surface, (2) the greenness index (GI) as a measurement value for the phenology of vegetation, and (3) the wetness index (WI; yellow stuff) as a measurement value for soil and vegetation (plant canopy moisture) interactions, which have attained a degree of acceptance in remote sensing studies (Huang et al. 2002; Sheng et al. 2011). This method is known as tasseled cap transformations (TCTs) of the multispectral scanner (MSS) and thematic mapper (TM) image data, which exemplify linear combination band features (Crist and Cicone 1984; Crist 1985; Zhai et al. 2022). The TCT-induced three indices have been used extensively in satellite remote sensing studies of agriculture, landscape, ecology, and forest (Dymond et al. 2002; Mbow et al. 2004; Wulder et al. 2004; Healey et al. 2005; Serra and Pons 2008; Baig et al. 2014).

Several studies have contributed to the shoreline dynamics of the Bangladesh delta; however, most of them were conducted for only limited portions of the coast (Matin and Jahid Hasan 2021), and TCT-based at-satellite reflectance was not applied to onshore change assessment. A small number of studies have explored coastal erosion and accretion in the coastal areas adjacent to Sundarbans in Bangladesh and calculated shoreline retreat rates (Rahman et al. 2011; Anwar and Takewaka 2014). Additionally, few studies have been devoted to detecting the shoreline change rate in the island areas of the Meghna Estuary (Hussain et al. 2014). The shoreline changes around Kutubdia Island of Cox's Bazar district (Islam et al. 2016) and Sandwip Island of Chittagong district (Emran et al. 2017), were observed in different studies (Mahamud and Takewaka 2016, 2018). However, the rates of shoreline change occurred between 1989 and 2009 along the coast of

Bangladesh, but overall, shoreline changes showed coastal erosion considering its 20-year gap (Sarwar and Woodroffe 2013). However, contributed research did not consider a baseline-based shift toward shorelines, which changes in time periods onshore. In shoreline change detection analysis, satellite remote sensing has advanced significantly with large numbers of historical datasets regionally, which can help to fix a baseline that did not change in a distant time to understand the changes or determination of shoreline alterations. Land accretion and erosion in the delta are required for coastal zone management (CZM) regarding climate change perspectives for agricultural land use and long-term regional settlement planning to execute a sustainable delta plan. The Government of Bangladesh (GoB) has taken necessary steps for implementing the country's "Bangladesh Delta Plan 2100" for the regional development of the Delta (coastal zone 27,738 $km^2$) to protect man and lands, including all biodiversity conservation, and to build sustainable marine resource management utilizing its blue economy (Alam et al. 2018). Therefore, the key purposes of this research were: (1) to conduct an onshore change assessment using a wide-ranging TCT analysis to project onshore alterations and (2) shoreline change assessment to focus on the land accretion and erosion of agricultural lands in the coastal belt of Bangladesh with a long time-interval for regional planning in the changing climates and seawater level changes.

## 4.2 Materials and Methods

### *4.2.1 Onshore Change Assessment*

The onshore change assessment was the first research objective of the study (Fig. 4.1). A comprehensive tasseled cap transformation analysis of onshore alterations was conducted for research on the entire coastal belt of the country using satellite remote sensing datasets.

#### 4.2.1.1 Geographical Extent of the Study Areas

The considered coastal belt of Bangladesh consists of a total of 14 administrative districts, including all islands located in the northern Bay of Bengal, allocated inside the maritime boundary of the country (Fig. 4.2). The study area of interest (AoI), with a 710-km-long coastline stretching from the Hariabhanga River in the west (along the Bangladesh–India border) to the Naf River in the east (along the Bangladesh–Myanmar border), running parallel to the Bay of Bengal (Karim and Mimura 2008), was also considered for the shoreline change assessment (Fig. 4.2). The coastal regions of the country are mainly distributed into three different regions, geomorphologically named as (1) the South-West Ganges Tidal Plain (western zone), (2) the South-Central Meghna Deltaic Plain (southern zone), and (3) the South-East Chittagong Coastal Belt (eastern zone) (Karim and Mimura 2008).

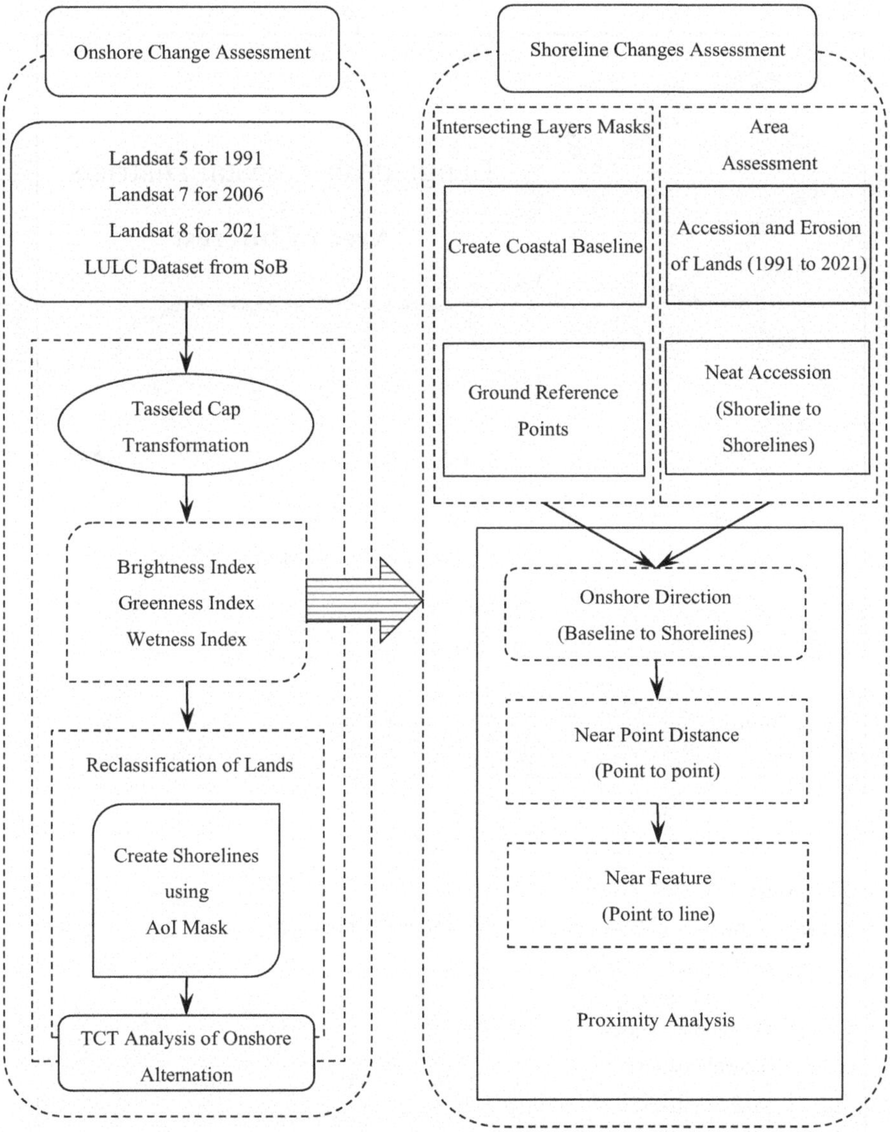


**Fig. 4.1** Research framework for determining onshore change and shoreline change assessments

Therefore, we illustrated the entire coastal belt as the AoI (Fig. 4.2). The country's data sheet (CDS), datum WGG84, was checked by the collected geodatabase (scale 1:25,000) with the associated topographic land use and land cover (LULC) feature classes of Bangladesh provided from the Survey of Bangladesh (SoB) under the Ministry of Defense (MoD), Government of Bangladesh (GoB). The eastern zone of Bangladesh is characterized by a muddy shore type in the upper stage of the Chittagong division, while the lower part is dominated by a sandy sea beach area

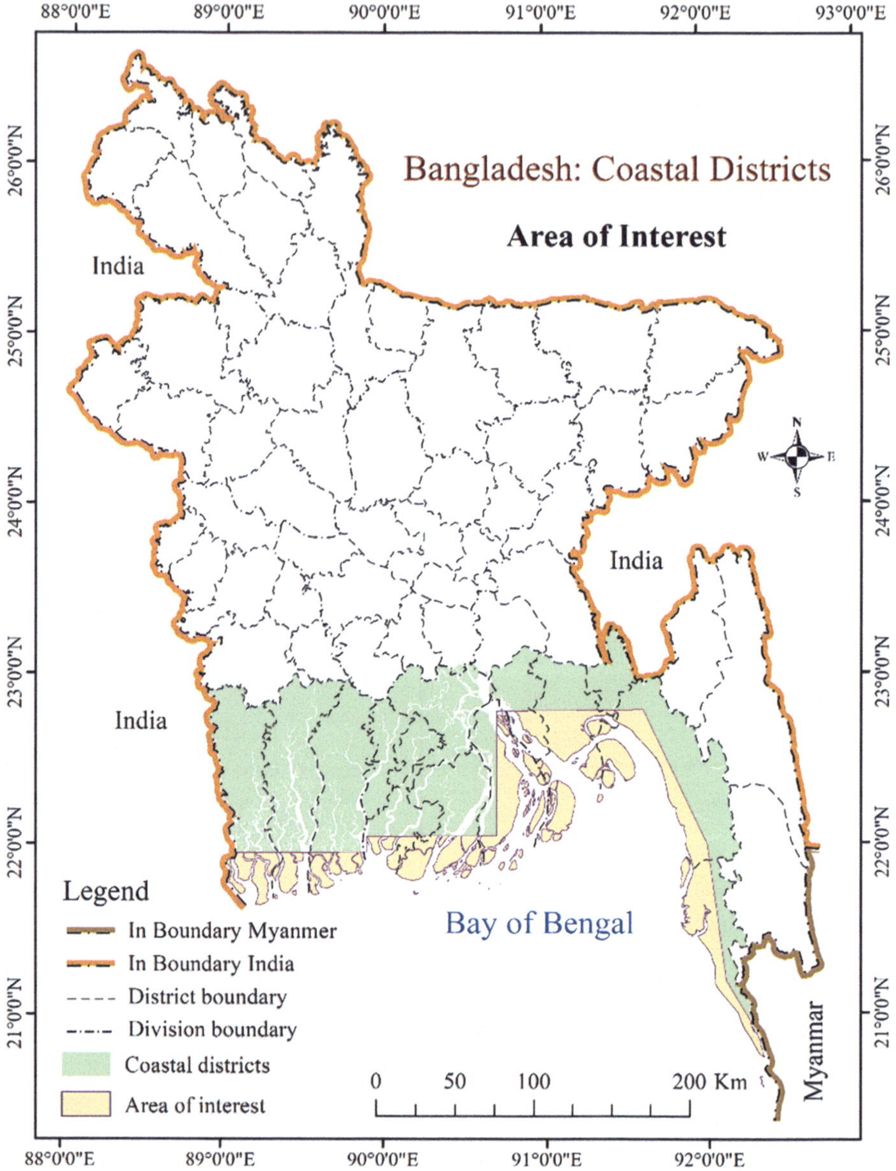


**Fig. 4.2** The study area (Datum WGG84): total 14 coastal districts and AoI

with the world's longest uninterrupted natural beach (Islam et al. 2016; Mahamud and Takewaka 2018; Hasan et al. 2021). The central zone is geomorphologically the most dynamic part of the active delta due to massive sediment flow from the Ganges–Brahmaputra–Meghna (GBM) river system and strong river currents close to the Meghna estuary (Alam 2014; Kabir et al. 2020). Several newly accreted islands, islets, landforms, and shoals are common features of the central estuary zone (McGranahan et al. 2007). Finally, the western zone is principally characterized by the world's largest mangroves, the Sundarbans Forest (Hossain et al. 2022), including tidal flatlands, swamps, floodplains, and natural levees. Every year, the Sundarbans mangrove forest saves the country as a buffer protection against severe erosion, tropical cyclones, and tidal surges, making the zone relatively stable (Sarwar and Woodroffe 2013; Shamsuzzoha and Ahamed 2023).

#### 4.2.1.2 Collection of Satellite Images

In this study, enhanced thematic mapper plus (ETM+) were collected from Landsat 5, 7 (ETM+), and operational land imager (OLI) image datasets were collected from 8 (OLI) satellites to appropriate spatial resolution (30-m spatial resolution). The ETM+ datasets for 1991 from Landsat 5, 2006 from Landsat 7, and OLI for 2021 from Landsat 8 were considered cloud-free coverage (Che et al. 2021; Shamsuzzoha et al. 2021; Arab et al. 2022). All Landsat image data were analyzed for the month of November due to hazard and cloud-free datasets separately from the Earth Explorer engine of the United States Geological Survey (USGS). The image acquisition dates for the study years were (1) November 19, 1991 (Landsat 5), (2) November 20, 2006 (Landsat 7), and (3) November 19, 2021 (Landsat 8). In this research, bands 1–5, band 7 (Landsat 5 and 7 [ETM+]), and bands 2–7 (Landsat 8 [OLI]) were employed for tasseled cap analysis (Fig. 4.1).

#### 4.2.1.3 Tasseled Cap Transformation (TCT)

The tasseled cap analysis was introduced by Kauth and Thomas in 1976 and recognized as the Kauth–Thomas transformation (Mondal et al. 2017). It gives a measure of the three indices of the pixel—(1) brightness, (2) greenness, and (3) wetness—and uses a linear combination of a total of six bands of the Landsat series (Huang et al. 2002; Baig et al. 2014). The Landsat Toolbox was utilized for image processing to make it usable in ArcGIS® 10.8.2 (ESRI, CA 92373, United States) for topographic correction, mosaicking, and tasseled cap analysis (Fig. 4.1). Bands 1–5 and 7 (visible and infrared) were: (1) blue ($B_1$ visible), (2) green ($B_2$ visible), (3) red ($B_3$ visible), (4) near-infrared ($B_4$ NIR), (5) shortwave infrared ($B_5$ SWIR-1), and (6) mid-infrared ($B_7$ SWIR-2), respectively. Landsat 5 and 7 (ETM+) imagery for the tasseled cap analysis was employed to generate 3 TCT indices of 14 coastal districts. The brightness, greenness, and wetness indices for ETM+ can be computed as the following equations, respectively (Mondal et al. 2017):

$$\mathrm{BI}_{\mathrm{ETM+}} = (\beta_1 \cdot B_1) + (\beta_2 \cdot B_2) + (\beta_3 \cdot B_3) + (\beta_4 \cdot B_4) + (\beta_5 \cdot B_5) + (\beta_7 \cdot B_7) \quad (4.1)$$

$$\mathrm{GI}_{\mathrm{ETM+}} = (\varepsilon_1 \cdot B_1) + (\varepsilon_2 \cdot B_2) + (\varepsilon_3 \cdot B_3) + (\varepsilon_4 \cdot B_4) + (\varepsilon_5 \cdot B_5) + (\varepsilon_7 \cdot B_7) \quad (4.2)$$

$$\mathrm{WI}_{\mathrm{ETM+}} = (\omega_1 \cdot B_1) + (\omega_2 \cdot B_2) + (\omega_3 \cdot B_3) + (\omega_4 \cdot B_4) + (\omega_5 \cdot B_5) + (\omega_7 \cdot B_7) \quad (4.3)$$

where the computed brightness ($BI_{ETM+}$), greenness ($GI_{ETM+}$), and wetness ($WI_{ETM+}$) indices based on the coefficients ($\beta$), ($\varepsilon$), ($\omega$), and bands ($B$) represent the top of atmosphere (TOA) reflectance, and the coefficients are constants accordingly given below (Huang et al. 2002):

$$\begin{aligned} BI_{ETM+} = & (0.3561 \cdot B_1) + (0.3972 \cdot B_2) + (0.3904 \cdot B_3) \\ & + (0.6966 \cdot B_4) + (0.2286 \cdot B_5) + (0.1596 \cdot B_7) \end{aligned} \tag{4.4}$$

$$\begin{aligned} GI_{ETM+} = & -(0.3344 \cdot B_1) - (0.3544 \cdot B_2) - (0.4556 \cdot B_3) \\ & + (0.6966 \cdot B_4) - (0.0242 \cdot B_5) - (0.2630 \cdot B_7) \end{aligned} \tag{4.5}$$

$$\begin{aligned} WI_{ETM+} = & (0.2626 \cdot B_1) + (0.2141 \cdot B_2) + (0.0926 \cdot B_3) \\ & + (0.0656 \cdot B_4) - (0.7629 \cdot B_5) - (0.5388 \cdot B_7) \end{aligned} \tag{4.6}$$

Similarly, for the Landsat 8 (OLI) dataset, bands 2–7 (band 1 skipped as coastal/aerosol) were named (1) blue ($B_2$ visible), (2) green ($B_3$ visible), (3) red ($B_4$ visible), (4) near-infrared ($B_5$ NIR), (5) shortwave infrared ($B_6$ SWIR-1), and (6) mid-infrared ($B_7$ SWIR-2), respectively. The brightness, greenness, and wetness indices for OLI can be computed as the following equations, respectively (Faliha Dzakiyah and Saraswati 2020):

$$BI_{OLI} = (\beta_2 \cdot B_2) + (\beta_3 \cdot B_3) + (\beta_4 \cdot B_4) + (\beta_5 \cdot B_5) + (\beta_6 \cdot B_6) + (\beta_7 \cdot B_7) \tag{4.7}$$

$$GI_{OLI} = (\varepsilon_2 \cdot B_2) + (\varepsilon_3 \cdot B_3) + (\varepsilon_4 \cdot B_4) + (\varepsilon_5 \cdot B_5) + (\varepsilon_6 \cdot B_6) + (\varepsilon_7 \cdot B_7) \tag{4.8}$$

$$WI_{OLI} = (\omega_2 \cdot B_2) + (\omega_3 \cdot B_3) + (\omega_4 \cdot B_4) + (\omega_5 \cdot B_5) + (\omega_6 \cdot B_6) + (\omega_7 \cdot B_7) \tag{4.9}$$

where the computed brightness ($BI_{OLI}$), greenness ($GI_{OLI}$), and wetness ($WI_{OLI}$) indices based on the coefficients ($\beta$), ($\varepsilon$), ($\omega$), and bands ($B$) represent the TOA reflectance, and the coefficients were constants accordingly given below (Baig et al. 2014):

$$\begin{aligned} BI_{OLI} = & (0.3029 \cdot B_2) + (0.2786 \cdot B_3) + (0.4733 \cdot B_4) \\ & + (0.5599 \cdot B_5) + (0.508 \cdot B_6) + (0.1872 \cdot B_7) \end{aligned} \tag{4.10}$$

$$\begin{aligned} GI_{OLI} = & -(0.2941 \cdot B_2) - (0.243 \cdot B_3) - (0.5424 \cdot B_4) \\ & + (7276 \cdot B_5) + (0.0713 \cdot B_6) - (0.1608 \cdot B_7) \end{aligned} \tag{4.11}$$

$$\begin{aligned} WI_{OLI} = & (0.1511 \cdot B_2) + (0.1973 \cdot B_3) + (0.3283 \cdot B_4) \\ & + (0.3407 \cdot B_5) - (0.7117 \cdot B_6) - (0.4559 \cdot B_7) \end{aligned} \tag{4.12}$$

#### 4.2.1.4 Marking of Shorelines

From the TCT processed in the ArcGIS® environment, three new bands—(1) brightness ($B_B$), (2) greenness ($B_G$), and (3) wetness ($B_W$)—were formed. Among the three new bands, the $B_W$ wetness index was found to be best fitted for the next step named reclassification of lands (Fig. 4.1). Reclassification of the wetness raster ($B_W$) was obtained from the wetness index raster into two major land cover classes: (1) land and (2) sea (as waterbodies). The waterbody (river and sea part) class was given a value of 0, and the land class was given a value of 10 by default referring to the Landsat Toolbox. After generating the reclassified land raster for 1991, 2006, and 2021, a reclassified binary image was used to create the shore boundary. The operation of the Landsat Toolbox was designed to create a polyline between the two earlier land use classes. In the process of creating shore boundaries, the AoI was masked to generate precise shore boundaries from the classified land raster set shorelines 1 (1991), 2 (2006), and 3 (2021) (Fig. 4.1).

#### 4.2.1.5 TCT Analysis of Onshore Alternation

TCT analysis was carried out for onshore alternations by masking (1) brightness ($B_B$), (2) greenness ($B_G$), and (3) wetness ($B_W$) band information using the shorelines of 1991 ($S_1$), 2006 ($S_2$), and 2021 ($S_3$), respectively. Therefore, a total of nine-line rasters of onshore information were prepared for an extensive TCT analysis of onshore changes (Fig. 4.1). The nine obtained line raster(s) of onshore information were considered shoreline brightness (SB), shoreline greenness (SG), and shoreline wetness (SW) of the study period 1991 ($SB_1$, $SG_1$, $SW_1$), 2006 ($SB_2$, $SG_2$, $SW_2$), and 2021 ($SB_3$, $SG_3$, $SW_3$), respectively.

### *4.2.2 Shoreline Change Assessment*

We created 3 shore boundaries for 1991, 2006, and 2021 using a 15-year interval. The shorelines (1) 1991, (2) 2006, and (3) 2021 were considered shorelines 1, 2, and 3, respectively, for the next step onshore line change assessment (Leatherman 2001). Therefore, exploring changes among the three shore boundaries, including mainland and island shores, was necessary. In this study, we considered the onshore concept for calculating baselines.

#### 4.2.2.1 Create Coastal Baseline

A baseline was generated from a novel intersecting layer mask approach of the existing shorelines. We selected three shoreline segments that best represent the study AoI (Fig. 4.1). For intersecting layer masks, the polygon at the specified shape file and size was selected at the intersection of the three input layers (shoreline 1991, 2006, and 2021). One reference scale was set as the default and set as the reference scale employed for calculating the masking geometry when masks are specified. After calculating the output feature class, the obtained feature class that contained the mask features was used as the baseline of the study. For calculating the coordinate system, we have the spatial reference of the map in which the masking polygons will be created. This is not the spatial reference that will be assigned to the output feature class.

It is the spatial reference of the map in which the masking polygons will be used since the position of symbology may change when features are projected.

#### 4.2.2.2 Ground Reference Points

A total of 16,550 ground reference points (GRPs) were generated by the "vertex to points" application from the study baseline as primary ground reference points (GRPs) for this research. The SoB landmass was used as the dataset and the red, green, and blue (RGB) Landsat images (prepared by band combination) of the respective years were used to confirm the onshore baseline was allocated onto land only. Similarly, for the baseline GRPs, we generated GRPs from the shorelines of 1991, 2006, and 2021. For the baseline GRPs, the vertex was found to be 16,550, but for the shorelines, the GRPs were 19,281, 21,945, and 16,548 for 1991, 2006, and 2021, respectively. For equal assessment and GRPs analysis, two vertices were additionally added for the shoreline of 2021, and unnecessary vertices were deleted from the shorelines of 1991 and 2006. During the deleting process, editing tools were used carefully in ArcGIS® Desktop. The shoreline-induced and edited 16,550 GRPs of 1991, 2006, and 2021 were also used to extract data from the TCT-generated raster(s).

#### 4.2.2.3 Accretion and Erosion of Lands (1991–2021)

For areal assessment, we applied features to polygon applications from data management tools to generate study baseline polygon features using the AoI. According to a similar process, we prepared AoIs for shoreline 1991, shoreline 2006, and shoreline 2021 polygon features. Thereafter, polygon features were intersected from geoprocessing tools to determine all accretion and erosion of lands into the mainland, islets, and islands of the Bangladesh delta.

#### 4.2.2.4 Neat Accretion of Lands

Neat accretion was calculated by the difference between accession and erosion of lands from (1) 1991 to 2006, (2) 1991 to 2021, and (3) 2006 to 2021 in $km^2$. Over the study of 30-year period, the difference between accession and erosion was considered the actual neat accession of lands of the Bangladesh delta. The actual neat accretion (in $km^2$) was also found to be equal to the difference in the neat accretion of lands from (1) 1991 to 2006, and (2) 1991 to 2021.

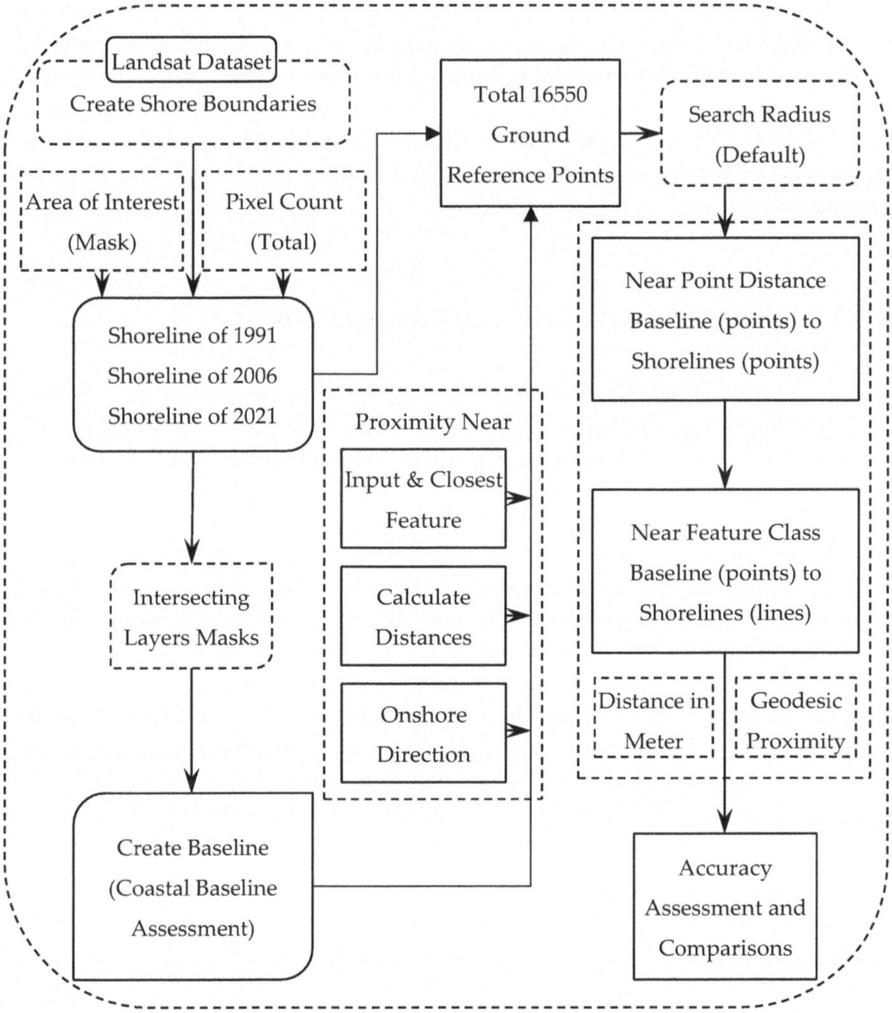


**Fig. 4.3** Proximity analysis of the study

#### 4.2.2.5 Proximity Analysis

Proximity tools in the ArcGIS® environment were distributed into two classes: (1) feature, or (2) raster. In the analysis, the feature type class was used for proximity analysis to calculate the near-point distance (NPD) from the baseline to shorelines (onshore direction) in 1991, 2006, and 2021 using 16,550 GRPs (Fig. 4.1). In the 16,550 GRPs, the point feature file was considered for calculating the point distance. Therefore, the proximity "Near tool" operational feature added the calculated distance measurement attribute to the input features from the onshore direction (Fig. 4.3). Again, the raster-based "Euclidean distance" tools measured distances from the “center of source” cells to the “center of destination” cells. The GRPs on the shoreline were used to calculate near-point distance (NPD) in meters and near-feature class (NFC) (Fig. 4.3). We calculated NPD 1991, 2006, and 2021 from the study baseline for the years 1991, 2006, and 2021 simultaneously. We reclassified the NPD 1991, 2006, and 2021 GRPs using the natural Jenks classification method into a total of ten classes to help the general readers understand near-proximity analysis very clearly.

### *4.2.3 Accuracy Assessment and Comparisons*

A total of 16,550 GRPs were used to make the study more precise to identify all types of landforms (Fig. 4.3). Previous studies took part in a small portion, excluding islands, due to the complex setting of the deltaic shorelines. We considered all land areas by utilizing the study GRPs on the SoB country data sheet accurately. The shoreline persuaded and corrected 16,550 GRPs of 1991, 2006, and 2021, which were also utilized to extract point data from the brightness, greenness, and wetness raster(s) named $SB_1$, $SG_1$, $SW_1$ of the year 1991, $SB_2$, $SG_2$, $SW_2$ raster(s) of the year 2006, and $SB_3$, $SG_3$, $SW_3$ raster(s) of the year 2021 accordingly to check the accuracy of the study. All GRPs were also converted from decimal degrees to meters in ArcGIS®. Then, we exported the meter GRPs in an Excel® spreadsheet and observed that the shoreline shape was also found in the Excel® sheet in meters perfectly. In the proximity analysis, all calculations were conducted in linear distance (meter). The analysis also checked Excel® separately for confirmation using all pixel brightness, greenness, and wetness data.

## 4.3 Results

### *4.3.1 Onshore Change Assessment*

The notion of onshore herein was that the “last boundaries” between “land” and “water” in the Bay of Bengal belong to Bangladesh. The onshore pixel data changed over the decades, including their locational shifts through the deltaic processes of “landmass formation” and “coastal bank erosion” as “accretion and erosion of lands,” respectively.

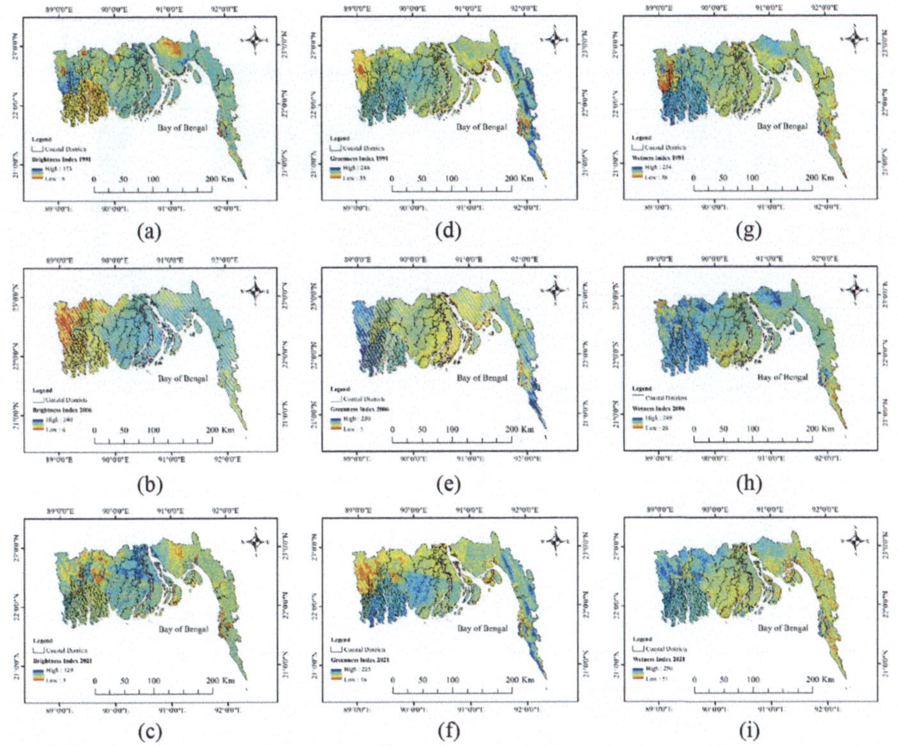

Fig. 4.4 Brightness index: (**a**) 1991, (**b**) 2006, and (**c**) 2021; Greenness index: (**d**) 1991, (**e**) 2006, and (**f**) 2021; Wetness index: (**g**) 1991, (**h**) 2006, and (**i**) 2021 of coastal districts, Bangladesh delta

#### 4.3.1.1 Tasseled Cap Transformation (TCT) of Coastal Districts

The tasseled cap transformation (TCT) method was applied as a remote sensing technique to explore onshore changes in the Bangladesh delta. A total of 9 raster(s) datasets with onshore brightness (BI), greenness (GI), and wetness (WI) indices were prepared for the 14 coastal administrative districts of Bangladesh. The BI values in 1991, 2006, and 2021 ranged from 8 to 173 (Fig. 4.4a), 6 to 240 (Fig. 4.4b), and 3 to 129 (Fig. 4.4c), respectively. The GI in 1991, 2006, and 2021 ranged from 33 to 246 (Fig. 4.4d), 5 to 230 (Fig. 4.4e), and 16 to 225 (Fig. 4.4f), respectively. The WI in 1991, 2006, and 2021 ranged from 38 to 234 (Fig. 4.4g), 20 to 249 (Fig. 4.4h), and 3 to 129 (Fig. 4.4i), respectively.

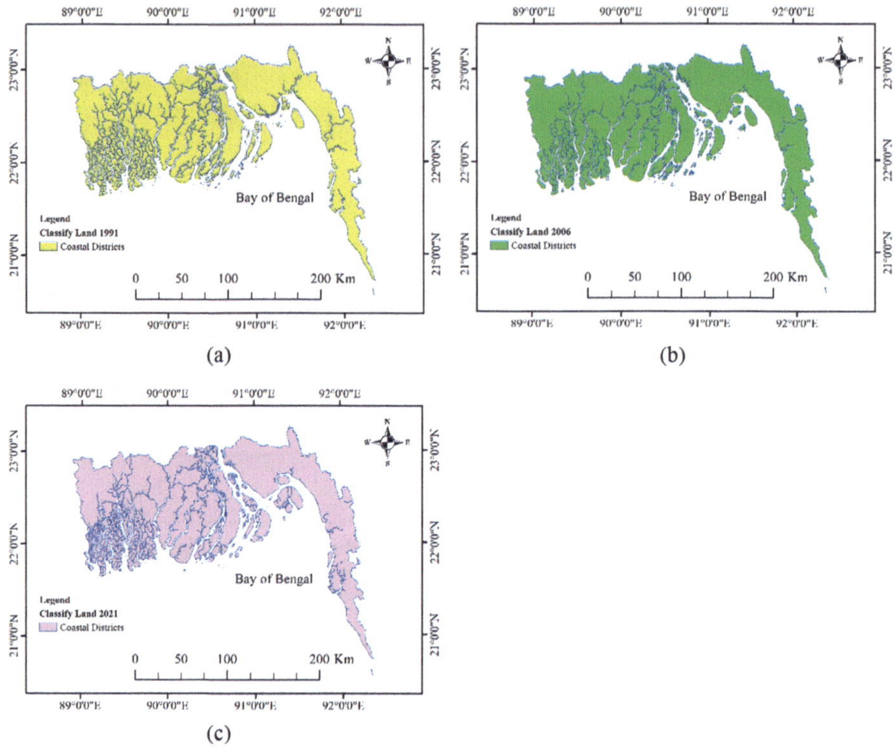


**Fig. 4.5** Reclassification of lands of coastal districts: (**a**) 1991, (**b**) 2006, and (**c**) 2021

### 4.3.1.2 Land Cover Map

Reclassified brightness, greenness, and wetness index raster(s) were experimented to prepare two land cover classes: (1) land and (2) waterbodies. In this research, the wetness raster(s) gave the best land cover map to prepare land classification. Therefore, the wetness index (WI) raster transformed into two land cover classes: land and sea (water). The sea part class was found by the value of 0, and the land class type was calculated by the value of 10 according to the default parameter. In the classification process, the land rasters of 1991, 2006, and 2021 were prepared (Fig. 4.5a–c). In the 14 coastal districts, 31,282 (Fig. 4.5a), 31,636 (Fig. 4.5b), and 31,693 (Fig. 4.5c) km$^2$ of land were observed in 1991, 2006, and 2021, respectively. The entire longitudinal length, in decimal degrees, was from 89.032 East to 92.358 East. The coast east of Bangladesh, adjacent to the Bay of Bengal, was considered for 1991. It was found from 89.032 East to 92.352 East in 2006 and 89.033 East to 92.346 East in 2021. Every study period, it was moved in east longitudinal coordinates given the changing vibrations of delta shorelines.

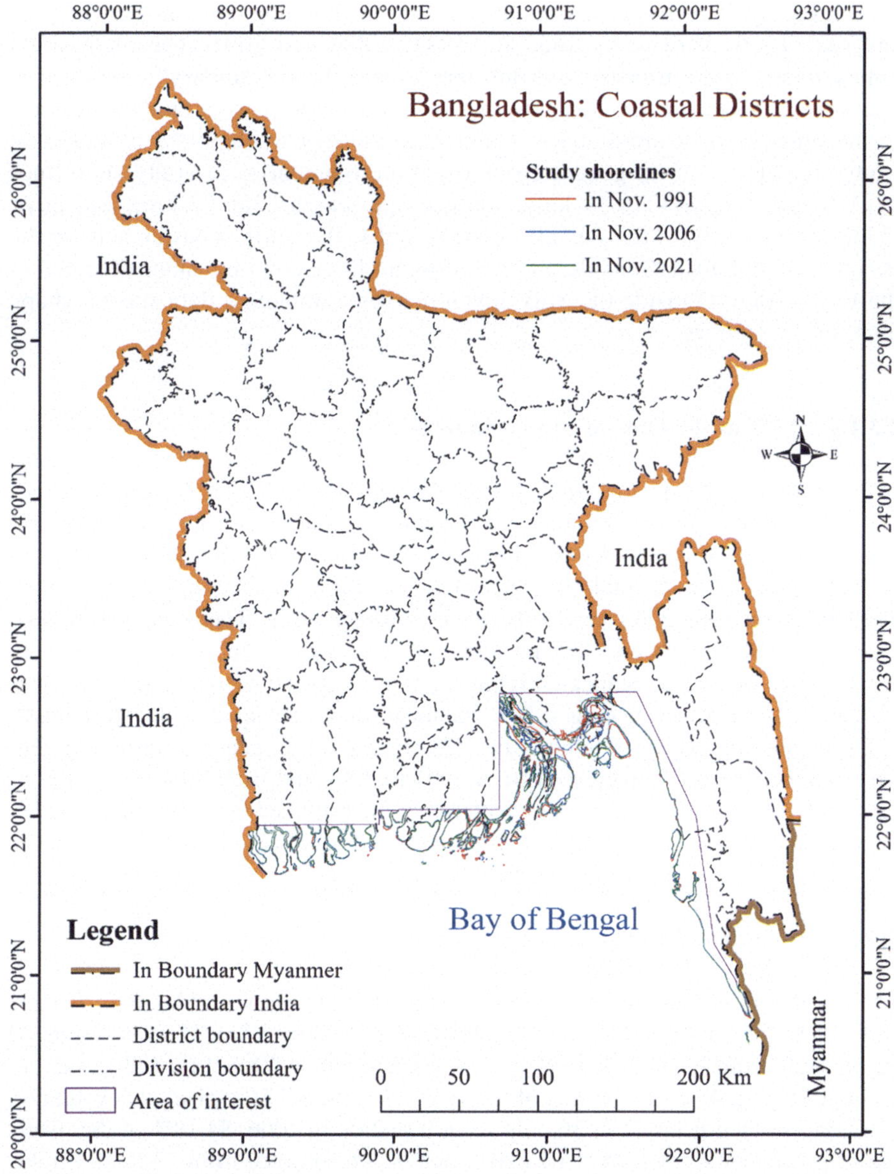


**Fig. 4.6** Study shoreline combinations of Bangladesh from 1991 to 2021 (15-year interval)

#### 4.3.1.3 Create Shorelines

The shorelines for the respective study years were calculated from the classified land and sea map using the AoI polygon. The classified land area was masked and edited within the AoI area using the cartography toolbox, and the polygon was erased by feature application, except for the shoreline boundaries. The land and sea rasters were reclassified for 1991, 2006, and 2021. The reclassified binary images were transformed into the "create shore boundary" operation of the Landsat Toolbox to create a polyline between the two earlier land use classes. In the process of creating shore boundaries, the AoI was applied to mask the study area (Fig. 4.2) to generate precise shore boundaries as study shorelines 1991 (shape length 3270.59 km), 2006 (shape length 3386.03 km), and 2021 (3338.60 km) (Fig. 4.6). For all shape lengths, the entire coastal belt of the Bangladesh delta was considered, including all large and tiny islands located in the Bay of Bengal inside the SoB, given maritime boundary data sheets (shape files) of the country.

#### 4.3.1.4 TCT of the Onshore Environment

In the observation of the longitudinal changes (unit decimal degree), wetness values were greater than greenness values, and brightness values were smaller than greenness values. This was repeated for all study years of the research in 1991, 2006, and 2021 (Fig. 4.7a–c). A total of 135,930 pixels were counted in onshore 1991, 111,838 in onshore 2006, and 140,225 in onshore 2021. The mean brightness, greenness, and wetness were 29, 102, and 187, respectively, in 1991 (Fig. 4.7a). In 2006, it was found that the mean brightness, greenness, and wetness changed by 56, 113, and 224, respectively (Fig. 4.7b). The mean values changed to 20 (brightness), 136 (greenness), and 238 (wetness) in 2021 (Fig. 4.7c). In general, onshore greenness and wetness (GW) increased from 1991 to 2006 and from 2006 to 2021 consecutively. Only a single decreasing pattern was investigated in the brightness index from 2006 to 2021. Additionally, the brightness index increased from 1991 to 2006.

##### 4.3.1.4.1 Onshore Brightness Index

The onshore brightness index was prepared as shore brightness (SB): $SB_1$ of 1991 (Fig. 4.8a), $SB_2$ of 2006 (Fig. 4.8b), and $SB_3$ of 2021 (Fig. 4.8c). The values ranged accordingly from 13 to 86 in 1991 (Fig. 4.8a), 5 to 110 in 2006 (Fig. 4.8b), and 2 to 53 in 2021 (Fig. 4.8c). In the $SB_1$ of the year 1991, the 85.78%-pixel data valuation was concentrated from 23 to 36 only from 0 to 255. In 2006, 84.18% of the pixels were agglomerated from 45 to 67 ($SB_2$), and 91.11% ranged from 13 to 28 brightness index values in 2021 ($SB_3$). Therefore, onshore brightness increased in 2006 from 1991 and decreased in 2021 from 2006 (Fig. 4.9a).

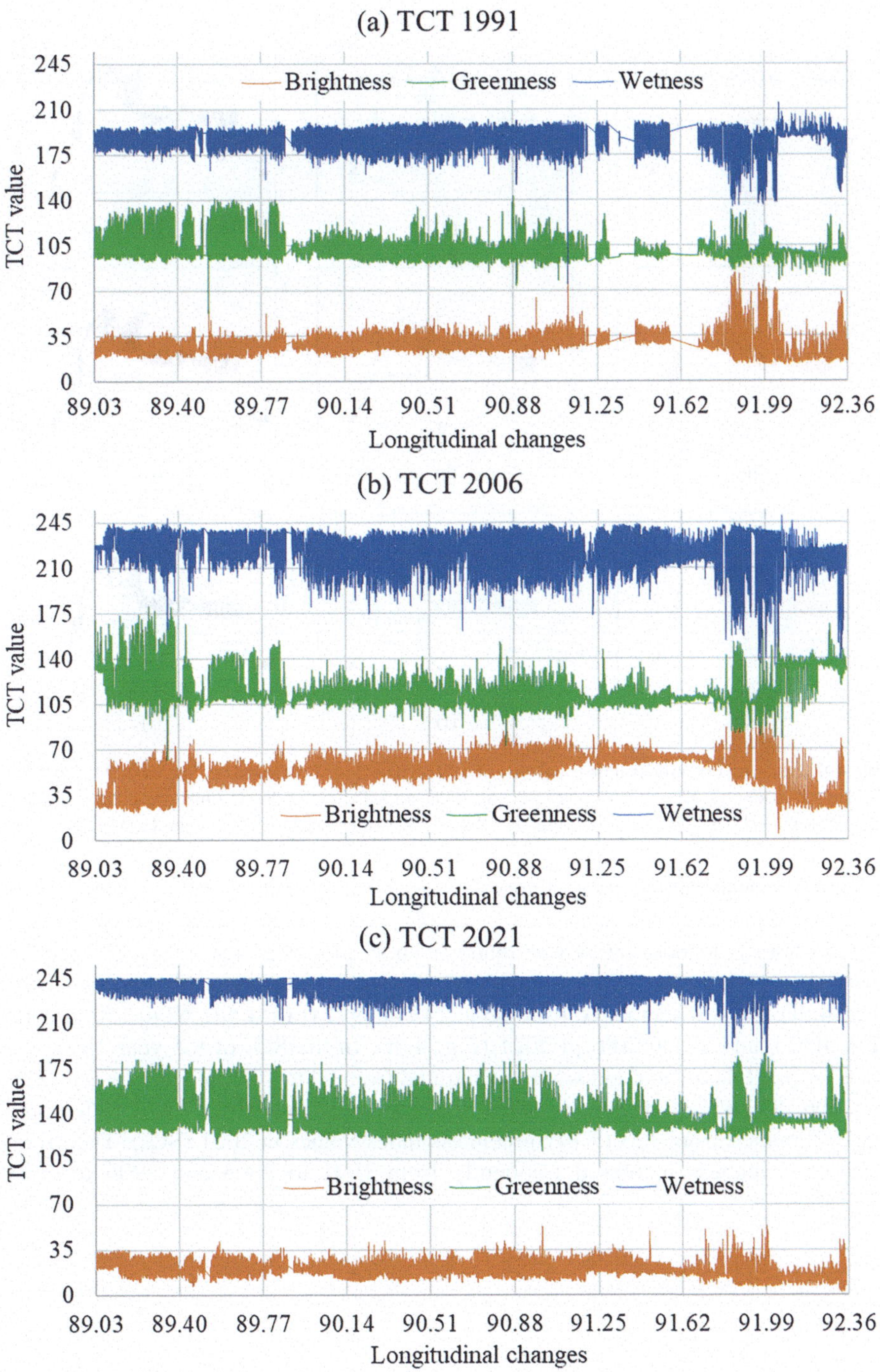


**Fig. 4.7** Onshore TCT and longitudinal changes (decimal degree): (**a**) 1991, (**b**) 2006, and (**c**) 2021

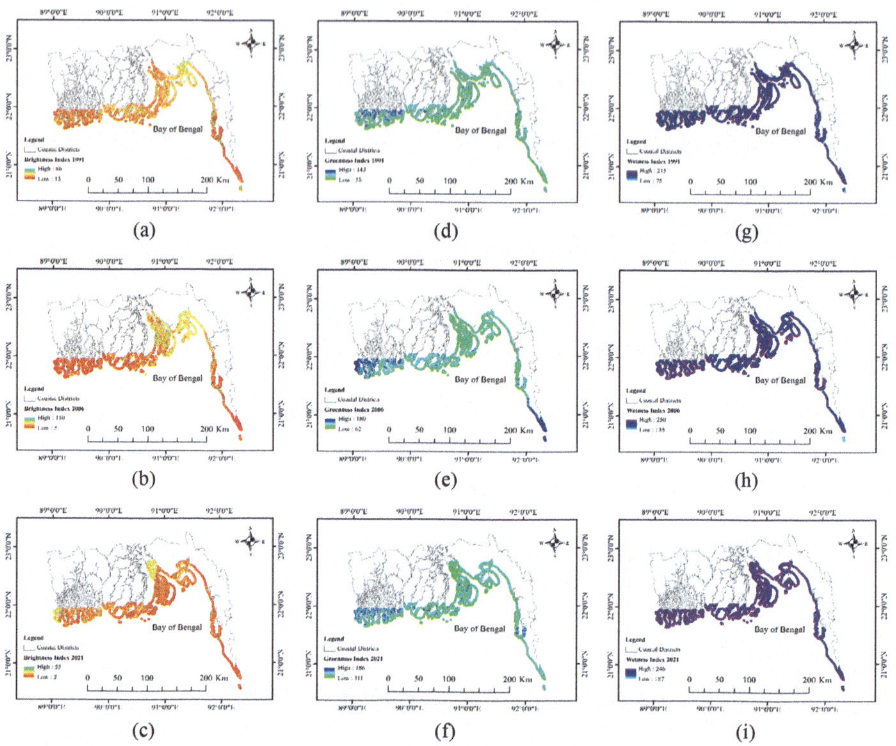


**Fig. 4.8** Onshore brightness index: (**a**) $SB_1$ of 1991, (**b**) $SB_2$ of 2006, and (**c**) $SB_3$ of 2021; Onshore greenness index: (**d**) $SG_1$ of 1991, (**e**) $SG_2$ of 2006, and (**f**) $SG_3$ of 2021; Onshore wetness index: (**g**) $SW_1$ of 1991, (**h**) $SW_2$ of 2006, and (**i**) $SW_3$ of 2021 of coastal districts, Bangladesh delta

#### 4.3.1.4.2 Onshore Greenness Index

The onshore greenness index was made as shore greenness (SG): $SG_1$ of the year 1991 (Fig. 4.8d), $SG_2$ of the year 2006 (Fig. 4.8e), and $SG_3$ of the year 2021 (Fig. 4.8f). The values ranged from 53 to 143 in 1991 (Fig. 4.8d), 62 to 180 in 2006 (Fig. 4.8e), and 111 to 186 in 2021 (Fig. 4.8f). In the $SG_1$ of the year 1991, the 88.83%-pixel data valuation was concentrated from 93 to 108 only among 0 to 255. In 2006, 83.16% of the pixels were agglomerated from 101 to 120 ($SG_2$), and 88.10% ranged from 123 to 147 greenness index values in 2021 ($SG_3$). Therefore, onshore greenness increased frequently from 1991 to 2006 and 2006 to 2021 (Fig. 4.9b).

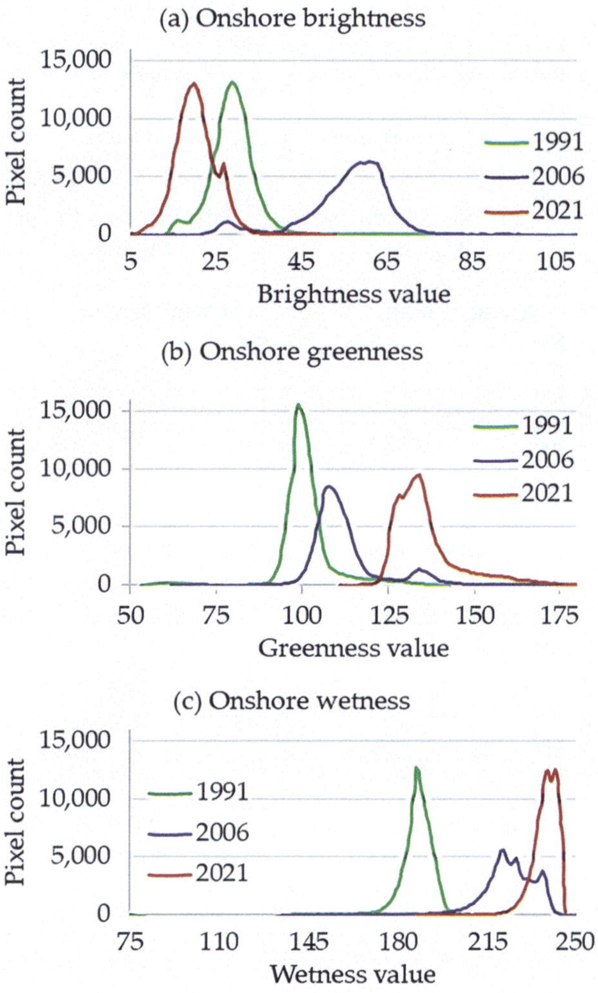


**Fig. 4.9** Onshore TCT change statistics: (**a**) brightness, (**b**) greenness, and (**c**) wetness

4.3.1.4.3 Onshore Wetness Index

The onshore wetness index was arranged as follows: $SW_1$ of the year 1991 (Fig. 4.8g), $SW_2$ of the year 2006 (Fig. 4.8h), and $SW_3$ of the year 2021 (Fig. 4.8i). The values ranged accordingly from 75 to 215 in 1991 (Fig. 4.8g), 135 to 250 in 2006 (Fig. 4.8h), and 187 to 246 in 2021 (Fig. 4.8i). In the $SW_1$ of the year 1991, the 87.73%-pixel data valuation was concentrated from 181 to 196 only among 0 to 255. In 2006, 82.22% of the pixels were agglomerated from 215 to 238 ($SW_2$), and 90.58% ranged from 231 to 244 wetness index values in 2021 ($SW_3$). Therefore, onshore wetness increased similarly to the greenness index of Bangladesh onshore from 1991 to 2006 and 2006 to 2021 (Fig. 4.9c).

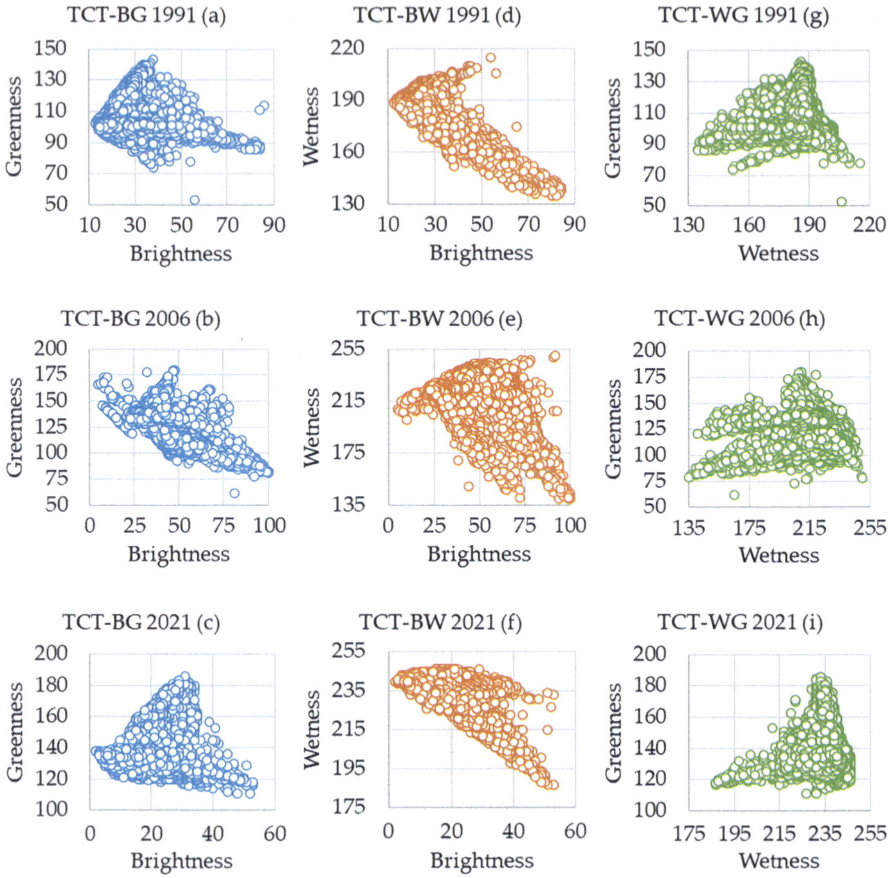


**Fig. 4.10** Onshore TCT change comparisons: (**a**) TCT-BG of 1991, (**b**) TCT-BG of 2006, and (**c**) TCT-BG of 2021; (**d**) TCT-BW of 1991, (**e**) TCT-BW of 2006, and (**f**) TCT-BW of 2021; and (**g**) TCT-WG of 1991, (**h**) TCT-WG of 2006, and (**i**) TCT-WG of 2021

#### 4.3.1.5 Onshore TCT Change Comparisons

TCT analysis provides linear relationships among (1) brightness, (2) greenness, and (3) wetness combinations. This study exemplified the linear combination of three band features: TCT-BG (brightness and greenness) (Fig. 4.10a–c), TCT-BW (brightness and wetness) (Fig. 4.10d–f), and TCT-WG (greenness and wetness) (Fig. 4.10g–i). From the TCT at-satellite reflectance of the three main uncorrelated indices, it was observed that the brightness index as a measurement value for the onshore ground surface was reduced. Therefore, greenness increased gradually in parallel with the wetness index. The greenness index, as a measurement value for the phenology of vegetation, significantly provided information positively that the Bangladesh delta will also be greener in the future. The yellow stuff belongs to the wetness index, which is the representation of soil and vegetation, as the formation

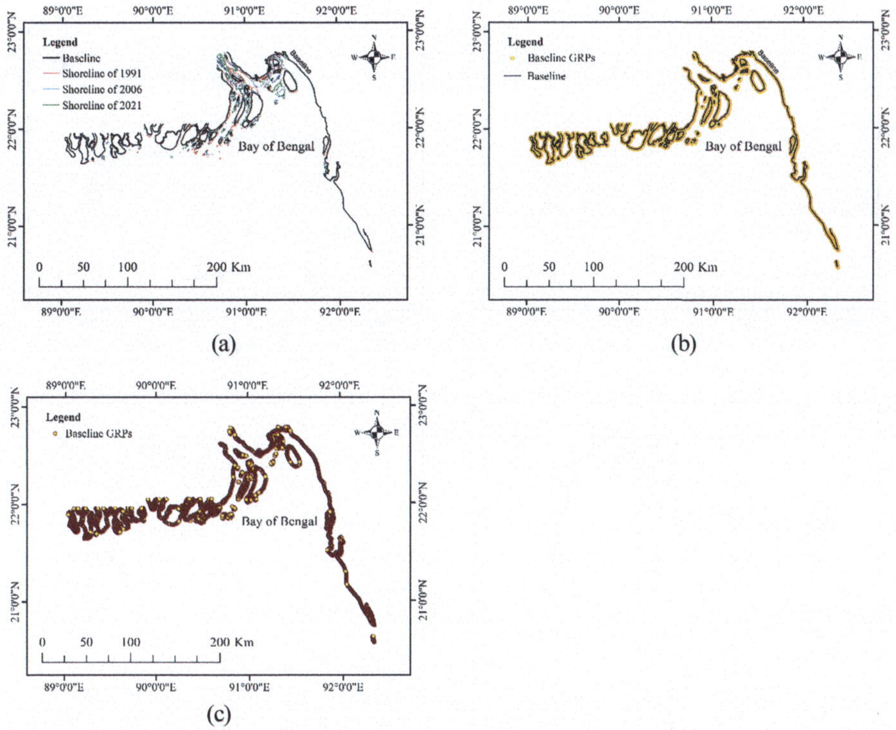


**Fig. 4.11** Baseline assessment and validation by ground reference points (GRPs) using the SoB-provided country data sheet of Bangladesh: (**a**) created coastal baseline, (**b**) point and line features intersection of baseline, and (**c**) total 16,550 baseline GRPs

of plant canopy moisture interacted gradually to improve greenness, usually to attain a degree of acceptance of this TCT application in onshore remote sensing.

## *4.3.2 Shoreline Change Assessment*

In the onshore change assessment, pixel TCT properties changed over the periods of 1991, 2006, and 2021 (Fig. 4.10a–i). The shorelines were shifted geographically, and the second portion of this study was focused on the shoreline change assessment of the Bangladesh delta in 2006 from 1991 and 2021 from 1991 and 2006, and from the study baseline of the Bangladesh coast.

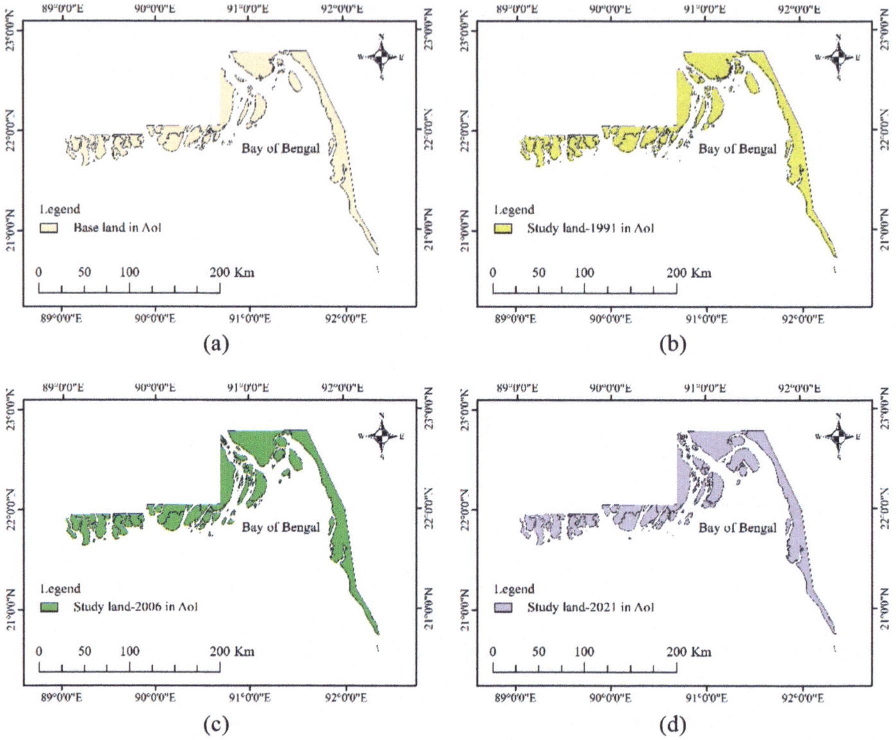


**Fig. 4.12** Area assessment using AoI Mask: (**a**) base land, (**b**) study land (1991), (**c**) study land (2006), and (**d**) study land (2021)

#### 4.3.2.1 Baseline Assessment and GRPs

Three shorelines for 1991, 2006, and 2021 for a period of 15-year interval were utilized for the change assessment (Figs. 4.6 and 4.11a). The shorelines of 1991, 2006, and 2021 were explored using onshore changes among the three shore boundaries, including mainland and island shores (Fig. 4.11a). In intersecting layer mask processing, the study baseline was created (Fig. 4.11a). In the study, the baseline was taken onshore (Fig. 4.11b). For areal analysis, the upper line of the AoI was combined with the acquired, baseline, shoreline of 1991, 2006, and 2021 (Fig. 4.11a). In this research, the point, line, and polygon shape features were embedded with the AoI (Fig. 4.2) to assess shoreline changes. The baseline length was calculated to be 2633.12 km (Fig. 4.11b). Point features were based on a total of 16,550 GRPs (Fig. 4.11c), line features were from the created shore boundaries (Fig. 4.11a), and thirdly, polygonal features were generated from the area assessment stage (Fig. 4.12a–d) as part of the AoI area assessment, land accretion from the baseline, land accretion, and erosion rates in the respective study years in the Bay of Bengal of the Bangladesh territory.

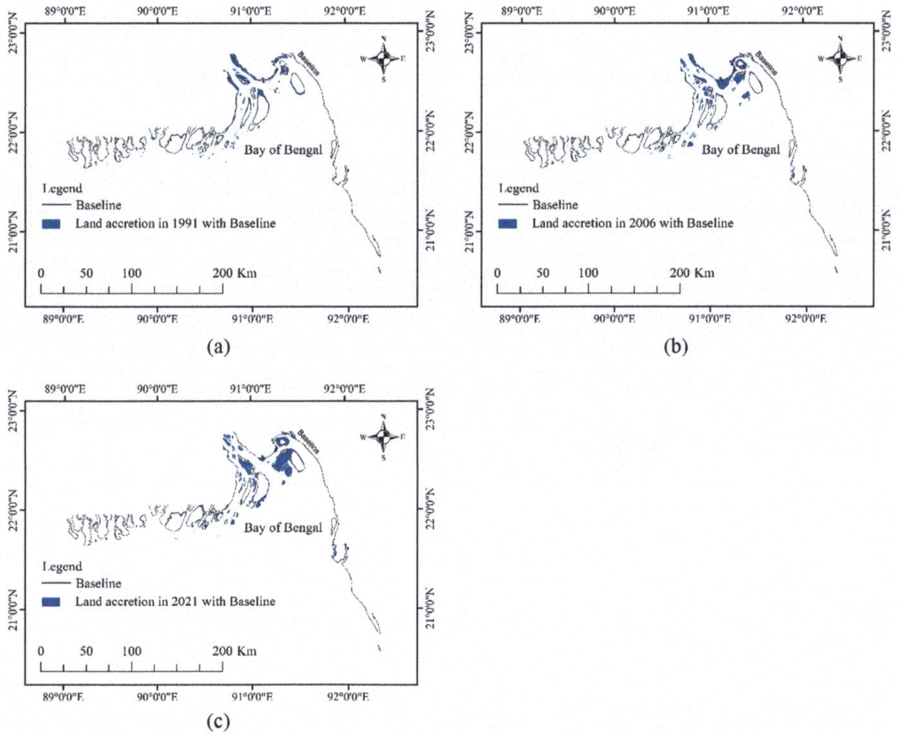


**Fig. 4.13** Land accretion along with the study baseline in the coastal region of Bangladesh (baseline to shoreline): (**a**) 1991, (**b**) 2006, and (**c**) 2021

**Table 4.1** Accession of lands from baseline to shoreline

| From baseline | Accretion (km$^2$) |
|---|---|
| To shoreline of 1991 | 876.01 |
| To shoreline of 2006 | 1225.28 |
| To shoreline of 2021 | 1299.22 |

**Table 4.2** Changes in accession and erosion of lands along shorelines

| Shoreline | Accretion (km$^2$) | Erosion (km$^2$) | Neat accretion (km$^2$) |
|---|---|---|---|
| 1991–2006 | 825.15 | 475.87 | 349.28 |
| 1991–2021 | 1223.94 | 800.72 | 423.22 |
| 2006–2021 | 756.69 | 682.75 | 73.94[a] |

[a] Actual neat accretion over 30 years from 1991 to 2021

#### 4.3.2.2 Area Assessment of Study Lands Using the AoI Mask

In the study, the baseline-induced base land AoI was 7572.81 km$^2$ (Fig. 4.12a). The AoI-based study lands were assessed in 1991 (8448.81 km$^2$), 2006 (8798.09 km$^2$), and 2021 (8872.03 km$^2$) accordingly (Fig. 4.12b–d). Therefore, the polygonal

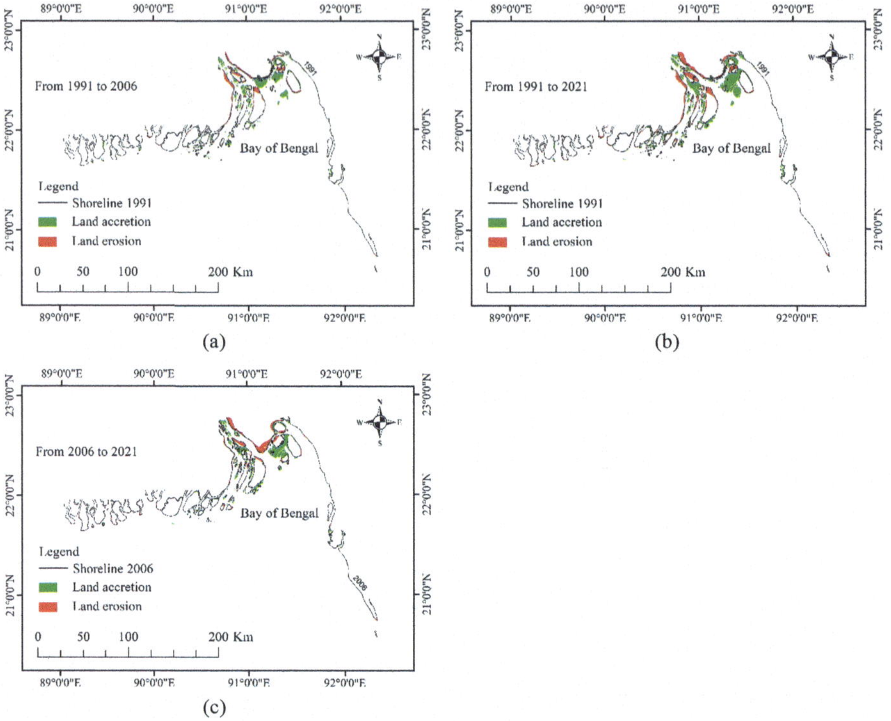


**Fig. 4.14** Cumulative accretion and erosion of lands in the coastal region of Bangladesh (shoreline to shoreline): (**a**) 1991–2006, (**b**) 1991–2021, and (**c**) 2006–2021

shapes represented that the shoreline adjacent areas were changed, and the overall lands were accredited for the area-increasing pattern of the coastal areas in Bangladesh.

#### 4.3.2.3 Accretion of Lands (Baseline to Shorelines)

The total land accretion with a baseline for 1991, 2006, and 2021 was observed through the distribution among the coastal areas (Fig. 4.13a–c). Additionally, land accretion was listed as 876.01 km$^2$, 1225.28 km$^2$, and 1299.22 km$^2$ in 1991, 2006, and 2021, respectively (Table 4.1). The study baseline was obtained as intersected layer masks among all shorelines, and, therefore, no erosion of the baseline was found (Fig. 4.11a).

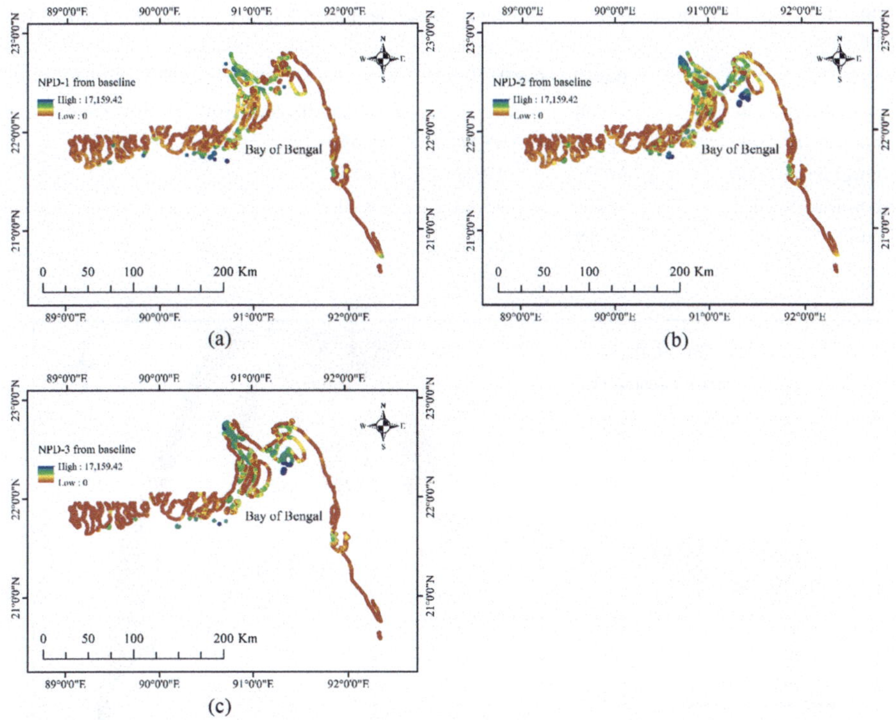


**Fig. 4.15** Near-point distance (NPD) in meters using proximity analysis (baseline to the shoreline): (**a**) NPD-1 of 1991, (**b**) NPD-2 of 2006, and (**c**) NPD-3 of 2021

**Table 4.3** Count (number) and percentage of baseline GRPs by near-point distance (NPD) classes to shorelines

| | | NPD-1 (1991) | | NPD-2 (2006) | | NPD-3 (2021) | |
|---|---|---|---|---|---|---|---|
| Class number | NPD class (m) | Count | % | Count | % | Count | % |
| 1 | 0–136.65 | 8301 | 50.16 | 7886 | 47.65 | 8568 | 51.77 |
| 2 | 136.66–379.51 | 2827 | 17.08 | 3611 | 21.82 | 1249 | 7.55 |
| 3 | 379.52–746.59 | 1547 | 9.35 | 1363 | 8.24 | 1214 | 7.34 |
| 4 | 746.60–1334.52 | 1128 | 6.82 | 1034 | 6.25 | 1204 | 7.27 |
| 5 | 1334.53–2086.62 | 790 | 4.77 | 847 | 5.12 | 938 | 5.67 |
| 6 | 2086.63–3050.70 | 840 | 5.08 | 566 | 3.42 | 772 | 4.66 |
| 7 | 3050.71–4432.38 | 529 | 3.20 | 587 | 3.55 | 1009 | 6.10 |
| 8 | 4432.39–6738.46 | 298 | 1.80 | 247 | 1.49 | 1032 | 6.24 |
| 9 | 6738.47–9540.68 | 242 | 1.46 | 115 | 0.69 | 364 | 2.20 |
| 10 | 9540.69–17,159.42 | 48 | 0.29 | 294 | 1.78 | 200 | 1.21 |
| | Total | 16,550 | 100 | 16,550 | 100 | 16,550 | 100 |

**Table 4.4** Near-feature analysis of the study baseline ground reference points (GRPs) for validation

| Near-shoreline feature | Baseline GRPs | | Average distance from baseline (in m) |
|---|---|---|---|
| | Count | % | |
| Shoreline of 1991 | 5935 | 35.86 | 245.45 |
| Shoreline of 2006 | 3305 | 19.97 | 474.84 |
| Shoreline of 2021 | 7310 | 44.17 | 379.19 |
| Total | 16,550 | 100 | |

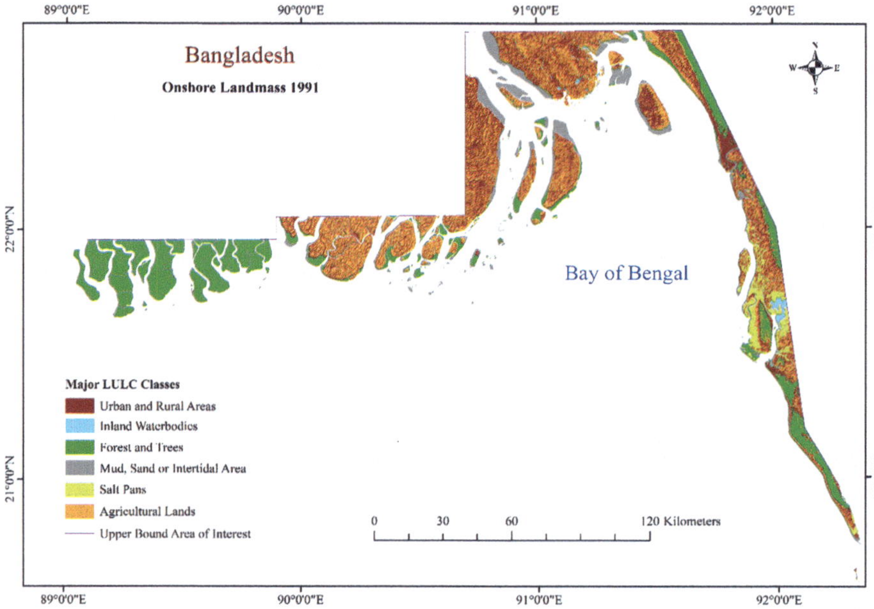


**Fig. 4.16** Agricultural land use (1991) among other major land use and land cover (LULC) classes in the study upper bound area of interest (AoI) at the onshore coastal region of Bangladesh delta

#### 4.3.2.4 Neat Accretion and Erosion of Lands (Shoreline to Shoreline)

The baseline was constantly used for the study to obtain accessions and erosion from the shoreline to the shoreline accordingly (Table 4.2). In Fig. 4.14a, a total of 825.15 km$^2$ of land was accredited in 2006 from the 1991 shoreline with 475.87 km$^2$ of coastal erosion (Table 4.2). Figure 4.14b shows that a total of 1223.94 km$^2$ of land was accredited in 2021 from the 1991 shoreline with 800.72 km$^2$ of coastal erosion (Table 4.2). Therefore, the study revealed that from 1991 to 2021, the neat accession was 73.94 km$^2$ (Table 4.2). By Fig. 4.14c, the land accession and erosion were observed to be 756.69 km$^2$ and 682.75 km$^2$, respectively, from 2006 to 2021 (Table 4.2).

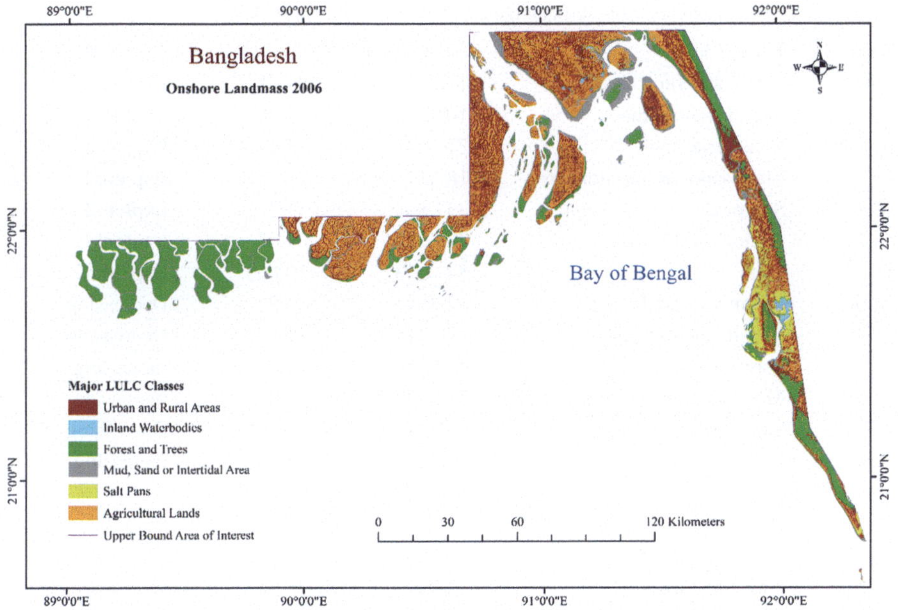


**Fig. 4.17** Agricultural land use (2006) among other major land use and land cover (LULC) classes in the study upper bound area of interest (AoI) at the onshore coastal region of Bangladesh delta

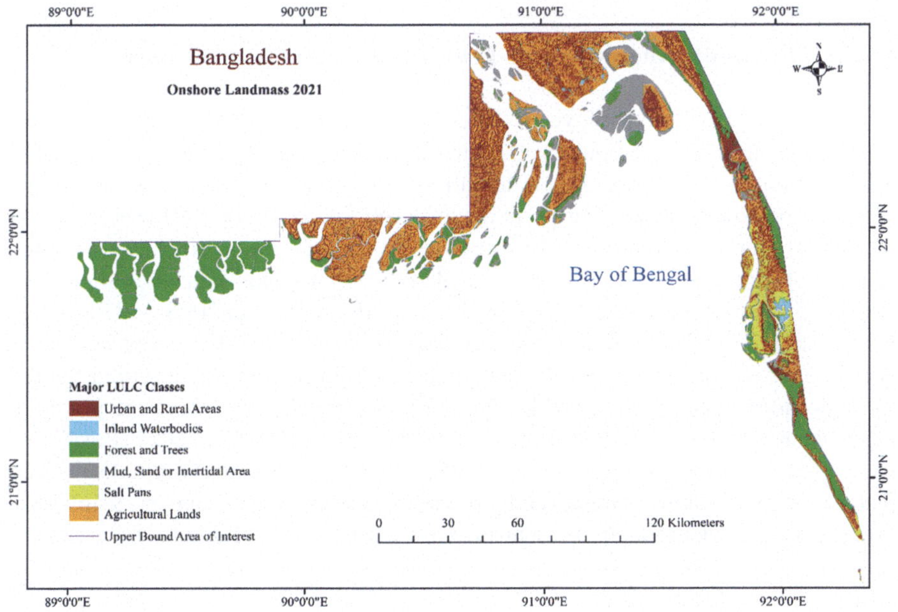


**Fig. 4.18** Agricultural land use (2021) among other major land use and land cover (LULC) classes in the study upper bound area of interest (AoI) at the onshore coastal region of Bangladesh delta

**Table 4.5** Major LULC classes and change assessment in the study AoI

| Year | Major LULCs | Area (km$^2$) | Area (%) | Pattern status |
|---|---|---|---|---|
| 1991 | Urban and rural areas | 1767.17 | 21.01 | Explored |
| | Inland waterbodies | 1091.96 | 12.98 | Explored |
| | Forest and trees | 2173.81 | 25.84 | Explored |
| | Mud, sand, or intertidal area | 182.17 | 2.17 | Explored |
| | Salt pans | 333.86 | 3.97 | Explored |
| | Agricultural lands | 2862.56 | 34.03 | Explored |
| | Total | 8411.53 | 100.00 | Explored |
| 2006 | Urban and rural areas | 1780.43 | 20.28 | Decreasing |
| | Inland waterbodies | 976.27 | 11.12 | Decreasing |
| | Forest and trees | 2337.19 | 26.63 | Increasing |
| | Mud, sand, or intertidal area | 322.92 | 3.68 | Increasing |
| | Salt pans | 335.17 | 3.82 | Decreasing |
| | Agricultural lands | 3025.61 | 34.47 | Increasing |
| | Total | 8777.60 | 100.00 | |
| 2021 | Urban and rural areas | 1753.63 | 19.92 | Decreasing |
| | Inland waterbodies | 891.65 | 10.13 | Decreasing |
| | Forest and trees | 2346.49 | 26.65 | Increasing |
| | Mud, sand, or intertidal area | 526.13 | 5.98 | Increasing |
| | Salt pans | 334.34 | 3.80 | Decreasing |
| | Agricultural lands | 2951.33 | 33.52 | Decreasing |
| | Total | 8803.56 | 100.00 | |

#### 4.3.2.5 Proximity Analysis: Near-Point Distance and Near-Feature

Proximity analysis was employed to calculate the relationship between the selected 16,550 ground reference points and their neighbors (Figs. 4.11c and 4.15a–c). We also calculated the entire shape length of the study baseline and shorelines (Fig. 4.11b). Additionally, near-point distance (NPD) classes of 1991 (natural Jenks in meters) from baseline points to shoreline points were also calculated to understand shoreline change classification in the Bay of Bengal area of Bengal delta (Fig. 4.15a for the study year 1991, Fig. 4.15b for the study year 2006, and Fig. 4.15c for the study year 2021). It was also classified to prepare NPD classes (Table 4.3), which gives us a clear view of the Bangladesh shoreline movement within the last three decades from 1991 to 2021 (Fig. 4.15a–c). Finally, the shoreline points as near-distance features for 1991 (Fig. 4.15a), 2006 (Fig. 4.15b), and 2021 (Fig. 4.15c) were used for assessing near-feature GRPs by baseline points (Table 4.4). Near-features of the study were observed by point-to-point proximity analysis using location (optional) tick marks in the near-proximity assessment (Table 4.4).

#### 4.3.2.6 Assessment of Agricultural Land Use Changes in the Onshore Coastal Region

The agricultural land LULC class was ranked one among other major LULC classes. It was found 34.03%, 34.47%, and 33.52% was occupied by agricultural land use for the study years 1991, 2006, and 2021 respectively (Table 4.4). In Khulna division there was no agricultural land found within the studied AoI. Meanwhile, In Chittagong division, the agricultural lands were found 1615.90 km$^2$ in 1991, 1720.67 km$^2$ in 2006, and 1656.35 km$^2$ in 2021. In Barishal administrative division, the agricultural lands were found 1246.66 km$^2$ in 1991, 1304.93 km$^2$ in 2006, and 1294.98 km$^2$ in 2021. The total agricultural lands of the AoI-based study area were assessed in 1991 (2862.56 km$^2$), 2006 (3025.61 km$^2$), and 2021 (2951.33 km$^2$) accordingly (Figs. 4.16, 4.17, and 4.18, Table 4.5). Therefore, the polygonal shapes represented that the shoreline-adjacent agricultural lands were also changed in the onshore coastal areas of Bangladesh. Among all LULC patterns in the study, the agricultural lands were increased in 2006 from the year 1991. On the other hand, the cultivated agricultural land was found to have been decreased in the year 2021 from the study year 2006 (Table 4.5). Hence, it was significant that agricultural lands were in great threat of loss of lands due to the coastal erosional processes in the Bangladesh delta. Therefore, it is urgent to develop new croplands area in the accredited new mud, soil, and intertidal (MSI) land-covered area for mitigating the loss of agricultural lands in onshore coastal areas of Bangladesh to strengthen food security of the coastal region in Bangladesh. There was a continuous increasing scenario in MSI land-covered class, and it was found 2.17%, 3.68%, and 5.98%, respectively, in the study years 1991, 2006, and 2021 (Table 4.5).

### *4.3.3 Accuracy Assessment*

This study was conducted utilizing 16,550 vertices to points of the baseline as GRPs (Fig. 4.11a–c). These GRPs were settled into the landmass of the country that was verified by the SoB country data sheet, and these landmasses have never faced any land erosion activities since 1991 to date. Therefore, we projected the degree decimal points features to meter points features graphically using Microsoft Excel®. Similar results were obtained from the Excel® analysis for the research and were good-fitted evidence. We obtained comparisons among NPD 1991, 2006, and 2021 classes, and listed them in Table 4.3. The changes in Bangladesh shorelines in the years from 1991 to 2021 were observed with a 15-year interval in between 2006 and 2021 (Table 4.4).

## 4.4 Discussion

The shoreline of coastal areas has changed frequently over the decades. Although it is a natural process, it has a vital role in influencing coastal livelihoods in the country due to densely populated coastal regions. In 1991, the Landsat 5 dataset was excellent compared with the 2006 Landsat 7 dataset. The limitations of the study were the "Landsat Toolbox" provided "Repair Landsat 7 (ETM+) Band" had been used for necessary corrections of 2006's Landsat 7 datasets. The Landsat 8 (OLI) bands were excellent. All the calculations were performed in meters (m) for point and line distances and in square kilometers ($km^2$) for polygonal area assessment. Bangladesh has a 710-km-long coastline stretching from the Hariabhanga River in the west (along the Bangladesh–India border) to the Naf River in the east (along the Bangladesh–Myanmar border), running parallel to the Bay of Bengal (Matin and Jahid Hasan 2021). However, after considering all islands, isles, islets, and riverine shoreline shapes, including mainland shore boundaries, we found land and water (sea) demarcation shapes (in km) of 3270 km in the shoreline year 1991, 3386 km in 2006, and 3338 km in 2021 (Fig. 4.11a). This length calculation was only inside the AoI coastal belt of the entire Bangladesh delta (Fig. 4.12). The calculated baseline shape length (2633.12 km) was less than all shorelines for the intersecting loss of many small islands (Fig. 4.11b). In the study, the baseline-induced base land AoI was 7572.81 $km^2$ (Fig. 4.12a). In addition, we found it to increase among all AoI-based study land in 1991 (8448.81 $km^2$), 2006 (8798.09 $km^2$), and 2021 (8872.03 $km^2$) (Fig. 4.12b, c). Therefore, a clear view of the country observed that the shorelines are moving toward change, and overall lands were accredited for the area-increasing aspects of the Bangladesh delta.

Matin and Jahid Hasan (2021) investigated the net area loss and gain of the Bangladesh delta by three zonal classes: west zone 40.67 $km^2$ (neat erosion), central zone 61.69 $km^2$ loss (neat erosion), and eastern zone 30.20 $km^2$ (neat accession) from 1989 to 2019. Sarwar and Woodroffe (2013) found that the total area along the Bangladesh shoreline changed slightly across the whole shoreline from 1989 to 2009 (20 years), with 315 $km^2$ of accretion and approximately 307 $km^2$ of erosion. In this study, in detail, neat accretion and erosion were calculated. From 1991 to 2006, the total accredited land area was 825.15 $km^2$ (Fig. 4.14a). However, it was 756.69 $km^2$ from 2006 to 2021 (Fig. 4.14c) (Table 4.2). Thus, we obtained total accretion from 1991 to 2021 over the 30-year period of 1223.94 $km^2$ (Fig. 4.14b) (Table 4.2). In the case of coastal erosion assessment, the results gained for the period 1991 to 2006 and 2006 to 2021 were 475.87 $km^2$ (Fig. 4.14a) and 682.75 (Fig. 4.14c) $km^2$, respectively (Table 4.2). Therefore, the total coastal erosion was 800.72 $km^2$ from 1991 to 2021 (Fig. 4.14c) (Table 4.2). Therefore, the net accretion was 73.94 $km^2$ from 1991 to 2021 (Table 4.2). This can be proof that the Bangladesh delta is active and witnessing frequent change. If the erosion process is increasing, it can be threatening for coastal people to continue their livelihoods, given the climate change and sea level rise issues. In the study, erosion was also considered in terms of landmass sinking and accretion related to land level rise due to sediment load deposition on the continental shelf of the Bay of Bengal. Table 4.3 represents the near-feature of the study baseline GRPs from the calculated three shoreline features as another result of shoreline changes in the Bay of Bengal.

In the studied AoI, major LULC classes were also determined using the SoB-provided country geodatabase for the years 1991, 2006, and 2021. A continuous increasing scenario was observed in the forest and trees and MSI-covered classes (Table 4.4). These two major LULC classes (forest and trees and MSI) were increased due to accretion processes of the active Bangladesh delta. On the other hand, other major LULC classes were decreased for the coastal erosion as per the study year observations. Agricultural lands were focused LULC class of the study that increased in 2006 from 1991, but then obtained a decreasing scenario in the study years 2021 from 2006. Therefore, further research is also significantly required to understand the impacts of shoreline alternations on major LULC and livelihood changes. In this stage of this research, we could predict the alternation process from 30 years of datasets. In our next stages, the impact of shoreline alternations will be assessed considering extreme climatic events.

## 4.5 Conclusion

Bangladesh is a country with a dense population and diversified riverine and deltaic landforms. A detailed shoreline assessment and change assessment study is required to support the delta plan implementation activities of the country's government including major LULC's change information. Bangladesh coastal areas are increasing, but the rates were slower after the period 1991–2006 in 2006–2021. In the 15-year period from 2006 to 2021, the erosion rate also increased. New lands could be added with the country data sheet that was collected from the Survey of Bangladesh office of the government. This study was important to prepare accurate mapping of the southern parts. There was a large gap found in the government offices in that they do not have updated or finalized shoreline boundaries (shape files) that are necessary for the respective years. In this study, we prepared shoreline boundaries for 1991, 2006, and 2021 as the country's shoreline boundary in the Bay of Bengal. This research can be utilized for future studies of Bangladesh delta, and, similarly, other researchers can build up shore boundaries, including changes over time, for future projections of land use planning and analysis of major LULC class transformation in the similar coastal regions of the world.

**Acknowledgments** The authors express their gratitude to the open-access journal *Remote Sensing* published by Multidisciplinary Digital Publishing Institute (MDPI) for their policy to support the authors for reusing the published article. We would like to thank *Remote Sensing* journal in publishing this article (Md Shamsuzzoha, Tofael Ahamed. Shoreline Change Assessment in the Coastal Region of Bangladesh Delta Using Tasseled Cap Transformation from Satellite Remote Sensing Dataset. *Remote Sensing*. *15*(2), 295; https://doi.org/10.3390/rs15020295). Moreover, the authors would like to thank the Bangabandhu Science and Technology Fellowship (BSTF) program, Ministry of Science and Technology (MoST), Government of Bangladesh (GoB) for providing a scholarship for Doctoral Program at the University of Tsukuba, Japan; the Survey of Bangladesh (SoB), Ministry of Defence (MoD), Government of Bangladesh (GoB) for providing country's data sheet; and United States Geological Survey (USGS) for Landsat Toolbox software support for analysis. The authors are also grateful to the Environmental Systems Research Institute (ESRI) for ArcGIS® 10.8.2 Software utilized for this research.

**Author Contributions** Research Investigation, Methodology, Data Curation, Analysis, Interpretation of Results, and Writing of Original Draft, M.S.; Research Conceptualization, Editing, and Supervision, T.A. All authors have read and agreed to the published version of the manuscript.

**Funding** This research did not receive any external funding.

**Data Availability Statement** The dataset that was generated and analyzed during this study is available from the corresponding author upon reasonable request, but restrictions apply to the data reproducibility for reusing and commercially confident details.

**Conflicts of Interest** The authors declare no conflict of interest regarding the research.

## References

Alam R (2014) Characteristics of hydrodynamic processes in the Meghna estuary due to dynamic whirl action. IOSR J Eng 4:39–50. https://doi.org/10.9790/3021-04633950

Alam S, de Heer J, Choudhury G (2018) Bangladesh Delta Plan 2100, Dhaka

Al-Maruf A, Craig Jenkins J, Bernzen A, Braun B (2021) Measuring household resilience to cyclone disasters in coastal Bangladesh. Climate 9:97. https://doi.org/10.3390/cli9060097

Anwar MS, Takewaka S (2014) Analyses on phenological and morphological variations of mangrove forests along the southwest coast of Bangladesh. J Coast Conserv 18:339–357. https://doi.org/10.1007/s11852-014-0321-4

Arab ST, Monirul Islam M, Shamsuzzoha M, Alam KF, Muhsin N, Noguchi R, Ahamed T (2022) A review of remote sensing applications in agriculture and forestry to establish big data analytics. In: Remote sensing application, pp 1–24. https://doi.org/10.1007/978-981-19-0213-0_1

Ataol M, Murat Kale M (2022) Shoreline changes in the river mouths of the Ceyhan Delta. Arab J Geosci 15:201. https://doi.org/10.1007/s12517-022-09516-2

Baig MH, Ali LZ, Shuai T, Tong Q (2014) Derivation of a tasselled cap transformation based on Landsat 8 at-satellite reflectance. Remote Sens Lett 5:423–431. https://doi.org/10.1080/2150704X.2014.915434

Bala BK, Hossain MA (2010) Modeling of food security and ecological footprint of coastal zone of Bangladesh. Environ Dev Sustain 12:511–529. https://doi.org/10.1007/s10668-009-9208-1

Boak EH, Turner IL (2005) Shoreline definition and detection: a review. J Coast Res 214:688–703. https://doi.org/10.2112/03-0071.1

Che X, Zhang HK, Liu J (2021) Making Landsat 5, 7 and 8 reflectance consistent using MODIS nadir-BRDF adjusted reflectance as reference. Remote Sens Environ 262:112517. https://doi.org/10.1016/j.rse.2021.112517

Costantino D, Pepe M, Dardanelli G, Baiocchi V (2020) Using optical satellite and aerial imagery for automatic coastline mapping. Geogr Tech 15:171–190. https://doi.org/10.21163/GT_2020.152.17

Crawford TW, Rahman MK, Giashuddin Miah M, Rafiqul Islam M, Paul BK, Curtis S, Sariful Islam M (2021) Coupled adaptive cycles of shoreline change and households in deltaic Bangladesh: analysis of a 30-year shoreline change record and recent population impacts. Ann Am Assoc Geogr 111:1002–1024. https://doi.org/10.1080/24694452.2020.1799746

Crist EP (1985) A TM Tasseled Cap equivalent transformation for reflectance factor data. Remote Sens Environ 17:301–306. https://doi.org/10.1016/0034-4257(85)90102-6

Crist EP, Cicone RC (1984) Comparisons of the dimensionality and features of simulated Landsat-4 MSS and TM data. Remote Sens Environ 14:235–246. https://doi.org/10.1016/0034-4257(84)90018-X

Dominici D, Zollini S, Alicandro M, Torre FD, Buscema P, Baiocchi V (2019) High resolution satellite images for instantaneous shoreline extraction using new enhancement algorithms. Geosciences 9:123. https://doi.org/10.3390/geosciences9030123

Dube SK, Chittibabu P, Sinha PC, Rao AD, Murty TS (2004) Numerical modelling of storm surge in the Head Bay of Bengal using location specific model. Nat Hazards 31:437–453. https://doi.org/10.1023/B:NHAZ.0000023361.94609.4a

Dymond CC, Mladenoff DJ, Radeloff VC (2002) Phenological differences in Tasseled Cap indices improve deciduous forest classification. Remote Sens Environ 80:460–472. https://doi.org/10.1016/S0034-4257(01)00324-8

Emran A, Abdur Rob M, Humayun Kabir M (2017) Coastline change and erosion-accretion evolution of the Sandwip Island, Bangladesh. Int J Appl Geospat Res 8:33–44. https://doi.org/10.4018/IJAGR.2017040103

Faliha Dzakiyah I, Saraswati R (2020) Drought area of agricultural land using Tasseled Cap Transformation (TCT) method in Ciampel Subdistrict Karawang Regency. E3S Web Confer 211:02005. https://doi.org/10.1051/e3sconf/202021102005

Ford M (2013) Shoreline changes interpreted from multi-temporal aerial photographs and high resolution satellite images: Wotje Atoll, Marshall Islands. Remote Sens Environ 135:130–140. https://doi.org/10.1016/j.rse.2013.03.027

Hasan M, Mahmudul RI, Saifur Rahman M, Ibrahim M, Shamsuzzoha M, Khanam R, Mostafa Zaman AKM (2021) Analysis of land use and land cover changing patterns of Bangladesh using remote sensing technology. Am J Environ Sci 17:64–74. https://doi.org/10.3844/ajessp.2021.64.74

Healey S, Cohen W, Zhiqiang Y, Krankina O (2005) Comparison of Tasseled Cap-based Landsat data structures for use in forest disturbance detection. Remote Sens Environ 97:301–310. https://doi.org/10.1016/j.rse.2005.05.009

Hossain A, Kanan MM, Pirotti F (2022) Land cover change across 45 years in the world's largest mangrove forest (Sundarbans): the contribution of remote sensing in forest monitoring. Eur J Remote Sens 1–17. https://doi.org/10.1080/22797254.2022.2097450

Huang C, Wylie B, Yang L, Homer C, Zylstra G (2002) Derivation of a tasselled cap transformation based on Landsat 7 at-satellite reflectance. Int J Remote Sens 23:1741–1748. https://doi.org/10.1080/01431160110106113

Hussain MA, Tajima Y, Gunasekara K, Rana S, Hasan R (2014) Recent coastline changes at the eastern part of the Meghna Estuary using PALSAR and Landsat images. IOP Confer Ser Earth Environ Sci 20:012047. https://doi.org/10.1088/1755-1315/20/1/012047

Islam MA, Hossain MS, Hasan T, Murshed S (2016) Shoreline changes along the Kutubdia Island, south East Bangladesh using digital shoreline analysis system. Bangladesh J Sci Res 27:99–108. https://doi.org/10.3329/bjsr.v27i1.26228

Kabir MA, Salauddin M, Hossain KT, Tanim IA, Saddam MMH, Ahmad AU (2020) Assessing the shoreline dynamics of Hatiya Island of Meghna estuary in Bangladesh using multiband satellite imageries and hydro-meteorological data. Reg Stud Mar Sci 35:101167. https://doi.org/10.1016/j.rsma.2020.101167

Karim M, Mimura N (2008) Impacts of climate change and sea-level rise on cyclonic storm surge floods in Bangladesh. Glob Environ Chang 18:490–500. https://doi.org/10.1016/j.gloenvcha.2008.05.002

Leatherman SP (2001) Chapter 8: Social and economic costs of sea level rise. pp 181–223. https://doi.org/10.1016/S0074-6142(01)80011-5

Mahamud U, Takewaka S (2016) Temporal and spatial characteristics of shoreline variability at Cox's Bazar, Bangladesh. J Jpn Soc Civil Eng Ser B2 (Coastal Engineering) 72:I_715–I_720. https://doi.org/10.2208/kaigan.72.I_715

Mahamud U, Takewaka S (2018) Shoreline change around a River Delta on the Cox's Bazar coast of Bangladesh. J Mar Sci Eng 6:80. https://doi.org/10.3390/jmse6030080

Matin N, Jahid Hasan GM (2021) A quantitative analysis of shoreline changes along the coast of Bangladesh using remote sensing and GIS techniques. Catena 201:105185. https://doi.org/10.1016/j.catena.2021.105185

Mbow C, Goïta K, Bénié GB (2004) Spectral indices and fire behavior simulation for fire risk assessment in savanna ecosystems. Remote Sens Environ 91:1–13. https://doi.org/10.1016/j.rse.2003.10.019

McGranahan G, Balk D, Anderson B (2007) The rising tide: assessing the risks of climate change and human settlements in low elevation coastal zones. Environ Urban 19:17–37. https://doi.org/10.1177/0956247807076960

Mondal A, Khare D, Kundu S, Mondal S, Mukherjee S, Mukhopadhyay A (2017) Spatial soil organic carbon (SOC) prediction by regression kriging using remote sensing data. Egypt J Remote Sens Space Sci 20:61–70. https://doi.org/10.1016/j.ejrs.2016.06.004

Natarajan L, Sivagnanam N, Usha T, Chokkalingam L, Sundar S, Gowrappan M, Roy PD (2021) Shoreline changes over last five decades and predictions for 2030 and 2040: a case study from Cuddalore, southeast coast of India. Earth Sci Inf 14:1315–1325. https://doi.org/10.1007/s12145-021-00668-5

Pardo-Pascual JE, Almonacid-Caballer J, Ruiz LA, Palomar-Vázquez J (2012) Automatic extraction of shorelines from Landsat TM and ETM+ multi-temporal images with subpixel precision. Remote Sens Environ 123:1–11. https://doi.org/10.1016/j.rse.2012.02.024

Rahaman MA, Mursheduzzaman GAM, Reza A, Chowdhury AM, Avi AR, Chakraborty TR, Shamsuzzoha M (2020) Nature-based solutions to promote climate change adaptation and disaster risk reduction along the Coastal Belt of Bangladesh. In: The Palgrave handbook of climate resilient societies. https://doi.org/10.1007/978-3-030-32811-5_49-1

Rahman AF, Dragoni D, El-Masri B (2011) Response of the Sundarbans coastline to sea level rise and decreased sediment flow: a remote sensing assessment. Remote Sens Environ 115:3121–3128. https://doi.org/10.1016/j.rse.2011.06.019

Sarwar MGM, Woodroffe CD (2013) Rates of shoreline change along the coast of Bangladesh. J Coast Conserv 17:515–526. https://doi.org/10.1007/s11852-013-0251-6

Serra P, Pons X (2008) Monitoring farmers' decisions on Mediterranean irrigated crops using satellite image time series. Int J Remote Sens 29:2293–2316. https://doi.org/10.1080/01431160701408444

Shamsuzzoha M, Ahamed T (2023) Shoreline change assessment in the coastal region of Bangladesh Delta using Tasseled Cap transformation from satellite remote sensing dataset. Remote Sens 15:295. MDPI. https://doi.org/10.3390/rs15020295

Shamsuzzoha M, Al-Maruf A (2012) Post SIDR life strategy: adaptation scenario of settlements of the south. Inst Bangladesh Stud J 19:207–222

Shamsuzzoha M, Noguchi R, Ahamed T (2021) Damaged area assessment of cultivated agricultural lands affected by cyclone bulbul in coastal region of Bangladesh using Landsat 8 OLI and TIRS datasets. Remote Sens Appl Soc Environ 23:100523. Elsevier B.V. https://doi.org/10.1016/j.rsase.2021.100523

Shamsuzzoha M, Noguchi R, Ahamed T (2022) Rice yield loss area assessment from satellite-derived NDVI after extreme climatic events using a fuzzy approach. Agric Inf Res 31:32–46. https://doi.org/10.3173/air.31.32

Sheng L, Huang J-f, Tang X-l (2011) A tasseled cap transformation for CBERS-02B CCD data. J Zhejiang Univ Sci B 12:780–786. https://doi.org/10.1631/jzus.B1100088

Talukder MF, Shamsuzzoha M, Hasan I (2018) Damage and agricultural rehabilitation scenario of post cyclone Mahasen in Coastal Zone of Bangladesh. J Sociol Anthropol 2:36–43. https://doi.org/10.12691/jsa-2-1-7

Wang X-Z, Zhang H-G, Fu B, Shi A-q (2013) Analysis on the coastline change and erosion-accretion evolution of the Pearl River Estuary, China, based on remote-sensing images and nautical charts. J Appl Remote Sens 7:073519. https://doi.org/10.1117/1.JRS.7.073519

Wulder MA, Skakun RS, Kurz WA, White JC (2004) Estimating time since forest harvest using segmented Landsat ETM+ imagery. Remote Sens Environ 93:179–187. https://doi.org/10.1016/j.rse.2004.07.009

Zhai Y, Roy DP, Martins VS, Zhang HK, Yan L, Li Z (2022) Conterminous United States Landsat-8 top of atmosphere and surface reflectance tasseled cap transformation coefficients. Remote Sens Environ 274:112992. https://doi.org/10.1016/j.rse.2022.112992

# Chapter 5
# A Damage-Based Crop Insurance System for Flash Flooding: A Satellite Remote Sensing and Econometric Approach

**Md. Monirul Islam, Tofael Ahamed, Shusuke Matsushita, and Ryozo Noguchi**

**Abstract** This research attempted to develop a damage-based crop insurance model using satellite remote sensing imagery and field surveys of flash flood regions in Bangladesh. A normalized difference water index (NDWI) was prepared from Landsat-8 Operational Land Imager (OLI) data, and inundated areas were delineated by using a very high-resolution land use land cover (LULC) map of a Survey of Bangladesh (SoB) to generate an accurate damage map for boro rice. The boro rice area damaged by flash floods in 2017 was estimated for the three most vulnerable flash flood-prone areas and sorted into high, moderate, and marginal damage categories. The results of the accuracy assessment showed excellent classification performance for all classes of LULC. Analysis of the classification of the damaged area showed that the Gowainghat (44.60%) and Kulaura (69.80%) subdistricts were in the highest marginal damage category and that Tahirpur (52.92%) was in the moderate damage category. The future value of expected losses was calculated to be \$536.25, \$442.00, and \$416.00 per hectare (ha) for high, moderate, and marginal damage areas, respectively. Moreover, findings concerning the damage-based crop insurance premium rate suggested that the higher the coverage levels were, the lower the insurance premium, and the lower the damaged class was, the lower the insurance premium rate. The lowest insurance premium rate was observed for areas with high coverage and moderate damage and was \$23.82/ha, and the highest insurance premium rate was observed for areas with marginal coverage and high damage and was \$39.49/ha. Evidence from the classification of damaged areas and from

M. M. Islam
Department of Agricultural Economics, Bangladesh Agricultural University, Mymensingh, Bangladesh

Agriculture & Food, CSIRO, Adelaide, SA, Australia

T. Ahamed (✉)
Institute of Life and Environmental Sciences, University of Tsukuba, Tsukuba, Japan
e-mail: tofael.ahamed.gp@u.tsukuba.ac.jp

S. Matsushita · R. Noguchi
Graduate School of Agriculture, Kyoto University, Kyoto, Japan



T. Ahamed (ed.), *Remote Sensing Application II*, New Frontiers in Regional Science: Asian Perspectives 77, https://doi.org/10.1007/978-981-97-1188-8_5

regression findings suggests that farmers' socioeconomic features and environmental awareness (occupation, educational level, total income, bank account ownership and awareness of climate change) are relevant to the decision to adopt a damage-based crop insurance system. The overall results show an empirical model for flash flood occurrence over both the temporal and spatial domains, which offers an effective measure for adopting a damage-based crop insurance model.

**Keywords** Crop damage assessment · Crop insurance · NDWI · Logistic and multiple regression · GIS · Satellite remote sensing

## 5.1 Introduction

In the current era, extreme climatic events, such as flash floods, droughts, floods, hurricanes, earthquakes, and landslides, have a negative effect on agricultural output, food and nutritional security for the growing population, and their livelihood, especially in developing nations (Pandey and Bardsley 2015; Zwane 2019). Among all natural calamities, flash floods have become one of the most devastating natural hazards worldwide, leading to significant economic and social damage (Svetlana et al. 2015; Benson and Edward 2002). Flash floods are rapid-onset floods of short durations accompanied by agricultural crop damage, landslides, human and animal life losses, and destruction of infrastructure and communication networks (Diakakis et al. 2016; Rahmati et al. 2016). The frequency and consequences of extreme flood events have increased rapidly worldwide in recent decades (Kron 2009).

Among developing countries, Bangladesh is one of the world's most susceptible nations to climate change due to its location at the confluence of the Ganga, Meghna, and Brahmaputra rivers and the Ganga-Brahmaputra delta (Yu et al. 2010). Bangladesh, which is ranked eighth in world population, faces tremendous challenges to secure agricultural production significantly affected by flash floods in the Haor area (Roy et al. 2017; Ahmed 2012). Extreme rainfall and large river discharge are predicted to rise due to climate change, increasing the possibility of frequent flash floods with greater intensity (Biswas et al. 2008; Kron 2009). Moreover, boro rice is the main contributor among all other rice types, and the flash flood stricken Haor region has produced 15% (in some cases 24%) of the total boro rice production in Bangladesh (BBS 2017). Therefore, any disturbance (such as flash flooding) on boro rice production potentially impacts the market prices of cereals. As extreme events are increasing and damage occurs in major crops due to the changing climate throughout the world, an effective intervention is required to support farmers' livelihoods.

Governments and other stakeholders in these vulnerable regions should strengthen agricultural growth and eliminate poverty through sound policymaking and investments in better risk management (Dawson et al. 2011; Rana et al. 2020). One potential strategy to cope with the devastating effects of crop damage is to introduce a crop insurance scheme (Dalhaus et al. 2018). However, a lack of reliable data, awareness of farmers, administrative costs, and experience are the key constraints for the successful execution of crop insurance programs in many developing countries, including Bangladesh (Al-Maruf et al. 2021; Hazell 2001; Anderson 2001). Bangladesh's small farmers tend to underinvest in new high-yield crops and agricultural equipment, withholding funds to protect themselves against weather and crop risk by following traditional coping strategies (Habiba and Shaw 2013). Traditional risk-reduction measures in Bangladesh, such as borrowing from family, selling land, employing microcredit, and saving, are frequently insufficient to address climate-related risk (Al-Maruf et al. 2021). From 1970 to 2020, the country has seen many microinsurance programs aimed at health and agriculture, with varying degrees of success and failure (Sultana and Kaisar 2021). In Bangladesh, rice farmers' subsidies on a variety of agricultural inputs (power & fuel for irrigation, fertilizer, cash incentive, etc.) are used to keep their prices within the rice farmers' purchasing power rather than as a crop insurance scheme (Tobias et al. 2012). Although the Bangladesh Climate Change Strategy and Action Plan (BCCSAP) 2008 highlighted "insurance" as a helpful disaster management strategy (WB 2010), until now, there has been no significant initiative taken by the government of Bangladesh.

In this regard, establishing appropriate risk management strategies for climate-induced vulnerable agriculture will be the topmost concern. However, advocates argue that crop insurance can play a vital role as an alternative risk coping instrument to enable poor farmers in developing economies to cope with weather-related production risks (Akter et al. 2008). Some researchers have developed index-based weather insurance for developing countries (Clarke 2016); willingness to pay (WTP) for crop insurance against extreme weather events such as floods and droughts (Arshad et al. 2015); demand assessments and testing of the commercial viability of crop insurance in Bangladesh (Akter et al. 2009); rainfall-based index insurance in Morocco (Skees et al. 2001); and assessments of the demand for flood insurance (Kousky 2011). Moreover, neither an index-based insurance product (Al-Maruf et al. 2021; Yuzva et al. 2018) nor an indemnity-based insurance product (Fadhliani et al. 2019) has been shown that can solve the challenge of proper risk management instruments in agriculture. In this connection, our research attempted to develop a new and unique damage-based crop insurance system for flash flood-affected wetland areas by integrating geographic information system (GIS), satellite remote sensing, and econometric models simultaneously.

However, it has been proven that to encounter flash flood damage and ensure farmers' livelihoods, satellite remote sensing and GIS can provide solutions and insurance adaptations to recover damage with quick assessments (Feng et al. 2019; Liu et al. 2019). In addition, Sentinel, Landsat, and NOAA's polar-orbiting satellites all have equipment that can be used to identify crop types and conditions and quantify crop damage acreage more accurately (Benedetti and Rossini 1993). Moreover, rapid flood inundation mapping and delivery of flood inundation maps during a flood event can provide valuable information for decision-makers to put necessary measures in place (Amarnath 2013). In addition, satellite-based flood inundation maps can be an essential input for assessing the progression of floodwaters and the severity of the flood situation (Du et al. 2014; Bubeck et al. 2012). In this regard, multidisciplinary approaches, including remote sensing, GIS, and machine learning algorithms, were used to effectively predict and manage floods due to quick and free access to satellite datasets (Waqas et al. 2021; Tien Bui et al. 2019; Khosravi et al. 2018). Therefore, freely available multispectral images such as Landsat 8 Operational Land Imager (OLI) images are suitable for land surface inundation (water content) mapping (Du et al. 2014; Xu 2006). In addition, various techniques have been developed to classify water pixels by processing surface reflectance captured by different spectral bands of optical sensors (Feyisa et al. 2014). The normalized difference water index (NDWI) for extracting surface water bodies and the development of inundation maps has been used in many studies and found to be the most important index for inundation map analysis worldwide (Du et al. 2014). Hence, our study also employed the NDWI index for detecting the surface water of flash floods in 2017 through inundation mapping.

In addition, flash flood vulnerability combines the inundation extent with social data, which is used to determine flood-prone communities with the greatest propensity for loss of property and life (Smith et al. 2019). Microlevel socioeconomic and demographic information is also very important for assessing the adoption viability of crop insurance and its premium policy (Adusumilli et al. 2019; Islam et al. 2018). However, socioeconomic spatial and nonspatial datasets are also required to examine the decision to adopt crop insurance using numerous econometric analyses. It was also significant to determine the break-even analysis for indemnity payouts. Once an indemnity payout is solicited, then the social adoption of crop insurance was practiced with different coverage levels and varying WTP. In this social adoption, practicing crop insurance and WTP were related to several social and economic indicators of stakeholder behavior (Abebe and Bogale 2014; Budhathoki et al. 2019). As the bivariate logistic regression model is becoming popular for analyzing the factors affecting farmers' adoption behavior for crop insurance (Aidoo et al. 2014; Abdullah et al. 2014), our study is also modeled use of the logistic regression method. In addition, a multiple regression analysis was applied to verify that the essential factors affecting the amount farmers are WTP for the proposed damaged-based crop insurance model (Aidoo et al. 2014; Sadati et al. 2010).

Therefore, while a significant number of crop insurance products are available based on individualistic subjective areas, none of them integrate satellite remote sensing and econometric approaches in a single platform. Additionally, to improve the sustainable livelihoods of the affected farmers, site-specific economic approaches can be adapted for optimizing agricultural subsidies according to the damage levels. Developing a cost-effective insurance product that assists farmers in risk management and establishing long-term insurance markets for small farmers is obligatory in Bangladesh. Once farmers have access to insurance against agricultural risk, they are more motivated to increase their investment irrespective of risks and uncertainty (Cole et al. 2017; Berber and Patrick 2018). In addition, all crop insurance policies consider the weather and, in some cases, natural disasters. The current study aimed to develop a novel damage-based crop insurance model, considering the limits of data, knowledge, farmer awareness, cost, and experience. Therefore, the significant potential of damaged-based crop insurance systems will help farmers avoid flash flood-related risks. To the best of our knowledge, the damage-based crop insurance system that combines GIS and remote sensing applications with an econometric approach in a single platform is one of the novel attempts. Therefore, the specific objectives of this research were as follows: (i) to estimate the damage areas for rice acreage due to flash floods using GIS and remote sensing; (ii) to determine a crop insurance premium rate based on different coverage levels; and (iii) to assess the adoption responses of affected farmers through the analyses of willingness to adopt (WTA) crop insurance and amount farmers are WTP as premiums.

## 5.2 Research Methodology

### *5.2.1 Selection of the Study Area*

Global warming causes the greatest harm to agriculture in developing countries, primarily because many farms in the low latitudes already endure increasing climatic variability (Mendelsohn 2008). Among developing countries, Bangladesh is one of the most climatically vulnerable, as measured by floods, droughts, cyclones, sea-level rise, salinity, flash floods, and soil erosion (Yu et al. 2010). The study areas consist of three subdistricts and were selected based on the consideration of the most affected areas from Haor regions, located in the northeastern part of Bangladesh with latitudes of 23°59′04.75"N to 25°12′49"N and longitudes of 90°27′22.13″E to 92°30′10.19″E (Ahmed 2012). To select the most vulnerable flash flood affected Haor areas, historical data were collected from daily newspapers, the Bangladesh Bureau of Statistics (BBS), the Haor Master Plan, the Bangladesh Meteorological Department (BMD), and the Bangladesh Water Development Board (BWDB). A very recent (last 5 years of 2016–2020) flash flood history in wetlands in Bangladesh is presented in Appendix 1. From that assessment, we initially selected three subdistricts from three districts. However, to confirm our review, we also checked the inundation extent for flash flood 2017 using Landsat 8 datasets through NDWI analysis and found that these were the most inundated areas. Finally, we confirmed and selected the following three most flash flood vulnerable subdistricts as our study area from all Haor regions in Bangladesh. The study areas were the Tahirpur, Gowainghat, and Kulaura subdistricts from the Sunamganj, Sylhet, and Moulvibazar districts, respectively (Fig. 5.1).

Agriculture is the principal livelihood of farmers in the study areas who practice mono agriculture (boro rice). This single crop remains under the constant threat of partial to complete damage from the early rush of flash floods (Haor Master Plan 2012). Notably, there are mainly two phases of flooding in the Haor basin, i.e., the early flood (pre-monsoon) phase and the deeply flooded (monsoon) phase. Early floods cause substantial damage to the standing crop of boro rice, while deep floods cause little crop damage in the field (Roy et al. 2017). Flash floods caused by the sudden flow of rainwater from adjacent Indian hills of Assam and Meghalaya pose

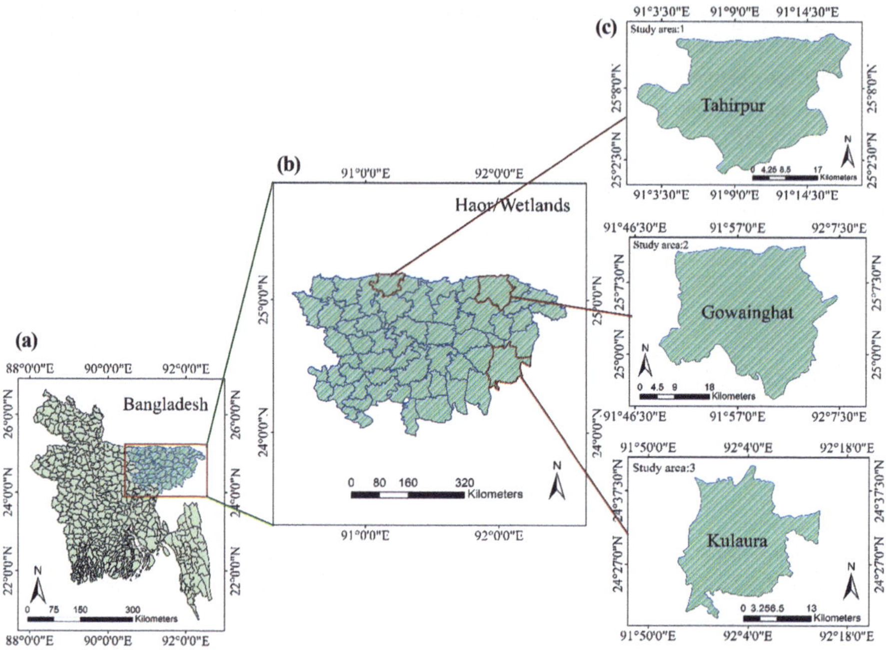


**Fig. 5.1** Flash flood-affected study areas in the northeastern part of Bangladesh showing (**a**) a map of Bangladesh, (**b**) the Haor/wetland area that covers 6 Haor districts (58 subdistricts), and (**c**) three selected most vulnerable flash flood-affected areas, namely the Tahirpur, Gowainghat, and Kulaura subdistricts

a high risk of damage to the standing winter boro rice crop just 2–3 weeks before harvesting (Ahmed 2012). A recent study found that the frequency of flash floods in the northern part of Bangladesh is increasing alarmingly, and flash flooding has hit the Haor region six times in the months between March and April during the period 2000–2017 (Roy et al. 2017). In addition, boro rice is the main contributor of all rice types, accounting for approximately 55% of the total rice production, and the flash flood-stricken region produces approximately 15% of the total boro rice (BBS 2017).

## 5.2.2 *Research Framework*

The research framework consisted of three major domains. In the first domain, satellite remote sensing and GIS techniques were applied to calculate the damage amount of boro rice due to flash floods from the high-resolution Survey of Bangladesh (SoB) cartographic datasets. Second, a damage-based crop insurance premium rate (monthly basis) was determined considering different coverage levels and interest rates. In the third section, an econometric approach was applied to assess adoption behavior by identifying the important factors affecting the adoption of crop insurance and WTP using the collected ground reference information through a field survey. Landsat 8

OLI datasets (10 April 2017) were used to map the flash flood extent in Haor regions in Bangladesh. In addition, at this stage, the selected vegetation indices show maximum sensitivity to changes in the biophysical conditions of the rice crop under observation (Whitcraft et al. 2015). NDWI, which is a well-established means of detecting surface water, was applied to high-resolution satellite image data to develop an inundation map in a GIS environment (ArcGIS 10.4.1®, ESRI, California, USA) (Du et al. 2014). Then, the damage-based premium rate was calculated following mathematical modeling (Chen and Tse 2017; Islam et al. 2018). In the third stage, to identify the important factors for WTA crop insurance, a binary logistic regression method was applied with maximum likelihood estimates (MLE), and a multiple regression approach was performed for the amount farmers were WTP as insurance premiums (Fig. 5.2).

## *5.2.3 Cartographic and Yield Datasets*

SoB cartographic datasets at the 1:25,000 scale were used in this research to develop the thematic layers, which included the administrative boundaries, buildings and structures, hydrographic features, vegetation and forests, roads, and geodetic control points from the Sylhet division of Bangladesh. Apart from this, secondary data

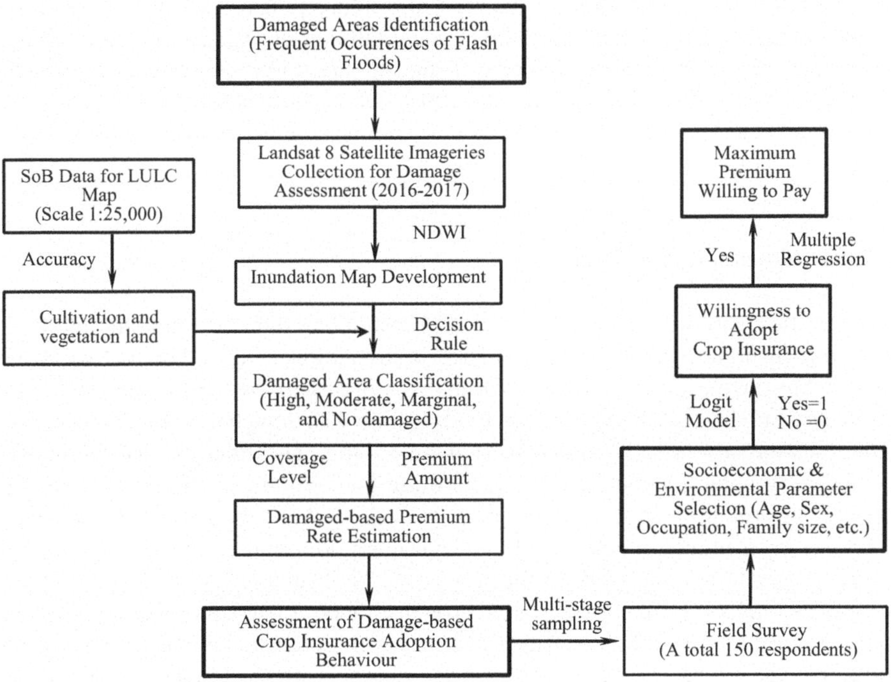


**Fig. 5.2** The conceptual framework for the geospatial crop insurance model based on damage proportions

related to boro rice yield at the subdistrict level for 3 years (2017–2019) were collected from the Department of Agricultural Extension (DAE), Ministry of Agriculture (MoA), Bangladesh (Fig. 5.2).

## *5.2.4 Ground Reference Information*

To incorporate socioeconomic indicators in the most vulnerable and frequently flash flood-affected subdistricts, ground reference information was collected to assess the adoption responses regarding the damage-based crop insurance system at the micro-level. A field survey was carried out in households to understand the effects of flash flooding on farmers and the damage to boro rice production that occurred in 2017. A total of 150 households from four damaged classifiers (50 from each selected subdistrict) were interviewed, and the selection was performed following a multi-stage sampling technique with the help of DAE officials. First, we collected a list of subdistricts affected by flash floods in 2017. Then, we selected the most affected unions from the most affected subdistricts (a small administrative unit of subdistrict). In the next stage, we chose the most affected villages (last administrative unit of subdistrict). After that, we collected a complete list of flash flood-affected farmers from the local DAE office. Finally, following a computer-generated random number table, we chose 50 flash flood-affected farmers from each subdistrict for ground reference information. In the case of a nonresponse, the interviewers proceeded to the next household until the targeted number of respondents for a village was reached. The data were collected using the structured questionnaire through face-to-face interviews in May and June 2019. To relate with the GIS approaches, the latitudes and longitudes of the respective farmers' fields were recorded using a global positioning system (GPS) unit (Fig. 5.2).

### 5.2.4.1 Questionnaire Design

The questionnaire for the field survey was prepared based on the literature of different crop insurance adaptation-related research (Aidoo et al. 2014; Falola et al. 2013). Moreover, the final version (draft) was pretested with 30 respondents (10 from each subdistrict), and necessary modification was done before the final survey. The questionnaire sought information in two parts. The first part asked for the sociodemographic characteristics and farming background of the respondents as general information. In the second part, we asked about the expected damaged-based crop insurance premium rate and farmer's adoption response behavior by identifying determinants of WTA the proposed crop insurance model. It is important to note that the WTP for damage-based crop insurance-related questions were asked only of those farmers who confirmed that they were interested in the proposed crop insurance model. Moreover, a sample questionnaire is attached in the Appendix 1 to understand the research question.

#### 5.2.4.2 Designing of WTA and WTP Crop Insurance

After introducing damage-based crop insurance, a farm household survey was used to elicit farmers' WTA for crop insurance. The damaged-base crop insurance scheme was described to farmers in as much detail as possible, focusing on how the system works. The basic principle was communicated to farmers as insurance guaranteeing a minimum payout for a specific risk over a predetermined period in exchange for a charge known as a premium. This study assumes the farmer decides to adopt crop insurance as a risk management tool against flash floods. This decision is made when risks and uncertainties are associated with the probability of a flash flood event. As a result, we used a binary logistic regression method to determine what variables influence farmers' WTA damage-based crop insurance.

We analyzed the risks associated with Boro rice production and their implications, along with providing details on the distinction between acquiring and not purchasing the proposed damage-based crop insurance prior to collecting data. Moreover, during data collection, farmers were presented with the benefits and drawbacks of crop insurance depending on the production cost or predicted return. By applying the guaranteed revenue formula, researchers calculated the premium amount farmers are willing to pay for the damage-based crop insurance model. Furthermore, to estimate the coverage level for the proposed insurance premium, a set of coverage levels (0.4–1.0) was asked of the respondents during data collection. A summary of farmers' WTA crop insurance strategy and WTP amount is presented below (Fig. 5.3).

### *5.2.5 Inundation Map Analysis Through NDWI*

Usually, NDWI monitors the changes related to water content in water bodies using the green and near-infrared (NIR) wavelengths; the index was defined by McFeeters (1996). The following formula was used to develop the inundation map:

$$\text{NDWI} = \frac{\left(X_{\text{green}} - X_{\text{nir}}\right)}{\left(X_{\text{green}} + X_{\text{nir}}\right)} \tag{5.1}$$

where NDWI is the normalized difference water index, *X*green is Band 3, and *X*nir is Band 5 for Landsat-8 OLI datasets. Generally, the NDWI value of water bodies is greater than 0.5. Vegetation has much smaller values, which easily distinguish vegetation from water bodies. Build-up features having positive values lie between 0 and 0.2. However, change detection is the process of identifying differences in the state of an object or phenomenon by observing the detection at different times (Long et al. 2014) and can be expressed as follows:

$$D_{x_{ij}^k} = x_{ij}^k(t_2) - x_{ij}^k(t_1) + C \tag{5.2}$$

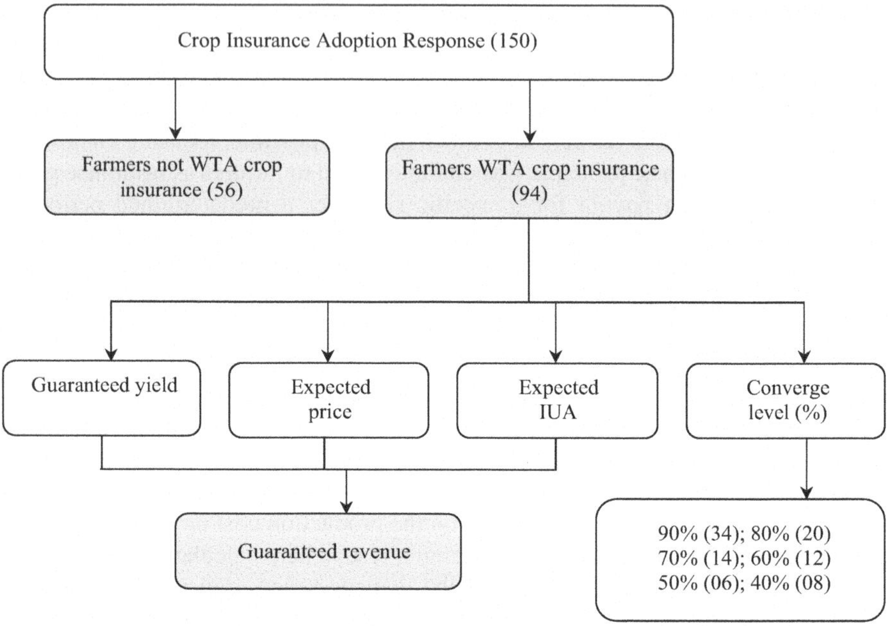


**Fig. 5.3** Schematic chart of farmers' adoption response behavior for the proposed crop insurance model (number in parentheses indicates the number of respondents)

where $x_{ij}^{k}$ is the pixel value for band $k$, $I$, and $j$ are line and pixel numbers in the image, $t_1$ is the first date, $t_2$ is the second date, and $C$ is a constant to produce positive digital numbers.

We followed the classification algorithms from relevant references (Long et al. 2014; Islam et al. 2010). Many studies have been conducted utilizing the spectroscopic characterization of SWIR to detect water content (McFeeters 1996). In this study, the production of inundation maps used a decision tree to associate each pixel with the following categories: flash flood, mixed, nonflash flood, and water-related pixels. The decision tree is illustrated in Fig. 5.4 and can be summarized as follows:

The first step was to estimate normalized difference vegetation index (NDVI) and NDWI and the different values of NDVI and NDWI and DVEL for each land class cover type. The separation of water-related pixels from non-flash flood pixels was carried out in this investigation using the threshold values (Bhatta 2012; Islam et al. 2010, Xiao et al. 2005; Acharya et al. 2018). If the NDWI is $<0.05$, it can be classified as a nonflash flood-related pixel. NDWI $\geq 0.05$ and DVEL$<0.01$ were considered water-related pixels. Water-related pixels should have a DVEL of $<0.01$. If the NDWI value ranged between 0 and 0.05, then it was considered a mixed pixel. However, excluding the "Lake" land-use type, the DVEL value is not always $<0.01$. To overcome this problem, another criterion is used to identify water-related pixels. In such cases, NDWI $\geq 0.05$ and NDVI $\leq 0.00$ were considered water-related pixels (Bhatta 2012).

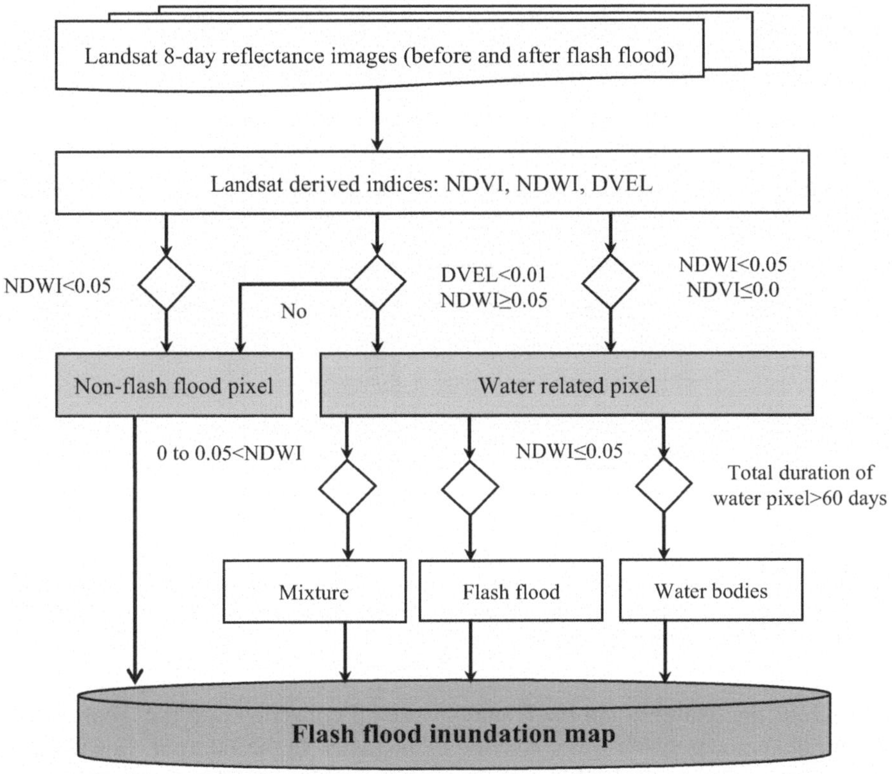


**Fig. 5.4** Flowchart for developing the flash flood inundation map using Landsat 8 OLI data

### *5.2.6 Land Use Land Cover Map*

Three land use land cover (LULC) maps were developed using the maximum likelihood classification method with the SoB datasets in the ArcGIS environment (Binte Mostafiz et al. 2021). In practice, several LULC classification techniques are available, namely supervised (artificial neural network, binary decision tree, image segmentation, maximum likelihood, minimum distance, parallelepiped, Mahalanobis distance, K-nearest neighbor, decision tree, semantic-based) and unsupervised (ISODATA, support vector machine and K-Means) (Anderson et al. 1976). However, an ideal classification technique is limited and unlikely to ever be developed (Abburu and Golla 2015). The maximum likelihood method is a statistical supervised approach for recognizing patterns (Munyati 2004). It allocates pixels to appropriate classes based on the probability values of the pixels. The nearest neighborhood method of maximum likelihood classification was applied to evaluate the classification's accuracy expediently, as the nearest neighbor is the best used for categorical data such as land-use classification or slope classification (Al-Ahmadi and Hames 2009). For processing these geospatial datasets, the merge function was used first to create a single vector dataset using spatial analysis tools. A single vector was made

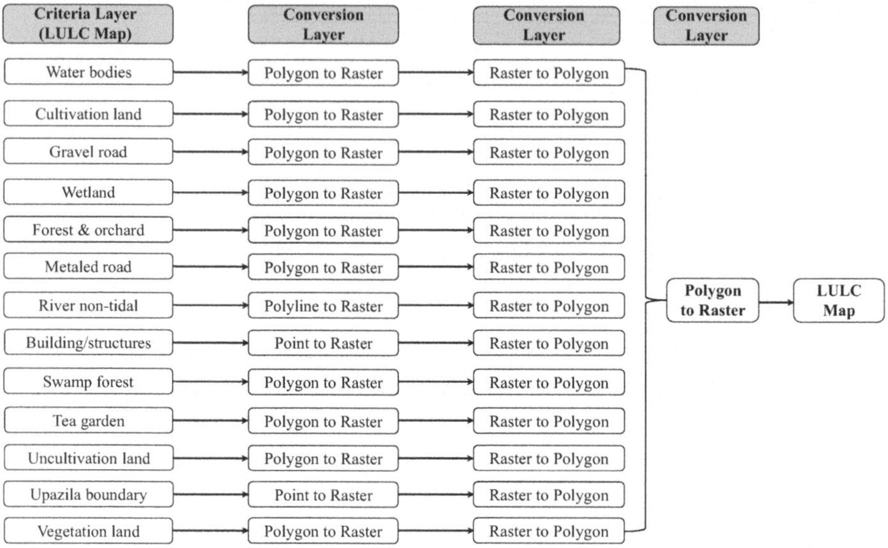


**Fig. 5.5** The schematic diagram for the land use/land cover map. The maximum likelihood classification method was applied with the integration of SoB cartographic datasets to develop highly accurate LULC maps for damage assessment

for each of the features, and the features were all converted into polygons. In the second stage, the polygon was converted to the raster. In the Tahirpur, Gowainghat, and Kulaura subdistricts, LULC maps were developed using 11, 14, and 14 classes, respectively. Finally, 11–14 features were used to develop the LULC map and to calculate the damaged area (Fig. 5.5).

#### 5.2.6.1 Accuracy Assessment of LULC

The users of LULC maps need to know how accurate the maps are to use the data more correctly and efficiently (Plourde and Congalton 2003). The most widely promoted classification accuracy is in the form of an error matrix, which can be used to derive a series of descriptive and analytical statistics (Liu et al. 2007). Producer accuracy (PA), user accuracy (UA), overall accuracy (OA), and Kappa statistics are generally reported, and these terms have been explained in detail in many studies (Liu et al. 2007; Plourde and Congalton 2003; Congalton 1991). In this study, accuracy assessment was performed for the three selected study area LULC maps, which were developed from the SoB layers. A stratified random sampling method was employed for selecting 200 points on the map with a minimum of 50 sample points for each considered category, as recommended by Congalton (1991). The PA refers to the number of correctly classified pixels in each category (on the major diagonal) by the number of reference pixels "known" to be of that category (the column total). The UA was computed by dividing the number of correctly classified pixels in each category by the total number of pixels that were classified in that category (the row total). On the other hand, OA refers to the ratio between the number of correct plots and the number of total plots. Moreover, to measure the agreement, the Kappa coefficient was applied in this study. Kappa measures the percentage of data values in the main diagonal of the table and then adjusts these values for agreement that could be expected due to chance alone. The Kappa needs to calculate the observed level of agreement and then compares calculated value to the value that is expected. The Kappa coefficient (K) is then calculated using Eq. (5.3):

$$K = \frac{N\sum_{i=1}^{r} X_{ii} - \sum_{i=1}^{r} X_{i+}X_{+i}}{N^2 - \sum_{i=1}^{r} X_{i+}X_{+i}} \tag{5.3}$$

where, $r$ denotes the number of rows/columns in the confusion matrix; $X_{ii}$ is the number of observations in row $i$ and column $i$; $X_{i+}$ is the total number of rows $i$; $X_{+i}$ is the total number of columns $i$, and $N$ stands for the number of observations.

The Kappa statistics provide a statistically valid assessment of the quality of classification and were used to assess overall class accuracy. The agreement of Kappa coefficients was classified as follows: values >0.79 are excellent, values between 0.6 and 0.79 are substantial, and values of 0.59 or less indicate moderate or poor agreement (Landis and Koch 1977).

#### 5.2.6.2 Clustering of Damaged Map

In the clustering stage, the respective criteria layers for the different subdistricts were chosen from the LULC map. Then, the separate boro rice acreage (cultivation and vegetation) was estimated with the raster calculator. After doing this, two layers were merged to create a combined layer. A value of 0 (zero) was assigned to the noncultivated area and 1 (one) for the boro rice cultivated area. Then, the damage map developed from the NDWI was multiplied by the combined layer to calculate the pixel-based damaged area. A similar method was followed to determine the high, moderate, marginal, and nondamaged classifiers (Fig. 5.6).

In addition, the presence of moisture in crop leaves and water in agricultural fields during the monsoon season is quite normal. Therefore, to avoid ambiguity in separating cropland from the inundated area, a mask of the nonvegetated area was prepared from a normalized difference vegetation index (NDVI) using the following equation.

$$\mathrm{NDVI} = \frac{(X_{\mathrm{nir}} - X_{\mathrm{red}})}{(X_{\mathrm{nir}} + X_{\mathrm{red}})} \tag{5.4}$$

where NDVI is the normalized difference vegetation index, $X_{\mathrm{nir}}$ is the value of Band 5, and $X_{\mathrm{red}}$ is the value of Band 4 for Landsat-8 OLI datasets. The threshold value was used for the binary classification of the NDVI layer into vegetation (above

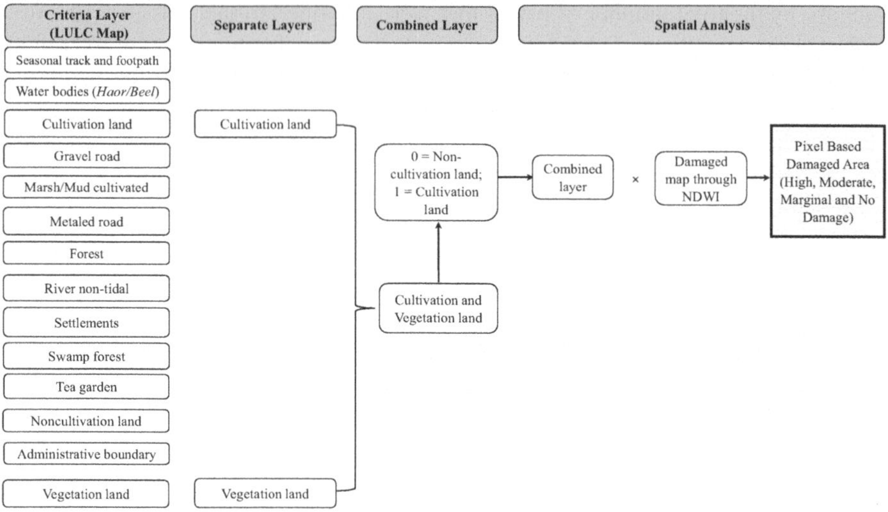


**Fig. 5.6** Schematic diagram for the separation of the damaged areas from the nondamaged areas

threshold) and nonvegetation (below threshold) classes. Finally, the inundated area was identified from the NDWI layer using a raster query that finds a pixel as inundated if the NDWI value is higher than the threshold but must have an NDVI value below the NDVI threshold (i.e., nonvegetated pixel) (Fig. 5.6). Generally, the NDVI value varies from −1 to 1. A higher NDVI value reflects a high NIR value, indicating dense greenery. Generally, NDVI = −1 to 0 represents water bodies; NDVI = −0.1 to 0.1 represents barren rocks, sand, or snow; NDVI = 0.2 to 0.5 represents shrubs and grasslands or senescing crops; and NDVI = 0.6 to 1.0 represents dense vegetation or tropical rainforest (Bhatta 2012).

## *5.2.7 Coverage Level for the Insurance Premium*

The insurance premium rates started according to the estimated loss (Skees et al. 1997). The estimated losses can be calculated based on the break-even yield for indemnity, excluding service costs (Martin et al. 2001). The estimation of the yield gap/damaged area was based on the forecasted yield and actual yield (Skees et al. 1997). To estimate losses, insurance actuaries use historical experience with the insurance product. In yield insurance, historical yield experience is used to calibrate the expected distribution of future crop yields (Godwin and Ker 1998). It is widely recognized that crop yields are skewed and thus not distributed normally (Just and Weninger 1999). Therefore, using only the mean and variance to determine premium rates may generate inappropriate estimations of losses. To address the problems created by the uncertainty over the parameters of the yield distribution, the commonly accepted procedure is to use guaranteed yield from historical yield to calculate premium rates. The premium rate estimation procedure can be expressed as follows:

$$f_{y_{(t+1)}} = \alpha y_t + (1-\alpha) f_{yt} \tag{5.5}$$

where $f_{y_{(t+1)}}$ is the forecasted area yield for year $t + 1$, $f_{yt}$ is the forecasted area yield for year $t$, $y_t$ is the actual area yield for year $t$, and $\alpha$ is the smoothing constant, which lies between 0 and 1. The expected break-even yield for indemnity (*EPR*) was calculated by dividing total indemnity payments ($I_t$) by the total number of years examined ($n$) and can be expressed as follows:

$$\text{EPR} = \sum_{t=1}^{n} \frac{I_t}{n} \tag{5.6}$$

where $I_t$ is the indemnity and calculated as the percentage shortfall in the boro rice yield in an area and can be expressed as follows:

$$I_t = \max\left[\left(y_{\text{g}} - \frac{y}{y_{\text{g}}}\right)(f_{\text{y}}), 0\right] \tag{5.7}$$

Equation (5.7) indicates that a farmer receives an indemnity payment whenever the area yield ($y$) is less than the guaranteed level of yield ($y_{\text{g}}$) for a given year $t$. Variable $y_{\text{g}}$ is calculated by multiplying the forecasted yield ($f_{\text{y}}$), and the different coverage levels chosen by the farmers. Different coverage levels (40–100%) were checked and verified to select the minimum coverage level for accepting damage-based crop insurance (Clover and Nieuwoudt 2003). We classified our coverage levels (40–100%) into two aspects: cost of production and expected return during data collection. Furthermore, to explore the extent of coverage level respondents are willing to accept, an additional classification was done, namely maximum and minimum coverage levels. Most of the respondents were ready to accept the proposed crop insurance model for regular return basis and maximum amount. In contrast, on the cost of production basis, respondents were less willing to choose as the cost of production increased significantly (Mottaleb and Mohanty 2015).

To determine the expected break-even yield for indemnity, the 5-year lifespan of the insurance products was considered. This procedure was applied for three (high, moderate, and marginal) damage classes in the study. Actual yield was considered from the average yield found during the field survey, while forecasted yield information was collected from the DAE office at the subdistrict level for the last 3 years (2017–2019). In this study, the simple moving average method was applied to calculate the forecasted yield from 3 years' average yield data for selected subdistricts using the following equation:

$$\text{MA} = Y_1 + Y_2 + Y_3 / N \tag{5.8}$$

where MA is the simple moving average, $Y_n$ is the average yield at the subdistrict level ($n = 1,2,3$), and $N$ stands for 3 years.

### 5.2.8 Calculation of Insurance Premium Rate

In this research, the insurance premium mainly considers the damaged proportions of boro rice acreage to cover the damages of farmers. To determine the premium rate, different low-interest rates were applied that could have the ability or "WTP" based on the farmer's socioeconomic condition. The range was 2–6% of the insured revenue, which was considered the WTP the premium for boro rice. The expected damage/loss amount was calculated with the help of the ground reference yield loss time-series data at three selected subdistricts and the government-boro rice procurement price in 2017 @ \$0.31 (BDT 26.0) per kg. The insurance coverage level was classified as high coverage (90%), moderate coverage (80%), and marginal coverage (70%) based on the yield gap (Skees et al. 1997; Clover and Nieuwoudt 2003). The minimum insurance coverage level was considered marginal coverage (70%) to recover production costs, while the maximum coverage was estimated as high coverage (90%) to ensure the expected maximum revenue from the boro rice. Similar insurance coverage levels were charged for different damaged classes of boro rice acreage. No premium rate was calculated for the undamaged areas due to physical and economic viability. A farmer could buy an insurance product from the three categories and could pay the premium in 12 installments (every month). The future value of the annuity ($V$) was calculated using the moving average yield of the selected subdistrict and the price of boro rice in 2017 as follows (Chen and Tse 2017).

$$V = 12P\frac{\left\{1+\left(\frac{i}{12}\right)\right\}^{12n} - 1}{i} \tag{5.9}$$

where $P$ is the premium for different coverage levels. The future expected loss from boro rice production can be expressed as follows:

$$V_j = \sum_{i=1}^{n} \mathrm{APL}_i \times \mathrm{PRT} \tag{5.10}$$

where $V_j$ is the cumulative loss of boro rice for different coverage levels ($j$) in one calendar year (here, only a single crop). The APL is the average production loss for boro rice (ton/ha; $i$ is the seasonal variety: if applicable; and $n$ is the number of years), and PRT is the average price of boro rice (\$/ton). Therefore, the premium for different coverage levels ($j$) is based on the expected loss and production function, which can be expressed as follows:

$$P_j = \frac{V_j \times i}{12\left\{1+\left(\frac{i}{12}\right)\right\}^{12n} - 1} \times C_{\mathrm{j}} \times C_{\mathrm{f}} \tag{5.11}$$

where $P_j$ is the premium for the specific coverage level ($j$) of a single rice practice in one calendar year. The parameter $C_{\mathrm{f}}$ is the conversion factor for the pixel resolution to scale down the pixel level (30 m × 30 m, Landsat 8, OLI). The parameter $C_{\mathrm{j}}$ is the coverage rate (highly, moderately, and marginally).

### 5.2.9 Model Specifications for Factors Influencing the Decision to Adopt Crop Insurance

The utility maximization assumption underpins farmers' choice of a specific approach to deal with adverse effects (Nchinda et al. 2010). A farmer adopts a strategy if the marginal utility of adoption outweighs the marginal utility of nonadoption (Suvedi et al. 2017). The value gained by a farmer who adopts the technology is not observable and is dependent on a set of exogenous factors that can be observed (Nchinda et al. 2010). The decision to use a damage-based crop insurance model is what is observed. Adoption, on the other hand, is based on several factors, including social, natural, and economic factors that influence product demand (Akter et al. 2008). A binary choice model is appropriate for the analysis given that farmers' WTA a damage-based crop insurance system fits into a dichotomous choice (i.e., whether a farmer adopts or does not adopt), essentially taking on values of zero and one (Cox 1958). The logistic regression model is a nonlinear transformation of the linear regression model (Cox 1958). The logistic distribution, like the ordinary normal distribution, is an S-shaped distribution function (cumulative density function) that restricts the estimated probabilities to a range of 0 to 1. Given the random effects ($Y = 0$ or 1), a dichotomous or binary logistic random-effects model regresses the log odds of the outcome probability on several predictors to estimate the likelihood that $Y = 1$ occurs. Moreover, the estimation of this type of model assumes that the distribution function model type results in logit models, which are widely applied (Nahayo et al. 2017; Aidoo et al. 2014; Abdullah et al. 2014). As a result, the MLE are made up of a set of parameters that would yield the observed sample more frequently (McFadden 1975). As a result, we used a binary logistic regression approach to find characteristics that influence farmers' WTA damage-based crop insurance. Due to data limitations, we were unable to include more variables, such as damage experiences, rice field distance from the flash flood sources/flood sources, and size of the land in the model. On the other hand, to identify the factors that affect the premium amount (guaranteed revenue) farmers are WTP for a damage-based crop insurance system, a multivariate regression analysis was conducted using the ordinary least square (OLS) estimation approach (Aidoo et al. 2014;

Sadati et al. 2010). To calculate the amount, farmers are WTP for the proposed model, sum insured, or guaranteed revenue or liability was determined as follows:

$$\mathrm{GR} = \mathrm{GY} \times \mathrm{EP} \times \mathrm{EIUA} \tag{5.12}$$

where, GR stands for guaranteed revenue, GY is the guaranteed yield, EP is the expected price, and EIUA denotes expected insured unit area.

Multiple regression is a statistical analysis technique that extends linear regression by including multiple independent variables in an equation to better understand how they relate to a dependent variable. Multiple regression is one of many extensions of linear regression in general linear model statistical series (e.g., analysis of variance, analysis of covariance, t-test, Pearson's product-moment correlation) (Allen 2017). There are two significant benefits to utilizing a multiple regression model to analyze data for a set of independent variables against a single dependent output. The first step is to determine how one or more predictor variables affect the criterion value. The ability to spot outliers or anomalies is the second spot. Thus, multiple regression analysis has been used in many studies to build better models for predicting the dependent variables (Allen 2017; Aidoo et al. 2014; Sadati et al. 2010). Considering the following advantages, we performed a multiple regression analysis to identify important determinants in relation to the amount farmers are willing to pay.

#### 5.2.9.1 Assessment of Damage-Based Crop Insurance Adoption Response

To check the farmers' responses regarding the damage-based crop insurance model, two econometric approaches were performed. First, to identify the factors affecting farmers' WTA the damage-based crop insurance model, a binary logistic regression model was applied. Dichotomous or binary logistic random-effects models have binary outcomes ($Y_i = 0$ or 1) and regress the log odds of the outcome probability against various predictors to estimate the probability that $I = 1$ will occur, given the random effects (Li et al. 2011). The simplest dichotomous model is given by the following equation:

$$\begin{aligned} Y_i = \ln\left[\frac{P_i}{1-P_i}\right] &= \beta_0 + \beta_1 X_1 + \beta_2 X_2 + \beta_3 X_3 + \beta_4 X_4 + \beta_5 X_5 + \beta_6 X_6 \\ &+ \beta_7 X_7 + \beta_8 X_8 + \beta_9 X_9 + \beta_{10} X_{10} + \beta_{11} X_{11} + \beta_{12} X_{12} + \beta_{13} X_{13} + U_i \end{aligned} \tag{5.13}$$

where $P_i$ is the probability of willingness to adopt or not adopt crop insurance; a $P_i$ of 1 indicates a WTA crop insurance; and a $P_i$ of 0 indicates an unwillingness to adopt crop insurance or otherwise. In addition, $\beta_0$ is the intercept; $\beta_1$ to $\beta_{13}$ are the regression coefficients of the binary dependent variable; and $U_i$ is the error term.

Second, to identify the factors that affect the premium amount farmers are WTP for damaged-based crop insurance, a multivariate regression analysis was conducted using the OLS estimation approach. The implicit form of the insurance premium model is proposed as follows:

$$Z_i = \beta_0 + \beta_1 Q_1 + \beta_2 Q_2 + \beta_3 Q_3 + \beta_4 Q_4 + \beta_5 Q_5 + \beta_6 Q_6 + \beta_7 Q_7 + \beta_8 Q_8 + \beta_9 Q_9 + \beta_{10} Q_{10} + \beta_{11} Q_{11} + \beta_{12} Q_{12} + V_i \tag{5.14}$$

Here, $Z_i$ denotes the amount farmers are WTP for damaged-based insurance premiums, $\beta_0$ to $\beta_{12}$ represent the coefficients, and $V_i$ is the error term. Apart from dichotomous variables, the natural logarithms (ln) of all variables were computed before the multiple regression model was estimated.

In the covariates (independent variables), $X_1$ is the age of the respondent (years), $X_2$ is the gender of the respondent (male = 1 female = 0), $X_3$ is their major occupation (agriculture = 1; otherwise = 0), $X_4$ is the family size (number), $X_5$ is the farming experience (years), $X_6$ is the education level (years of schooling), $X_7$ is the total income ($/month), $X_8$ is the total expenditure ($/month), $X_9$ is the ownership of a bank account (yes = 1, no = 0), $X_{10}$ is the land tenure system (1 = owner; 0 = otherwise). These ten variables were common for both logistic regression and multiple regression analysis for the study, and covariates of multiple regression were defined as $Q_1$, $Q_2$, ...$Q_{10}$. However, $X_{11}$ is the earning members (number), $X_{12}$ is the awareness of climate change (yes = 1, no = 0), and $X_{13}$ is the distance to agricultural offices (km) were only taken into consideration for the willingness to adopt the crop insurance model in the logit regression. On the other hand, $Q_{11}$ is introduced here as the number boro acreage (decimal), and $Q_{12}$ is the access to credit (yes = 1, no = 0) added to obtain the accurate model for the analysis of crop insurance premium amount in the multiple regression approach. These variables were selected based on previous studies and expert interviews and then related to WTA crop insurance and the amount of premiums. Our hypotheses for different socioeconomic variables are highlighted in Table 5.1. The analysis was conducted using STATA 14.0® (StataCorp, Texas, USA).

**Table 5.1** Explanation of the variables in the empirical analysis of damage-based crop insurance adoption responses and the amount farmers are WTP

| Variables | Indicators | Descriptions of the variables | Parameter | Likely sign | References |
|---|---|---|---|---|---|
| Dependent | $Y_i$ | Willingness to adopt damage-based crop insurance | N/A | N/A | |
| | $Z_i$ | Amount farmers are willing to pay as premiums for insurance | N/A | N/A | |
| Independent | $X_1/Q_1$ | Age of respondent (in years) | $\beta_1$ | – | Emmanuella (2017); Abebe and Bogale (2014) |
| | $X_2/Q_2$ | Gender of the respondent (male = 1, female = 0) | $\beta_2$ | + | Aidoo et al. (2014); Emmanuella (2017) |
| | $X_3/Q_3$ | Major occupation of the respondents (agriculture = 1; otherwise = 0) | $\beta_3$ | + | Aidoo et al. (2014); Emmanuella (2017) |
| | $X_4/Q_4$ | Family size (no.) | $\beta_4$ | + | Aidoo et al. (2014); Emmanuella (2017) |
| | $X_5/Q_5$ | Farming experience of the respondent (in years) | $\beta_5$ | – | Aidoo et al. (2014); Emmanuella (2017) |
| | $X_6/Q_6$ | Education level (years of schooling) | $\beta_6$ | + | Aidoo et al. (2014); Emmanuella (2017) |
| | $X_7/Q_7$ | Total income ($/month) | $\beta_8$ | + | Emmanuella (2017); Ghazanfar et al. (2015) |
| | $X_8/Q_8$ | The total expenditure of the respondent ($/month) | $\beta_9$ | – | Emmanuella (2017) |
| | $X_9/Q_9$ | Ownership of a bank account (yes = 1, no = 0) | $\beta_{10}$ | + | Aidoo et al. (2014) |
| | $X_{10}/Q_{10}$ | Land tenure system (1 = owner; 0 = otherwise) | $\beta_{11}$ | + | Aidoo et al. (2014); Ghazanfar et al. (2015) |
| | $X_{11}$ | The number of earning members (no.) | $\beta_7$ | + | Wairimu et al. (2016); Abebe and Bogale (2014) |
| | $X_{12}$ | Awareness of climate change (yes = 1, no = 0) | $\beta_{12}$ | + | Long et al. (2013) |
| | $X_{13}$ | The distance of agricultural offices (km) | $\beta_{13}$ | – | Wairimu et al. (2016); Mutaqin and Usami (2019) |
| | $Q_{11}$ | Number of boro rice acreage (decimal) | $\beta_7$ | + | Aidoo et al. (2014); Emmanuella (2017) |
| | $Q_{12}$ | Access to credit (yes = 1, no = 0) | $\beta_{12}$ | + | Emmanuella (2017); Wairimu et al. (2016) |

*Source: Author's compilation based on secondary review, 2019*
*Note: N/A stands for not applicable*

## 5.3 Results

### *5.3.1 Accuracy Assessment of LULC*

The results of the accuracy assessment of the maximum supervised classification of land use land cover classes for all three selected study areas showed excellent classification for all classes. The results revealed excellent classification with an overall accuracy of 83.14, 79.78, and 79.57% for Tahirpur, Gowainghat, and Kulaura, respectively. Similarly, good agreement for Kappa coefficients was also observed in the study areas, indicating 0.81, 0.76, and 0.75 for Tahirpur, Gowainghat, and Kulaura, respectively, implying a statistically valid classification, and strong agreement between classified and referenced images was observed for the studied LULC analysis (Table 5.2).

### *5.3.2 Land Use Land Cover (LULC) Analysis*

LULC is an important factor in identifying the vegetative condition for any crop production. To increase the accuracy of the maximum likelihood classifiers, a significant number of classes were documented following the very high-resolution datasets obtained from the SoB (Fig. 5.7). In the LULC map analysis, two-thirds of the Tahirpur subdistrict was covered by wetlands (marsh/mud), and the boro rice

**Table 5.2** Accuracy assessments for the LULC map using the producer's accuracy (PA), user's accuracy (UA), overall accuracy (OA), and Kappa coefficients

| | Tahirpur | | Gowainghat | | Kulaura | |
|---|---|---|---|---|---|---|
| Particulars | PA (%) | UA (%) | PA (%) | UA (%) | PA (%) | UA (%) |
| Seasonal track and footpath | 87 | 89 | 79 | 81 | 82 | 76 |
| Administrative boundary | 84 | 87 | 75 | 83 | 85 | 77 |
| Metaled road | 82 | 83 | 78 | 79 | 86 | 72 |
| Gravel road | 87 | 86 | 81 | 85 | 81 | 84 |
| Settlements | 78 | 83 | 84 | 86 | 79 | 81 |
| River non-tidal | 77 | 88 | 79 | 82 | 81 | 82 |
| Marsh/Mud cultivated | 72 | 81 | 76 | 77 | 77 | 85 |
| Water bodies (Haor/Beel/Lake) | 80 | 84 | 77 | 83 | 82 | 84 |
| Cultivation land | 79 | 82 | 82 | 87 | 79 | 80 |
| Noncultivation | 80 | 89 | 78 | 79 | 73 | 82 |
| Vegetation land | 81 | 88 | 79 | 76 | 77 | 77 |
| Forest | 88 | 85 | 77 | 82 | 79 | 79 |
| Tea garden | 85 | 81 | 74 | 77 | 75 | 80 |
| Swamp forest | 83 | 79 | 79 | 79 | 72 | 81 |
| Overall accuracy (%) | 83.14 | | 79.78 | | 79.57 | |
| Kappa Coeefficient | 0.79 | | 0.76 | | 0.75 | |

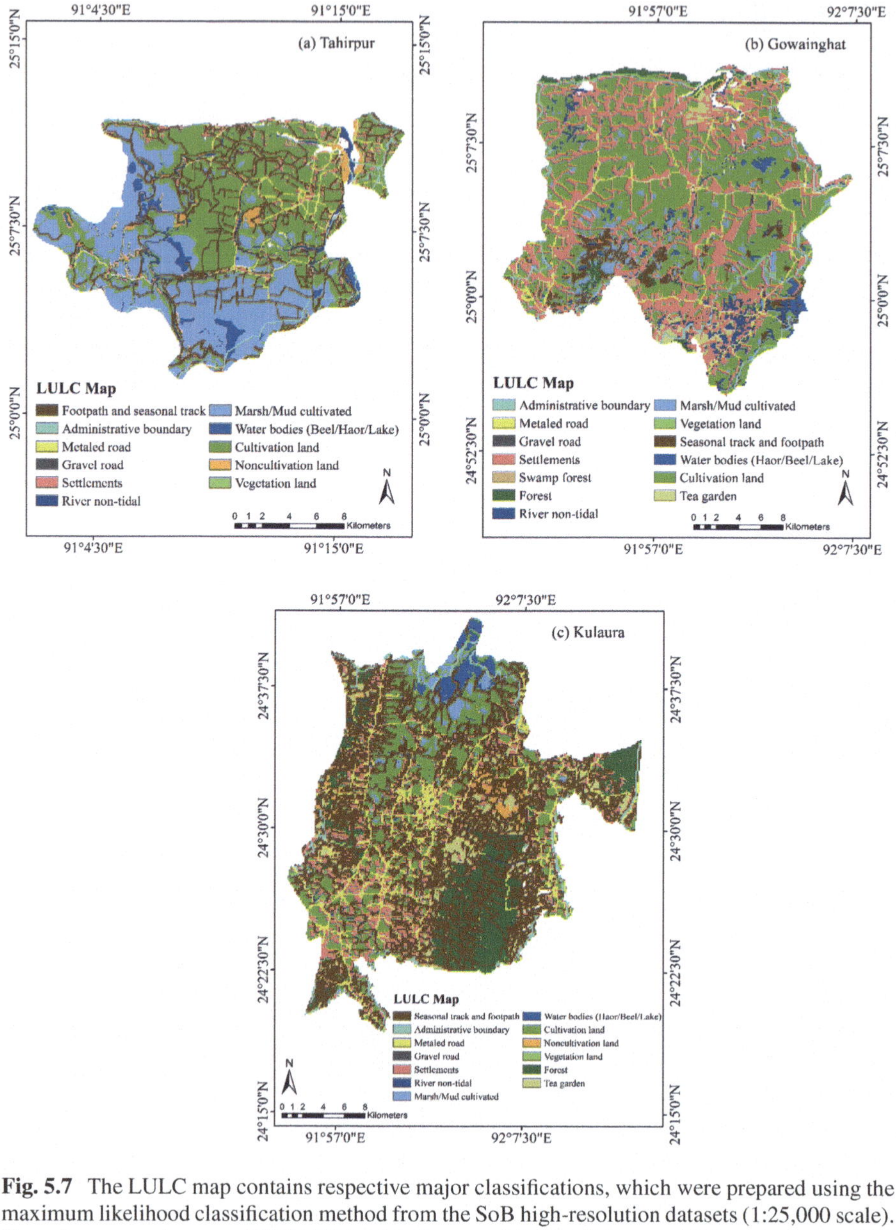


**Fig. 5.7** The LULC map contains respective major classifications, which were prepared using the maximum likelihood classification method from the SoB high-resolution datasets (1:25,000 scale). The reference system was UTM-GS84. LULC maps of the (**a**) Tahirpur, (**b**) Gowainghat, and (**c**) Kulaura subdistricts

coverage was 10,243.7 ha out of the total area of 30,437.6 ha (Table 5.3 and Fig. 5.7a). The boro rice cultivation in the Gowainghat subdistrict was the largest and accounted for approximately half (24,827.0 ha) of the total boundary (Fig. 5.7b). The building and structure layers also made up a significant portion (16,725.6 ha) of

**Table 5.3** The extent of the land use land cover map (both in pixels and ha) for selected study areas

| | Tahirpur | | Gowainghat | | Kulaura | |
|---|---|---|---|---|---|---|
| Particulars | Area (ha) | % | Area (ha) | % | Area (ha) | % |
| Seasonal track and footpath | 4437.80 | 14.58 | 2210.40 | 3.95 | 28,657.60 | 40.30 |
| Administrative boundary | 497.90 | 1.64 | 1569.60 | 2.81 | 1886.80 | 2.65 |
| Metaled road | 974.60 | 3.20 | 2407.70 | 4.30 | 6515.90 | 9.16 |
| Gravel road | 81.60 | 0.27 | 83.50 | 0.15 | 363.00 | 0.51 |
| Settlements | 1164.70 | 3.83 | 16,725.60 | 29.89 | 7338.60 | 10.32 |
| River non-tidal | 703.60 | 2.31 | 1892.20 | 3.38 | 95.50 | 0.13 |
| Marsh/Mud cultivated | 10,243.70 | 33.65 | 2571.80 | 4.60 | 1464.40 | 2.06 |
| Water bodies (Haor/Beel/Lake) | 966.00 | 3.17 | 1182.20 | 2.11 | 1529.60 | 2.15 |
| Cultivation land | 10,355.90 | 34.02 | 24,827.00 | 44.37 | 12,567.90 | 17.68 |
| Noncultivation | 751.10 | 2.47 | – | | 927.20 | 1.30 |
| Vegetation land | 260.60 | 0.86 | 460.80 | 0.82 | 224.80 | 0.32 |
| Forest | – | | 1091.50 | 1.95 | 6775.60 | 9.53 |
| Tea garden | – | | 869.80 | 1.55 | 2758.60 | 3.88 |
| Swamp forest | – | | 64.80 | 0.12 | – | |
| Total | 30,437.60 | 100.00 | 55,957.00 | 100.00 | 71,105.30 | 100.00 |

*Note: (–) indicates the non-available items*

the Gowainghat subdistrict. For the Kulaura subdistrict, an area of approximately 28,657.6 ha was covered by footpaths (Fig. 5.7c). The map also identified a 12,567.9 ha area covered by boro rice, which was a significant proportion of the Kulaura subdistrict.

### *5.3.3 Inundation Analysis*

A ground scenario of inundation was identified from the Landsat-8 OLI image for the after flash flood period on March 28, 2017. After testing several band combinations for NDWI calculations, we found that the green and NIR bands were very useful for extracting water pixels due to their capability to remove noise, such as built-up and shallow cloud areas. The red color in the Fig. 5.8 indicates higher inundation intensity in the images taken after flash flooding. Green represents the vegetative stages of boro rice. In addition, the NDWI images over green vegetation areas show several red streaks related to the drainage patterns. These streaks indicate that NDWI is sensitive to liquid water in vegetation. The minimum and maximum NDWI values for the Kulaura subdistrict were 0.13 and −0.39, respectively. NDWI was also calculated for the Gowainghat (minimum 0.10 and maximum −0.35) and Tahirpur subdistricts (minimum 0.03 and maximum −0.33) (Fig. 5.8).

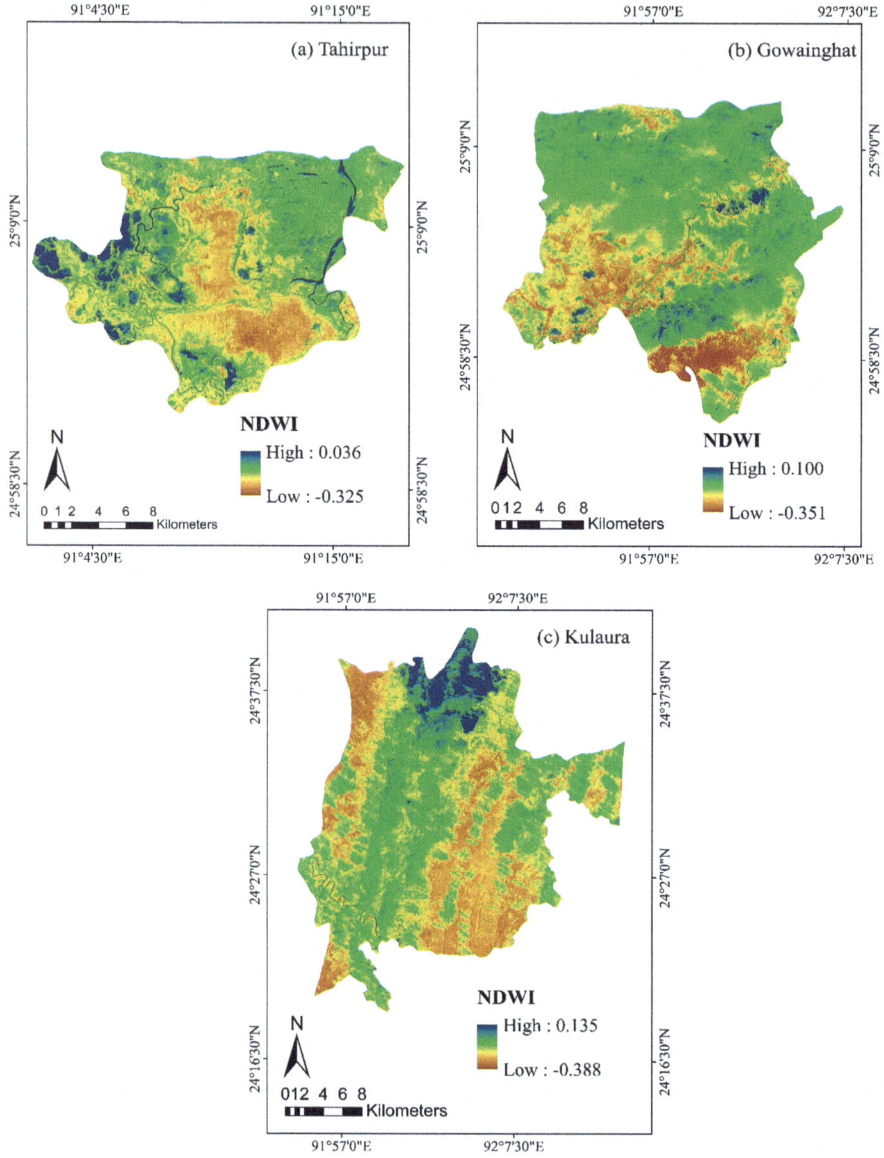


**Fig. 5.8** NDWI maps for damage assessment for the (**a**) Tahirpur, (**b**) Gowainghat, and (**c**) Kulaura subdistricts

Then, reclassification was performed to classify these range values into the four categories of high, moderate, marginal, and no damage according to the inundation scale obtained from the analysis (Fig. 5.9).

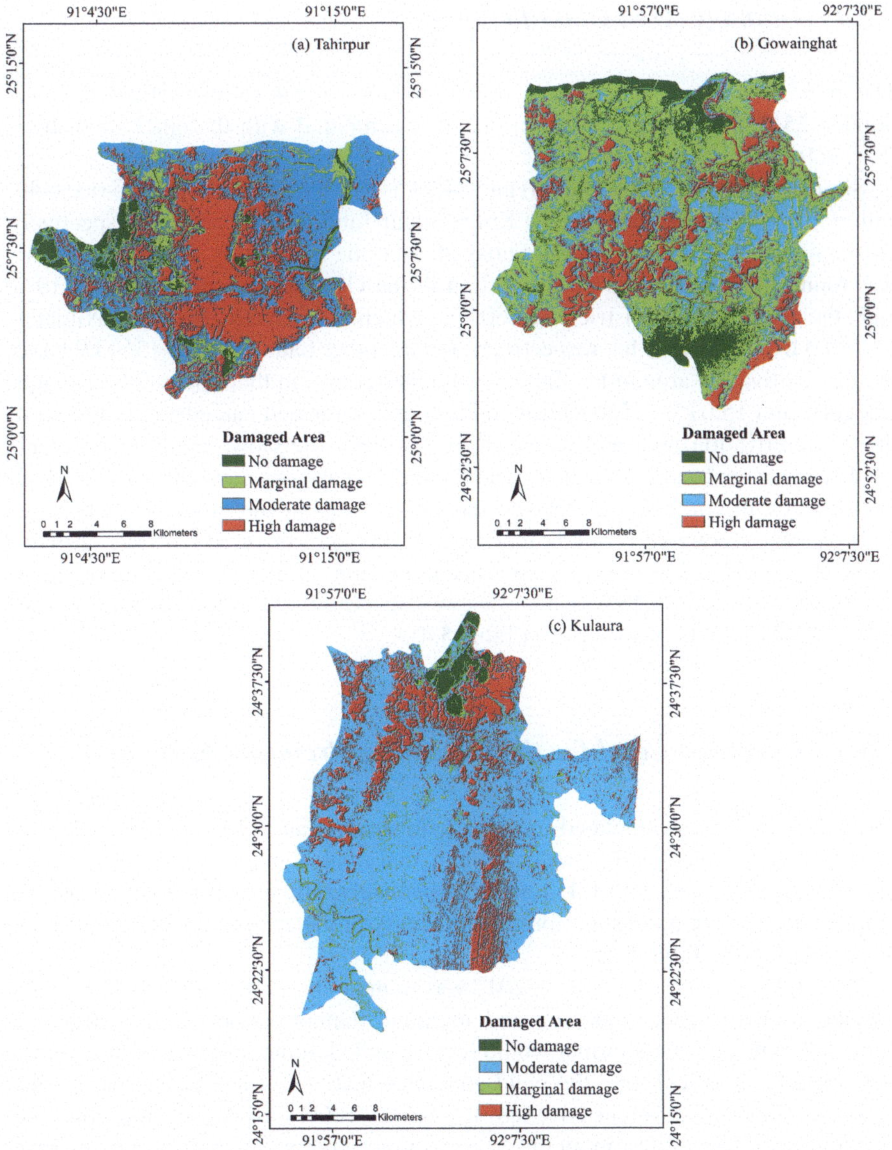


**Fig. 5.9** Assessment of overall damage area with proper classification (i.e., high, moderate, marginal, and no damage) for flash flood in 2017 using LULC map and NDWI for the (**a**) Tahirpur, (**b**) Gowainghat, and (**c**) Kulaura subdistricts

### 5.3.4 Damage Area Classification

The inundation map analysis showed that the Gowainghat subdistrict had the most flooded 25,204.4 ha (45.28%) boro rice area compared with the other subdistricts (Fig. 5.9).

Similarly, 18.05% (12,719.3 ha) and 34.94% (10,588.2 ha) of the boro rice cultivation areas were inundated in the Kulaura and Tahirpur subdistricts, respectively. Of the total boro rice cultivation-damaged area, the marginally damaged category was found to have the highest proportion in the Gowainghat subdistrict (20.19%) and the Kulaura subdistrict (12.60%), which accounted for approximately 11,240.0 ha and 8878.1 ha, respectively. On the other hand, 18.49% (5603.9 ha) of the rice cultivation area in the Tahirpur subdistrict was in the moderately damaged category, and 15.07% (4566.8) was in the highly damaged category. In the case of the rice cultivation area and damaged proportions, the Gowainghat and Kulaura subdistricts were found in the marginally damaged category at 44.60% and 69.80%, respectively. In contrast, the Tahirpur subdistrict was in the moderately damaged category and accounted for more than half (52.92%) of the total boro rice-growing areas (Fig. 5.10). Moreover, a significant portion of the damage in moderate (34.96%) and high (43.13%) categories was also observed in the Gowainghat and Tahirpur subdistricts, respectively (Table 5.4).

### 5.3.5 Coverage Level for the Insurance Premium Rate

The actual yield was calculated from the field survey data. The study observed that for respondents with high, moderate, and marginal damage, the actual average yields were 1.04, 1.33, and 1.41 ton/ha, respectively. The forecasted yield was 2.69 ton/ha, and we kept it constant for the uniformity of the estimation among different damaged classes (Table 5.5).

Yield loss due to flash floods in 2017 was found to be 1.65, 1.36, and 1.28 ton/ha for classes with high, moderate, and marginal damage, respectively. Moreover, Table 5.5 indicates that a farmer would receive an indemnity payment whenever the area yield ($y$) was less than the yield guarantee level ($y_g$) for a given year. At this stage, we checked different coverage levels (40–100%) to identify the minimum coverage level required to adopt the crop insurance policy. The findings showed that at the minimum coverage level of 70%, farmers could select the expected break-even yield for the indemnity to adopt damaged-based crop insurance. Until the 70% coverage level, farmers were not eligible to reach the expected break-even yield for indemnity. Hence, we considered a 70% coverage level as the benchmark (marginal) for the proposed crop insurance model to protect the farmer's production cost. Similarly, 80% and 90% were considered moderate and high coverage levels, respectively, to cover the production risk of ready-to-harvest boro rice in the study

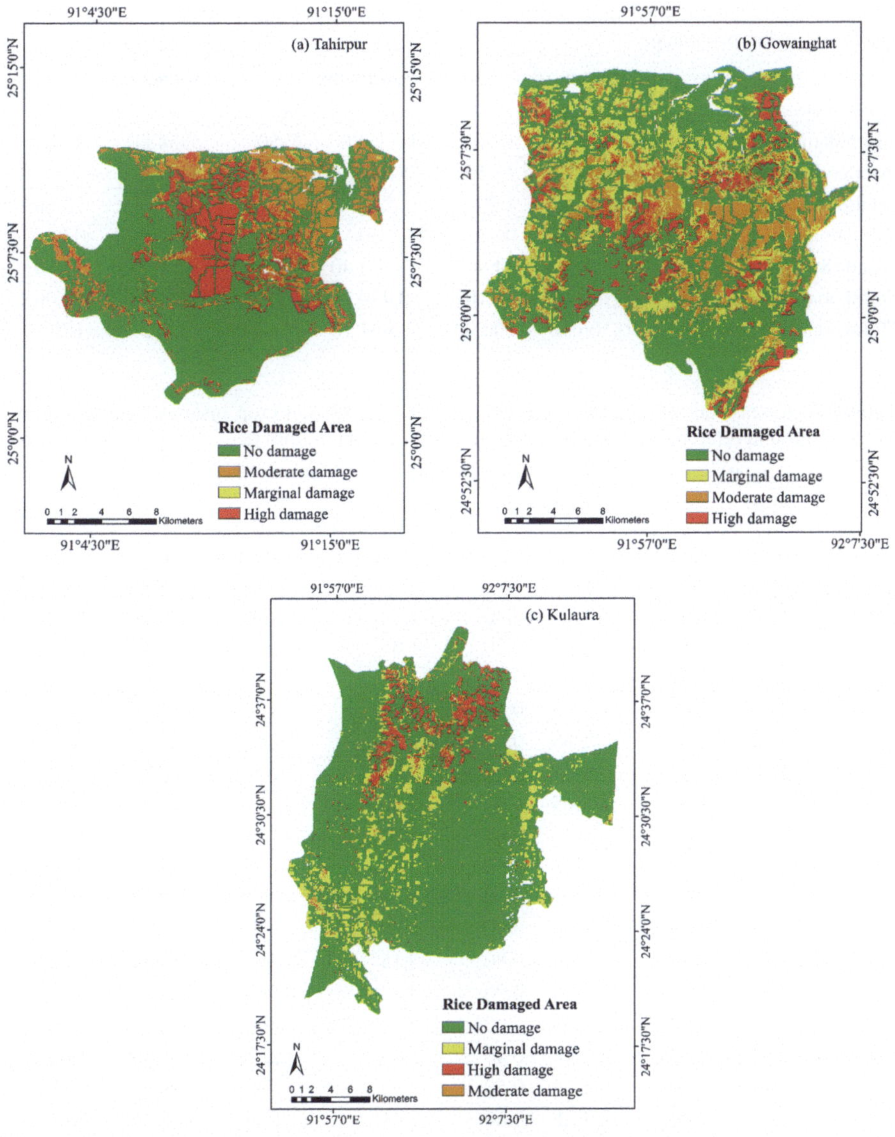


**Fig. 5.10** Assessment of boro rice damage area with proper classification (i.e., high, moderate, marginal, and no damage) due to flash flood in 2017 using LULC map and NDWI for the (**a**) Tahirpur, (**b**) Gowainghat, and (**c**) Kulaura subdistricts

areas. The findings also suggested that the higher the coverage level, the greater the expected break-even yield for indemnity for farmers in flash flood affected Haor areas. Therefore, coverage level selection is an important factor for damage-based crop insurance systems (Clover and Nieuwoudt 2003).

**Table 5.4** Calculations of damaged proportions for four classes based on the total number of pixels, ha, and in percentage

| Particulars | Tahirpur | | Gowainghat | | Kulaura | |
|---|---|---|---|---|---|---|
| | Area (ha) | % | Area (ha) | % | Area (ha) | % |
| High damage | 4566.80 | 15.10 | 4181.20 | 7.50 | 3354.60 | 4.80 |
| Moderate damage | 5603.90 | 18.50 | 8810.90 | 15.80 | 348.50 | 0.50 |
| Marginal damage | 374.30 | 1.20 | 11,240.00 | 20.20 | 8878.10 | 12.60 |
| No damage | 43.30 | 0.10 | 972.30 | 1.80 | 138.20 | 0.20 |
| Total rice area damaged | 10,588.20 | 34.90 | 25,204.40 | 45.30 | 12,719.30 | 18.10 |
| Total nonrice area | 19,718.60 | 65.10 | 30,459.70 | 54.70 | 57,734.30 | 82.00 |
| Total area | 30,306.90 | 100.00 | 55,664.10 | 100.00 | 70,453.50 | 100.00 |

**Table 5.5** Assessment of coverage levels for the damaged-based crop insurance premium rate using the ground reference moving average yield information of boro rice

| Damaged classes | Actual yield (yt) (ton/ha) | Forecasted yield (fy) (ton/ha) | Yield loss (ton/ha) | Coverage level (%) | Guaranteed yield (yg) (ton/ha) | Indemnity (ton/ha) | EPR (5 years) | Remarks |
|---|---|---|---|---|---|---|---|---|
| High | 1.04 | 2.69 | 1.65 | 0.40 | 1.08 | −1.52 | −0.30 | N/A |
| | 1.04 | 2.69 | 1.65 | 0.50 | 1.35 | −0.74 | −0.15 | N/A |
| | 1.04 | 2.69 | 1.65 | 0.60 | 1.61 | −0.12 | −0.02 | N/A |
| | 1.04 | 2.69 | 1.65 | 0.70 | 1.88 | 0.40 | 0.08 | A |
| | 1.04 | 2.69 | 1.65 | 0.80 | 2.15 | 0.85 | 0.17 | A |
| | 1.04 | 2.69 | 1.65 | 0.90 | 2.42 | 1.27 | 0.25 | A |
| | 1.04 | 2.69 | 1.65 | 1.00 | 2.69 | 1.65 | 0.33 | A |
| Moderate | 1.33 | 2.69 | 1.36 | 0.40 | 1.08 | −2.25 | −0.45 | N/A |
| | 1.33 | 2.69 | 1.36 | 0.50 | 1.35 | −1.32 | −0.26 | N/A |
| | 1.33 | 2.69 | 1.36 | 0.60 | 1.61 | −0.60 | −0.12 | N/A |
| | 1.33 | 2.69 | 1.36 | 0.70 | 1.88 | −0.02 | 0.00 | N/A |
| | 1.33 | 2.69 | 1.36 | 0.80 | 2.15 | 0.49 | 0.10 | A |
| | 1.33 | 2.69 | 1.36 | 0.90 | 2.42 | 0.94 | 0.19 | A |
| | 1.33 | 2.69 | 1.36 | 1.00 | 2.69 | 1.36 | 0.27 | A |
| Marginal | 1.41 | 2.69 | 1.28 | 0.40 | 1.08 | −2.45 | −0.49 | N/A |
| | 1.41 | 2.69 | 1.28 | 0.50 | 1.35 | −1.48 | −0.30 | N/A |
| | 1.41 | 2.69 | 1.28 | 0.60 | 1.61 | −0.74 | −0.15 | N/A |
| | 1.41 | 2.69 | 1.28 | 0.70 | 1.88 | −0.13 | −0.03 | N/A |
| | 1.41 | 2.69 | 1.28 | 0.80 | 2.15 | 0.39 | 0.08 | A |
| | 1.41 | 2.69 | 1.28 | 0.90 | 2.42 | 0.85 | 0.17 | A |
| | 1.41 | 2.69 | 1.28 | 1.00 | 2.69 | 1.28 | 0.26 | A |

*Note: Expected break-even yield for indemnity (EPR); N/A stands for not applicable; and A for applicable*

## 5.3.6 Damage-Based Crop Insurance Premium Rate

### 5.3.6.1 Future Value of the Annuity

First, three selected study areas and their corresponding high, moderate, and marginal damaged areas were summed and then converted into expected yield losses based on average yield losses due to flash floods in 2017. Then, the future value of expected losses was calculated based on three damaged classes, and the result showed that the expected loss in the highly damaged area was $536.25/ha. Similarly, $ 442.0 and $ 416.0 per ha were observed for the moderately and marginally damaged classes, respectively (Table 5.6).

### 5.3.6.2 Insurance Premium Rate

The insurance coverage levels were categorized as high coverage (90%), moderate coverage (80%), and marginal coverage (70%). The insurance coverage level was the same for different damage classes, but different premium rates were found compared with different interest rates (2–6%). The findings suggested that the minimum coverage was marginal coverage (70%), which covered the production costs, while maximum coverage was designed using high coverage (90%) to insure production against this adverse climate change and high flash flood intensity. The higher the coverage levels are, the lower the insurance premium and vice versa, but the lower the damage class is, the lower the insurance premium (Table 5.7).

On the other hand, the greater the interest rate charged by the insurance company, the higher the per month insurance premium under this damage-based crop insurance model. The lowest insurance premium was observed for the high coverage and moderately damaged class, at the 4% interest rate, was $23.82/ha, and the highest premium was for the marginal coverage and high damage class at the 4% interest rate ($ 39.49/ha).

## 5.3.7 Adoption Response for Damage-Based Crop Insurance

The damage-based crop insurance scheme was described to farmers in as much detail as possible, focusing on how the system works. The basic principle was communicated to farmers as insurance guaranteeing a minimum payout for a specific risk over a predetermined period in exchange for a charge known as a premium. Production risk and its effects and information on the difference between purchasing or not purchasing the product in terms of the costs and benefits were presented to farmers during data collection. Among 150 respondents, 94 respondents were

**Table 5.6** Calculation of the future value of the annuity (*V*) for high, moderate, and marginal damaged classes

| Particulars | Tahirpur (pixel) | Gowainghat (pixel) | Kulaura (pixel) | Total (pixel) | Total area (ha) | Yield loss (ton/ha) | Total yield loss (ton) | Price ($/ton) | Total expected loss ($/ha) |
|---|---|---|---|---|---|---|---|---|---|
| High damage | 50,742 | 46,458 | 37,273 | 134,473 | 12,102.57 | 1.65 | 19,969.24 | 325 | 536.25 |
| Moderate damage | 62,265 | 97,899 | 3872 | 164,036 | 182.26 | 1.36 | 247.87 | 325 | 442.0 |
| Marginal damage | 4159 | 124,889 | 98,645 | 227,693 | 252.99 | 1.28 | 323.83 | 325 | 416.0 |

**Table 5.7** Monthly premium rate based on the proposed coverage levels and interest rates for the damaged-based crop insurance model

| | | High damage | | | Moderate damage | | | Marginal damage | | |
|---|---|---|---|---|---|---|---|---|---|---|
| | | Percentage (%) | | | Percentage (%) | | | Percentage (%) | | |
| Level of coverage (% of yield) | Premium rate | 2% | 3% | 4% | 3% | 4% | 5% | 4% | 5% | 6% |
| Marginally coverage (70%) | Per Hectare ($) | 31.00 | 35.26 | 39.49 | 25.55 | 29.06 | 32.55 | 24.05 | 27.35 | 30.63 |
| | Per Pixel ($) | 2.79 | 3.17 | 3.55 | 2.30 | 2.62 | 2.93 | 2.16 | 2.46 | 2.76 |
| Moderately coverage (80%) | Per Hectare ($) | 30.85 | 35.10 | 39.31 | 25.43 | 28.93 | 32.40 | 23.93 | 27.23 | 30.49 |
| | Per Pixel ($) | 2.78 | 3.16 | 3.54 | 2.29 | 2.60 | 2.92 | 2.15 | 2.45 | 2.74 |
| Highly coverage (90%) | Per Hectare ($) | 30.71 | 34.94 | 39.12 | 25.31 | 28.80 | 32.25 | 23.82 | 27.10 | 30.35 |
| | Per Pixel ($) | 2.76 | 3.14 | 3.52 | 2.28 | 2.59 | 2.90 | 2.14 | 2.44 | 2.73 |

WTA damage-based crop insurance systems. However, a significant number of respondents were not interested in joining the proposed damage-based crop insurance model. The key reasons for not participating in our proposed model are lack of confidence regarding crop insurance, no previous crop insurance-related information, lack of education, lack of knowledge on satellite datasets, etc. Key determinants for the adoption responses for the damage-based crop insurance model are highlighted in the following paragraphs.

#### 5.3.7.1 Determinants of Willingness to Adopt Damage-Based Crop Insurance

The estimated outcome using logit regression was used to analyze farmers' WTA crop insurance strategy. The analytical statistics showed that the estimated model had a good fit with a chi-square value that was statistically significant at the 1% level. This finding indicated that farmers' socioeconomic characteristics were relevant in explaining farmers' adaptation decisions in the study areas. However, the study observed that the relationship between the major occupation and the willingness to adopt a crop insurance strategy was negative and statistically significant at the 10% significance level. As the odds ratio results suggest, as the farmers' major occupation (farming) increases, their WTA damaged-based crop insurance strategy decreases by 0.862 units and vice versa, with other factors remaining constant. This result indicates a lower probability for the only agriculture-dependent respondents (boro rice producers) to adopt crop insurance strategies than the farmers with alternate occupations providing different sources of income.

Farmers' education levels were found to have a significant positive relationship with the WTA the developed damaged-based crop insurance system. As the farmers' educational levels increased, their WTA crop insurance strategy also increased by 0.111, which was significant at the 10% probability level. The positive effect on the WTA crop insurance implied that better-educated farmers were more likely to receive and understand the insurance policy and were more willing to purchase insurance than those with lower educational levels. This result followed the result obtained by Ali (2013). On the other hand, the logit estimates show that farmers' total incomes had significant negative relationships with farmers' WTA a damaged-based crop insurance model. This finding was contrary to the a priori expectation. The result implied that farmers with higher household incomes were less vulnerable to production risks and the subsequent effects on their welfare and therefore had a lower WTA crop insurance (Falola et al. 2013). Similarly, ownership of a bank account was significant at the 5% significance level and negatively affected farmers' WTA crop insurance strategies. According to a prior expectation, ownership of a bank account would result in a higher probability of engaging in a crop insurance strategy. However, the logit regression results from the field surveys showed that ownership of a bank account negatively affected the respondent's WTA crop insurance. Therefore, for the crop insurance company, it is recommended that farmers not be required to own a bank account to obtain crop insurance. Most of the farmers ranked climate change impacts from flash floods as the major source of losses. Awareness of climate change had a positive effect on farmers' WTA crop insurance and was statistically significant at the 10% level. This finding suggested that farmers who were more knowledgeable about climate change had a higher tendency to adopt crop insurance strategies (Long et al. 2013) (Table 5.8).

#### 5.3.7.2 Determinants of Premiums for Crop Insurance

The main determinants of premiums for crop insurance were the family size and total income of the respondents. The natural logarithm was taken to normalize each variable for the multivariate regression. Moreover, our findings observed that family size had a positive correlation with the number of farmers WTP insurance premiums for boro rice and was statistically significant at the 1% level. This finding indicates that the larger the household size is, the higher the responsibility to reduce potential farm losses because rural households depend on farming as a major source of livelihood and are more WTP to insure their crops. Likewise, household income had a positive and significant association with the amount farmers were WTP to insure their crops. This result means that the higher-income farmers were WTP more to insure their crops. Therefore, farmers with lower incomes were more WTP a lower amount for insurance since they tended to have lower payment capacities (Abebe and Bogale 2014). As a

**Table 5.8** Respondents' WTA crop insurance and the premium amount. Binary logistic regression and multiple linear regression methods were applied

| Variables | Logit estimation | | Multiple regression | | |
|---|---|---|---|---|---|
| Willingness to adopt crop insurance | Odds Ratio | *z* | Guaranteed revenue | Coefficient | *t*-value |
| Age ($X_1$) | −0.011 | −0.44 | Ln_Age ($Q_1$) | −0.19 | −0.76 |
| Sex ($X_2$) | 0.55 | 1.19 | Ln_Sex ($Q_2$) | 0.14 | 1.09 |
| Occupation ($X_3$) | −0.86* | −1.68 | Ln_Occupation ($Q_3$) | −0.05 | −0.48 |
| Family size ($X_4$) | 0.07 | 0.89 | Ln_Family size ($Q_4$) | 0.46*** | 2.88 |
| Farming experience ($X_5$) | 0.012 | 0.51 | Ln_Education level ($Q_5$) | 0.03 | 0.71 |
| Education level ($X_6$) | 0.11* | 1.72 | Ln_Farming experience ($Q_6$) | 0.10 | 0.91 |
| No. of earning members ($X_7$) | 0.33 | 1.20 | Ln_ Boro area (Decimal) ($Q_7$) | −0.06 | −0.85 |
| Total income ($X_8$) | −0.00*** | −3.21 | Ln_Total income (annual) ($Q_8$) | 0.48*** | 3.65 |
| Total expenditure ($X_9$) | −0.00 | −1.38 | Ln_Total expenditure (annual) ($Q_9$) | −0.02 | −0.22 |
| Ownership of a bank account ($X_{10}$) | −0.76** | −1.91 | Ln_Ownership of bank account ($Q_{10}$) | 0.16 | 1.40 |
| Land tenure system ($X_{11}$) | −0.11 | −0.25 | Ln_Land tenure system ($Q_{11}$) | −0.05 | −0.43 |
| Awareness of climate change ($X_{12}$) | 0.85* | 1.88 | Ln_Access to credit ($Q_{12}$) | −0.13 | −1.09 |
| The distance of agriculture office ($X_{13}$) | −0.007 | −0.23 | Constant | 2.71 | 1.20 |
| Constant | 2.56 | 1.37 | RMSE: 0.6710 | | |
| Model summary Statistics | $R\,\chi2 = 31.53$; Prob>$\chi2 = 0.0028$; Pseudo $R2 = 0.1736$; Log likelihood = −75.0744 Dependent variable: WTA crop insurance (1 = yes; 0 = otherwise) | | *R*-squared: 0.285 *F*-value: 4.285 *P*-value: 0.000 Dependent variable: Ln_ amount farmers are WTP for a damage-based crop insurance premium (USD) | | |

*Note: ***, **, and * indicates the level of significance at 1, 5, and 10%, respectively*

result, it can be said that although lower-income farmers would be WTA damage-based crop insurance to manage risk better and secure their welfare, they may not be able to afford the insurance (Table 5.8).

## 5.4 Discussion

Flash floods are one of the most frequent climate-related disasters that occur worldwide and often have devastating consequences for the food security of agricultural households. These challenges are more crucial for developing countries, including Bangladesh. Over the last several decades, the Bangladesh government, with the help of organizations, has taken a number of measures against these flash floods and agricultural losses. In addition, in the last few years, different scholars have conducted research to determine a long-lasting solution for flash flood affected Haor areas in Bangladesh. In their research, flood hazard mapping was evaluated (Gazi et al. 2019); flash flood area mapping was performed by Aldhshan et al. 2019; seasonal land cover changes and their effects on essential services of Haor and non-Haor areas were examined (Nahar et al. 2018); and the impacts of floods on boro crop production in Haor areas were examined (Ahmed et al. 2017; Khan et al. 2012). However, the success of mitigating or preventing this disaster was not satisfactory, and the challenges continue to increase. Therefore, it is important to develop a model to secure and enhance rice production by adopting advanced cutting-edge technology. In this regard, an initiative was taken to develop a damage-based crop insurance model by optimizing the insurance premium rate according to the damage levels in each region. It has been proven that introducing a community-based micro crop insurance model can play a vital role and may significantly contribute to poverty alleviation (Mechler et al. 2006; Hazell 2001). Hence, the plan to develop a damage-based crop insurance model was initiated for flash flood-affected areas to reduce their livelihood vulnerability. In addition, farmers' adoption behavior to accept crop insurance strategy was examined through statistical analysis (Emmanuella 2017; Aidoo et al. 2014). However, microlevel socioeconomic and demographic data are essential for evaluating the viability of crop insurance adoption and premium amount determination (Adusumilli et al. 2019; Islam et al. 2018).

Through the inundation map analysis, boro rice damaged acreage was estimated accurately in the high, moderate, and marginal damage categories using GIS and satellite remote sensing datasets. This result was also supported by the data obtained from the Department of Disaster Management (DDM) in 2017, and up to 19% of the total land inundation area was calculated to be in flashflood areas in the 39 subdistricts (Situation Report, Nirapod 2017). In the rice cultivation area and damage proportions, the Gowainghat and Kulaura subdistricts were found in the marginally damaged category at 44.60% and 69.80%, respectively. In contrast, the Tahirpur subdistrict was moderately damaged and accounted for more than half (52.92%) of the total boro rice-growing areas. It is evident that NDWI is a good indicator for analyzing inundation maps and accurately identifying the most flash flood-affected areas. Several remote sensing measures of vegetation moisture based on water absorption have been applied in different studies and found significant findings for inundation analysis (Acharya et al. 2018; Bhatta 2012; Islam et al. 2010), including indices based on NIR absorption and shortwave infrared (SWIR) absorption (Sims and Gamon 2003; Steven et al. 2003).

Moreover, during the primary data collection, the respondents mentioned that these phenomena tend to occur every year and negatively impact their livelihood with food insecurity. To enable an intervention, the study proposed damage-based crop insurance model to support their livelihoods. In this regard, ground reference information was collected, and it was estimated that yield loss due to flash floods in 2017 was 1.65, 1.36, and 1.28 ton/ha for high, moderate, and marginally damaged classes, respectively. From the break-even yield for indemnity analysis, it was observed that the minimum 70% coverage level is required to adopt the crop insurance policy. Moreover, the future value of expected losses was calculated based on three damaged classes, and the findings showed that the high, moderate, and marginal damaged losses were $536.25, $442.0, and $416.0 per ha, respectively. In addition, the damage-based crop insurance findings suggested that the higher the coverage levels, the lower the insurance premium, and the lower the damaged class, the lower the insurance premium. Similarly, the greater the interest rate applied to insurance, the more positive the response to monthly installments under this developed model. Therefore, the developed crop insurance model could help farmers stabilize their farm income and investment and guard against flash floods that cause losses in rice production. However, similar research on the development of land suitability, harvest index, and subsidy dependence factor-based crop insurance premiums for Bangladesh was done, and the study found that producing rice up to marginally suitable land, a low harvest index, could gain support through insurance (Islam et al. 2018).

Moreover, the developed models can be transferred to any similar environment for any crop with proper classification of the LULC map and separation of damaged areas from nondamaged areas. Furthermore, the calculation of the premium rate could also be possible for other sensors (i.e., Sentinel), as the calculation of the damaged amount depends on the conversion factor for the pixel resolution to scale down the pixel level (30 m × 30 m, Landsat 8, OLI for the present research) and the ground reference yield estimation. However, after analyzing the field survey results, some essential determinants were identified that could help farmers adopt damage-based crop insurance in flash flood-affected areas. The insurance company could consider these determinants (family size and total monthly income) to introduce a damage-based crop insurance program for farmers. As most farmers live under the poverty line, this small and reasonable coping mechanism will significantly impact the sustainability of their livelihoods.

Furthermore, the outcomes of farmers' responses regarding WTA damage-based crop insurance showed that approximately 62.67% of the participants accepted the concept as the mechanism for flash flood risk reduction. Researchers have shown that payouts from index-based insurance products are poorly correlated with the actual damage experienced by farmers in their fields (Clarke 2016; Dercon et al. 2014). A pilot project was conducted on agricultural insurance products to address climate-related risks in Sub-Saharan Africa, and the results showed that yield index insurance, index-based crop insurance, and index-based livestock insurance were positively correlated with on-farm income/savings, literacy, and family size, with estimated coefficients of 0.211, 0.292 and 0.018, respectively, and negatively correlated with the premium rate (−0.183), age of farmer (−0.058), land tenure (−0.800), and farm size (−0.167) (Ntukamazina et al. 2017). Index-based insurance has been recognized to potentially reduce insurance premiums and make insurance affordable to farmers (Al-Maruf et al. 2021; Yuzva et al. 2018). In Asia, countries such as China, India, and Thailand are at different phases of adopting index-based crop insurance with varying government support and private sector engagement (Sinha and Tripathi 2014). From our findings, the premium amount indicated that the monthly installment is quite acceptable and cost-effective in the developing country context, as most farmers are poor and lack the financial resources to cope with bad weather events. Therefore, the developed model can be applied for highly flash flood-susceptible countries around the world, namely floods that frequently occur due to yearly monsoon rainfall (Bangladesh, China, Vietnam) or sudden flash floods after torrential rains in mountainous areas (Bangladesh, China, Indonesia, Japan) (Svetlana et al. 2015). Finally, the model can be useful for farm risk management, insurance premium determinations, and relevant stakeholder decision-making to mitigate the effects of extreme natural hazard events. Further research needs to be undertaken that considers site-specific crop insurance policies along with government intervention to assess the sustainability of the developed insurance model globally.

## 5.5 Conclusions and Policy Implications

Remote sensing and GIS-based methods were used to assess the damaged area of boro rice production through inundation map analysis in Bangladesh. We demonstrated the effectiveness of using Landsat-8 images for mapping the inundated Haor areas in 2017 using NDWI. The LULC map was developed with higher resolution ground reference datasets to identify boro rice acreage more accurately. The damage areas were classified as high, moderate, and marginal, and the results showed that on average 32% of the boro rice acreage was damaged in the three most vulnerable subdistricts of Bangladesh. To confront the adverse impacts of flash floods on and risks to agriculture, a damage-based crop insurance model was proposed as a risk mitigation strategy. The estimated premium rate (approximately $23–40/ha monthly basis) was obtained for marginal, moderate, and highly damaged classes in the study areas. However, the premium amount suggested that the monthly installment is very reasonable and cost-effective in the developing country aspect as most of the farmers are very poor with insufficient income to cope with adverse climatic phenomena. Moreover, adoption responses for the damage-based crop insurance system findings suggested that creating alternative income-generating activities, intensive formal education of farmers, and awareness campaigns for adopting and practicing crop insurance would be necessary. This study represents the first attempted research for crop insurance that combines GIS applications with an econometric approach in a single platform considering damage-based crop insurance premium rate estimation. Finally, given the relatively low premium rates, farmers were willing to contribute to crop insurance schemes, and government subsidies might be required if future crop insurance schemes are to be sustainable. Nevertheless, today, there is no successful crop insurance model for Bangladesh due to the government's lack of a risk-sharing program. Therefore, the developed damage-based crop insurance model has great potential to support flash flood-affected farmers by introducing a damage-based agricultural insurance policy through subsidizing premiums as part of a standard subsidy program. Finally, given the relatively low premium rates, farmers are willing to adopt crop insurance schemes, and government subsidies might be required if future crop insurance schemes are sustainable. This research suggests potential policy implications for meteorological agencies, agricultural support extension systems, and disaster and risk mitigation systems for the provision of insurance premiums along with agricultural subsidies to improve the livelihoods of farmers in flash flood-prone areas globally. Certain restrictions apply to the generalization of this study's results. Because of time and budget constraints, the study areas were limited to just three subdistricts in Bangladesh's Haor region. Although the sample size may not be statistically sufficient, it may help understand the factors that influence the adoption of a damage-based crop insurance system in flash flood-prone areas. Even though the outcomes of logit and multiple regression models are only rough indicators, we could not accumulate all important sociodemographic variables simultaneously. However, as a supporting tool, it may contribute to the discussion on the drivers of farmers' WTA for the proposed crop insurance model in connection to qualitative and quantitative as well as social, economic, and environmental factors, mainly at the subdistrict level. Therefore, policymakers have the flexibility to modify the model and alter the variables as needed. Moreover, more in-depth research on willingness to pay based on the cost of production and expected return can be done in the future that will consider additional crop insurance-related variables.

**Acknowledgments** The authors would like to thank the Japan Section of the Regional Science Association International for granting the copyright to include this published article—Md. Monirul Islam, Matsushita Shusuke, Ryozo Noguchi, Tofael Ahamed. A damage-based crop insurance system for flash flooding: a satellite remote sensing and econometric approach. Asia-Pacific Journal of Regional Science, 1–43. https://doi.org/10.1007/s41685-021-00220-9, 2022. Some minor modifications have been made in this book chapter. Moreover, the authors would like to express their sincere gratitude to the University of Tsukuba, Japan for providing GIS facilities to carry out the research. We are grateful to the Department of Agricultural Extension (DAE), Ministry of Agriculture, Bangladesh for their cordial support by providing ground reference boro rice yield data. We also express our sincere thanks to the United States Geological Survey (USGS) and the European Space Agency (ESA) for geographical and satellite data. Furthermore, we also express our gratitude to the Ministry of Education, Culture, Sports, Science and Technology (MEXT), Japan for providing scholarships to pursue this research in Japan.

# Appendix 1. Flash Flood History in the Affected Haor Areas of Bangladesh

| Year and date | Districts | Affected sub-districts | | | |
|---|---|---|---|---|---|
| 2020/06/27 | Sylhet | Jaintiapur | **Gowainghat** | Companiganj | Kanaighat |
| | Sunamganj | Sunamganj Sadar | Dowarabazar | Shulla | Chhatak |
| | | **Tahirpur** | Bishwamvarpur | Jamalganj | Dharmapasha |
| 2019/06/04 | Sunamganj | Sadar | Bishwamvarpur | Tahirpur | Dowarabazar |
| | | Dharmapasha | | | |
| 2019/05/06 | Sylhet | Gowainghat | | | |
| | Sunamganj | Jamalganj | **Tahirpur** | | |
| | Netrakona | 2 upazila | | | |
| 2018/06/20 | Moulvibazar | **Kulaura** | Kamalganj | | |
| | Sylhet | Gowainghat | Kanaighat | Fenchuganj | |
| | Sunamganj | Chhatak | Bishwamvarpur | **Tahirpur** | |
| 2017/03/28 | Sylhet | 13 (including **Gowainghat**) | | | |
| | Moulvibazar | 7 (including Kulaura) | | | |
| | Sunamganj | 11 (including **Tahirpur**) | | | |
| | Habiganj | 8 | | | |
| | Netrokona | 9 | | | |
| | Kishoreganj | 6 | | | |
| 2016/04/25 | Sylhet | **Gowainghat** | Jointapur | Kanaighat | |
| 2016/06/24 | Moulvibazar | Barlekha | | | |
| | Sunamganj | Sunamganj Sadar | **Tahirpur** | | |

Source: Author's development from different sources

# References

Abburu S, Golla SB (2015) Satellite image classification methods and techniques: a review. Int J Comput Appl 119(8):20–25. https://doi.org/10.5120/21088-3779

Abdullah MA, Auwal AG, Darham S, Radam A (2014) Farmers willingness to pay for crop insurance in Northwest Selangor Irrigated Agricultural Development Area (IADA), Malaysia. Int Soc Southeast Asian Agric Sci 20(2):19–30

Abebe TH, Bogale A (2014) Willingness to pay for rainfall-based insurance by smallholder farmers in central rift valley of Ethiopia: the case of Dugda and Mieso Woredas. Asia-Pac J Energy Environ 1(2):121–155

Acharya TD, Subedi A, Lee DH (2018) Evaluation of water indices for surface water extraction in a Landsat 8 scene of Nepal. Sensors (Basel, Switzerland) 18(8):2580. https://doi.org/10.3390/s18082580

Adusumilli N, Wang H, Dodla S, Deliberto M (2019) Estimating risk premiums for adapting no-till and cover crops management practices in soybean production system using stochastic efficiency approach. Agric Syst 178:102744. https://doi.org/10.1016/j.agsy.2019.102744

Ahmed AU (2012) Bangladesh environmental and climate change assessment. The main report was prepared for IFAD's Country Strategic Opportunities Programme 2012–2018

Ahmed MR, Rahaman KR, Kok A, Hassan QK (2017) Remote sensing-base quantification of the impact of flash flooding on the rice production: a case study over Northeastern Bangladesh. Sensors 17(10):2347. https://doi.org/10.3390/s17102347

Aidoo R, James OM, Prosper W, Awunyo-Vitor D (2014) Prospects of crop insurance as a risk management tool among arable crop farmers in Ghana. Asian Econ Soc Soc 4(3):341–354. https://EconPapers.repec.org/RePEc:asi:aeafrj:2014:p:341-354

Akter S, Brouwer R, Chowdhury S, Aziz S (2008) Determinants of participation in a catastrophe insurance programme: empirical evidence from a developing country. Presented at the Australian agricultural and resource economics society, 52nd annual conference, 5–8 February, Canberra, Australia

Akter S, Brouwer R, Choudhury S, Aziz S (2009) Is there a commercially viable market for crop insurance in rural Bangladesh? Mitig Adapt Strateg Glob Chang 14:215–229. https://doi.org/10.1007/s11027-008-9161-6

Al-Ahmadi FS, Hames AS (2009) Comparison of four classification methods to extract land use and land cover from raw satellite images for some remote arid areas, Kingdom of Saudi Arabia. J King Abdulaziz Univ Earth Sci 20(1):167–191

Aldhshan SRS, Mohammed OZ, Mohd Shafri HZ (2019) Flash flood area mapping using sentinel-1 SAR data: a case study of eight upazilas in Sunamganj district, Bangladesh. IOP Confer Ser Earth Environ Sci 357(1):012034. https://doi.org/10.1088/1755-1315/357/1/012034

Ali A (2013) Farmers' willingness to pay for index-based crop insurance in Pakistan: a case study on food and cash crops of rain-fed areas. Agric Econ Res Rev 26(2):241–248

Allen M (2017) The sage encyclopedia of communication research methods, vol 1–4. Sage Publications, Inc, Oaks. https://doi.org/10.4135/9781483381411

Al-Maruf A, Mira SA, Rida TN, Rahman MS, Sarker PK, Jenkins JC (2021) Piloting a weather-index-based crop insurance system in Bangladesh: understanding the challenges of financial instruments for tackling climate risks. Sustainability 13(15):8616. https://doi.org/10.3390/su13158616

Amarnath G (2013) An algorithm for rapid flood inundation mapping from optical data using a reflectance differencing technique. J Flood Risk Manag 7:239–250

Anderson JR (1976) A land use and land cover classification system for use with remote sensor data. U.S. Government Printing Office

Arshad M, Amjath-Babu TS, Kächele H, Müller K (2015) What drives the willingness to pay for crop insurance against extreme weather events (flood and drought) in Pakistan? A hypothetical market approaches. Clim Dev 8(3):234–244. https://doi.org/10.1080/17565529.2015.1034232

Bangladesh Bureau of Statistics (BBS) (2017) Yearbook of agricultural statistics of Bangladesh. Statistics Division, Ministry of Planning, Government of the People's Republic of Bangladesh, 2017. www.bbs.gov.bd. Accessed 12 May 2018

Benedetti R, Rossini P (1993) On the use of NDVI profiles as a tool for agricultural statistics: the case study of wheat yield estimate and forecast in Emilia Romagna. Remote Sens Environ 45:311–326. https://doi.org/10.1016/0034-4257(93)90113-C

Benson C, Edward JC (2002) Bangladesh: disasters and public finance. Disaster risk management, working paper series no. 5. World Bank, Washington, DC. http://www.proventionconsortium.org/files/Bangladesh.pdf. Accessed 18 July 2019

Berber K, Patrick W (2018) Research for agricultural insurance in South Asia: a regional dialogue. International Food Policy Research Institute, Washington, DC, pp 1–5

Bhatta B (2012) Remote sensing and GIS, 2nd edn. Oxford University Press, New Delhi

Binte Mostafiz R, Noguchi R, Ahamed T (2021) Calorie-based seasonal multicrop land suitability analysis for regional food nutrition security in Bangladesh. Asia-Pac J Reg Sci 5:757. https://doi.org/10.1007/s41685-021-00197-5

Biswas JK, Hossain MS, Mamin MSI, Muttaleb MA (2008) Manipulation of seeding date and seedling age to avoid flash flood damage of Boro rice at the Northeastern haor area of Bangladesh. Bangladesh Rice J 13:57–61

Bubeck P, Botzen W, Aerts J (2012) A review of risk perceptions and other factors that influence flood mitigation behavior. Risk Anal 32:1481–1495. https://doi.org/10.1111/j.1539-6924.2011.01783.x

Budhathoki NK, Lassa JA, Pun S, Zander KK (2019) Farmers' interest and willingness-to-pay for index-based crop insurance in the lowlands of Nepal. Land Use Policy 85:1–10. https://doi.org/10.1016/j.landusepol.2019.03.029

Chen WS, Tse YK (2017) Financial mathematics for actuaries, 2nd edn. World Scientific Publishing Pte Ltd, Hackensack, pp 4–45. ISBN 978-981-3224-67-4

Clarke DJ (2016) A theory of rational demand for index insurance. Am Econ J Microecon 8(1):283–306. https://www.aeaweb.org/articles?id=10.1257/mic.20140103

Clover TA, Nieuwoudt WL (2003) An economic evaluation of area yields insurance for small-scale cane growers. Dev South Afr 20:293–305

Cole S, Giné X, Vickery J (2017) How does risk management influence production decisions? Evidence from a field experiment. Rev Financ Stud 30(6):1935–1970. https://doi.org/10.1093/rfs/hhw080

Congalton RG (1991) A review of assessing the accuracy of classification of remotely sensed data. Remote Sens Environ 37(1):35–46. https://doi.org/10.1016/0034-4257(91)90048-B.\

Cox DR (1958) The regression analysis of binary sequences. J R Stat Soc 20:215–242. https://doi.org/10.1111/j.2517-6161.1958.tb00292.x

Dalhaus T, Musshoff O, Finger R (2018) Phenology information contributes to reduce temporal basis risk in agricultural weather index insurance. Sci Rep 8:1–10. https://doi.org/10.1038/s41598-017-18656-5

Dawson RJ, Ball T, Werritty J, Werritty A, Hall JW, Roche N (2011) Assessing the effectiveness of non-structural flood management measures in the Thames Estuary under conditions of socioeconomic and environmental change. Glob Environ Chang 21(2):628–646. https://doi.org/10.1016/j.gloenvcha.2011.01.013

Dercon S, Vargas-Hill R, Clarke D, Outes-Leon I, Taffesse AS (2014) Offering rainfall insurance to informal insurance groups: evidence from a field experiment in Ethiopia. J Dev Econ 106:132–143. (C)

Diakakis M, Deligiannakis G, Katsetsiadou K, Lekkas E, Melaki M, Antoniadis Z (2016) Mapping and classification of direct effects of the flood of October 2014 in Athens. Bull Geol Soc Greece 50:681–690. https://doi.org/10.12681/bgsg.11774

Du Z, Li W, Zhou D, Tian L, Ling F, Wang H, Sun B (2014) Analysis of Landsat-8 OLI imagery for land surface water mapping. Remote Sens Lett 5:672–681. https://doi.org/10.1080/2150704X.2014.960606

Emmanuella E (2017) Farmers willingness to pay for crop insurance: evidence from Eastern Ghana. Int J Agric Manag Dev 7(4):447–463

Fadhliani Z, Luckstead J, Wailes EJ (2019) The impacts of multiperil crop insurance on Indonesian rice farmers and production. Agric Econ 50:15–26

Falola A, Ayinde AE, Agboola BO (2013) Willingness to take agricultural insurance by Cocoa farmers in Nigeria. Int J Food Agric Econ 1(1):97–107. https://doi.org/10.22004/ag.econ.156837

Feng P, Wang B, Liu DL, Yu Q (2019) Machine learning-based integration of remotely-sensed drought factors can improve the estimation of agricultural drought in South-Eastern Australia. Agric Syst 173:303–316. https://doi.org/10.1016/j.agsy.2019.03.015

Feyisa GL, Meilby H, Fensholt R, Proud SR (2014) Automated water extraction index: a new technique for surface water mapping using Landsat imagery. Remote Sens Environ 140:23–35. https://doi.org/10.1016/j.rse.2013.08.029

Gazi MY, Islam MA, Hossain S (2019) Flood-hazard mapping in a regional scale—way forward to the future Hazard atlas in Bangladesh. Malaysian J Geosci 3(1):1–11. https://doi.org/10.26480/mjg.01.2019.01.11. Zibeline International Publishing

Ghazanfar S, Qiwen Z, Abdullah M, Ahmad Z, Lateef M (2015) Farmers' perception and awareness and factors affecting awareness of farmers regarding crop insurance as a risk coping mechanism evidence from Pakistan. J Northeast Agric Univ 22:76–82

Godwin BK, Ker AP (1998) Nonparametric estimation of crop yield distributions: implications for rating group-risk crop insurance contracts. Am J Agric Econ 80:139–153

Haor Master Plan (2012) Bangladesh Haor and Wetland Development Board. Ministry of Water Resources, Government of the People's Republic of Bangladesh, Vol. II. Accessed 20 Jan 2020

Habiba U, Shaw R (2013) Crop insurance as risk management strategy in Bangladesh. In: Disaster risk reduction approaches in Bangladesh. Springer, Tokyo, pp 281–305

Hazell P (2001) Potential role for insurance in managing catastrophic risk in developing countries. Occasional Paper, IFPRI

Islam AS, Bala SK, Haque MA (2010) Flood inundation map of Bangladesh using MODIS time-series images. J Flood Risk Manag 3:210–222. https://doi.org/10.1111/j.1753-318X.2010.01074.x

Islam MM, Ahamed T, Noguchi R (2018) Land suitability and insurance premiums: a GIS-based multicriteria analysis approach for sustainable rice production. Sustainability 10(6):1759. https://doi.org/10.3390/su10061759

Just RE, Weninger Q (1999) Are crop yields normally distributed. Am J Agric Econ 81:287–304

Khan MNH, Mia MY, Hossain MR (2012) Impacts of flood on crop production in Haor areas of two Upazillas in Kishoregonj. J Environ Sci Nat Resour 5:193–198

Khosravi K, Pham BT, Chapi K, Shirzadi A, Shahabi H, Revhaug I, Prakash I, Bui DT (2018) A comparative assessment of decision trees algorithms for flash flood susceptibility modeling at Haraz Watershed. Northern Iran Sci Total Environ 627:744–755. https://doi.org/10.1016/j.scitotenv.2018.01.266

Kousky C (2011) Understanding the demand for flood insurance. Nat Hazards Rev 12(2):96–110. https://doi.org/10.1061/(ASCE)NH.1527-6996.0000025

Kron W (2009) Flood insurance: from clients to global financial markets. J Flood Risk Manag 2(1):68–75. https://doi.org/10.1111/j.1753-318X.2008.01015.x

Landis JR, Koch G (1977) The measurement of observer agreement for categorical data. Biometrics 33(1):159–174. https://doi.org/10.2307/2529310

Li B, Lingsma HF, Steyerberg EW (2011) Logistic random-effects regression models: a comparison of statistical packages for binary and ordinal outcomes. BMC Med Res Methodol 11:77. https://doi.org/10.1186/1471-2288-11-77

Liu C, Frazier P, Kumar L (2007) Comparative assessment of the measures of thematic classification accuracy. Remote Sens Environ 107(4):606–616. https://doi.org/10.1016/j.rse.2006.10.010

Liu J, Xu ZW, Chen FL, Chen F, Zhang L (2019) Flood hazard mapping and assessment on the Angkor world heritage site, Cambodia. Remote Sens 11:98. https://doi.org/10.3390/rs11010098

Long QT, Minh TB, Manh CN, Thanh TV (2013) Farm households' willingness to pay for crop (micro) insurance in rural Vietnam: an investigation using contingent valuation method. East Asian Development Network (EADN) Working Paper No. 64

Long S, Fatoyinbo TE, Policelli F (2014) Flood extent mapping for Namibia using change detection and thresholding with SAR. Environ Res Lett 9:035002

Martin SW, Barnet BJ, Coble KH (2001) Developing and pricing precipitation insurance. J Agric Resour Econ 26(1):261–274. https://www.jstor.org/stable/40987107

McFadden D (1975) The measurement of urban travel demand. University of California, Berkeley

McFeeters SK (1996) The use of the Normalized Difference Water Index (NDWI) in the delineation of open water features. Int J Remote Sens 17:1425–1432. https://doi.org/10.1080/01431169608948714

Mechler R, Linnerooth-Bayer J, Peppiatt D (2006) Disaster insurance for the poor. A review of micro-insurance for natural disaster risks in developing countries. Provention/IIASA Study. Provention Consortium, Geneva

Mendelsohn R (2008) The impact of climate change on agriculture in developing countries. J Nat Resour Policy Res 1(1):5–19. https://doi.org/10.1080/19390450802495882

Mottaleb KA, Mohanty S (2015) Farm size and profitability of rice farming under rising input costs. J Land Use Sci 10(3):243–255

Munyati C (2004) Use of principal component analysis (PCA) of remote sensing images in wetland change detection on the Kafue Flats, Zambia. Geocarto Int 19:11–22. https://doi.org/10.1080/10106040408542313

Mutaqin DJ, Usami K (2019) Smallholder farmers' willingness to pay for agricultural production cost insurance in rural West Java, Indonesia: a contingent valuation method (CVM) approach. Risks 7(2):69. https://doi.org/10.3390/risks7020069

Nahar N, Sultana N, Miah J (2018) Seasonal land cover changes and its effects on essential services of haor and non-haor areas of Kishoreganj district, Bangladesh. Asia-Pac J Reg Sci 2:399–429. https://doi.org/10.1007/s41685-017-0067-8

Nahayo A, Omondi MO, Zhang XH, Li LQ, Pan GX, Joseph S (2017) Factors influencing farmers' participation in crop intensification program in Rwanda. J Integr Agric 16:1406–1416. https://doi.org/10.1016/S2095-3119(16)61555-1

Nchinda VP, Ambe TE, Holvoet N, Leke W, Che MA, Nkwate SP, Ngassam SB, Njualem DK (2010) Factors influencing the adoption intensity of improved yam (Dioscorea spp.) seed technology in Western Highlands and Guinea Savannah of Cameroon. J Appl Biosci 36:2389–2402

Ntukamazina N, Onwonga RN, Sommer R, Rubyogo JC, Mukankusi CM, Mburu J, Kariuki R (2017) Index-based agricultural insurance products: challenges, opportunities, and prospects for uptake in sub-Sahara Africa. J Agric Rural Dev Trop 118(2):171–185. https://www.jarts.info/.../913

Pandey I, Bardsley DK (2015) Social-ecological vulnerability to climate change in the Nepali Himalaya. Appl Geogr 64:74–86. https://doi.org/10.1016/j.apgeog.2015.09.008

Plourde L, Congalton RG (2003) Sampling method, and sample placement: how do they affect the accuracy of remotely sensed maps? Photogramm Eng Remote Sens 69(3):289–297. https://doi.org/10.14358/PERS.69.3.289

Rahmati O, Haghizadeh A, Pourghasemi HR, Noormohamadi F (2016) Gully erosion susceptibility mapping: the role of GIS-based bivariate statistical models and their comparison. Nat Hazards 82(2):1231–1258. https://doi.org/10.1007/s11069-016-2239-7

Rana S, Kiminami L, Furuzawa S (2020) Analysis on the factors affecting farmers' performance in disaster risk management at community level: focusing on a Haor locality in Bangladesh. Asia-Pac J Reg Sci 4:737–757. https://doi.org/10.1007/s41685-020-00171-7

Roy P, Chowdhury D, Deshwara M (2017) Havoc in haor: early flashfloods shock farmers. The Daily Star, 14 April 2017. http://www.thedailystar.net/frontpage/havoc-haor-1391089. Accessed 15 June 2017

Sadati SA, Ghobadi F, Mohamadi Y, Sharifi O, Asakereh A (2010) Survey of effective factors on adoption of crop insurance among farmers: a case study of Behbahan County. Afr J Agric Res 5:2237–2242

Sims DA, Gamon JA (2003) Estimation of vegetation water content and photosynthetic tissue area from spectral reflectance: a comparison of indices based on liquid water and chlorophyll absorption features. Remote Sens Environ 84:526–537

Sinha S, Tripathi NK (2014) Assessment of crop insurance international practices, policies and technologies as risk mitigation tools in India and Thailand. Int J Adv Res 2:769–788

Situation Report 03 May 2017, Nirapod (2017) Flash Flood Situation Update, 3 May 2017. http://reliefweb.int/sites/reliefweb.int/files/resources/Update%20Report%20of%20Flash%20Flood_MAY%2003%2C%202017.pdf. Accessed 15 June 2017

Skees RJ, Black JR, Barnett BJ (1997) Designing and rating an area yield crop insurance contract. Am J Agric Econ 79:430–438

Skees J, Stephanie G, Panos V, Rodney L, Kalavakonda V (2001) Developing rainfall-based index insurance in Morocco, Policy, Research Working Paper No. WPS 2577. World Bank, Washington, DC

Smith A, Bates PD, Wing O, Sampson C, Quinn N, Neal J (2019) New estimates of flood exposure in developing countries using high-resolution population data. Nat Commun 10:1814. https://doi.org/10.1038/s41467-019-09282-y

Steven MD, Malthus TJ, Baret F, Xu H, Chopping MJ (2003) Inter-calibration of vegetation indices from different sensor systems. Remote Sens Environ 88(4):412–422

Sultana D, Kaisar F (2021) Evolution of micro insurance in Bangladesh: financial cushion for the bottom of the pyramid population. Light Castle Partners. 2021. https://www.lightcastlebd.com/insights/2021/05/evolution-of-micro-insurance-in-bangladesh-financial-cushion-for-the-bottom-of-the-pyramidpopulation?utm_source=rss&utm_medium=rss&utm_campaign=evolution-of-micro-insurance-in-bangladesh-financial-cushion-for-the-bottom-of-the-pyramid-population. Accessed 24 Aug 2021

Suvedi M, Ghimire R, Kaplowitz M (2017) Farmers' participation in extension programs and technology adoption in rural Nepal: a logistic regression analysis. J Agric Educ Ext 23:351–371. https://doi.org/10.1080/1389224X.2017.1323653

Svetlana S, Dobrovič R, Dobrovič J (2015) The economic impact of floods and their importance in different regions of the world with emphasis on Europe. Proc Econ Finance 34:649–655. https://doi.org/10.1016/S2212-5671(15)01681-0

Tien Bui D, Khosravi K, Shahabi H, Daggupati P, Adamowski JF, Melesse AM, Thai Pham B, Pourghasemi HR, Mahmoudi M, Bahrami S (2019) Flood spatial modeling in northern Iran using remote sensing and GIS: a comparison between evidential belief functions and its ensemble with a multivariate logistic regression model. Remote Sens 11(13):1589. https://doi.org/10.3390/rs11131589

Tobias A, Molina I, Valera HG, Mottaleb KA, Mohanty S (2012) Handbook on rice policy for Asia. International Rice Research Institute, Los Baños, p 47

Wairimu E, Obare G, Odendo M (2016) Factors affecting weather index-based crop insurance WB, (2010). The World Bank. Agricultural Insurance in Bangladesh: Promoting Access to Small and Marginal Farmers. Report No. 53081-BD

Waqas H, Lu L, Tariq A, Li Q, Baqa MF, Xing J, Sajjad A (2021) Flash flood susceptibility assessment and zonation using an integrating analytic hierarchy process and frequency ratio model for the Chitral District, Khyber Pakhtunkhwa, Pakistan. Water 13:1650. https://doi.org/10.3390/w13121650

Whitcraft AK, Becker-Reshef I, Justice CO (2015) A framework for defining spatially explicit earth observation requirements for a global agricultural monitoring initiative (GEOGLAM). Remote Sens 7:1461–1481. https://doi.org/10.3390/rs70201461

World Bank (2010) Weather index insurance for agriculture: guidance for development practitioners; agriculture and rural development discussion paper. The World Bank, Washington, DC

Xiao X, Boles S, Liu J, Zuang D, Frolking S, Li C (2005) Mapping paddy rice agriculture in southern China using multitemporal MODIS images. Remote Sens Environ 95:480–492

Xu H (2006) Modification of normalised difference water index (NDWI) to enhance open water features in remotely sensed imagery. Int J Remote Sens 27:3025–3033. https://doi.org/10.1080/01431160600589179

Yu WH, Alam M, Hassan A, Khan AS, Ruane AC, Rosenzweig C, Major DC, Thurlow J (2010) Climate change risk and food security in Bangladesh. EarthScan, London

Yuzva K, Botzen WW, Aerts J, Brouwer R (2018) A global review of the impact of basis risk on the functioning of and demand for index insurance. Int J Disaster Risk Reduct 28:845–853. https://doi.org/10.1016/j.ijdrr.2018.01.001

Zwane EM (2019) Impact of climate change on primary agriculture, water sources, and food security in Western Cape, South Africa. J Disaster Risk Stud 11(1):1–7. https://doi.org/10.4102/jamba.v11i1.562

# Chapter 6
# Application of Introducing Damaged Area Index for Cultivated Agricultural Lands Affected by Recurrent Tropical Cyclones Bulbul and Amphan Using Satellite Remote Sensing

**Md Shamsuzzoha and Tofael Ahamed**

**Abstract** South and Southeast Asian countries are prone to cyclones throughout the year due to climate changes. The damage is also severe due to tidal surges and intrusion of saline water in agricultural lands. Therefore, in this research, a damaged area index (DAI) of rice crop-based agricultural lands, affected by two consecutive tropical cyclones, was developed after assessing damaged type areas in the coastal region of Bangladesh. Repeatedly the cyclone Bulbul and Amphan struck India and Bangladesh's coastline on November 10, 2019, and May 20, 2020, and both cyclones caused huge damage to rice crops and dominant agricultural lands. The study was conducted in 12 unions of Kalapara subdistrict under Patuakhali district, Bangladesh, to measure DAI for the crop calendar 2019–2020. It was found that cyclone Bulbul was more destructive than cyclone Amphan by using Landsat-8 OLI and TIRS datasets. A total of 10-DAI was calculated over the 30,907.59 hectares (ha) of agricultural lands. This study assessed two damaged areas-sets for the cyclone Bulbul and Amphan, respectively, those were defined as (1) not damaged (ND); (2) slightly damaged (SD); (3) moderately damaged (MD); (4) very damaged (VD); and (5) extremely damaged (ED) using damaged assessment area (DAA) method. The non-agricultural land use/land cover (LULC) areas were considered DAI-1 for the study. The DAI-6 to 9 areas for both cyclones were found at 6705.94 ha (21.70%); 8386.06 ha (27.13%); 7043.07 ha (22.79%); and 3994.51 ha (12.92%) respectively.

M. Shamsuzzoha
Graduate School of Life and Environmental Sciences, University of Tsukuba, Tsukuba, Ibaraki, Japan

Department of Emergency Management, Faculty of Environmental Science and Disaster Management, Patuakhali Science and Technology University, Patuakhali, Bangladesh

T. Ahamed (✉)
Institute of Life and Environmental Sciences, University of Tsukuba, Tsukuba, Ibaraki, Japan
e-mail: tofael.ahamed.gp@u.tsukuba.ac.jp



T. Ahamed (ed.), *Remote Sensing Application II*, New Frontiers in Regional Science: Asian Perspectives 77, https://doi.org/10.1007/978-981-97-1188-8_6

A total of 420 ground reference points (GRPs) data were also investigated separately to validate the proposed DAI method of the research. A similar observation was found for the DAI-6 (ND to ED), 7 (SD to ED), 8 (MD to ED), and 9 (VD to ED) significantly from the GRPs validation consistently. Therefore, this study could help prepare a damage area index to support farmers with a new adaptive policy in recurrent cyclones for tropical cyclone-prone global south countries.

**Keywords** Agricultural crops · Crop calendar · Cyclone Amphan · Cyclone Bulbul · Damaged area assessment · DAI · Satellite remote sensing

## 6.1 Introduction

The top rice-producing countries in South Asia are being affected by tropical cyclones in the early summer (April and May) and the late rainy season (October and November) over the Bay of Bengal (Khalil 1992; Hafsa et al. 2021; Shamsuzzoha et al. 2022; Htun et al. 2023). Bangladesh is one of the global south hotspots due to tropical cyclones outbreaks in South and Southeast Asian countries (Shamsuzzoha and Al-Maruf 2012; Al-Maruf et al. 2021; Hafsa et al. 2021). Bangladesh is the fourth-largest rice-harvested country in the world (IRRI 2020) (Fig. 6.1). The cultivated agricultural lands of this country are damaged by frequent and recurrent tropical cyclones almost every year (Haq et al. 2012; Talukder et al. 2018; Shamsuzzoha

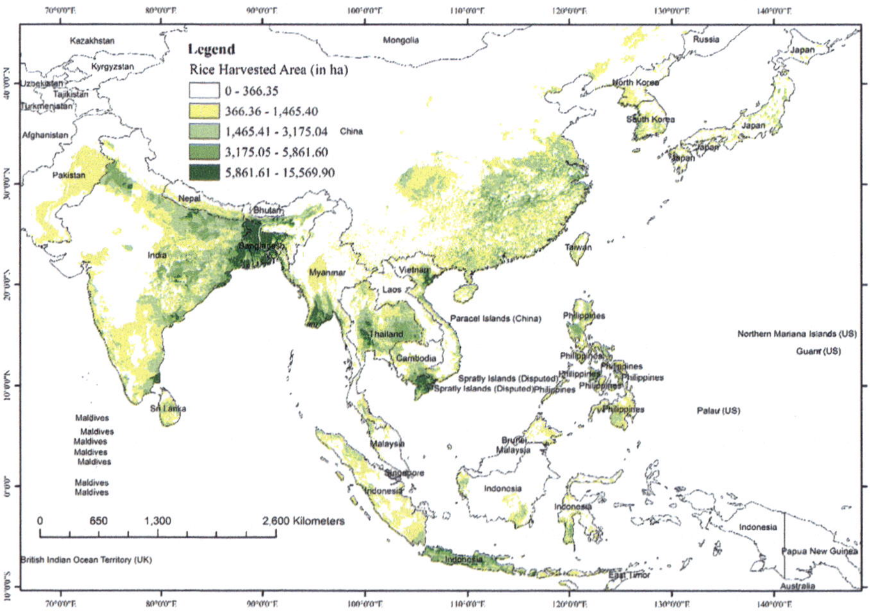


**Fig. 6.1** Rice harvested area in South and Southeast Asian countries (IRRI 2020; Earth Stat 2023)

et al. 2021) (Figs. 6.2 and 6.3). The physiography of the country's southern part is an active delta with changing shorelines and tidal rivers (Alam and Ahamed 2022, 2023a, b; Shamsuzzoha and Ahamed 2023). Therefore, due to the geographical setting, the country is prone to climate-induced natural disasters, and tropical cyclones cause massive loss of human lives and resources, destroy settlements, and damage agricultural lands with impacts of huge crop yield losses (Matsuda 1993; Hossain et al. 2008; Nadiruzzaman and Wrathall 2015; Parks et al. 2021; Du et al. 2023).

Various studies have been accomplished using different methods from satellite remote sensing approaches on tropical cyclones-induced disaster impact (Kim et al. 2019; Mishra and Vanganuru 2020; Nandi et al. 2020; Zhang et al. 2020; Banik et al. 2021; Zhou and Matyas 2021; Roy et al. 2023). Though cyclones are repeated every year in southern and southeast Asian countries, the studies could not cover the index-based assessment of damaged areas impacted by recurrent tropical cyclones or typhoons. The assessment of damaged areas in cyclone-prone regions has been conducted in several studies to gain the emergency response and recovery stages after the extreme climatic event (e.g., tropical cyclones, floods, and tidal surges) to reduce disaster risks in divergent sectors (e.g., agricultural sector, urban and rural settlements, and transport sector), to provide humanitarian agriculture support for local livelihoods and production purposes (e.g., medical support, relief, rehabilitation, aid, subsidy, insurance, donation, and exchange) (Di et al. 2018; Talukder et al. 2018; Malmgren-Hansen et al. 2020; Chen et al. 2022; Karaer et al. 2022). The agricultural sector is one

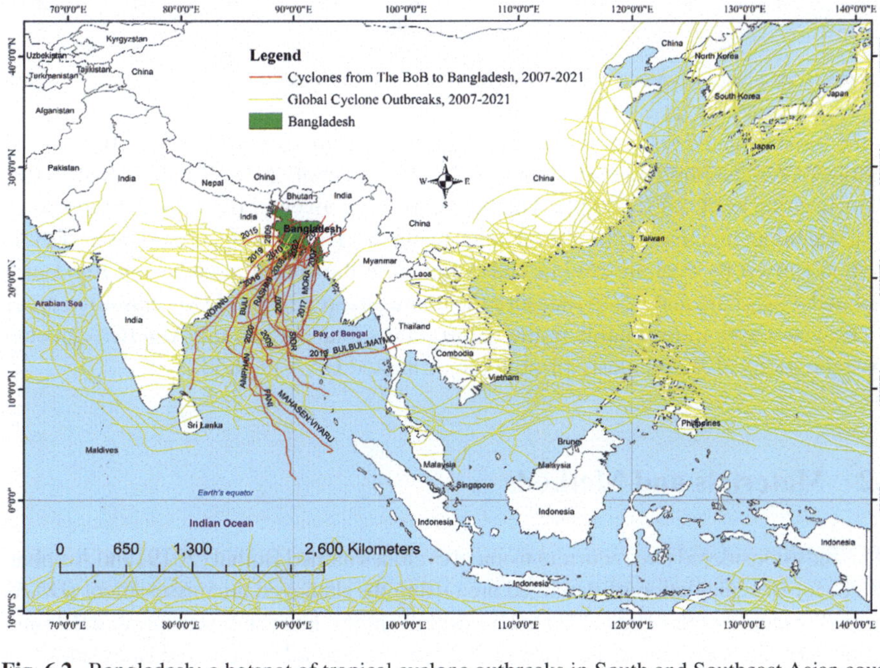


**Fig. 6.2** Bangladesh: a hotspot of tropical cyclone outbreaks in South and Southeast Asian countries (NOAA 2023)

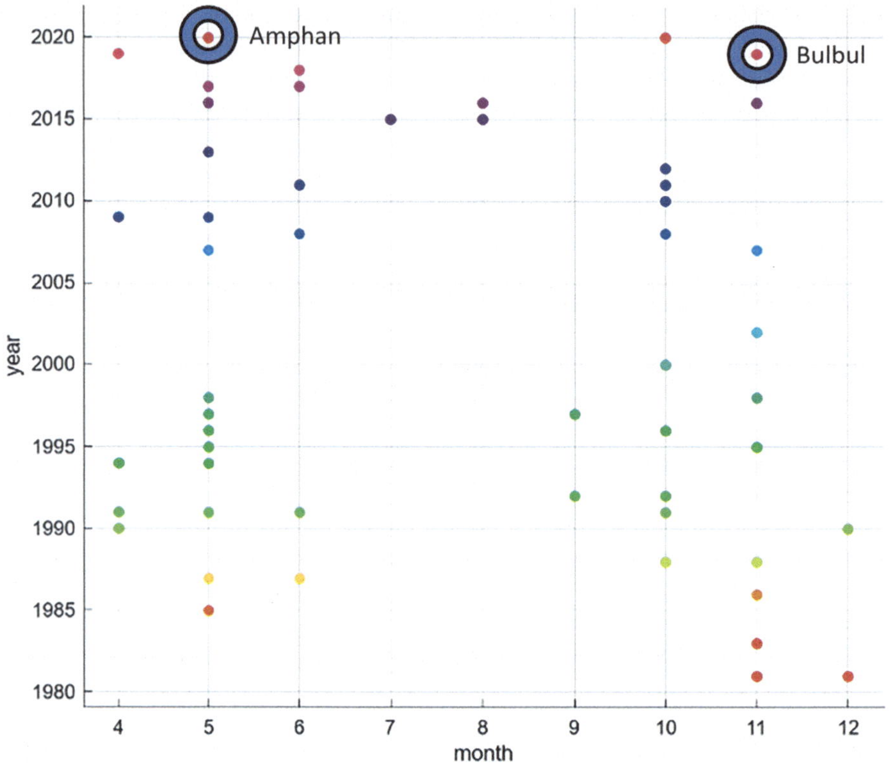


**Fig. 6.3** Calander of cyclones in Bangladesh, 1980–2020 (NOAA 2023)

of the most affected sectors that are damaged by cyclones landfalling in the coastal regions of Asian countries (Masutomi et al. 2012; Di et al. 2018; Rahman et al. 2018; Shamsuzzoha et al. 2021, 2022). Therefore, the objectives of the present study were (1) to calculate damaged agricultural lands using the damaged area assessment (DAA) method caused by cyclone Bulbul and Amphan in the coastal areas using Landsat 8 OLI and TIRS data and (2) to measure the damaged area index (DAI) classified areas of cultivated agricultural lands in the recurrent cyclone Bulbul and Amphan-affected coastal regions of Bangladesh through the estimated DAA method.

## 6.2 Materials and Methods

This study considered two recurrent tropical cyclones named Bulbul (2019) and Amphan (2020) for the calculation of damaged area using the damaged area assessment (DAA) method (Shamsuzzoha et al. 2021). The Bulbul (in the late rainy season of 2019) and Amphan (in the early summer season of 2020) made landfall and entered Bangladesh on November 10th, 2019, and May 20th, 2020, respectively, in the same hazard-crop calendar (2019–2020) (Shamsuzzoha et al. 2021; Arab et al. 2022) (Fig. 6.3).

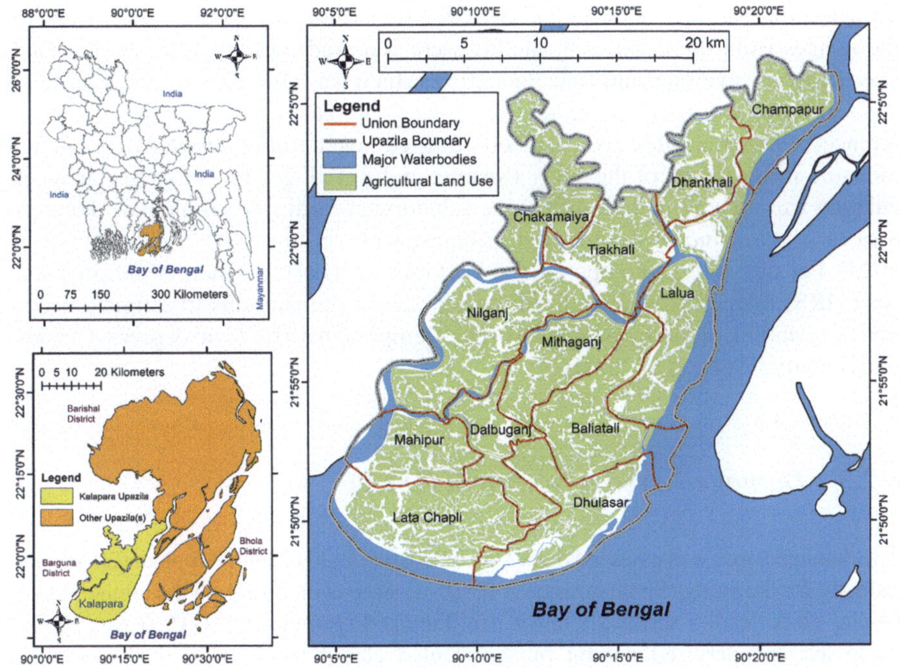


**Fig. 6.4** Study area map (Data source: SoB, 2019)

## 6.2.1 *Study Area*

The Kalapara *Upazila* (subdistrict) adjacent to the Bay of Bengal of the Patuakhali district in Bangladesh was considered as the study area of the research. This study area is also bounded by the Barguna district from the west and the Bay of Bengal from the south with an active delta physiographic condition (Shamsuzzoha and Ahamed 2023) (Fig. 6.4). The total area of the subdistrict is 492.102 km$^2$ and is encompassed by four substantial tidal rivers (Hoque et al. 2019; Khalil 2020; Shamsuzzoha et al. 2022). The necessary agricultural land use map of the study subdistrict was extracted for the research by leveraging an enormous number of data layers (scale 1:25,000) from the attributed topographic geo-database with feature classes from the Survey of Bangladesh (SoB) office under the Ministry of Defense in Dhaka, Bangladesh in November 2019. The agricultural land use of the Kalapara subdistrict was used for the renovation of Landsat 8 OLI and TIRS satellite image data, the selection of the total 420 GRPs, and the accuracy assessment of the land use class with the satellite imagery. All datasets of the study were georeferenced to UTM 45 N of WGS 1984.

## 6.2.2 *Data Preparation (Before and After Cyclones)*

Landsat 8 image data from November 7th, 2019, before the cyclone Bulbul, and from November 23rd, 2019, after the cyclone was acquired separately from the

Earth Explorer archive of the USGS (USGS 2019). Concerning the cyclone Amphan, the images from the Landsat 8 satellite were acquired on May 17th, 2020, which was before the cyclone, and June 2nd, 2020, which was after the cyclone. Landsat 8 dataset was attained for the preparation of necessary calculations of normalized difference vegetation index (NDVI), soil adjusted vegetation index (SAVI), and soil moisture index (SMI) of the study (Shamsuzzoha et al. 2021). Necessary area calculation was done with their difference valuation accordingly to calculate damaged area in hectare (ha) in the research (Shamsuzzoha et al. 2021, 2022; Htun et al. 2023). For the study, both operational land imager (OLI) and thermal infrared sensor (TIRS) imageries were utilized for the dataset preparations to estimate necessary vegetation indices and land surface temperature (LST in degrees Celsius) individually (Weng et al. 2004).

### *6.2.3 Damaged Area Assessment (DAA)*

The damaged area was assessed due to cyclone Bulbul (2019) and Amphan (2020) individually for the study (Shamsuzzoha et al. 2021; Arab et al. 2022). In the applied DAA, similar classification set of a total of five damaged type classes (DTCs) was applied using the unsupervised natural breaks (Jenks) classification method (Jenks 1967; Shamsuzzoha et al. 2021). The DAA method was developed to calculate a composite layer by pix-code sum procedures among the corresponding weighted pixel-code values with classified pixels of NDVI-change type class (CTC), SAVI-CTC, and SMI-CTC rasters using a pseudocode algorithm (Shamsuzzoha et al. 2021). In this research, the damaged areas were classified using the natural breaks (Jenks) method, and classified as not damaged (ND), slightly damaged (SD), moderately damaged (MD), very damaged (VD), and extremely damaged (ED), respectively (Shamsuzzoha et al. 2021).

### *6.2.4 Damaged Area Index (DAI)*

In the study, a novel damaged area index (DAI) for the session 2019–2020 (area calculated in hectares) due to the recurrent two cyclones Bulbul (2019) and Amphan (2020) is introduced. Five DAA classes were utilized by the "Geoprocessing" tool in ArcGISArcGIS® 10.8.2 (ESRI, CA 92373, United States) software. A geometric intersection was computed of the cyclone Bulbul and Amphan-generated DAA features as input of the "Intersect" operation. The Bulbul DAA and Amphan DAA features overlap in all layers by their feature classes. The output feature class was written as the damaged area index (DAI) for the study cyclone hazard-crop calendar, 2019–2020 (Fig. 6.5). A total of 10 DAI area classes were obtained from the intersect operation and transversely followed the natural breaks (Jenks) classification method for labeling the results for drawing the expected DAI map of the study area (Table 6.1). In the study area, for collecting necessary GRPs information, a GPS60® handheld device (KS 66062-3426, United States) was used from March 28th to April 10th, 2020. For validation of the DAI results, a total of 420 GRPsbased DAI was also computed individually in the study.

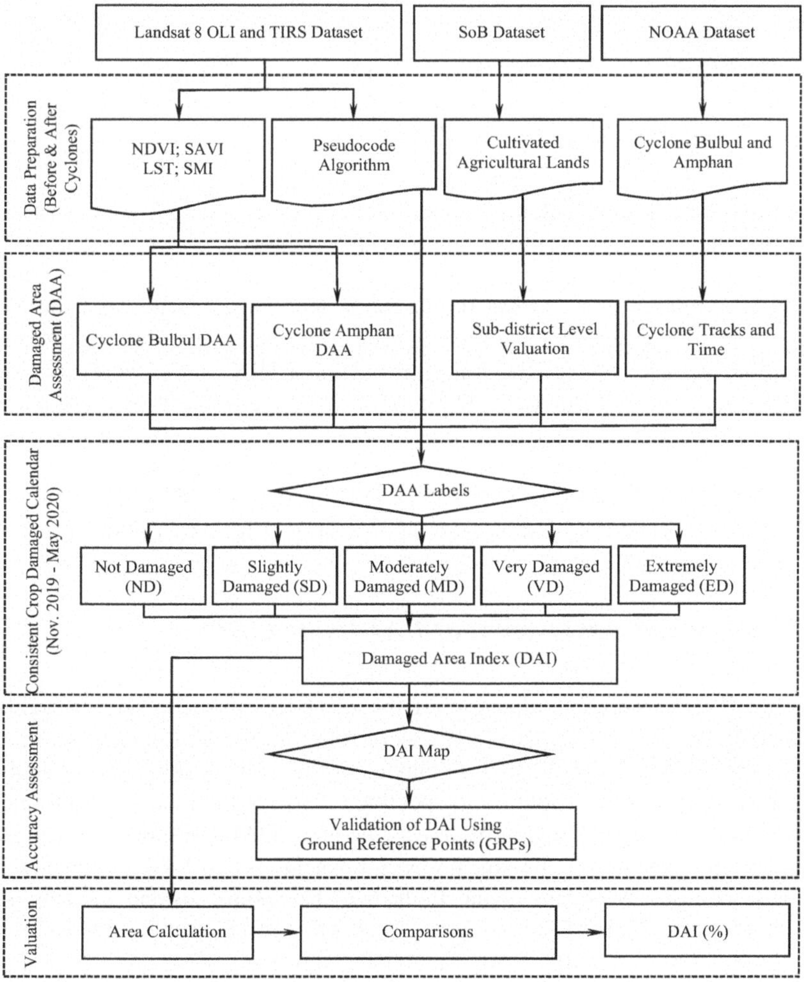


**Fig. 6.5** The research framework of the study

**Table 6.1** Labeled classes description of the proposed damaged area index (DAI) due to recurrent tropical cyclones named cyclone Bulbul (2019) and Amphan (2020)

| DAI label | Description | Labeled classes |
|---|---|---|
| 1 | Non-agricultural lands | 1 (Adjacent/Other LULC) |
| 2 | Not damaged | 2 (ND, ND) |
| 3 | Not, slightly damaged | 3 (ND, SD) |
| 4 | Not, slightly, and moderately damaged | 4 (ND, SD, MD) |
| 5 | Not, slightly, moderately, and very damaged | 5 (ND, SD, MD, VD) |
| 6 | Not, slightly, moderately, very, and extremely damaged | 6 (ND, SD, MD, VD, ED) |
| 7 | Slightly, moderately, very, and extremely damaged | 7 (SD, MD, VD, ED) |
| 8 | Moderately, very, and extremely damaged | 8 (MD, VD, ED) |
| 9 | Very, extremely damaged | 9 (VD, ED) |
| 10 | Extremely damaged | 10 (ED, ED) |

## 6.3 Results

### *6.3.1 Damaged Area Index (DAI)*

A total of 10-DAI were calculated over the 30,907.59 ha agricultural lands by following the intersect methods using double raster sets obtained as Bulbul-induced DAA and Amphan-induced DAA (Fig. 6.6). In the study, two damaged cultivated cropland areas during the October–November period and the April–May period of the tropical cyclone landfall calendar were defined. After the appropriate geoprocessing, the non-agricultural LULC areas were considered DAI-1 for the study (Table 6.1). The DAI-2 to 10 areas for both cyclones were found accordingly in Table 6.2. A total of 420 GRPs data were also investigated separately to validate the proposed DAI method of the research. In the Kalapara subdistrict area, the DAI-6 (ND to ED), 7 (SD to ED), 8 (MD to ED), and 9 (VD to ED) were observed to be damaged significantly (Fig. 6.6, Table 6.2).

### *6.3.2 Accuracy Assessment of DAI Using GRPs*

A total of 10-DAI were calculated using the 420 GRPs over the agricultural lands of the study area (Fig. 6.7). No GRPs were found in the non-agricultural LULC areas as the DAI-1, 2, and 3 category classes of the study (Table 6.2). In damaged area index using GRPs validation, respective GRPs for the recurrent extreme climatic events were distributed similarly as the areal DAI assessment of the study. The calculated results by the study GRPs from DAI-4 to DAI-10 classes were obtained accordingly in the given Table 6.2. Therefore, among all four study areas, the DAI-6 (ND to ED), 7 (SD to ED), 8 (MD to ED), and 9 (VD to ED) were also observed significantly from the GRPs validation consistently (Fig. 6.7,

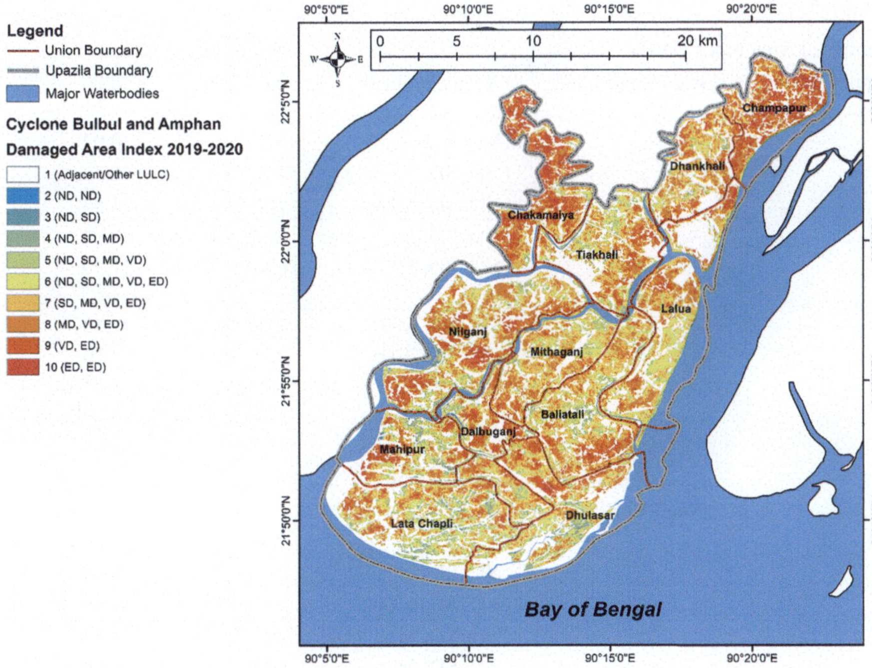


**Fig. 6.6** Damaged area index map

Table 6.2). In Fig. 6.7, it was also found that the respective GRPs were also dissolved into the areal features with similar patterns of DAI classifications. Therefore, it ensured the accuracy of the DAI class distribution in both geospatially and GRPs analysis.

We have found through the GPS survey of 420 GRPs information that during the cyclone Bulbul (in November), the major standing crops were transplanted *Aman* rice, and during the cyclone Amphan (in May), the major standing cropping pattern was a mixed type (pulse, oilseed, transplant *Aus* rice, *Boro* rice, melon, watermelon, vegetable, maize, etc.). In May, 24.0% of cropland was found as fallow lands according to local farmers' responses. Therefore, it was found that the agricultural lands of the study area were damaged extremely due to the cyclone Bulbul in November 2019 rather than Amphan in May 2020 (Shamsuzzoha et al. 2021; Arab et al. 2022).

**Table 6.2** Damaged area index, 2019–2020 (ha) due to cyclone Bulbul (2019) and Amphan (2020)

| Damaged Area Index (DAI) | | | Total (ha) | Total % |
|---|---|---|---|---|
| Damaged Area Index 2019–2020 | Area (in hectare) | 1 (Adjacent/Other LULC) | 0.00 | 0.00 |
| | | 2 (ND, ND) | 5.05 | 0.02 |
| | | 3 (ND, SD) | 112.23 | 0.36 |
| | | 4 (ND, SD, MD) | 786.21 | 2.54 |
| | | 5 (ND, SD, MD, VD) | 2978.46 | 9.64 |
| | | 6 (ND, SD, MD, VD, ED) | 6705.94 | 21.70 |
| | | 7 (SD, MD, VD, ED) | 8386.06 | 27.13 |
| | | 8 (MD, VD, ED) | 7043.07 | 22.79 |
| | | 9 (VD, ED) | 3994.51 | 12.92 |
| | | 10 (ED, ED) | 896.08 | 2.90 |
| | | Total | 30,907.59 | 100.00 |
| | Ground reference points (GRPs) | 1 (Adjacent/Other LULC) | 0 | 0.00 |
| | | 2 (ND, ND) | 0 | 0.00 |
| | | 3 (ND, SD) | 0 | 0.00 |
| | | 4 (ND, SD, MD) | 4 | 1.00 |
| | | 5 (ND, SD, MD, VD) | 20 | 4.75 |
| | | 6 (ND, SD, MD, VD, ED) | 67 | 16.00 |
| | | 7 (SD, MD, VD, ED) | 117 | 27.75 |
| | | 8 (MD, VD, ED) | 109 | 26.00 |
| | | 9 (VD, ED) | 77 | 18.25 |
| | | 10 (ED, ED) | 26 | 6.25 |
| | | Total | 420 | 100.00 |

## 6.4 Discussion

The study region was threatened with more adverse conditions, mainly the cultivated agricultural lands due to cyclones. The cyclone Bulbul hit the study region on November 9, 2019, and crossed over the study area on November 10, 2019. It was 6 months, 10 days intervals only when Amphan impacted the area again on May 20, 2020. Cyclone Bulbul and Amphan were consecutive recurrent extreme climatic events that simultaneously made massive destruction in coastal areas of Bangladesh. After assessing damaged areas using the DAA method due to Bulbul, 2.46, 10.35, 25.08, 43.25, and 18.87% of cultivated agricultural lands were distributed as not (ND), slightly (SD), moderately (MD), very (VD), and extremely damaged (ED), respectively, in November 2019 (Shamsuzzoha et al. 2021). Within almost 6 months intervals, the same cultivated croplands were damaged again by another tropical cyclone named Amphan in May 2020. The total agricultural lands were damaged due to Amphan not damaged 3.54%, slightly 16.87%, moderately by 33.25%, very 29.01%, and extremely by 17.33% (Arab et al. 2022). Previous studies were based on mono-cyclone hazard-based analytics for Bangladesh to find out an annual

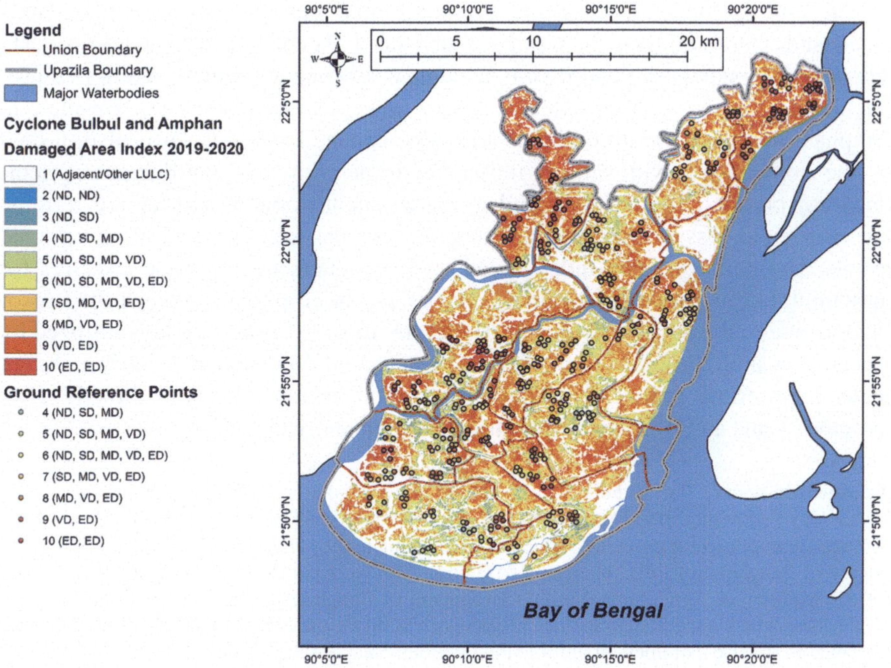


**Fig. 6.7** Validation of DAI using GRPs

response to support the local farmers mainly the rice grower farmers (NAWG 2019; Mishra and Vanganuru 2020; Shamsuzzoha et al. 2021, 2022; Arab et al. 2022; Nasrin et al. 2023). Therefore, this study was focused on the repetition of damaged occurrences on the same agricultural lands that belong to comparable crop calendar or cropping practices. It was clearly identified that in the time of cyclone landfall coastal region of Bangladesh, the late rainy season (October and November) time is more risk prone to formulate damaged incidents in the cultivated agricultural lands. The limitation of the research is that this study was done for the single crop dominant agricultural land use pattern. Other land use and land cover (LULC)'s damaged information was not considered according to the research objective of the study.

## 6.5 Conclusion

Every year, tropical cyclones are coming to make the coastal areas of the Bangladesh delta that affirm the country as the part of the cyclone prone global south hotspot. Tropical cyclones were repeated and always frequent by making landfall over the study country affecting through the extreme climatic events. This study was projected to develop a satellite remote sensing-based damaged area index to determine the yearly classifiers for agricultural lands due to cyclonic events in the coastal region of Bangladesh. In this study, cyclone Bulbul and Amphan impacts on cultivated lands were studied in the Kalapara subdistrict, Patuakhali District, Bangladesh. The Kalapara subdistrict had mainly a single crop conquered by T. *Aman* rice production during the cyclone Bulbul impact time in November 2020. Besides, in Amphan time the agricultural lands area was damaged again that was also flashed by the storm surges with salinity intrusions in the croplands that degrade agricultural crop productivity. The damaged area assessment would be helpful to determine the crop yield loss assessment in future for tropical cyclones and other extreme events. The damaged area index based on DAA intersects could be effective for agricultural crop growers to evaluate the expense of crop yields against the loss of income due to the adverse cyclones. Therefore, this DAI process can be utilized by researchers and agronomists to determine year-round damages, identify risks, to make appropriate policies, and support coastal region's farmers directly after extreme events and from the different governmental organizations.

**Acknowledgments** The authors would like to acknowledge the Bangabandhu Science and Technology Fellowship (BSTF) program, Ministry of Science and Technology (MoST), Government of Bangladesh (GoB) for providing a doctoral scholarship to the University of Tsukuba in Japan, the Survey of Bangladesh (SoB), Ministry of Defense (MoD), Department of Agricultural Extension (DAE), Ministry of Agriculture (MoA), Government of Bangladesh (GoB) for necessary country geodatabase, and United States Geological Survey (USGS) for the Landsat 8 satellite image dataset. The authors are also extremely grateful to the Environmental Systems Research Institute (ESRI) for providing the ArcGIS® 10.8.2 software support to do the research based on satellite remote sensing.

**Author Contributions** M.S. conducted research investigation, methodology, data curation, and analysis. He obtained the results and drafted the manuscript; T.A. supervised the research for the study conceptualization and editing of the final manuscript. All authors have read and agreed to the published version of the manuscript.

**Funding** This research did not receive any external funding.

**Data Availability Statement** The dataset that was generated and analyzed during this study is available from the corresponding author upon reasonable request, but restrictions apply to the data reproducibility for reusing and commercially confident details.

**Conflict of Interest** The authors declare no conflict of interest regarding the research.

## References

Alam KF, Ahamed T (2022) Assessment of land use land cover changes for predicting vulnerable agricultural lands in River basins of Bangladesh using remote sensing and a fuzzy expert system. Remote Sensing 14:5582. MDPI. https://doi.org/10.3390/rs14215582

Alam KF, Ahamed T (2023a) Erosion vulnerable area assessment of Jamuna River system in Bangladesh using a multi-criteria-based geospatial fuzzy expert system and remote sensing. Asia Pac J Reg Sci 7:433. https://doi.org/10.1007/s41685-023-00292-9

Alam KF, Ahamed T (2023b) Climate-adaptive potential crops selection in vulnerable agricultural lands adjacent to the Jamuna River basin of Bangladesh using remote sensing and a fuzzy expert system. Remote Sens 15:2201. https://doi.org/10.3390/rs15082201

Al-Maruf A, Craig Jenkins J, Bernzen A, Braun B (2021) Measuring household resilience to cyclone disasters in coastal Bangladesh. Climate 9:97. https://doi.org/10.3390/cli9060097

Arab ST, Monirul Islam M, Shamsuzzoha M, Alam KF, Muhsin N, Noguchi R, Ahamed T (2022) A review of remote sensing applications in agriculture and forestry to establish big data analytics. In: Remote sensing application, pp 1–24. https://doi.org/10.1007/978-981-19-0213-0_1

Banik T, Thandlam V, De BK, Kundu SS, Gogoi RB, Raju PLN, Guha A (2021) Understanding dynamics of tropical cyclones in the Bay of Bengal using lightning data. Meteorol Atmos Phys 133:1505–1522. Springer. https://doi.org/10.1007/s00703-021-00824-y

Chen X, Avtar R, Umarhadi DA, Louw AS, Shrivastava S, Yunus AP, Khedher KM, Takemi T, Shibata H (2022) Post-typhoon forest damage estimation using multiple vegetation indices and machine learning models. Weather Clim Extremes 38. Elsevier B.V. https://doi.org/10.1016/j.wace.2022.100494

Di L, Yu E, Shrestha R, Lin L (2018) DVDI: a new remotely sensed index for measuring vegetation damage caused by natural disasters. In: IGARSS 2018–2018 IEEE international geoscience and remote sensing symposium, IEEE, pp 9067–9069. https://doi.org/10.1109/IGARSS.2018.8518022

Du X, Chen H, Li Q, Ge X (2023) Urban impact on landfalling tropical cyclone precipitation: a numerical study of Typhoon Rumbia (2018). Adv Atmos Sci 40:988. Science Press. https://doi.org/10.1007/s00376-022-2100-8

Earth Stat (2023) Dataset now available: crop allocation to food, feed, nonfood. In: Harvested area and yield for 4 crops (1995–2005)

Hafsa B, Ahmed R, Kumar R (2021) Spatio-temporal pattern of tropical cyclones: the case of cyclones of The Bay of Bengal. J Geogr Inst Jovan Cvijic SASA 71(71):213–229. Geographical Institute "Jovan Cviji" of the Serbian Academy of Sciences and Arts. https://doi.org/10.2298/IJGI2103213H

Haq MZ, Robbani M, Mohammad Ali M, Mainul Hasan M, Mahmudul Hasan M, Uddin J, Begum M, Teixeira da Silva JA, Pan XY, Karim MR (2012) Damage and management of cyclone Sidr-affected homestead tree plantations: a case study from Patuakhali, Bangladesh. Nat Hazards 64:1305–1322. https://doi.org/10.1007/s11069-012-0299-x

Hoque M, Tasfia S, Ahmed N, Pradhan B (2019) Assessing spatial flood vulnerability at Kalapara Upazila in Bangladesh using an analytic hierarchy process. Sensors 19:1302. https://doi.org/10.3390/s19061302

Hossain MZ, Islam MT, Sakai T, Ishida M (2008) Impact of tropical cyclones on rural infrastructures in Bangladesh. Agric Eng Int CIGR Ejournal. Invited Overview

Htun M, Shamsuzzoha AM, Ahamed T (2023) Rice yield prediction model using normalized vegetation and water indices from Sentinel-2A satellite imagery datasets. Asia Pac J Reg Sci 7:491–519. https://doi.org/10.1007/s41685-023-00299-2

IRRI (2020).Rice profile in Bangladesh

Jenks GF (1967) The data model concept in statistical mapping. Int Yearb Cartogr 7:186–190

Karaer A, Chen M, Gazzea M, Ghorbanzadeh M, Abichou T, Arghandeh R, Ozguven EE (2022) Remote sensing-based comparative damage assessment of historical storms and hurricanes in Northwestern Florida. Int J Disaster Risk Reduct 72:102857. https://doi.org/10.1016/j.ijdrr.2022.102857

Khalil GM (1992) Cyclones and storm surges in Bangladesh: some mitigative measures. Nat Hazards 6:11–24. https://doi.org/10.1007/BF00162096

Khalil MI (2020) Coastal groundwater aquifer characterization from geoelectrical measurements—a case study at Kalapara, Patuakhali, Bangladesh. J Appl Geol 5:1. https://doi.org/10.22146/jag.55009

Kim M, Park MS, Im J, Park S, In Lee M (2019) Machine learning approaches for detecting tropical cyclone formation using satellite data. Remote Sens 11:1195. MDPI AG. https://doi.org/10.3390/rs11101195

Malmgren-Hansen D, Sohnesen T, Fisker P, Baez J (2020) Sentinel-1 change detection analysis for cyclone damage assessment in urban environments. Remote Sens 12:2409. https://doi.org/10.3390/rs12152409

Masutomi Y, Iizumi T, Takahashi K, Yokozawa M (2012) Estimation of the damage area due to tropical cyclones using fragility curves for paddy rice in Japan. Environ Res Lett 7. Institute of Physics Publishing. https://doi.org/10.1088/1748-9326/7/1/014020

Matsuda I (1993) Loss of human lives induced by the cyclone in Bangladesh. GeoJournal 31:319–325. https://doi.org/10.1007/BF00812781

Mishra AK, Vanganuru N (2020) Monitoring a tropical super cyclone Amphan over Bay of Bengal and nearby region in May 2020. Remote Sens Appl Soc Environ 20:100408. https://doi.org/10.1016/j.rsase.2020.100408

Nadiruzzaman M, Wrathall D (2015) Participatory exclusion—cyclone Sidr and its aftermath. Geoforum 64:196–204. Elsevier Ltd. https://doi.org/10.1016/j.geoforum.2015.06.026

Nandi G, Neogy S, Roy AK, Datta D (2020) Immediate disturbances induced by tropical cyclone Fani on the coastal forest landscape of eastern India: a geospatial analysis. Remote Sens Appl Soc Environ 20:100407. https://doi.org/10.1016/j.rsase.2020.100407

Nasrin T, Mohd Ramiz M, Sarif N, Hashim M, Siddiqui MA, Siddiqui L, Mohibul S, Mankotia S (2023) Modeling of impact assessment of super cyclone Amphan with machine learning algorithms in Sundarban Biosphere Reserve, India. Nat Hazards 117:1945–1968. https://doi.org/10.1007/s11069-023-05935-w

NAWG (2019) Bangladesh cyclone Bulbul 2019 joint rapid assessment

NOAA (2023) International-best-track-archive-for-climate-stewardship-ibtracs. National Oceanic and Atmospheric Administration. April 5

Parks RM, Brooke Anderson G, Nethery RC, Navas-Acien A, Dominici F, Kioumourtzoglou MA (2021) Tropical cyclone exposure is associated with increased hospitalization rates in older adults. Nat Commun 12:1545. Nature Research. https://doi.org/10.1038/s41467-021-21777-1

Rahman M, Rejaur HL, Nazrul Islam M (2018) Climate of Bangladesh: temperature and rainfall changes, and impact on agriculture and groundwater—a GIS-based analysis. In: Bangladesh I: climate change impacts, mitigation and adaptation in developing countries, pp 27–65. https://doi.org/10.1007/978-3-319-26357-1_2

Roy C, Rejaur Rahman M, Ghosh MK, Biswas S (2023) Tropical cyclone intensity forecasting in the Bay of Bengal using a biologically inspired computational model. Model Earth Syst Environ 10:523–537. https://doi.org/10.1007/s40808-023-01786-3

Shamsuzzoha M, Ahamed T (2023) Shoreline change assessment in the coastal region of Bangladesh Delta using Tasseled Cap transformation from satellite remote sensing dataset. Remote Sens 15:295. MDPI. https://doi.org/10.3390/rs15020295

Shamsuzzoha M, Al-Maruf A (2012) Post SIDR life strategy: adaptation scenario of settlements of the south. J Inst Bangladesh Stud 19:207–222

Shamsuzzoha M, Noguchi R, Ahamed T (2021) Damaged area assessment of cultivated agricultural lands affected by cyclone bulbul in coastal region of Bangladesh using Landsat 8 OLI and TIRS datasets. Remote Sens Appl Soc Environ 23. Elsevier B.V. https://doi.org/10.1016/j.rsase.2021.100523

Shamsuzzoha M, Noguchi R, Ahamed T (2022) Rice yield loss area assessment from satellite-derived NDVI after extreme climatic events using a fuzzy approach. Agric Inf Res 31:32–46. https://doi.org/10.3173/air.31.32

Talukder MF, Shamsuzzoha M, Hasan I (2018) Damage and agricultural rehabilitation scenario of post cyclone Mahasen in coastal zone of Bangladesh. J Sociol Anthropol 2:36–43. https://doi.org/10.12691/jsa-2-1-7

USGS (2019) Landsat 8 (L8) data users handbook. United States Geological Survey (USGS)

Weng Q, Lu D, Schubring J (2004) Estimation of land surface temperature–vegetation abundance relationship for urban heat Island studies. Remote Sens Environ 89:467–483. https://doi.org/10.1016/j.rse.2003.11.005

Zhang G, Perrie W, Zhang B, Yang J, He Y (2020) Monitoring of tropical cyclone structures in ten years of RADARSAT-2 SAR images. Remote Sens Environ 236:111449. Elsevier Inc. https://doi.org/10.1016/j.rse.2019.111449

Zhou Y, Matyas CJ (2021) Regionalization of precipitation associated with tropical cyclones using spatial metrics and satellite precipitation. GISci Remote Sens 58:542–561. Bellwether Publishing, Ltd. https://doi.org/10.1080/15481603.2021.1908675

# Chapter 7
# Land Suitability Analysis for Potential Vineyards Extension Using a Fuzzy Expert System and Remote Sensing Datasets at a Regional Scale

**Sara Tokhi Arab and Tofael Ahamed**

**Abstract** Grapes are one of the world's most widely distributed crops and are cultivated in more than 100 countries in the global scheme. Due to climate change and improper vine growth, variable selection, production has significantly decreased across countries. Combining multiple criteria such as climatic, topographic, vegetation, and soil properties based on satellite remote sensing datasets can play an important role in finding the potential classified areas for suitable lands to increase grapevine cultivation. While using multicriteria analysis, fuzzy set theory can serve to prevent misclassification of vineyard in the suitability analysis. Therefore, the primary purpose of this study was to develop a land suitability analysis method using a fuzzy expert system at a regional scale. In this regard, fuzzy-based multicriteria decision-making system was used with geographical information system (GIS) and fuzzy membership functions. The criteria maps such as normalized difference vegetation index (NDVI), rainfall, topographic variables (elevation, and slope), soil property related variables, roads, and rivers were collected from Google Earth Engine (GEE), Climate Hazards Group InfraRed Precipitation with Station (CHIRPS), United States Geological Survey (USGS), Food and Agriculture Organization (FAO), Afghanistan road network and United Nations Disaster Assessment (OCHA) websites. The fuzzy membership function was used in the ArcGIS® environment to perform the spatial analysis, and the overlay function was used to generate the final suitability map for Afghanistan considering policy planning. The results indicated that 23% (15,760,144 ha) of the areas were potential and located in the highly suitable region for grape production; however, 11% (7,370,025 ha) of the regions were not suitable for vineyards throughout the country

S. T. Arab
Division of Intelligent Agricultural Machinery Research, Institute of Agricultural Machinery, National Agriculture and Food Research Organization, Tsukuba, Ibaraki, Japan

T. Ahamed (✉)
Institute of Life and Environmental Sciences, University of Tsukuba, Tsukuba, Ibaraki, Japan
e-mail: tofael.ahamed.gp@u.tsukuba.ac.jp



T. Ahamed (ed.), *Remote Sensing Application II*, New Frontiers in Regional Science: Asian Perspectives 77, https://doi.org/10.1007/978-981-97-1188-8_7

of Afghanistan. In the present study, it was observed that most of the vineyards were in highly suitable areas (90%, 80,466 ha), while 0.01% (5 ha) of the vineyards were in less suitable areas. The present analysis demonstrated that the significant extension of grape vines can be possible in highly suitable areas. The results of this research can support decision-makers, farm managers, and land developers to find more prospective acreage for expanding vineyards in Afghanistan.

**Keywords** Land suitability analysis · Vineyards · Regional scale and fuzzy

## 7.1 Introduction

Land suitability analysis is an essential tool for preserving the long-term viability of agricultural fields. It is also an important management strategy to identify the ideal farming locations for various crops and fruits. Thus, the primary objective of most nations is to produce sustainable products for an increasing global population. In this regard, assessing the potential of land for grapevine extension is very important to increase grape production on a regional scale. The assessment can also support farmers in increasing their income and ensuring their livelihood. In recent land suitability analysis, researchers considered geographical information system (GIS)-based multicriteria, satellite remote sensing vegetation indices, and UAV (unmanned aerial vehicle) images to increase the resolutions for higher accuracy in interpretations (Worqlul et al. 2019; Gilliams et al. 2005; Grassano et al. 2011).

Land suitability analysis using remote sensing data at the regional scale may increase the complexity of data collection and image processing due to the high data volume and diversity that it generates. With regard to incorporating big data analytics with cloud computing, large-scale scientific applications have demonstrated the advantages of high computational and storage constraints that are simple to implement (Wang et al. 2018). The Google Earth engine (GEE) makes this task possible for everyone to access and manipulate these data without cost and large computational facilities (Xie et al. 2019). In addition, rainfall datasets are also difficult to obtain on a regional scale. Across most of the globe, there is neither a reliable system of weather data nor a homogeneous distribution of those that exist. Consequently, it is essential to study other data sources for rainfall information, such as satellite-based near real-time rainfall information and radar data. In addition, there are products with low latency and extensive records, such as the Climate Hazards Group Infrared Precipitation with Stations (CHIRPS) dataset. This package contains integrated models of terrain-induced precipitation, precipitation estimates from measurement satellites that cover the majority of the globe and have low latency and polarization, and achieve precipitation estimates from in situ stations. The CHIRPS dataset contains a lengthy recording period (1981 to the present) with a fine spatial resolution of 0.05° (Funk et al. 2015; Ghozat et al. 2022). Vineyard management depends not only on precipitation but also on vegetation and soil properties. which makes it important to find more agroecology-friendly vineyard sites.

In agroecological zoning, satellite remote sensing-based topographical, vegetation, climate, and soil features can be included for land suitability analysis (Akpoti et al. 2019). The Food and Agriculture Organization (FAO) categorized land suitability into four categories: highly suitable ($S_1$), moderately suitable ($S_2$), marginally suitable ($S_3$), and not suitable ($N_1$ and $N_2$) (FAO 1976) (Table 7.1). In order to establish and identify land suitability categories, a number of elements that influence land quality have been identified. Since several variables were considered, the method is known as a multicriteria decision-making process (Romano et al. 2015). To develop a suitability map in different classifications and overlay methods, AHP (analytical hierarchy process)-weighted overlay, fuzzy logic, and weighted linear combination (WLC) are commonly used. Furthermore, a combination of two methods, fuzzy and AHP, is also used in the land assessment process (Zabel et al. 2014; Badr et al. 2018; Purnamasari et al. 2019).

The AHP is the simplest suitability overlay method for weight generation. However, there is evidence of bias in the conclusions drawn by experts (Saaty 1980; Pramanik 2016) (Fig. 7.1). WLC is another method that combines the suitableness ratings of multiple evaluation elements into a single composite score using the weighted average operation. Due to its simplicity, adaptability, and effectiveness, WLC has become one of the most extensively used land suitability analysis techniques (Deng et al. 2014; Radočaj et al. 2020) (Fig. 7.2).

Additionally, the Fuzzy-AHP method is used to generate suitability maps. In this method, classification was performed using the fuzzy membership function and AHP-determined weights for each criterion, and the suitability map was created using the overlaying technique (Elaalem et al. 2011; Perrot et al. 2015). Fuzzy suitability analysis is an efficient approach with greater accuracy for evaluating land suitability. To standardize the criteria, use fuzzy set membership (Romano et al. 2015). Fuzzy logic produces more realistic results than other approaches, such as Boolean classification, since it can categorize variables in a continuous way and address all uncertainties in a more accurate way (Badr et al. 2018; Perrot et al. 2015).

All the overlay methods have been used for different crop suitability analyses, such as rice, maize, cassava, grapes, and other crops (Tatem 2005; Roy and Saha

**Table 7.1** Land suitability classification based on FAO

| Suitability classes | Description |
|---|---|
| $S_1$ (highly suitable) | These types of land have no significant limitations for production |
| $S_2$ (moderately suitable) | These types of land have moderate limitations for production. It will reduce productivity by increasing the input by a certain amount, which can be changed to $S_1$ |
| $S_3$ (marginally suitable) | These types of lands have marginal limitations. These limitations reduce productivity because by increasing the input, the expenditure on land will increase |
| $N_1$ (not suitable currently) | These types of lands have significant constraints that may be overcome in time but cannot be solved with present knowledge at an affordable cost |
| $N_2$ (not suitable permanently) | These types of lands have severe limitations, and with the use of technique and technology, we cannot make them suitable |

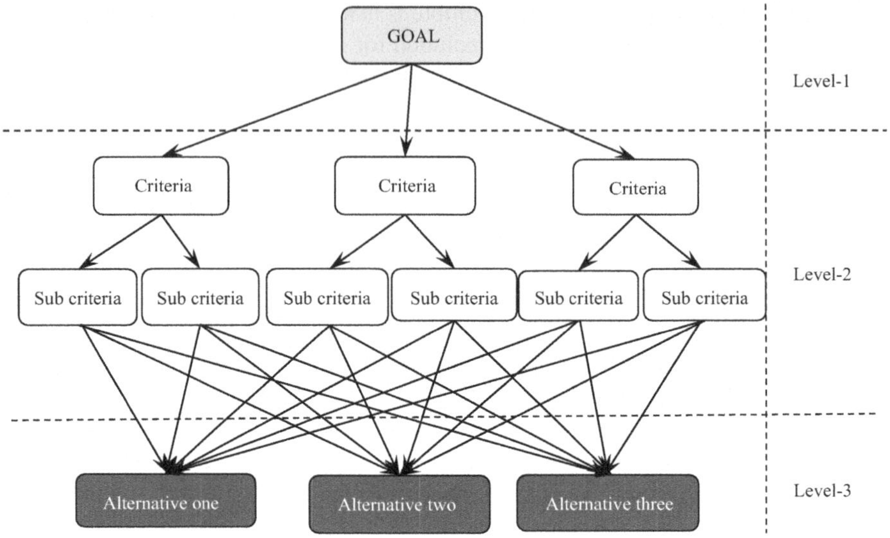


**Fig. 7.1** AHP's multicriteria hierarchy structure

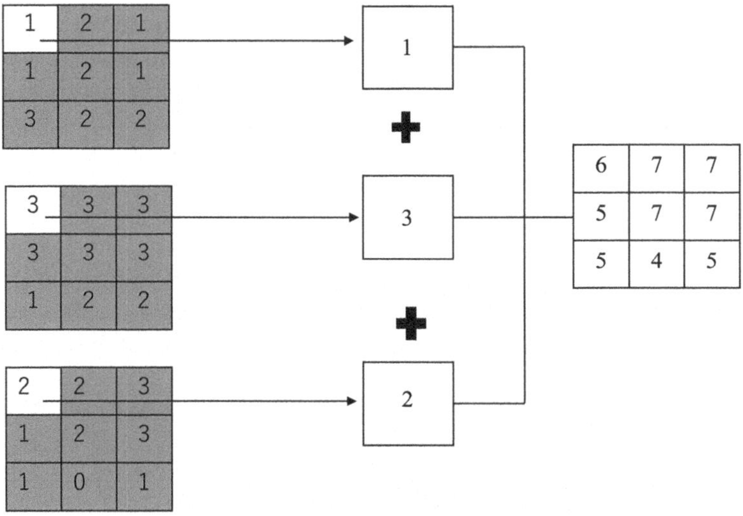


**Fig. 7.2** Overlay analysis using weighted linear combination (WLC)

2018; Tashayo et al. 2020). There is a significant advantage to employing CHIRPS rainfall for land suitability analysis, which can overcome the limitations of stations and radar data. Most studies are used station- and Japan Aerospace Exploration Agency (JAXA) rainfall-based (Habibie et al. 2020; Moisa et al. 2022). In addition, the GEE is also an effective platform for climate-weather and geophysical datasets. Moreover, fuzzy sets reduce subjectivity and increase effectiveness in land suitability analysis (Guo et al. 2022). There is very limited research in the world on land suitability analysis for table grapes at the regional scale where climatic variability is severe and environmental factors are uncertain. Former suitability assessment mostly focused on the relationship between vintage quality and other environmental variables for wine grapes (Perrot et al. 2015; Grelier et al. 2007; Coulon-Leroy et al. 2014). Coulon-Leroy et al. (2014) developed a vine vigor model using a fuzzy set based on data related to soil, rootstock, and inter-row management strategies and spatial suitability assessment was done for vineyard site selection based on biophysical models and fuzzy logic for spatial land assessment in micro-scale (Coulon-Leroy et al. 2014; Badr et al. 2018).

Accordingly, in Afghanistan, land suitability analysis was performed based on the AHP method for saffron and grapes on a micro-scale (Arab et al. 2022; Wali et al. 2016). This study focused on all of Afghanistan on a regional scale and to reduce the subjectivity of suitability analysis, the fuzzy method was performed. The fuzzy set addresses such uncertainty of climatic variables and biophysical and soil properties with justification on a continuous scale precisely. Thus, for the extension of vineyards, the government needs a comprehensive land assessment system to determine the potential region for vineyards. Aside from ground assessment, suitability analysis for finding the potential vineyard area is one of the best options. Therefore, the primary objective of this study is to develop a land suitability analysis method for potential vineyard extension based on biophysical, infrastructural, and climate variable using a fuzzy-based expert system and satellite remote sensing at the regional scale.

## 7.2 Materials and Methods

### *7.2.1 Study Area*

Afghanistan is a landlocked country that is located between 33.9391° to the north and 67.7100° to the east. Afghanistan's scenery is dominated by high mountains and arid deserts. The craggy mountain summits are dangerous and snow-covered for most of the year. The country consists of 34 provinces with a total population of 40.2 million (Fig. 7.3) (ACSO 2020). Most of the population lives in fertile valleys. The summers are very hot and dry, and the winters are very cold, especially at high elevations. The area is divided into three parts: the eastern, central, and western regions. The eastern regions are Badakhshan, Takhar, Kunduz, Baghlan, Panjsher, Nuristan, Parwan, Kapisa, Kunar, Wardak, Kabul, Laghman, Nangarhar, Ghazni, Logar, Paktya, Khost, and Paktika Provinces. The central regions are the Balkh, Jawzjan, Sari Pul, Samangan, Bamyan, Ghor, Daykundi, Uruzgan, Zabul, and Kandahar Provinces, and the western regions are the Faryab, Badghis, Hirat, Farah, Nimroz, and Hilmand Provinces. The country's climate is favorable for different vegetables and fruits. Among all fresh fruits, grapes are the most economically

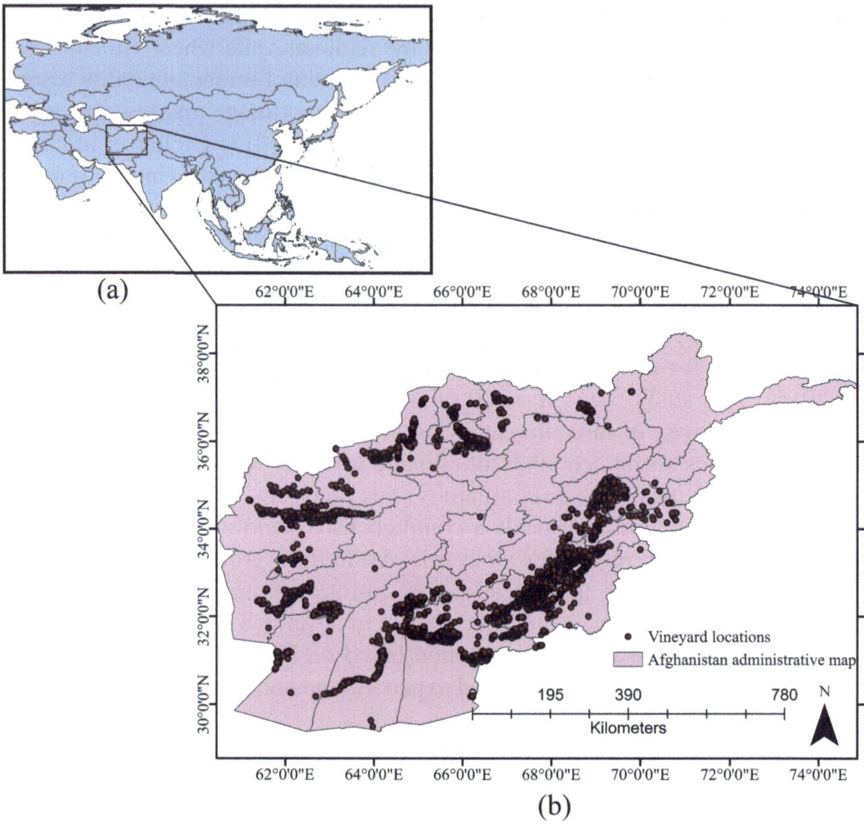


**Fig. 7.3** Geographical extent of study area (**a**) Afghanistan on the world map (**b**) Afghanistan administrative map with vineyard location

important horticultural crop, with a total production of 993,382 tons in 2020 in the country (ACSO 2020). The total vineyard area is 87,593 ha, which makes up 0.13% of Afghanistan's land (FAO 2016). A total of 46.97% of Afghanistan consists of rangeland and 34.45% of non-arable land. This means that the area for agricultural activities is very limited.

### *7.2.2 Vineyard Suitability Framework*

This research was conducted in three steps. In the first step, infrastructure, soil, vegetation, and climate variables were organized for vineyard suitability analysis, as portrayed in the research flowchart (Fig. 7.4 and Table 7.2). Variable maps, such as (a) elevation, (b) slope, (c) river, (d) road, (e) soil depth, (f) soil pH, (g) soil texture, (h) soil salinity, (i) NDVI, (j) LULC, and (k) rainfall, were developed (Fig. 7.5).

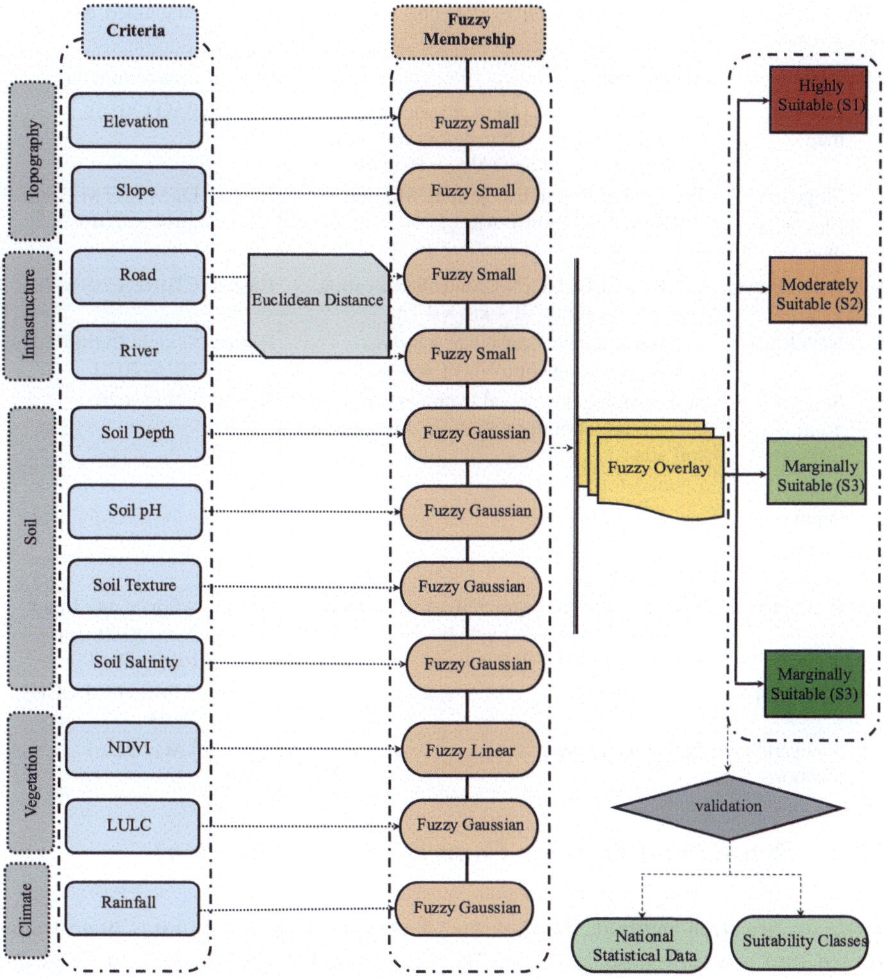


**Fig. 7.4** Research framework for vineyard suitability analysis based on a fuzzy algorithm

In this research, different sources were used to obtain the datasets, such as FAO, readily available sources, and GEE for downloading big datasets of Landsat 8 Operational Land Imager (OLI) and rainfall. All primary suitability variables were resampled to the same resolution (30 m) as Landsat 8 OLI. Second, the fuzzy membership function was applied to each variable based on previous literature (Fig. 7.6 and Table 7.3) (Zabel et al. 2014; Badr et al. 2018; Salata et al. 2022; Stanchi et al. 2013; Keshavarzi et al. 2010). Third, in the ArcGIS® environment, fuzzy gamma was applied to overlay all the variables. The suitability classes were identified (Fig. 7.7) (Bellman and Zadeh 1970). Finally, validation was performed with ground reference dataset.

**Table 7.2** Variable selections and their explanation for specifications that included data types and sources

| No | Data | Explanation | Types | Source of data |
|---|---|---|---|---|
| 1 | Land use map | Derived from Spot (10 m color), Google Earth (2.5 m 1 m, /0.6 m color) and Arial Photographs (1 m color/0.5 m B and W) | Raster | FAO (2016) |
| 2 | Slope map | The Shuttle Radar Topography Mission (SRTM), resolution 1-ARC | Raster | DEM SRTM USGS 2014 & 2015 |
| 3 | Elevation map | | | |
| 4 | Rainfall map | CHIRPS PERSIANN-Cloud classification system, resolution of 4 km × 4 km | Raster | CHIRPS 2016–2020 |
| 5 | NDVI map | Landsat 8 (Collection 1 Tire 1 8-days composite) composite NDVI scenes | Raster | Google Earth Engine 2016–2020 |
| 6 | Soil pH | Soil mapping developed from field soil survey and laboratory analysis. Afghanistan soil atlas | Raster | FAO (2020) |
| 7 | Topsoil texture | | | |
| 8 | Topsoil depth | | | |
| 9 | Topsoil salinity | | | |
| 10 | Road map | This dataset is an extraction of roads from OpenStreetMap data made by WFP following UNSDI-T standards | Vector | Afghanistan Road Network (main roads) 2018 |
| 11 | River map | Scale 1:50,000 | Vector | AIMS OSM OCHA 2019 |
| 12 | Vineyard's locations | Polygon and point | Vector | FAO (2016) |

### *7.2.3 Dataset and Variable Conversion (Fuzzification)*

In the fuzzification method, the datasets of various ranges and unites were transformed into a common scale (0–1). The fuzzy small, large, linear, and Gaussian were assigned to biophysical, climatic, infrastructure, topographic, and soil-related variables in this study, respectively. The fuzzy small transformation function was used when the small values of the input raster were more likely to be a member of the set (Table 7.3). The defined midpoint identifies the crossover point (assigned a membership of 0.5), with values greater than the midpoint having a lower chance of membership and values less than the midpoint having a higher chance of membership (Eq. (7.1). Fuzzy linear indicates that the linear relationship in datasets and minimum values were assigned to 0, and maximum values were assigned to 1 (Eq. 7.2). The fuzzy Gaussian demonstrated the normal distribution of datasets. The midpoint was assigned 1, and the remaining datasets moved in positive and negative directions. The input value membership was decreased when data moved from the midpoint (Eq. 7.3) (Purnamasari et al. 2019). All fuzzy membership functions have a midpoint ($f_2$) and a spread ($f_1$) (Purnamasari et al. 2019). Then each variable was classified according to the number and four categories, 1–0.85, 0.85–0.5, 0.5–0.25,

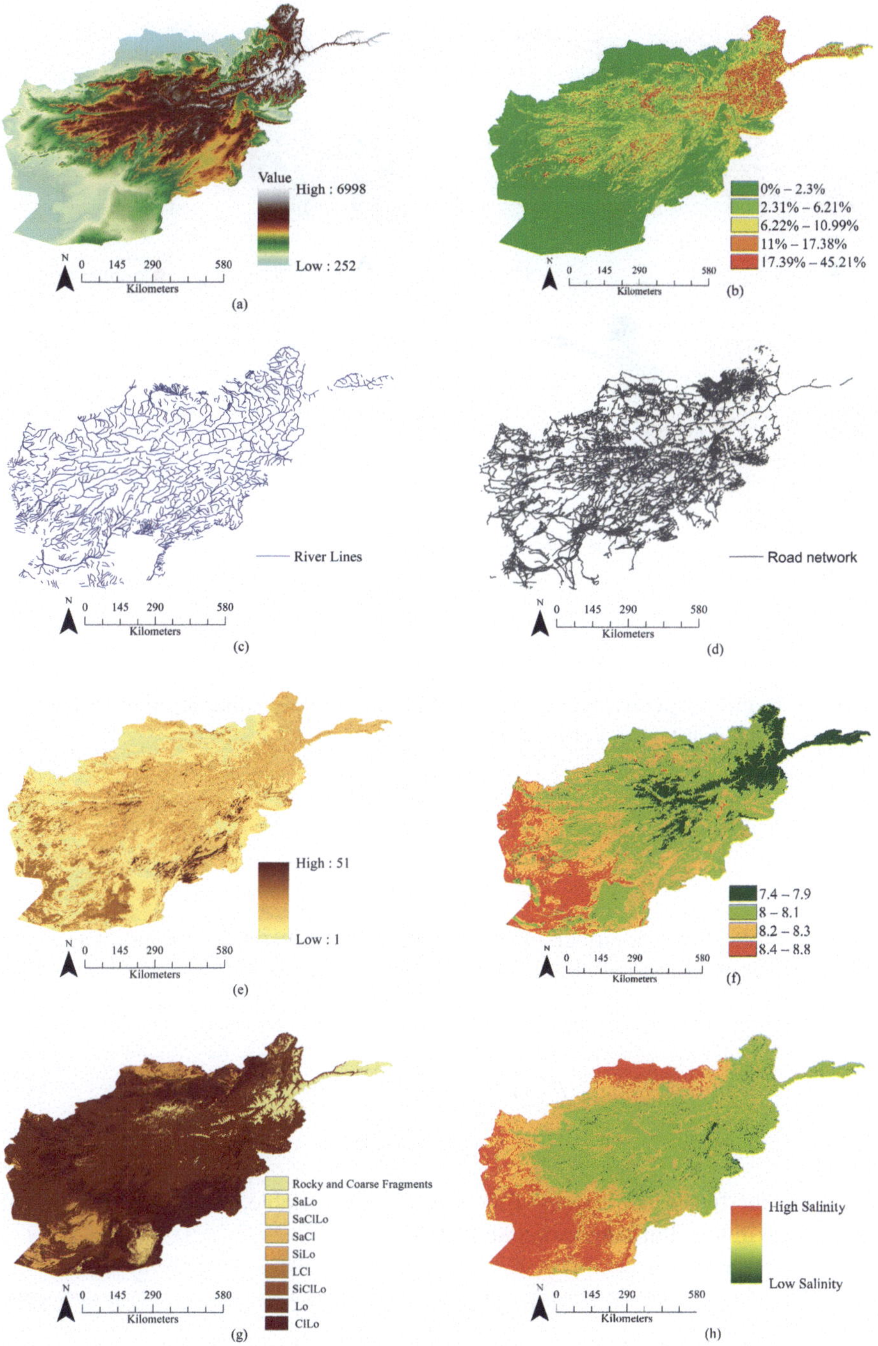


**Fig. 7.5** Multicriteria maps for land suitability analysis: (**a**) elevation (m), (**b**) slope (%), (**c**) river, (**d**) road, (**e**) soil depth, (**f**) soil pH, (**g**) soil texture, (**h**) soil salinity, (**i**) normalized difference vegetation index (NDVI), (**j**) land use/land cover (LULC), and (**k**) rainfall (mm/h)

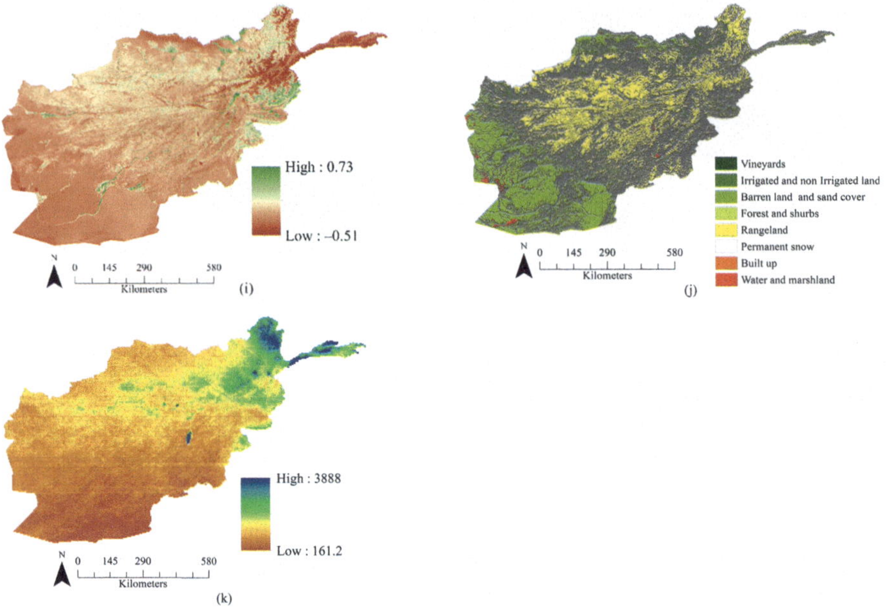


**Fig. 7.5** (continued)

and 0.25–0 and, respectively, highly suitable, moderately suitable, marginally suitable, and not suitable areas (Fig. 7.7).

$$\mu_{(x)} = \frac{1}{1 + \left(\frac{x}{f_2}\right)^{f_1}} \tag{7.1}$$

where $\mu$ is the fuzzy membership function, $x$ is the variable, midpoint ($f_2$), and spread ($f_1$) (Purnamasari et al. 2019).

$$\mu(x) = \begin{cases} 0\,\mathrm{x} \le \mathrm{a} \\ \frac{x-a}{b-a}\,\mathrm{a} < \mathrm{x} < \mathrm{b} \\ 1\,\mathrm{x} \ge \mathrm{b} \end{cases} \tag{7.2}$$

where $\mu$ is the fuzzy membership function, $x$ is the variable, $a$ is the minimum, and $b$ is the maximum (Purnamasari et al. 2019).

$$\mu(x) = \mathrm{e}^{-\left(f_1 \times (x - f_2)^2\right)} \tag{7.3}$$

where $\mu$ is the fuzzy membership function, $x$ is the variable midpoint ($f_2$) and spread ($f_1$) (Purnamasari et al. 2019).

**Table 7.3** Fuzzy suitability membership function for vineyard

| No. | Variable | Fuzzy membership function: Mid-point | Spread | Equation | Fuzzy membership type |
|---|---|---|---|---|---|
| 1 | Elevation | 2500 m | 5 | $\mu_{(x)} = \frac{1}{1+\left(\frac{x}{2500}\right)^5}$ | Small |
| 2 | Slope | 15.6 | 5 | $\mu_{(x)} = \frac{1}{1+\left(\frac{x}{15.6}\right)^5}$ | Small |
| 3 | Road | 1000 | 3 | $\mu_{(x)} = \frac{1}{1+\left(\frac{x}{1000}\right)^3}$ | Small |
| 4 | River | 1000 | 5 | $\mu_{(x)} = \frac{1}{1+\left(\frac{x}{1000}\right)^5}$ | Small |
| 5 | Soil depth | 1 | 0.1 | $\mu(x) = e^{-\left(0.1\times(x-1)^2\right)}$ | Gaussian |
| 6 | Soil pH | 8.13 | 0.1 | $\mu(x) = e^{-\left(0.1\times(x-8.13)^2\right)}$ | Gaussian |
| 7 | Soil texture | 4.5 | 0.1 | $\mu(x) = e^{-\left(0.1\times(x-4.5)^2\right)}$ | Gaussian |
| 8 | Soil salinity | 3.5 | 0.1' | $\mu(x) = e^{-\left(0.1\times(x-3.5)^2\right)}$ | Gaussian |
| 9 | LULC | 0.5 | 1 | $\mu(x) = e^{-\left(1\times(x-0.5)^2\right)}$ | Gaussian |
| 10 | Rainfall | 500 m | 0.1 | $\mu(x) = e^{-\left(0.1\times(x-500)^2\right)}$ | Gaussian |
| Variable | | Minimum | Maximum | | |
| 11 | NDVI | 0.513 | 0.716 | $\mu(x) = \begin{cases} 0 & x \le 0.513 \\ \frac{x-0.513}{0.716-0.513} & 0.513 < x < 0.716 \\ 1 & x \ge 0.716 \end{cases}$ | Linear |

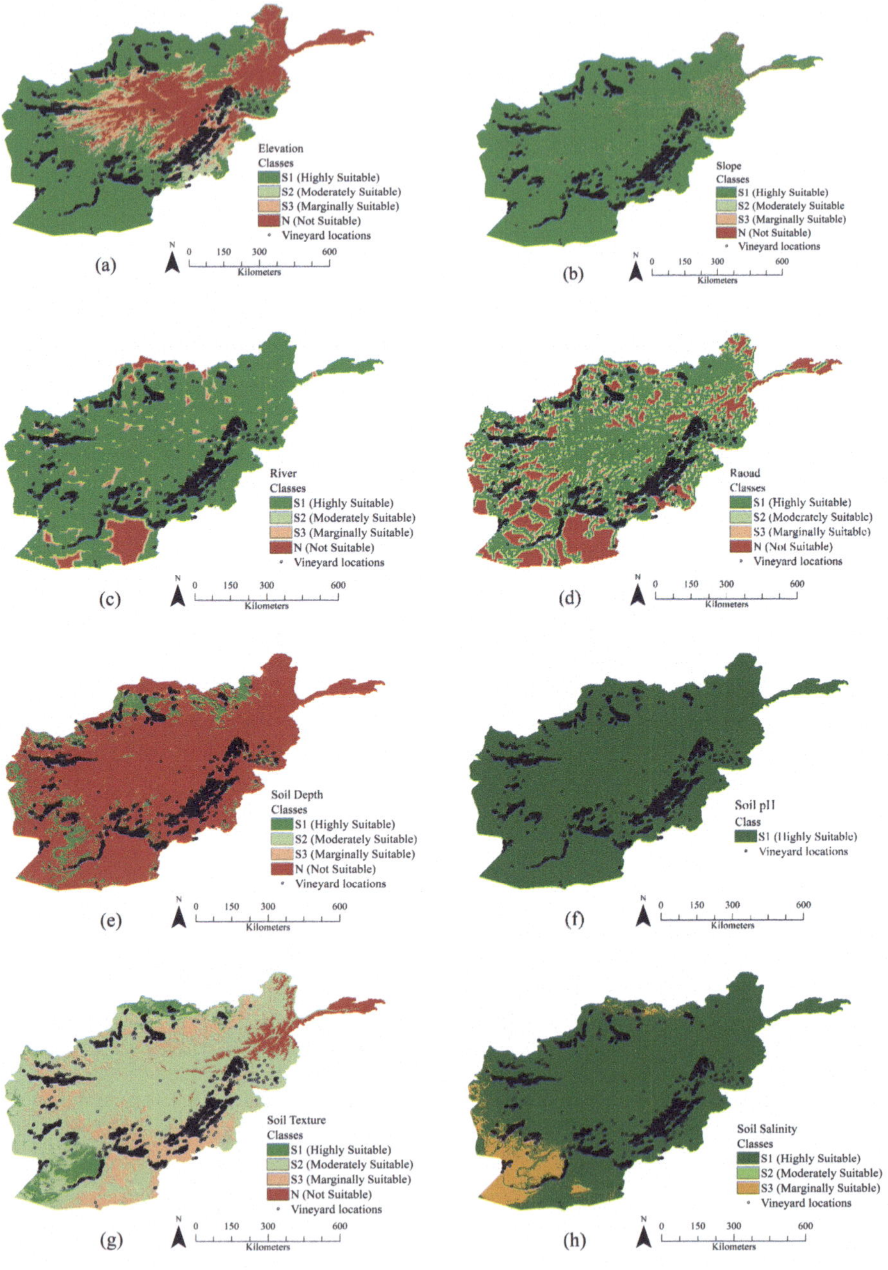


**Fig. 7.6** Reclassification of the criterion used for suitability analysis (**a**–**k**)

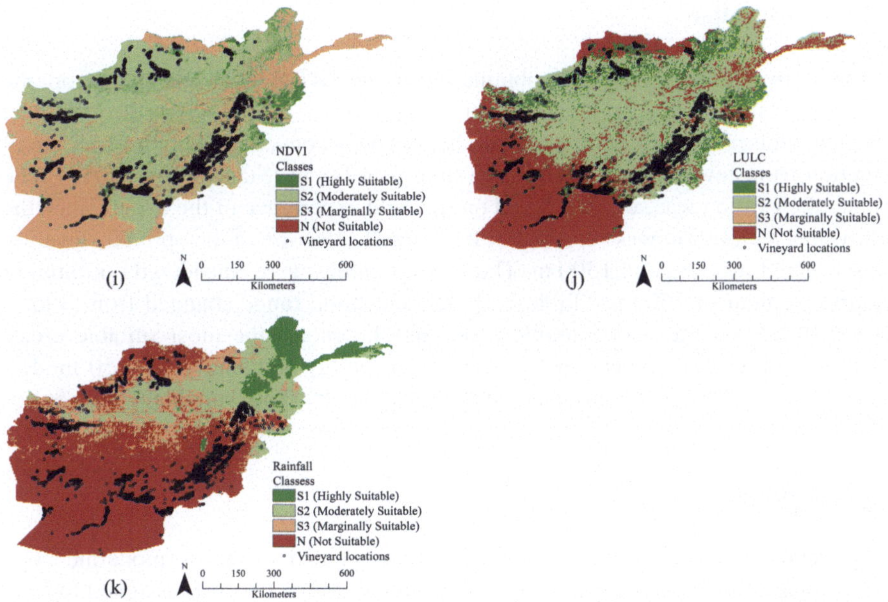


**Fig. 7.6** (continued)

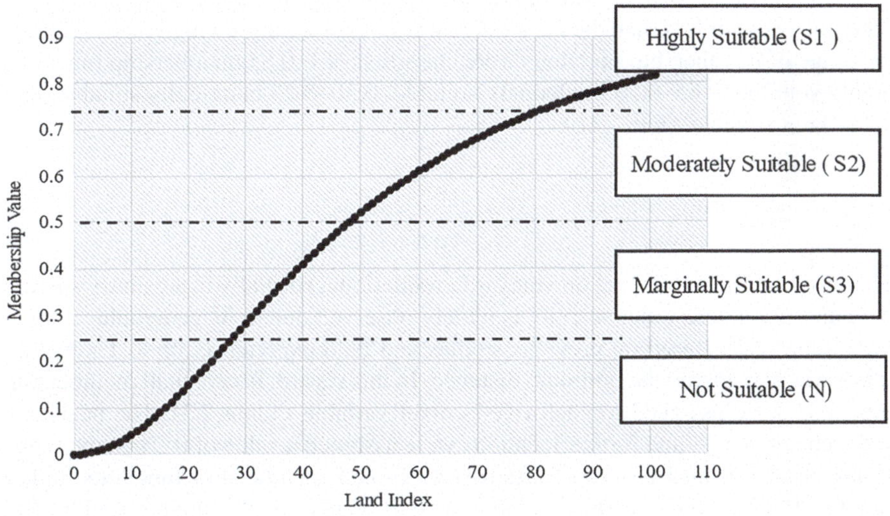


**Fig. 7.7** Selection of suitability classes for vineyard location using a fuzzy membership function and land index

#### 7.2.3.1 Elevation

In this study, elevation data were obtained from the SRTM DEM and taken from the USGS website (Table 7.2). This criterion was important because of its direct impact on grapevine phenology. The elevation dataset was converted using a fuzzy small function from low, ranging from 0 to a high of 6998 m (Stanchi et al. 2013). The range of elevation was selected based on the field experience of the expert since the study area's elevation was very high. Fuzzy small selection of the optimal elevation was selected in a range of 1500 m. The smaller values were considered optimum in an arrangement of 1500 m (Table 7.3). The elevation range changed from 0 to 1, where 0 denotes the least suitable areas and 1 denotes the most suitable areas. Therefore, 1–0.85 membership functions were highly suitable, 0.85–0.50 moderately suitable, 0.50–0.25 marginally suitable, and 0.25–0 not suitable (Fig. 7.6a).

#### 7.2.3.2 Slope

Slope relates to the vineyard's degree of inclination, and a slight-to-moderate slope can be favorable for grape production. The Gaussian function was assigned to each pixel of the slope (Table 7.3). In the study area, a 5–15% slope was considered the optimal slope based on previous studies [Arab et al. 2022; Badr et al. 2018; Stanchi et al. 2013]. In this process, the optimum elevation was assigned at the peak of the function. More or less than the ideal range was not good for growing grapes. Finally, the range of membership functions were classified as 1–0.85 membership functions highly suitable, 0.85–0.50 moderately suitable, 0.50–0.25 marginally suitable, and 0.25–0 not suitable. (Fig. 7.6b).

#### 7.2.3.3 River

There is a significant impact on vineyard production, especially in a country such as Afghanistan, where water scarcity is widely observed due to its geographical location (Table 7.2). Based on previous studies and field experience, up to 1 km from rivers was considered the optimum distance. In this regard, fuzzy small membership functions were assigned to each river's pixel variable (Table 7.3). The near area value changed to 1, and further locations were assigned a value of 0. Then the range of membership functions was classified as 1–0.85 membership functions highly suitable, 0.85–0.50 moderately suitable, 0.50–0.25 marginally suitable, and 0.25–0 not suitable (Fig. 7.6c) (Purnamasari et al. 2019).

#### 7.2.3.4 Road

Roads are an important parameter in facilitating grower's access to input and output markets. Based on previous studies and a field expert's experience, a distance of up to 1000 m from the road was selected as optimal for vineyard suitability. The fuzzy small membership function was assigned to roads (Table 7.3). The closer the road was assigned 1, and the farther distance was assigned 0. Then the range of membership functions was classified as 1–0.85 membership functions highly suitable, 0.85–0.50 moderately suitable, 0.50–0.25 marginally suitable, and 0.25–0 not suitable (Fig. 7.6d) (Purnamasari et al. 2019; Arab et al. 2022).

#### 7.2.3.5 Soil Datasets

Several important soil variables were chosen, such as soil pH, soil depth, soil texture, and soil salinity. All these variables were obtained from the FAO soil database (Table 7.2). Fuzzy Gaussian membership was assigned for all soil components. The soil pH range in the study area was 7.4–8.8. The optimum range was 6.5–8 (USAID 2016). Soil texture was categorized for study areas into nine classes based on FAO classification. Based on reference, sandy loam, loam, and coarse sandy loam were selected with higher scores because clay or silt soils have less water-holding capacity in the root zone of grapevines (FAO 2020). Poor irrigation and drainage are the primary causes of soil salinity in the study area because it is mostly prone to drought (Goes et al. 2016). The saline category is assigned a value of 0, and the less saline category is assigned a value of 1.2 ECe dS/m (De Pascale and Barbieri 1995). The fuzzy Gaussian was used to assign all soil parameters (Table 7.3). The optimal level is considered the peak of distribution, which is less than or higher than the range considered 0 (Badr et al. 2018; Park et al. 2011; Rameshkumar et al. 2006; Goldammer 2018). Therefore, 1–0.85 membership functions highly suitable, 0.85–0.50 moderately suitable, 0.50–0.25 marginally suitable, and 0.25–0 not suitable were classified (Fig. 7.6e–h).

#### 7.2.3.6 Normalized Difference Vegetation Index (NDVI)

A Google Earth engine environment was used to calculate composite NDVI scenes from Landsat 8 OLI (TIRE 1, 8-days composite NDVI) from April to October for 5 years (2016–2020) (Table 7.2). Finally, the average of all NDVIs from 2016 to 2020 was considered for the suitability analysis. Further analysis was carried out with ArcGIS® 10.8. The fuzzy linear membership function was assigned (Table 7.3). The higher NDVI is given a value of 1, and the lower NDVI is given a value of 0 (Tiwari and Mishra 2022). The membership functions 1–0.85 highly suitable, 0.85–0.50 moderately suitable, 0.50–0.25 marginally suitable, and 0.25–0 not suitable were classified (Fig. 7.6i).

#### 7.2.3.7 Land Use Land Cover (LULC)

The LULC datasets were obtained from FAO (Table 7.2). For instance, LULC consisted of different classes, such as vineyards, irrigated and non-irrigated agricultural land, forest and shrubs, rangeland, barren land and sand cover, permanent snow, build up, water, and marshland. Most of the Afghanistan's land is covered in sand and rock, with only 12% suitable for agricultural activities. All the LULC categories were reclassified to a new range of values. After the normalization, fuzzy Gaussian membership was given to LULC. The maximum vineyard and agricultural land membership is 1 (Table 7.3). However, the minimum membership function for buildings, roads, and water bodies is 0 (Rameshkumar et al. 2006). The membership functions of 1–0.85 as highly suitable, 0.85–0.50 moderately suitable, 0.50–0.25 marginally suitable, and 0.25–0 not suitable were classified (Fig. 7.6j).

#### 7.2.3.8 Rainfall

The rainfall data were collected yearly from the website of CHIRPS. Following that, the years 2016–2020 were used in this calculation (Table 7.2). Based on previous studies, the optimum rainfall for grapevines is approximately 500 mm, and the reference fuzzy Gaussian was used (USAID 2016; Rameshkumar et al. 2006) (Table 7.3). The optimum rainfall was assigned to 1, which was less than or greater than that assigned to 0. The membership functions of 1–0.85 highly suitable, 0.85–0.50 moderately suitable, 0.50–0.25 marginally suitable, and 0.25–0 not suitable were classified (Fig. 7.6k).

### 7.2.4 *Fuzzy Overlay*

The fuzzy overlay was used for multicriteria to indicate the likelihood of a feature belonging to various sets. Fuzzy gamma was used to develop the final suitability map. Fuzzy gamma is made of fuzzy product and fuzzy sum. The fuzzy product was calculated by multiplying each cell's fuzzy values by all the input criteria, and the fuzzy sum was calculated by adding the fuzzy values of each set to the cell location. The fuzzy gamma developed the links between the numerous input variables rather than merely returning the value of a single membership set, according to fuzzy OR and fuzzy AND (Eq. 7.4). The final suitability map was classified based on the land suitability index and membership value (Fig. 7.5).

$$\mu(x) = (\text{fuzzy sum})^{\gamma} \times (\text{fuzzy product})^{1-\gamma} \quad (7.4)$$

where $\mu(x)$ combination is the calculated fuzzy membership function, $\gamma$ is a parameter chosen in the range (0, 1), and the fuzzy algebraic sum and fuzzy algebraic product are calculated using the Eq. (7.5).

$$\mu(x) = \left[\prod_{i=1}^{n} \mu i\right]^{\gamma} \times \left[1 - \prod_{i=1}^{n} (1 - \mu i)\right]^{1-\gamma} \quad (7.5)$$

where $\mu i$ is the fuzzy membership function for the '$i$' th map, and $i$ = 1, 2, 3,.....$n$ maps are to be combined and $\gamma$ is the user input variable selected in the range 0–1. When $\gamma = 0$, the combination equals the fuzzy algebraic product and when $\gamma = 1$, the combination equals the fuzzy algebraic sum (Coulon-Leroy et al. 2014).

The suitable vineyards were classified based on the land index. The current and permanently unsuitable areas were considered in a similar class to not suitable, highly, moderately, and marginal areas. In this fuzzy method, a land index was calculated and converted from a 0–1 fuzzy value to be multiplied by 100. The suitability classes were determined by the value of the land index: S1 highly suitable land is 75–100, moderately suitable land is 75–50, S3 marginally suitable land is 50–25, and N not suitable land is 25–0 (USAID 2016) (Fig. 7.7).

### *7.2.5 Validation of Suitability Map with Ground Reference Data*

The total yield of 2020 for each of the province was collected from the statistical book of the Islamic Republic of Afghanistan (Fig. 7.8) (ACSO 2020). The average yield in each province was calculated from the total yield divided by the total vineyard area. Furthermore, the model validation was performed by evaluating and testing the proxy of ground data. In the context of this study, validation was performed using linear regression and polynomial regression function between the average vineyard area from the land suitability index and the average yield of grapes in each province level.

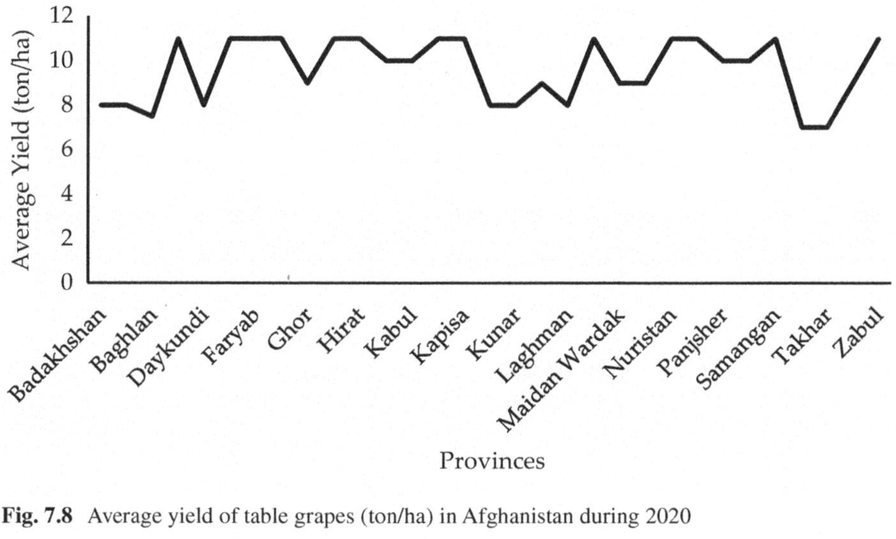


**Fig. 7.8** Average yield of table grapes (ton/ha) in Afghanistan during 2020

## 7.3 Results

### 7.3.1 *Fuzzy Overlay Analysis*

In this research, 11 fuzzy layers were used for vineyard suitability analysis (Fig. 7.7). The fuzzy suitability results indicated that in Afghanistan overall, 23% (15,760,144 ha) of lands were highly, 44% (30,307,470 ha) were moderately, 22% (15,403,607 ha) were marginally, and 11% (7,370,025 ha) were not suitable for grape production in Afghanistan. To confirm the present vineyard locations, the results indicated that 90.3% (80,466 ha) of the vineyards were in the highly, 7.3% (6533 ha) moderately, 2.4% (2124 ha) marginally, and 0.01% (5 ha) unsuitable areas in Afghanistan (Table 7.4 and Table 7.5) (Fig. 7.9).

Moreover, based on fuzzy suitability analysis (gamma overlay), suitability classes were defined in present practice and potential categories. The results indicated that in most provinces, the potential for extension is very high, such as Badghis's present practice of 81 ha and a potential of 11,442,659 ha, Balkh's present practice of 650 ha and a potential of 802,525 ha, Faryab's present practice of 7863 ha and a potential of 1,065,155 ha, Herat's present practice of 7499 ha and a potential of 1,503,025 ha, Khandahar's present practice of 18,784 ha and a potential of 615,253 ha, Samangan's present practice of 628 ha and a potential of 465,673 ha, Sar-e-pul's present practice of 8210 ha and a potential of 562,666 ha, and Takhar's present practice of 107 ha and a potential of 836,973 ha (Table 7.6).

**Table 7.4** Potential areas belonging to the suitable classes based on fuzzy algorithms for Afghanistan

| Suitability classes | Pixels | Vineyard area (ha) | Vineyard area (%) |
|---|---|---|---|
| S1 | 892,282 | 80,466 | 90.3 |
| S2 | 72,447 | 6533 | 7.3 |
| S3 | 23,553 | 2124 | 2.4 |
| N | 57 | 5 | 0.013 |
| Total | | | 100 |

**Table 7.5** Vineyard locations at different suitability classes distributed among the provinces of Afghanistan

| Suitability classes | Pixels | Area (ha) | Area (%) |
|---|---|---|---|
| S1 | 174,763,186 | 15,760,144 | 23 |
| S2 | 336,077,507 | 30,307,470 | 44 |
| S3 | 170,809,566 | 15,403,607 | 22 |
| N | 81,725,718 | 7,370,025 | 11 |
| Total | | | 100 |

**Table 7.6** Vineyard locations and potential for vineyard extension at different suitability classes distributed among the provinces of Afghanistan

| No | Provinces | Categories | S1 (ha) | S2 (ha) | S3 (ha) | N (ha) |
|---|---|---|---|---|---|---|
| 1 | Badghis | Present practice | 81 | 0 | 0 | 0 |
| | | Potential | 1,142,659 | 810,647 | 286,037 | 1484 |
| 2 | Balkh | Present practice | 650 | 6 | 3 | 0 |
| | | Potential | 802,525 | 444,395 | 407,443 | 194,948 |
| 3 | Farah | Present practice | 717 | 170 | 207 | 0 |
| | | Potential | 387,215 | 2,703,118 | 1,685,316 | 433,877 |
| 4 | Faryab | Present practice | 7863 | 2 | 29 | 0 |
| | | Potential | 1,065,155 | 653,186 | 428,675 | 119,013 |
| 5 | Ghazni | Present practice | 8900 | 1891 | 158 | 0 |
| | | Potential | 483,463 | 1,517,790 | 324,367 | 10,073 |
| 6 | Helmand | Present practice | 587 | 165 | 250 | 0 |
| | | Potential | 586,359 | 2,648,833 | 1,803,261 | 1,217,535 |
| 7 | Herat | Present practice | 7499 | 260 | 396 | 0 |
| | | Potential | 1,503,025 | 3,003,630 | 1,287,774 | 55,550 |
| 8 | Jawzjan | Present practice | 11,130 | 489 | 1 | 0 |
| | | Potential | 379,067 | 236,147 | 435,719 | 182,225 |
| 9 | Kabul | Present practice | 547 | 63 | 14 | 0 |
| | | Potential | 11,130 | 489 | 1 | 0 |
| 10 | Kandahar | Present practice | 18,784 | 1755 | 372 | 0 |
| | | Potential | 615,253 | 1,698,481 | 1,079,035 | 2,215,816 |
| 11 | Kapisa | Present practice | 1004 | 16 | 6 | 0 |
| | | Potential | 121,173 | 62,767 | 22,188 | 661 |
| 12 | Kunduz | Present practice | 235 | 1 | 4 | 0 |
| | | Potential | 442,893 | 150,901 | 229,115 | 39,966 |
| 13 | Laghman | Present practice | 8 | 9 | 0 | 0 |
| | | Potential | 193,205 | 175,397 | 40,687 | 11,162 |
| 14 | Logar | Present practice | 1010 | 130 | 0 | 0 |
| | | Potential | 113,391 | 300,938 | 62,851 | 367 |
| 15 | Nangarhar | Present practice | 192 | 34 | 80 | 0 |
| | | Potential | 363,646 | 343,736 | 63,826 | 2513 |
| 16 | Nemroz | Present practice | 102 | 165 | 79 | 0 |
| | | Potential | 90,494 | 1,698,144 | 1,712,493 | 712,428 |
| 17 | Paktia | Present practice | 626 | 125 | 33 | 0 |
| | | Potential | 537,537 | 952,389 | 430,318 | 71,961 |
| 18 | Patyka | Present practice | 278 | 24 | 2 | 0 |
| | | Potential | 224,665 | 299,047 | 41,469 | 0 |
| 19 | Parwan | Present practice | 6863 | 97 | 36 | 0 |
| | | Potential | 132,406 | 324,202 | 136,561 | 21,154 |
| 20 | Samangan | Present practice | 628 | 0 | 0 | 0 |
| | | Potential | 465,673 | 694,886 | 269,141 | 6645 |

(continued)

**Table 7.6** (continued)

| No | Provinces | Categories | S1 (ha) | S2 (ha) | S3 (ha) | N (ha) |
|---|---|---|---|---|---|---|
| 21 | Sar-e-pul | Present practice | 8210 | 13 | 0 | 0 |
| | | Potential | 562,666 | 713,517 | 400,439 | 14,081 |
| 22 | Takhar | Present practice | 107 | 0 | 38 | 0 |
| | | Potential | 836,973 | 195,768 | 226,469 | 107,492 |
| 23 | Urozgan | Present practice | 19 | 3 | 1 | 0 |
| | | Potential | 427,891 | 629,480 | 105,621 | 0 |
| 24 | Wardak | Present practice | 58 | 18 | 1 | 0 |
| | | Potential | 104,788 | 836,273 | 199,258 | 13,304 |
| 25 | Zabul | Present practice | 4364 | 1096 | 414 | 5 |
| | | Potential | 369,797 | 1,094,514 | 325,407 | 47,186 |

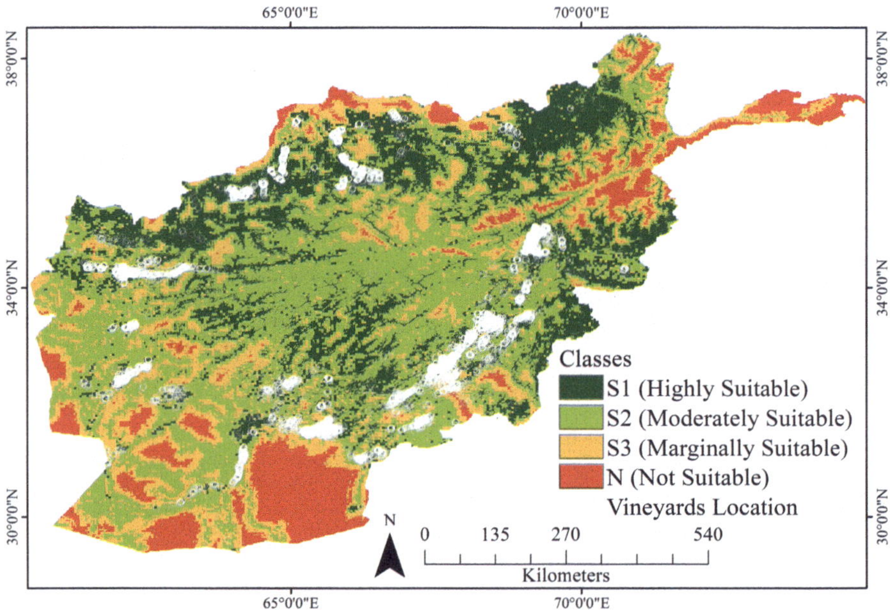


**Fig. 7.9** Suitable areas for table grape production in Afghanistan based on fuzzy multicriteria decision analysis

## 7.3.2 *Fuzzy Suitability Validation*

Ground validation of the suitability map was significant for confirming each suitable vineyard and extending the vineyard areas for the future in Afghanistan. However, obtaining many ground datasets in the country's current condition was very difficult. The results indicated good agreement between the land suitability index and average grape yield at the provincial level. The datasets were used for linear regression ($R^2$ = 0.74) and polynomial regression ($R^2$ = 0.84) (Fig. 7.10). Polynomial regression indicated higher accuracy in prediction than the linear trend.

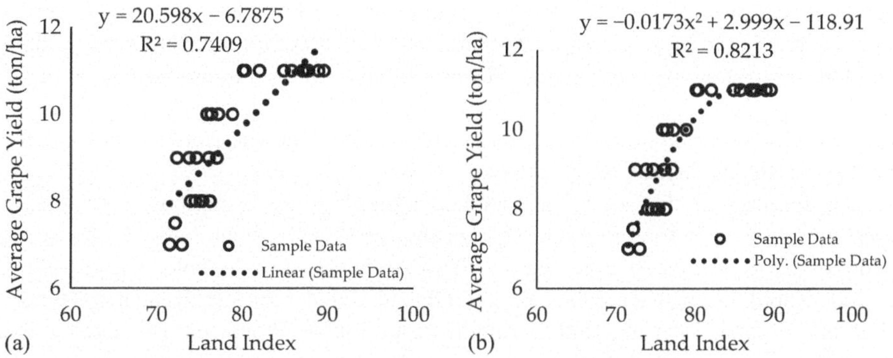


**Fig. 7.10** Validation of the fuzzy-based land suitability score referring to the average grape yield from different provinces in Afghanistan—(**a**) linear regression and (**b**) polynomial regression

## 7.4 Discussion

Fuzzy suitability analysis is a technique in land suitability analysis when the variables are more complicated and need a continuous output. In this regard, the fuzzy set model has many advantages, including the ability to convert all data to a range of 0–1, which is a great way to solve different magnitudes at different data layers, and the use of fuzzy membership functions to produce a thorough assessment of the suitability of the land for each crop production on a continuous scale across various land categories (Kılıc et al. 2022; Chrobak et al. 2020). Moreover, the implementation of fuzzy inference as a useful addition to the fundamental of GIS and remote sensing techniques were proposed in this paper to extend the areas of vineyard (Coulon-Leroy et al. 2014). Fuzzy sets were used in previous studies to determine the suitability analysis of land (Acharya and Yang 2015; Salata et al. 2022; Grelier et al. 2007; Braimoh et al. 2004; Pilevar et al. 2020; Purnamasari et al. 2019). Most of the prior studies used fuzzy models to construct agricultural suitability analysis framework considering multiple variables related to specific micro-regions for cereals, fruits, and wine grape production. Pilevar et al. (2020) used topographic variables such as slope and elevation and soil characteristics such as soil texture and soil pH for fuzzy suitability. In addition, other studies have used variables like rainfall, evapotranspiration, NDVI, LULC, soil, soil moisture, groundwater level, slope, and elevation to assess fuzzy suitability assessment (Ahamed et al. 2000; Chrobak et al. 2020). In this study, topographic variables (elevation and slop), soil variables (soil depth, soil pH, soil texture, soil salinity), vegetation (NDVI), infrastructure (road and river), and climatic (rainfall) variables were changed to fuzzy membership. After applying different fuzzy membership functions to each variable based on references and the literature, the suitability classes were defined (Badr et al. 2018; Mustafiz et al. 2022). Literatures indicated how well-developed fuzzy inference variables selection have significant advantages over the conventional method of mathematical modeling (Hong and Lee 1996; Chrobak et al. 2020). In the USA, a study was done on vineyard site selection and revealed that fuzzy set methods were the best method for land suitability classification with expert knowledge. Similarly, other studies mentioned expert knowledge is an influential factor (Purnamasari et al. 2019) for vineyard management and long-term strategy development. However, the advantageous application of fuzzy methodology has been confirmed by Tang et al. (1997). In this research, higher accuracy has been reported for the fuzzy method in comparison with Boolean methods for land suitability evaluation of different crops (Tang et al. 1997).

The findings of this study indicated that highly suitable regions were mostly located in the southern regions of Afghanistan for its agroclimatology. Furthermore, some parts of the western zone and northern zone were also suitable. These regions include Kabul, Logar, Khandahar, Kapisa, Ghazni, Zabul, Uruzgan, Kandahar, Herate, Badghis, Frayah, Jawzjan, Balkh, Kunduz, and Takhar Provinces. However, some parts of the northeast and central highlands, where the elevation is very high, are not suitable for grape production. Increasing elevation raises the likelihood of frost, cooler temperatures, and a decrease in plant growth months (Stanchi et al. 2013). The study also proved that there was potential for vineyard extension in Afghanistan for grape production. However, only 0.13% of the land area is currently under vineyard production practices. Some provinces have the highest potential for vineyard extension, such as Badghis, Faryab, Herat, and Takher. The government could act to increase vineyards since the suitable areas exceed reality because grapes are an industrial crop, and fresh grapes play an important role in growers' farm income and international trade in Afghanistan. The extension of vineyards can support farmers' livelihoods, particularly in the southern provinces (Table 7.6 and Fig. 7.9).

In this study, the suitability model was validated with the average yield of each provincial vineyard. Thus, the model accuracy was approximately 82% on the regional scale. Previous studies also reported the higher accuracy of the fuzzy method in comparison with Boolean methods for land suitability evaluation of different crops. They compared land index to the observed yield of rubber and maize. They discovered that the fuzzy approach had a correlation coefficient of 0.89, 0.87 (Braimoh et al. 2004; Van Ranst et al. 1996) for land suitability for cassava done in Indonesia. They compared land indices with observed yield and obtained a correlation coefficient of 0.55 (Purnamasari et al. 2019). Based on the high accuracy findings that came from this research, the fuzzy set technique could be a good model for land assessment in other environments. However, the average yield of each province was not adequate for validation due to the inaccessibility of ground reference data and information. This study can help researchers and decision-makers find more prospective acreage for expanding vineyards at the regional scale. Further research will be performed to expand the model with additional expertise knowledge with respect to economical points of view and logistics, including storage facilities for table grapes to reduce the post-harvest losses at the regional scale.

## 7.5 Conclusions

Selecting appropriate land for a vineyard is very important for sustainable grape production. Therefore, the primary objective of this study was to develop a land suitability analysis method using a fuzzy membership function on a regional scale to extend the best areas for table grape production. In this study, different climate, topographic, vegetation, and soil parameters were considered for vineyard suitability analysis. After this step, the final suitability map was developed based on the fuzzy overlay method. In further assessments, the validation of the suitability map was performed for table grape yield at the provincial level. The validation results of the suitability analysis indicated that the land suitability model had 82% accuracy at the regional scale. In the analysis, we found that less than 1% of the highly suitable area was used for vineyards. There is a significant opportunity to increase the production of table grapes in another 22% highly suitable areas. Therefore, the main finding of this study suggests that there is a potential for vineyard expansion in Afghanistan. This study contributes to the expansion of limited studies on vineyard suitability by bringing in multiple variables centering on the fuzzy method at the regional scales to increase table grape production.

**Acknowledgments** The authors express their gratitude to open-access journal *Remote Sensing* published by MDPI for their policy to support the authors for reusing the published article (Sara Tokhi Arab, Tofael Ahamed. Land Suitability Analysis for Potential Vineyards Extension in Afghanistan at Regional Scale Using Remote Sensing Datasets. Remote Sensing. *14*(18), 4450. https://doi.org/10.3390/rs14184450, 2022). Some minor modification has been conducted in this book chapter. Furthermore the authors are grateful for the LANDSAT/LC08/C01/T1 8DAY NDVI product of the United States Geological Survey and the Google Earth Engine developers, in addition to the CHIRPS precipitation dataset from the climate hazards group, the Afghanistan Geodesy and Cartography Head Office (AGCHO) for the map of Afghanistan's administrative boundaries, FAO for the land cover shapefile and soil characteristics datasets, and the Central Statistics Organization of the Islamic Republic of Afghanistan for yield datasets. The More Jobs, Better Lives (MJBL) Foundation, which gives a scholarship, is thankful to the initial author.

**Funding** This research received no external funding.

**Data Availability Statement** In this section, please provide details regarding where data supporting reported results can be found, including links to publicly archived datasets analyzed or generated during the study. Please refer to suggested Data Availability Statements in section "MDPI Research Data Policies" at https://www.mdpi.com/ethics. You might choose to exclude this statement if the study did not report any data.

**Conflicts of Interest** The authors declare no conflict of interest.

## References

Acharya TD, Yang IT (2015) Vineyard suitability analysis of Nepal. Int J Environ Sci 6:13. Acharya: Vineyard suitability analysis of Nepal—Google Scholar

Afghanistan Central Statistics Organization (ACSO) (2020) Central Statistics Organization of Afghanistan's Statistical Yearbook of 2018–2019. Islamic Republic of Afghanistan National Statistics and Information Authority, Kabul Afghanistan (42). 2021. https://invest.gov.af/theme3/wp-content/uploads/2021/06/Afghanistan-Statistical-Yearbook-first-Version.pdf. Accessed 12 Mar 2022

Ahamed TN, Rao KG, Murthy JSR (2000) GIS-based fuzzy membership model for crop-land suitability analysis. Agric Syst 63:75–95. https://doi.org/10.1016/S0308-521X(99)00036-0

Akpoti K, Kabobah AT, Zwart SJ (2019) Review–agricultural land suitability analysis: state-of the-art and outlooks for integration of climate change analysis. Agric Syst 173:172–208. https://doi.org/10.1016/j.agsy.2019.02.013

Arab ST, Salari T, Noguchi R, Ahamed T (2022) Land suitability analysis for grape (Vitis vinifera L.) production using satellite remote sensing, GIS, and analytical hierarchy process. In: Ahamed T (ed) Remote sensing application. New frontiers in regional science: Asian perspectives, vol 59. Springer, Singapore. https://doi.org/10.1007/978-981-19-0213-0_6

Badr G, Hoogenboom G, Moyer M, Keller M, Rupp R, Davenport J (2018) Spatial suitability assessment for vineyard site selection based on fuzzy logic. Precis Agric 19:1027–1048. https://doi.org/10.1007/s11119-018-9572-7

Bellman RE, Zadeh LA (1970) Decision-making in a fuzzy environment. Manag Sci 17:141–164. https://doi.org/10.1287/mnsc.17.4.B141

Braimoh AK, Vlek PL, Stein A (2004) Land evaluation for maize based on fuzzy set and interpolation. Environ Manag 33:226–238. https://doi.org/10.1007/s00267-003-0171-6

Chrobak K, Chrobak G, Kazak JK (2020) The use of common knowledge in fuzzy logic approach for vineyard site selection. Remote Sens 12:416. https://doi.org/10.3390/land11030416

Coulon-Leroy C, Charnomordic B, Thiollet-Scholtus M, Guillaume S (2014) Fuzzy modeling of a composite agronomical feature using FisPro: the case of vine vigor. In: Laurent A, Strauss O, Bouchon-Meunier B, Yager RR (eds) Information processing and management of uncertainty in knowledge-based systems. Springer, Berlin/Heidelberg, pp 127–137

De Pascale S, Barbieri G (1995) Effects of soil salinity from long-term irrigation with saline-sodic water on yield and quality of winter vegetable crops. Sci Hortic 64:145–157. https://doi.org/10.1016/0304-423800823-3

Deng F, Li X, Wang H, Zhang M, Li R, Li X (2014) GIS-based assessment of land suitability for alfalfa cultivation: a case study in the dry continental steppes of northern China. Span J Agric Res 12:364–375. https://doi.org/10.5424/sjar/2014122-4672

Elaalem M, Comber A, Fisher P (2011) A comparison of fuzzy AHP and ideal point methods for evaluating land suitability. Trans GIS 15:329–346. https://doi.org/10.1111/j.1467-9671.2011.01260.x

FAO (1976) A framework for land evaluation, soils bulletin no. 32. UNO-FAO, Rome. https://www.fao.org/3/X5310E/x5310e00.htm. Accessed 1 Dec 2020

FAO (2016) Land Cover ATLAS. The Islamic Republic of Afghanistan. https://www.fao.org/publications/card/en/c/cc0ac143-38ed-41f0-b7c6-2342ffa7f0e6/. Accessed 15 June 2020

FAO (2020) The Lasmic Republic of Afghanistan Soil ATLAS, Volume 1: maps derived from soil survey of twenty-six district of nine provinces. https://www.fao.org/publications/card/en/c/CA6928EN/. Accessed 15 June 2020

Funk C, Peterson P, Landsfeld M, Pedreros D, Verdin J, Shukla S, Husak G, Rowland J, Harrison L, Hoell A et al (2015) The climate hazards infrared precipitation with stations—a new environmental record for monitoring extremes. Sci Data 2:150066. https://doi.org/10.1038/sdata.2015.66

Ghozat A, Sharafati A, Hosseini SA (2022) Satellite-based monitoring of meteorological drought over different regions of Iran: application of the CHIRPS precipitation product. Environ Sci Pollut Res 29:36115–36132. https://doi.org/10.1007/s11356-022-18773-3

Gilliams S, Van Orshoven J, Muys B, Kros H, Heil GW, Van Deursen W (2005) AFFOREST sDSS: a metamodel based spatial decision support system for afforestation of agricultural land. New For 30:33–53. https://doi.org/10.1007/s11056-004-0761-z/

Goes BJM, Howarth SE, Wardlaw RB, Hancock IR, Parajuli UN (2016) Integrated water resources management in an insecure river basin: a case study of Helmand River Basin, Afghanistan. Int J Water Resour Dev 32:3–25. https://doi.org/10.1080/07900627.2015.1012661

Goldammer T (2018) Grape Grower's handbook. A guide to viticulture for wine production. http://www.wine-grape-growing.com/. Accessed 15 June 2021

Grassano N, Tedone L, Verdini L, De Mastro G (2011) Evaluation of rapeseed cultivation suitability in Apulia with GIS-multicriteria analysis. Ital J Agron 6:e16–e16. https://doi.org/10.4081/ija.2011.e16

Grelier M, Guillaume S, Tisseyre B, Scholasch T (2007) Precision viticulture data analysis using fuzzy inference systems. OENO One 41:19–31. https://doi.org/10.20870/oeno-one.2007.41.1.857

Guo Y, Xia H, Pan L, Zhao X, Li R (2022) Mapping the northern limit of double cropping using a phenology-based algorithm and Google Earth Engine. Remote Sens 14:1004. https://doi.org/10.3390/rs14041004

Habibie MI, Noguchi R, Matsushita S, Ahamed T (2020) Development of micro-level classifiers from land suitability analysis for drought-prone areas in Indonesia. Remote Sens Appl Soc Environ 20:100421. https://doi.org/10.1016/j.rsase.2020.100421

Hong T-P, Lee C-Y (1996) Induction of fuzzy rules and membership functions from training examples. Fuzzy Sets Syst 84:33–47. https://doi.org/10.1016/0165-0114(95)00305-3

Keshavarzi A, Sarmadian F, Heidari A, Omid M (2010) Land suitability evaluation using fuzzy continuous classification (a case study: Ziaran region). Mod Appl Sci 4:72. https://doi.org/10.1287/mnsc.17.4.B141

Kılıc OM, Ersayın K, Gunal H, Khalofah A, Alsubeie MS (2022) Combination of fuzzy-AHP and GIS techniques in land suitability assessment for wheat (Triticum aestivum) cultivation. Saudi J Biol Sci 29:2634–2644. https://doi.org/10.1016/j.sjbs.2021.12.050

Moisa MB, Tiye FS, Dejene IN, Gemeda DO (2022) Land suitability analysis for maize production using geospatial technologies in the Didessa watershed, Ethiopia. Artif Intell Agric 6:34–46. https://doi.org/10.1016/j.aiia.2022.02.001

Mustafiz RB, Noguchi R, Ahamed T (2022) Agricultural land suitability assessment using satellite remote sensing-derived soil-vegetation indices. In: Remote sensing application. Springer, Singapore, pp 65–98. https://doi.org/10.1007/978-981-19-0213-0_3

Park S, Jeon S, Kim S, Choi C (2011) Prediction and comparison of urban growth by land suitability index mapping using GIS and RS in South Korea. Landsc Urban Plan 99:104–114. https://doi.org/10.1016/j.landurbplan.2010.09.001

Perrot N, Baudrit C, Brousset JM, Abbal P, Guillemin H, Perret B, Goulet E, Guérin L, Barbeau G, Picque D (2015) A decision support system coupling fuzzy logic and probabilistic graphical approaches for the agri-food industry: prediction of grape berry maturity. PLoS One 10:e0134373. https://doi.org/10.1371/journal.pone.0134373

Pilevar AR, Matinfar HR, Sohrabi A, Sarmadian F (2020) Integrated fuzzy, AHP and GIS techniques for land suitability assessment in semi-arid regions for wheat and maize farming. Ecol Indic 110:105887. https://doi.org/10.1016/j.ecolind.2019.105887

Pramanik MK (2016) Site suitability analysis for agricultural land use of Darjeeling district using AHP and GIS techniques. Model Earth Syst Environ 2:56. https://doi.org/10.1007/s40808-016-0116-8

Purnamasari RA, Noguchi R, Ahamed T (2019) Land suitability assessments for yield prediction of cassava using geospatial fuzzy expert systems and remote sensing. Comput Electron Agric 166:105018. https://doi.org/10.1016/j.compag.2019.105018

Radočaj D, Jurišić M, Gašparović M, Plaščak I (2020) Optimal soybean (Glycine max L.) land suitability using gis-based multicriteria analysis and sentinel-2 multitemporal images. Remote Sens 12:1463. https://doi.org/10.3390/rs12091463

Rameshkumar S, Vadivelu S, Reddy R, Naidu L, Hegde R, Srinivas S (2006) Land suitability for grape cultivation and its economic evaluation in Rajanukunte Watershed, Karnataka. http://isslup.in/wp-content/uploads/2018/09/Land-suitability-for-grape-cultivation-and-its.pdf. Accessed 15 June 2021

Romano G, Dal Sasso P, Liuzzi GT, Gentile F (2015) Multi-criteria decision analysis for land suitability mapping in a rural area of Southern Italy. Land Use Policy 48:131–143. https://doi.org/10.1016/j.landusepol.2015.05.013

Roy J, Saha S (2018) Assessment of land suitability for the paddy cultivation using analytical hierarchical process (AHP): a study on Hinglo river basin, Eastern India. Model Earth Syst Environ 4:601–618. https://doi.org/10.1007/s40808-018-0467-4

Saaty TL (1980) The analytic hierarchy process (AHP). J Oper Res Soc 41:1073–1076. https://doi.org/10.1007/978-3-642-83555-1_5

Salata S, Ozkavaf-Senalp S, Velibeyoğlu K, Elburz Z (2022) Land suitability analysis for vineyard cultivation in the Izmir metropolitan area. Land 11:416. https://doi.org/10.3390/land11030416

Stanchi S, Godone D, Belmonte S, Freppaz M, Galliani C, Zanini E (2013) Land suitability map for mountain viticulture: a case study in Aosta Valley (NW Italy). J Maps 9:367–372. https://doi.org/10.1080/17445647.2013.785986

Tang H, Van Ranst E, Groenemans R (1997) Application of fuzzy set theory to land suitability assessment. Malays J Soil Sci 3:39–58. Application-of-fuzzy-set-theory-to-land-suitability-assessment.pdf (researchgate.net)

Tashayo B, Honarbakhsh A, Akbari M, Eftekhari M (2020) Land suitability assessment for maize farming using a GIS-AHP method for a semi-arid region, Iran. J Saudi Soc Agric Sci 19:332–338. https://doi.org/10.1016/j.jssas.2020.03.003

Tatem AJ (2005) Global climate matching: satellite imagery as a tool for mapping vineyard suitability. J Wine Res 16:19–32. https://doi.org/10.1080/1260500236682

Tiwari ND, Mishra D (2022) Geospatial analysis of land degradation by fuzzy membership algorithm in reference to Satna river basin in northern foreland of Peninsular India. Arab J Geosci 15:1128. https://doi.org/10.1007/s12517-022-10324-x

USAID, Commercial Horticulture and Agricultural Marketing Program and Roots of Peace (2016) Best practices for GRAPE production and marketing in Afghanistan. Guide offers tips on cultivation, harvesting and marketing techniques to improve sales of Afghan produce on international export markets. pp 1–37. chrome-extension://efaidnbmnnnibpcajpcglclefindmkaj/https://static1.squarespace.com/static/5b69fa24506fbeb93ef780e2/t/5c65bc46e5e5f088ebcad662/1550171215318/Best+Practices+for+GRAPE+Production+and+Marketing+in+Afghanistan+Roots+of+Peace+ROP.pdf. Accessed 15 June 2021

Van Ranst E, Tang H, Groenemam R, Sinthurahat S (1996) Application of fuzzy logic to land suitability for rubber production in peninsular Thailand. Geoderma 70:1–19. https://doi.org/10.1016/0016-7061(95)00061-5

Wali E, Datta A, Shrestha RP, Shrestha S (2016) Development of a land suitability model for saffron (Crocus sativus L.) cultivation in Khost Province of Afghanistan using GIS and AHP techniques. Arch Agron Soil Sci 62:921–934. https://doi.org/10.1080/03650340.2015.1101519

Wang L, Ma Y, Yan J, Chang V, Zomaya AY (2018) PipsCloud: high performance cloud computing for remote sensing big data management and processing. Futur Gener Comput Syst 78:353–368. https://doi.org/10.1016/j.future.2016.06.009

Worqlul AW, Dile YT, Jeong J, Adimassu Z, Lefore N, Gerik T, Srinivasan R, Clarke N (2019) Effect of climate change on land suitability for surface irrigation and irrigation potential of the shallow groundwater in Ghana. Comput Electron Agric 157:110–125. https://doi.org/10.1016/j.compag.2018.12.040

Xie Z, Phinn SR, Game ET, Pannell DJ, Hobbs RJ, Briggs PR, McDonald-Madden E (2019) Using Landsat observations (1988–2017) and Google Earth Engine to detect vegetation cover changes in rangelands-a first step towards identifying degraded lands for conservation. Remote Sens Environ 232:111317. https://doi.org/10.1016/j.rse.2019.111317

Zabel F, Putzenlechner B, Mauser W (2014) Global agricultural land resources—a high resolution suitability evaluation and its perspectives until 2100 under climate change conditions. PLoS One 9:e107522. https://doi.org/10.1371/journal.pone.0107522

# Chapter 8
# Integrating Analytical Hierarchy Process with GIS and Satellite Remote Sensing to Assess Land Suitability for Sustainable Tea Production in Bangladesh

**Animesh Chandra Das, Ryozo Noguchi, and Tofael Ahamed**

**Abstract** Land suitability evaluation is crucial for assessing environmental limitations that inhibit higher yield and productivity in tea. The aim of this research was to determine the suitable lands for sustainable tea production in the northeastern part of Bangladesh using phenological datasets from remote sensing, geospatial datasets of soil-plant biophysical properties, and expert opinions. Sentinel-2 satellite image datasets were processed to obtain layers for land use and land cover (LULC) as well as the normalized difference vegetation index (NDVI). Data from the Shuttle Radar Topography Mission (SRTM) were processed to generate the elevation layer. Other vector layers of edaphic and climatic parameters were processed in ArcGIS 10.7.1® software for the respective raster layers. Finally, land suitability classes were obtained from the reclassified raster layers of all parameters along with the results from multicriteria analysis using spatial analysis. The results of the study showed that only 41,460 hectares of land (3.37% of the total land) were in the highly suitable category in the Sylhet region of Bangladesh. The proportions of moderately suitable, marginally suitable, and not suitable land categories for tea cultivation in the Sylhet division were 9.01%, 49.87% and 37.75%, respectively.

---

A. C. Das
Graduate School of Life and Environmental Sciences, University of Tsukuba, Tsukuba, Ibaraki, Japan

Faculty of Agriculture, Sylhet Agricultural University, Sylhet, Bangladesh
e-mail: s1836018@s.tsukuba.ac.jp

R. Noguchi
Graduate School of Agriculture, Kyoto University, Kyoto, Japan
e-mail: noguchi.ryozo.gm@u.tsukuba.ac.jp

T. Ahamed (✉)
Institute of Life and Environmental Sciences, University of Tsukuba, Tsukuba, Ibaraki, Japan
e-mail: tofael.ahamed.gp@u.tsukuba.ac.jp



T. Ahamed (ed.), *Remote Sensing Application II*, New Frontiers in Regional Science: Asian Perspectives 77, https://doi.org/10.1007/978-981-97-1188-8_8

Thirty-one tea estates were located in highly suitable areas, 79 in moderately suitable areas, 24 in marginally suitable areas, and only 1 in a not suitable area. Yield estimation was performed with the NDVI ($R^2$ = 0.69, 0.66, and 0.67) and the LAI ($R^2$ = 0.68, 0.65, and 0.63) for 2017, 2018, and 2019. This research suggests that satellite remote sensing and GIS applications with the analytical hierarchy process (AHP) could be used by land policy makers and agricultural land use planners to select suitable lands for increasing tea production.

**Keywords** Tea · Land suitability assessment · Phenological indices · GIS · Remote sensing · Analytical hierarchy process · Yield validation

## 8.1 Introduction

Land suitability assessment is important for sustainable land resource planning and management (Jayasinghe et al. 2019). A range of parameters, e.g., topography, soil conditions, state of the climate, and vegetation indices, are considered to assess land suitability (Ahamed et al. 2000; Wang et al. 1990). Such assessment provides information about specific land use potentials and constraints. Effective management with proper land use decisions results in higher productivity of land as well as a sustained environment (McDowell et al. 2018). For sustainable land resource management, the Food and Agricultural Organization (FAO 1976) proposed guidelines for land suitability assessment. According to the guidelines, land is classified into four categories: highly suitable (S1), moderately suitable (S2), marginally suitable (S3), and not suitable (N) (Purnamasari et al. 2019; Habibie et al. 2019). The global warming as well as the rapidly growing population exerts considerable pressure on scarce land resources all over the world (Vitousek 1994). The yield of plantation crops such as tea is decreasing due to the effects of drought and land degradation caused by climate change, especially changes in temperature and rainfall (Gunathilaka et al. 2017). Therefore, determining suitable lands and climate for tea to obtain maximum yield and production is urgent. In addition, it is imperative to utilize fallow lands, islands, and hilly areas for sustainable land management.

Tea (*Camellia sinensis* L. (O.) Kuntze) is a popular beverage as well as a valuable cash crop and is renowned for its nutritional, antimicrobial, medicinal, and anticancer properties throughout the world (Lambert and Yang 2003; Cooper et al. 2005; Bhattacharya et al. 2014). Bangladesh is one of the world's leading tea-producing countries. Tea cultivation in Bangladesh began in the British colonial period in 1854. Most of Bangladesh's tea is produced in the Sylhet division, and approximately 96% is cultivated in three districts of the Sylhet division. Among these districts, Moulvibazar produces 63% of the tea, and Sylhet and Habiganj combinedly produce 33% of 93% of the total plantation areas (Islam and Al-Amin 2019). At present, the country has 167 tea estates that produce approximately 96.07 million kg of tea annually with an average yield of 1768.52 kg/ha (Bangladesh Tea Board 2020). The tea industry of Bangladesh annually earns roughly Tk. 1.775

billion (0.81% of GDP) in foreign currency, exporting nearly 18 million kg of tea (1.37% of the export of the global tea trade) (Kamruzzaman et al. 2015). According to world rankings, China ranked first, producing 2350 million kg of tea, followed by India (1267 million kg), Kenya (473 million kg), Sri Lanka (293 million kg), Turkey (253 million kg), Vietnam (180 million kg), Indonesia (125 million kg), Argentina (84 million kg), and Bangladesh (83 million kg). Thus, Bangladesh ranked ninth in tea production (International Tea Committee 2017). The global production of tea has increased tremendously over the last 50 years. In addition, the average yield per hectare of tea in Bangladesh is apparently lower than that in other leading tea-producing countries. The drawback to the higher tea yield in Bangladesh is the existence of marginally suitable lands with unfavorable climates (Bangladesh Tea Board 2020). For this reason, the selection of suitable lands is crucial in making proper use of available lands for tea cultivation. Tea growers establish estates based on traditional knowledge and experience without utilizing scientific information or methods of validation. They consider site suitability rather than using appropriate information. This also adversely impacts production in the long term, exacerbating environmental problems in tea-plantation areas (Jayasinghe et al. 2019).

With advances in information and communication technology, land suitability assessment has been performed using geographical information system (GIS) and satellite remote sensing techniques (Bandyopadhyay et al. 2009). In addition, important criteria for sustainable tea production must be considered in the analytical hierarchy process (AHP) for prioritizing experts' opinions in accordance with the weight obtained from consistent GIS results (Habibie et al. 2019). Satellite remote sensing with a GIS-based AHP is a robust tool for spatial decision-making processes for land suitability analysis (Purnamasari et al. 2019). The results of land suitability assessment should be validated with obtained yield information. Research on yield estimation of agricultural crops indicates that remote sensing techniques alone are not capable of accurate yield estimation (Clevers et al. 1994). To improve accuracy, the actual yield of individual tea estates should be incorporated with remote sensing data.

Multiple studies have been undertaken to assess land suitability for tea. A land suitability assessment was performed in Sri Lanka using a GIS-based multicriteria approach (Jayasinghe et al. 2019). Land suitability evaluation was performed for tea and orange in the Nghe An Province of Vietnam using land suitability evaluation (LSE) software by considering several ecological factors (Nguyen et al. 2020). A comprehensive suitability assessment for tea was carried out in Zhejiang Province of China using a geographic information system (GIS), and a modified land ecological suitability evaluation model demonstrated the scientific basis for land suitability and the planting distribution of tea crops (Bo et al. 2012). A GIS-based land use suitability assessment for forests and tea crops was performed by Chanhda et al. (2010) along the Laos-China border, while Gahlod et al. (2017) carried out research to evaluate the suitability of land for cardamom, rubber, and tea using geospatial techniques in Kerala, India, by considering various physicochemical factors. Those studies provided information regarding the constraints of land use for tea and opportunities for decision making as well as optimal utilization of land resources (Collette and Nauen 1983).

Suitability analysis facilitates the recognition of marginally suitable lands with limiting factors that aid decision makers in developing appropriate crop management strategies for increasing the productivity of land (Gahlod et al. 2017). Upon consideration of suitability assessment results, it is urgent to implement further initiatives for turning unproductive estates into productive estates by adopting better management practices. This will also allow tea growers on new plantations to work in highly suitable and moderately suitable fallow lands (Jayasinghe et al. 2019). Hence, it has been necessary to perform land suitability assessment by considering other crop requirements. Accordingly, no studies regarding land suitability evaluation for tea in Bangladesh to increase production have been performed. Therefore, a comprehensive study utilizing physical, climatic, and vegetative parameters for sustainable land use and higher productivity of tea was undertaken. This study has attempted to assess land suitability, considering multiple criteria to ensure long-term progress in the tea industry. This land suitability assessment could also improve land use policy for the sustainable management of lands in tea-plantation areas in order to increase tea production in Bangladesh.

## 8.2 Materials and Methods

Sentinel-2 multispectral instrument (MSI) satellite images were utilized as the remote sensing dataset, and certain vector-layered edaphic and climatic geodata were processed to develop the map for suitability analysis. The criteria were categorized into four types for land selection of tea according to FAO guidelines. Primary data as well as ground reference information were obtained through fieldwork using a global positioning system (GPS) receiver to locate the tea estates in the study area.

### *8.2.1 Study Area*

The study area is the Sylhet division, located in the northeastern part of Bangladesh, and consists of four districts, Habiganj, Moulvibazar, Sunamganj, and Sylhet, which include 38 subdistricts (Fig. 8.1).

The population size of this locality is approximately ten million, which is less than 7% of the total population of Bangladesh. The study area lies between the latitudes of 23°58′ and 25°12′ north and the longitudes of 90°56′ and 92°30′ east. The area is surrounded by the Indian states of Meghalaya, Assam, and Tripura to the north, east, and south, respectively, and divisions of Chattagram to the southwest and Dhaka and Mymensingh to the west. The land within the study area is 1,229,840 hectares, and the elevation is less than 335 m. The study area receives an adequate amount of rainfall in the monsoon season that is favorable for tea cultivation.

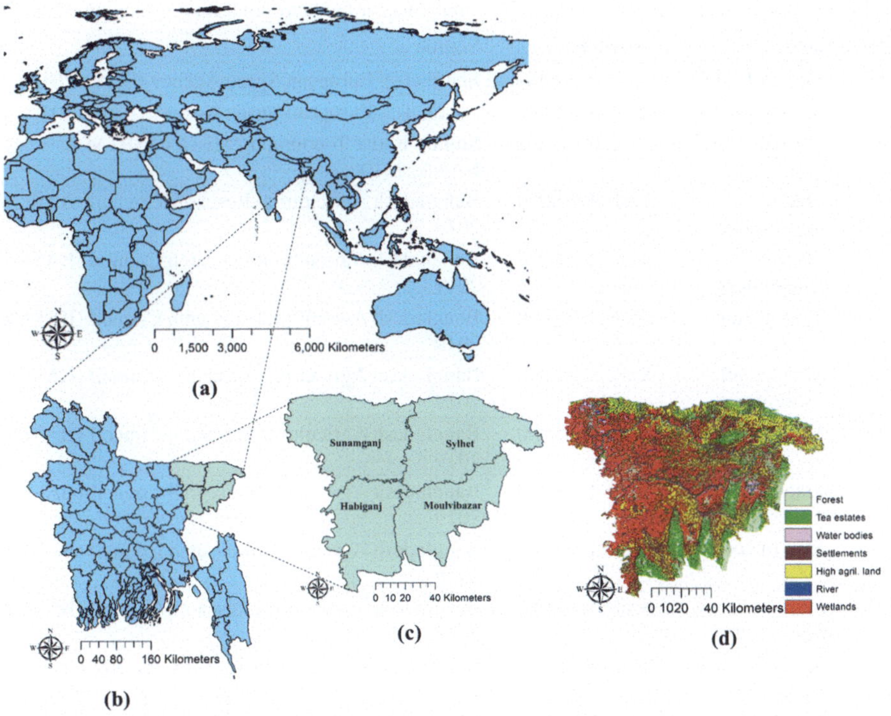


**Fig. 8.1** Geographical extent of the study area: (**a**) Bangladesh on the world map, (**b**) Bangladesh, (**c**) Sylhet Division, and (**d**) Physiographic image of Sylhet Division

## *8.2.2 Criteria for Suitability Analysis*

Twelve criteria, viz. land use and land cover (LULC), the normalized difference vegetation index (NDVI), elevation, precipitation, temperature, slope, soil texture, pH, drainage, soil type, distance from roads and distance from rivers, were considered to determine suitable land for tea cultivation (Table 8.1, Fig. 8.2 and Appendix 1)

### 8.2.2.1 Land Use and Land Cover

Land use and land cover (LULC) data were utilized to assess the lands for forests, tea estates, high agricultural lands, settlements, water bodies, rivers, and wetlands. According to land use and land cover, the majority of the study area was occupied by forests, tea estates, high agricultural lands, and wetlands for rice cultivation. LULC was built from Sentinel-2 datasets with 20 m resolution and processed using the supervised and maximum likelihood classification in ArcGIS®. The raster layer for LULC was categorized into seven classes: forests, tea estates, high agricultural lands, settlements, water bodies, rivers, and wetlands. The forest

**Table 8.1** Generated map and sources of original data for the land suitability evaluation of tea

| No. | Data | Description | Source |
|---|---|---|---|
| 1 | Map of LULC | 20 m resolution | Sentinel-2, European Space Agency (ESA), 2019 |
| 2 | Map of NDVI | 20 m resolution | Sentinel-2, European Space Agency (ESA), 2019 |
| 3 | Map of elevation | 30 m resolution | Shuttle Radar Topography Mission (SRTM), NASA, 2019 |
| 4 | Map of precipitation | Scale 1:50,000 | Bangladesh Agricultural Research Council (BARC), 2019 |
| 5 | Map of temperature | Scale 1:50,000 | Bangladesh Agricultural Research Council (BARC), 2019 |
| 6 | Map of slope | Scale 1:50,000 | Bangladesh Agricultural Research Council (BARC), 2019 |
| 7 | Map of soil texture | Scale 1:50,000 | Bangladesh Agricultural Research Council (BARC), 2019 |
| 8 | Map of soil pH | Scale 1:50,000 | Bangladesh Agricultural Research Council (BARC), 2019 |
| 9 | Map of drainage | Scale 1:50,000 | Bangladesh Agricultural Research Council (BARC), 2019 |
| 10 | Map of soil type | Scale 1:50,000 | Bangladesh Agricultural Research Council (BARC), 2019 |
| 11 | Map of distance from roads | Scale 1:50,000 | Bangladesh Agricultural Research Council (BARC), 2019 |
| 12 | Map of distance from rivers | Scale 1:50,000 | Bangladesh Country Almanac (BCA), 2019 |
| 13 | Location of tea estates | GPS data | Field survey, 2019 |
| 14 | Tea production | Statistical data | Bangladesh Tea Board (BTB), 2017–2019 |

class consisted mostly of national reserve forests; the settlements consisted of households, public offices, and other infrastructures; and the water bodies consisted of areas of water such as beels, ponds, and lakes; a larger portion of the Sylhet division is a low-lying area occupied by haor, a wetland ecosystem (Kamruzzaman and Shaw 2018) (Table 8.2).

An accuracy assessment was performed to calculate the accuracy of the LULC classification. According to the accuracy assessment for LULC, user accuracy (UA), producer accuracy (PA), and overall accuracy (OA) were determined along with the Kappa coefficient (T). UA was calculated from the total correct samples in rows divided by the total reference samples in each row. PA was determined from the total correct samples in a column divided by the total reference samples in each column. OA was calculated from the total samples along the matrix diagonal from the reference divided by the total samples.

$$\text{UA} = \frac{\text{Number of correctly classified pixels in each category}}{\text{Total number of classified pixels in that category}\left(\text{the row total}\right)} \times 100 \tag{8.1}$$

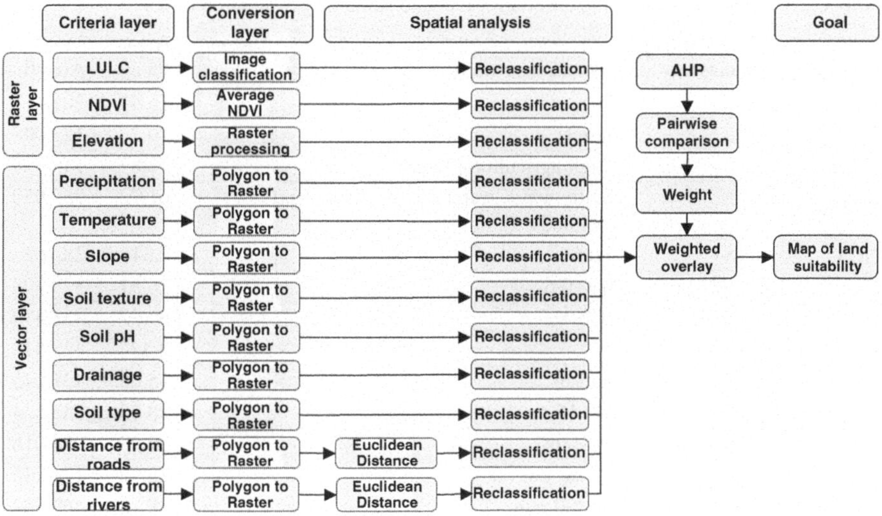


**Fig. 8.2** Conceptual framework for the land suitability evaluation of tea estates

$$\text{PA} = \frac{\text{Number of correctly classified pixels in each category}}{\text{Total number of reference pixels in that category (the column total)}} \times 100 \tag{8.2}$$

$$\text{OA} = \frac{\text{Total number of correctly classified pixels (diagonal)}}{\text{Total number of reference pixels}} \times 100 \tag{8.3}$$

$$\text{Kappa coefficient}\,(T) = \frac{(\text{TS} \times \text{TCS}) - \Sigma(\text{column total} \times \text{row total})}{\text{TS}^2 - \Sigma(\text{column total} \times \text{row total})} \times 100 \tag{8.4}$$

where TS denotes the total number of samples and TCS is the total number of correct samples.

#### 8.2.2.2 Normalized Difference Vegetation Index (NDVI)

The NDVI is a vegetation index correlated with various biophysical parameters and different crop indices (Ahamed et al. 2016). The proportion of green biomass sensed or captured in satellites is important for vegetation monitoring. Apart from this, the NDVI is used to measure the phenological variations in vegetation. Tea is a perennial crop that exhibits active vegetative growth in the monsoon season from March to November, and harvesting occurs within this time. In this study, the NDVI was calculated for tea plantations using Sentinel-2 satellite images. The map for the NDVI was developed using extraction by masking from Sentinel-2 images to distinguish the vegetation status of this area.

**Table 8.2** Reclassification of the criteria for the land suitability evaluation of tea

| Criteria | Suitability classes | Values/subcriteria | Area (%) | Area (ha) |
|---|---|---|---|---|
| LULC | S1 | Tea estates | 16.41 | 201,818 |
| | S2 | Forest | 10 | 123,008 |
| | S3 | High agricultural land | 11.87 | 145,964 |
| | N | Settlements, water bodies, rivers, and wetlands | 61.72 | 759,050 |
| NDVI | S1 | >0.6 | 2.31 | 28,362 |
| | S2 | 0.4–0.6 | 26.51 | 325,998 |
| | S3 | 0–0.4 | 67.77 | 833,430 |
| | N | <0 | 3.42 | 42,051 |
| Elevation | S1 | >15 m | 35.05 | 431,045 |
| | S2 | 10–15 m | 28.75 | 353,533 |
| | S3 | 7–10 m | 22.28 | 274,016 |
| | N | <7 m | 13.92 | 171,246 |
| Precipitation | S1 | >1800 mm | 38.18 | 469,553 |
| | S2 | 1600–1800 | 46.54 | 572,386 |
| | S3 | 1000–1600 | 15.28 | 187,901 |
| Temperature | S1 | 18–25 °C | 100 | 1,229,840 |
| Slope | S1 | 5–25° | 14.73 | 181,103 |
| | S2 | <5° | 85.12 | 1,046,828 |
| | S3 | >25° | 0.16 | 1909 |
| Soil texture | S1 | scl, l, cl, sl | 71.36 | 877,555 |
| | S2 | c, sicl, sic | 27.22 | 334,729 |
| | S3 | c(ss), ls, s | 1.43 | 17,556 |
| Soil pH | S1 | 4.5–5.5 | 13.99 | 172,008 |
| | S2 | 5.5–7.3 | 81.05 | 996,795 |
| | S3 | 7.3–8.4 | 4.96 | 61,037 |
| Drainage | S1 | Moderately well drained to well drained | 13.36 | 164,353 |
| | S2 | Imperfectly drained | 9.60 | 118,083 |
| | S3 | Poorly drained | 66.30 | 815,421 |
| | N | Very poorly drained | 10.73 | 131,983 |
| Soil type | S1 | Brown hill soils | 13.24 | 162,782 |
| | S2 | Gray piedmont soils | 11.84 | 145,659 |
| | S3 | Noncalcareous alluvium, brown flood plain soils,<br>Dark gray floodplain soils, gray floodplain soils, acid basin clays, deep-red brown terrace soils | 73.72 | 906,642 |
| | N | Peat, water bodies, urban | 1.20 | 14,757 |
| Distance from roads | S1 | 0–1.0 km | 13.97 | 171,759 |
| | S2 | 1.0–2.0 km | 14.51 | 148,404 |
| | S3 | 2.0–4.0 km | 21.81 | 268,246 |
| | N | >4.0 km | 49.72 | 611,431 |

**Table 8.2** (continued)

| Criteria | Suitability classes | Values/subcriteria | Area (%) | Area (ha) |
|---|---|---|---|---|
| Distance from rivers | S1 | 0–0.5 km | 6.23 | 76,601 |
| | S2 | 0.5–1.0 km | 11.68 | 143,618 |
| | S3 | 1.0–2.0 km | 18.03 | 221,779 |
| | N | >2.0 km | 64.06 | 787,842 |

#### 8.2.2.3 Elevation

Tea grows in a wide range of elevations from sea level to approximately 2200 m (Hajiboland 2017; Jayasinghe et al. 2019). Tea is planted in the flat valleys of Assam, India, at an elevation ranging from a few meters to approximately 200 m above sea level, while on the hill slopes of Darjeeling, it is cultivated up to an altitude of 2000 m. The tea-growing hillocks in Cachar, Assam, rarely extend above 150 m, whereas a few tea estates located at the base of the Himalayas ascend to 500 m (Bhagat et al. 2010). The elevation in the Sylhet division ranges from −55 to 335 m. Most of the highland area is free from water logging and suitable for tea cultivation, despite the higher topographic elevation (Bozdağ et al. 2016). The elevation data were extracted using the Shuttle Radar Topography Mission (SRTM), 2019, NASA, with a 30 m resolution.

#### 8.2.2.4 Precipitation

Moderate temperatures with high precipitation are favorable for tea cultivation, although tea is susceptible to water stress. Tea plants require an average minimum rainfall of 1000 mm per year, but 1800–2000 mm is optimal (Gahlod et al. 2017). The study area is characterized by higher rainfall ranges of 1000–2300 mm/year. Rainfall data were collected from the Bangladesh Agricultural Research Council and converted to a raster file. The raster file was processed according to the mean monthly rainfall data from March to November, when tea plants grow vigorously (Ahmed et al. 2014a, b).

#### 8.2.2.5 Temperature

The growth of tea plants is highly influenced by temperature. The yield of tea is also affected by increased average monthly temperature as well as constant higher temperature for longer periods. Temperature regimes below 13 °C and above 30 °C have been observed as detrimental for the shoot growth of tea plants (Jayasinghe et al. 2019). In the growing season, plants grow satisfactorily at temperatures ranging between 18 and 25 °C (Gahlod et al. 2017), which matched the temperatures of the study area.

#### 8.2.2.6 Slope

Slope influences the quality of land and is an important determinant for the land suitability evaluation for crops. Land slope affects erosion and surface runoff due to its microclimate variation. It also affects other soil properties, such as the soil moisture percentage, the proportion of clay materials, and the availability of other nutrients, such as nitrogen, calcium, and magnesium (Khormali et al. 2012). Land slopes ranging between 5° and 25° are optimal for tea cultivation. A slope gradient greater than 35° is considered unsuitable because it encourages soil erosion and landslides. Moreover, flattened slopes are also unsuitable for tea cultivation, as they may cause waterlogged conditions (Jayasinghe et al. 2019). Slope data were obtained from the Bangladesh Agricultural Research Council (BARC), 2019.

#### 8.2.2.7 Soil Texture

Soil texture is considered an important criterion, as it influences other soil properties such as bulk density, hydraulic conductivity, and the water holding capacity that regulate the flow dynamics of water and the nutrients in soils. Tea plants grown in loamy soils produce leaves containing higher proportions of polyphenols, caffeine, and amino acids that are related to an adequate supply of soil nutrients, increased microbial activity, and effectiveness of nitrogenous fertilizer. Soils with higher clay and sand contents are worse than loamy soil in terms of moisture and nutrient holding capacity along with favorable microbial activities (Zhang et al. 2017). Textural classes designed for potential tea soils were considered according to Gahlod et al. (2017).

#### 8.2.2.8 Soil pH

Tea usually grows well in soils with lower pH levels ranging from 4.5 to 5.5 (Natesan 1999). Soil pH may further decline with the use of nitrogenous fertilizers such as ammonium sulfate and urea for higher yields. For this, dolomite or lime is added to the tea soil as an ameliorating agent when the pH decreases to below 4.5. Moreover, tea plants take up larger quantities of Al ions from soil and thus require an adequate supply of exchangeable Al and Fe ions in the soil (Foy et al. 1978). On the other hand, the growth of tea plants in soils is stunted, and the mortality rate is due to a higher pH level and lower amounts of exchangeable Al, Fe, and Zn (Sultana et al. 2014). Classification with respect to soil pH was performed according to the guidelines of the Bangladesh Agricultural Research Council (Hussain et al. 2012).

#### 8.2.2.9 Drainage

Tea plants are sensitive to stagnant water and cannot survive in areas with persistent waterlogging (Deka et al. 2006; Mukhopadhyay and Mondal 2017). Under waterlogged conditions, tea bushes are thinner with retarded growth and strive to survive in the presence of water-tolerant weeds. Severely affected tea plantations under very poor drainage conditions display total crop failure with resulting bare soils. Moderately affected tea bushes are also thinner, showing the symptoms of yellowing and reddening of older leaves and failing to develop tillers (Rama Rao et al. 2007). Importantly, it has been reported that adequate drainage in tea soils can increase the yield by 30–35% (Deka et al. 2006). Tea plants are affected by excess water due to heavy precipitation in the Brahmaputra valley of Assam during the summer monsoons, causing a problem of surplus water disposal (Bhagat et al. 2010). Improved drainage is essential to provide adequate aeration in the root zone of tea plants for proper plant growth as well as to increase production. Drainage also prevents surface runoff to limit soil loss and maintain optimal soil moisture. Therefore, according to Nguyen et al. (2020), soil drainage was considered an important criterion for evaluating land suitability for tea.

#### 8.2.2.10 Soil Type

Tea grows well in the hilly regions of the northeastern part of Bangladesh, which consists of brown hill soils (Egashira et al. 2007). Piedmont soils are formed within the transition between hills and lowland plains due to the accumulation of sediments from hills. Gray piedmont soils are moderately suitable for tea with respect to their moderate drainage and acidic to almost neutral pH levels (Prokop and Ploskonka 2014; Akhtaruzzaman et al. 2014). Noncalcareous alluvium and brown, dark gray, and gray floodplain soils were formed due to sedimentation in the flood plains and are thus vulnerable with respect to drainage facilities and nutrient availability for tea plant growth (Egashira et al. 1998; Bhuiya 1987). Acidic basin clay soils are mostly observed in the Sylhet basin, but they are not important for agricultural production. Deep red-brown terrace soil is the core component of barren lands covered with grasses (Egashira et al. 2007). Peat soils are not suitable for tea production, as they are found in low-lying areas and wetland ecosystems in Bangladesh, where flooding and waterlogging are common features (Islam et al. 2017).

#### 8.2.2.11 Distance from Roads

There are three types of roads in the Sylhet division: highways, district roads, and local roads, including rural and urban roads. Highways are the main roads for large transportation across divisions. Distance from roads was considered because of the need to minimize the transportation costs associated with the tea cultivation input supply as well as the export of tea in the country and abroad using highways, districts- and local roads. Minimum distances between fields and roads facilitate the transport of inputs and collected tea leaves. Data for distance from roads were retrieved in polyline vector form and then converted into a raster. Spatial analysis was performed to measure the distances using the Euclidean distance. Reclassification of the distances from roads was performed according to Pramanik (2016).

#### 8.2.2.12 Distance from Rivers

The Sylhet division is crossed by the Surma, Kushiyara, Khowai, and Manu rivers as well as a large number of small rivers, which are part of the watershed in this area. Distances from rivers are a factor influencing the enhancement of tea production in the study area, as it may facilitate the transport of input materials as well as processed tea leaves. In addition, it may be the source of irrigation water during the drought season. The data for distances from rivers were retrieved as a polyline vector and then converted into raster data. Once the vector data were converted to raster form, they were analyzed using Euclidean distance to calculate the distance from rivers. The study area with respect to distance from rivers was extracted by masking followed by reclassification according to the shortest distances.

### *8.2.3 Digital Image Processing*

Geospatial data, including both image and feature datasets, were used in this research. Sentinel-2 images with a 20 m resolution were processed to generate the LULC and NDVI maps. Elevation data with a 30 m resolution from the Shuttle Radar Topography Mission (SRTM), NASA, were processed to form an elevation map. Vector data with a 1:50,000 scale for precipitation, temperature, slope, soil texture, soil pH, drainage, soil type, and distance from roads and distance from rivers were processed to form raster layers, followed by reclassification and weighted overlaying.

#### 8.2.3.1 NDVI Computation

The NDVI is the most popular index for monitoring vegetation health as well as crop yields (Rouse et al. 1973). The NDVI is generated from two important wave bands—near-infrared and red bands—and is measured as follows:

$$\mathrm{NDVI} = \frac{R_{\mathrm{NIR}} - R_{\mathrm{RED}}}{R_{\mathrm{NIR}} + R_{\mathrm{RED}}} \tag{8.5}$$

where $R_{\mathrm{NIR}}$ is the reflectance of near-infrared and $R_{\mathrm{RED}}$ is the reflectance of red. In this research, cloud-free Sentinel-2 image data were collected and utilized to measure the NDVI from band combinations of the NIR and red reflections using band 8 and band 4. The NDVI values were extracted according to the ground reference data. The ground reference data for tea were collected in 2019 during its active vegetative growth stage starting in March and ending in November.

#### 8.2.3.2 LAI Computation

The leaf area index (LAI) measured from remotely sensed data is an important parameter that can be effectively used for tea yield prediction (Tripathi et al. 2004). According to the correlation of the LAI with the NDVI observed in previous studies, the LAI was determined using the least square method and can be expressed as follows (Tewari et al. 2003; Tuvshinbayar et al. 2017):

$$\text{LAI} = 0.57 \times \exp\left(2.33 \times \text{NDVI}\right) \tag{8.6}$$

### *8.2.4 Reclassification of Criteria*

Reclassification was performed to interpret the data in raster form by substituting a new single value or by categorizing the ranges of values into a single value. The raster map for each criterion was reclassified into four classes: highly suitable (S1), moderately suitable (S2), marginally suitable (S3), and not suitable (N) (Fig. 8.4a–l).

### *8.2.5 Analytical Hierarchy Process (AHP)*

The AHP is a multicriteria decision-making process developed by Saaty (1990). In the first stage, the decision elements are expanded into a hierarchy that includes three classes consisting of the top class (goals), the middle class (criteria), and the bottom class (alternatives). The top class of the hierarchy is involved in the selection of goals. The middle class defines the criteria, and the bottom class defines alternative decisions. A survey questionnaire was used to obtain expert opinions on the relative significance of the criteria and factors. Comparisons for each factor pair were depicted as integer values of 1 (equal importance) to 9 (extreme difference), where a higher number denoted the alternative factor being more important than another (Table 8.3).

For instance, in the comparison between LULC and the NDVI, a score of 1 denoted that both factors were equally important for evaluating suitability, and a score of 9 indicated that LULC was more important than the NDVI. All the scores were collected in a pairwise comparison matrix, with the diagonal and reciprocal scores presented in the lower left-hand triangle. Reciprocal scores (1/3, 1/5, 1/7, and

**Table 8.3** Scale of preference for the AHP pairwise comparison (Saaty 1989)

| Scale | Degree of preference | Description |
|---|---|---|
| 1 | Equal importance | Two factors contribute equally |
| 3 | Moderate importance of one factor over other factor | Experience and judgment slightly favor one over another |
| 5 | Strong importance | Experience and judgment strongly favor one over another |
| 7 | Very strong importance | Experience and judgment very strongly favor one over another |
| 9 | Extreme importance | The evidence favoring one over another is of the highest possible order of affirmation |
| 2,4,6,8 | Intermediate values between two adjacent scales | When compromise is required |
| Reciprocals | Opposite of the above | Used for inverse comparisons |

1/9) were used in the row criterion that was observed as less important over the column criterion. In the second stage, the scoring of the criteria was performed via pairwise comparisons followed by scoring the scales of relative importance (Table 8.4).

The third stage involved the calculation of the matrix, ensuring consistency among the criteria in the pairwise comparison matrix. The AHP was also used to measure the normalized values for each criterion and alternative to determine the normalized principal eigenvectors and priority vectors. The pairwise comparison matrix was calculated according to the following expression:

$$\begin{bmatrix} C_{11}C_{12}C_{13} & \cdots & C_{1n} \\ C_{21}C_{22}C_{23} & \cdots & C_{2n} \\ \vdots\vdots\vdots & \cdots & \vdots \\ \vdots\vdots\vdots & \cdots & \vdots \\ C_{m1}C_{m2}C_{m3}\cdots C_{mn} \end{bmatrix} \tag{8.7}$$

The sum of each column of the pairwise matrix was expressed as follows:

$$C_{ij} = \sum_{n}^{i=1} C_{ij} \tag{8.8}$$

Each element of the matrix was divided by its column total to generate a normalized matrix:

$$X_{ij} = \frac{C_{ij}}{\sum_{i=1}^{n} C_{ij}} = \begin{bmatrix} X_{11}X_{12}X_{13} & \cdots & X_{1n} \\ X_{21}X_{22}X_{23} & \cdots & X_{2n} \\ \vdots\vdots\vdots & \cdots & \vdots \\ \vdots\vdots\vdots & \cdots & \vdots \\ X_{m1}X_{m2}X_{m3}\cdots X_{mn} \end{bmatrix} \tag{8.9}$$

**Table 8.4** Pairwise comparison for scoring the criteria for tea cultivation

| Criteria | LULC | NDVI | Elevation | Precipitation | Temperature | Slope | Soil texture | Soil pH | Drainage | Soil type | Distance from roads | Distance from rivers |
|---|---|---|---|---|---|---|---|---|---|---|---|---|
| LULC | 1 | 1 | 0.33 | 0.14 | 0.20 | 0.20 | 0.33 | 0.20 | 0.14 | 0.33 | 0.33 | 0.33 |
| NDVI | 1 | 1 | 0.33 | 0.11 | 0.14 | 0.20 | 0.33 | 0.20 | 0.14 | 0.33 | 0.33 | 0.33 |
| Elevation | 3 | 3 | 1 | 0.20 | 0.33 | 0.33 | 1 | 0.33 | 0.20 | 1 | 1 | 1 |
| Precipitation | 7 | 9 | 5 | 1 | 3.00 | 5.00 | 5 | 3 | 1 | 5 | 7 | 7 |
| Temperature | 5 | 7 | 3 | 0.33 | 1 | 3.00 | 5 | 5 | 0.20 | 5 | 5 | 5 |
| Slope | 5 | 5 | 3 | 0.20 | 0.33 | 1 | 3 | 1 | 0.33 | 3 | 3 | 3 |
| Soil texture | 3 | 3 | 1 | 0.20 | 0.20 | 0.33 | 1 | 0.33 | 0.20 | 1 | 3 | 3 |
| Soil pH | 5 | 5 | 3 | 0.33 | 0.20 | 3.00 | 1 | 1 | 0.33 | 3 | 5 | 5 |
| Drainage | 7 | 7 | 5 | 1 | 5.00 | 3.00 | 5 | 3.00 | 1 | 5 | 7 | 7 |
| Soil type | 3 | 3 | 1 | 0.20 | 0.20 | 0.33 | 1 | 0.33 | 0.20 | 1 | 1 | 1 |
| Distance from roads | 3 | 3 | 1 | 0.14 | 0.20 | 0.33 | 0.33 | 0.20 | 0.14 | 1 | 1 | 1 |
| Distance from rivers | 3 | 3 | 1 | 0.14 | 0.20 | 0.33 | 0.33 | 0.20 | 0.14 | 1 | 1 | 1 |

The sum of the normalized matrix column was divided by the number of criteria used ($n$) to calculate the weighted matrix of the priority criteria:

$$W_{ij} = \frac{\sum_{j=1}^{n} X_{ij}}{n} \begin{bmatrix} W_{11} \\ W_{12} \\ \vdots \\ W_{1n} \end{bmatrix} \tag{8.10}$$

The initial consistency vectors were generated by multiplying the pairwise matrix by the vector of weights:

$$\begin{bmatrix} C_{11}C_{12}C_{13} & \cdots & C_{1n} \\ C_{21}C_{22}C_{23} & \cdots & C_{2n} \\ \vdots\vdots\vdots & \cdots & \vdots \\ \vdots\vdots\vdots & \cdots & \vdots \\ C_{m1}C_{m2}C_{m3}\cdots C_{mn} & & \end{bmatrix} x \begin{bmatrix} W_{11} \\ W_{12} \\ \vdots \\ W_{1n} \end{bmatrix} = \begin{bmatrix} C_{11}W_{11} & C_{12}W_{12} & \cdots & C_{1n}W_{1n} \\ C_{21}W_{21} & C_{22}W_{22} & \cdots & C_{2n}W_{2n} \\ \vdots & \vdots & \vdots & \vdots \\ \vdots & \vdots & \vdots & \vdots \\ C_{m1}W_{m1} & C_{m2}W_{m2} & \cdots & C_{mn}W_{mn} \end{bmatrix}$$

$$= \begin{bmatrix} V_{11} \\ V_{12} \\ \vdots \\ V_{1n} \end{bmatrix} \tag{8.11}$$

The principal eigenvector ($\lambda_{max}$) was calculated by averaging the values for the consistency vector:

$$\lambda_{max} = \sum_{n}^{i} CV_{ij} \tag{8.12}$$

Eigenvalues were measured to determine the relative weights by averaging the rows of each matrix. The largest value for the eigenvector was equal to the number of criteria, and when $\lambda_{max} = n$, the judgments were consistent. Normalized eigenvalues were calculated to determine the weights of the priority criteria. The principal value suggested that all criteria were consistent in the pairwise comparison matrix (Table 8.5).

The judgments were also verified to measure the consistency index (CI), which was calculated as follows:

$$CI = \frac{\lambda_{max} - n}{n - 1} \tag{8.13}$$

Here, n is the total number of criteria. Saaty (1989) also suggested the consistency ratio (CR), which was compared with the consistency index and the random index (RI) (Aldababseh et al. 2018) (Table 8.6).

The consistency ratio was calculated as follows:

**Table 8.5** Normalized matrix of the criteria for tea cultivation

| Criteria | LULC | NDVI | Elevation | Precipitation | Temperature | Slope | Soil texture | Soil pH | Drainage | Soil type | Distance from roads | Distance from rivers |
|---|---|---|---|---|---|---|---|---|---|---|---|---|
| LULC | 0.022 | 0.020 | 0.014 | 0.036 | 0.018 | 0.012 | 0.014 | 0.014 | 0.035 | 0.013 | 0.010 | 0.010 |
| NDVI | 0.022 | 0.020 | 0.014 | 0.028 | 0.013 | 0.012 | 0.014 | 0.014 | 0.035 | 0.013 | 0.010 | 0.010 |
| Elevation | 0.065 | 0.060 | 0.041 | 0.050 | 0.030 | 0.020 | 0.043 | 0.023 | 0.050 | 0.038 | 0.029 | 0.029 |
| Precipitation | 0.152 | 0.180 | 0.203 | 0.250 | 0.272 | 0.293 | 0.214 | 0.203 | 0.248 | 0.188 | 0.202 | 0.202 |
| Temperature | 0.109 | 0.140 | 0.122 | 0.083 | 0.091 | 0.176 | 0.214 | 0.338 | 0.050 | 0.188 | 0.144 | 0.144 |
| Slope | 0.109 | 0.100 | 0.122 | 0.050 | 0.030 | 0.059 | 0.129 | 0.068 | 0.083 | 0.113 | 0.087 | 0.087 |
| Soil texture | 0.065 | 0.060 | 0.041 | 0.050 | 0.018 | 0.020 | 0.043 | 0.023 | 0.050 | 0.038 | 0.087 | 0.087 |
| Soil pH | 0.109 | 0.100 | 0.122 | 0.083 | 0.018 | 0.176 | 0.043 | 0.068 | 0.083 | 0.113 | 0.144 | 0.144 |
| Drainage | 0.152 | 0.140 | 0.203 | 0.250 | 0.454 | 0.176 | 0.214 | 0.203 | 0.248 | 0.188 | 0.202 | 0.202 |
| Soil type | 0.065 | 0.060 | 0.041 | 0.050 | 0.018 | 0.020 | 0.043 | 0.023 | 0.050 | 0.038 | 0.029 | 0.029 |
| Distance from roads | 0.065 | 0.060 | 0.041 | 0.036 | 0.018 | 0.020 | 0.014 | 0.014 | 0.035 | 0.038 | 0.029 | 0.029 |
| Distance from rivers | 0.065 | 0.060 | 0.041 | 0.036 | 0.018 | 0.020 | 0.014 | 0.014 | 0.035 | 0.038 | 0.029 | 0.029 |

**Table 8.6** Consistency random index (RI) (Saaty 1989)

| *n* | 1 | 2 | 3 | 4 | 5 | 6 | 7 | 8 | 9 | 10 | 11 | 12 | 13 | 14 | 15 |
|---|---|---|---|---|---|---|---|---|---|---|---|---|---|---|---|
| RI | 0 | 0 | 0.58 | 0.9 | 1.12 | 1.24 | 1.32 | 1.41 | 1.45 | 1.49 | 1.51 | 1.48 | 1.56 | 1.57 | 1.59 |

$$\mathrm{CR} = \frac{\mathrm{CI}}{\mathrm{RI}} \tag{8.14}$$

### 8.2.6 *Land Suitability Evaluation*

The land suitability evaluation for tea was conducted according to the classification guidelines proposed by the FAO. The guidelines for suitability classification were utilized to assess the suitability of each land unit for a particular use. According to the FAO's guidelines for land evaluation, it was initially determined whether the land was suitable (S) or not suitable (N). The suitable class (S) was further divided as required. In practice, three categories—S1, S2, and S3—were used to evaluate the lands for tea cultivation. Thus, the land suitability evaluation was performed for the prioritized criteria that were reclassified into four categories. Eventually, the suitability classes for tea were determined using the weighted overlay based on the fraction weights obtained from expert opinions.

$$\text{Weighted Overlay} = \sum_{i=1}^{n} C_i * W_n \tag{8.15}$$

where $C_i$ denotes the criterion ($i$) that was reclassified and $W_n$ denotes the number of criteria ($n$) that were weighted.

### 8.2.7 *Ground Reference Information and Field Survey*

Primary data were collected during the 2019 field survey. GPS waypoints for tea estate locations were collected around the Sylhet division using a handheld GPS locator (eTrex, Garmin, USA). These waypoints were used as references to determine the locations of tea estates. According to the statistical references, among the 135 tea estates in the study area, 91 were located in Moulvibazar, 25 in Habiganj, and 19 in Sylhet.

### 8.2.8 *Validation of Yield*

Yield data for three consecutive years (from 2017 to 2019) were collected from the Bangladesh Tea Board (BTB). The NDVI values were also extracted from Sentinel-2 satellite imagery for the years 2017 to 2019. The monthly average NDVI values were extracted for the active growing season (March to November) over the tea estates located in the study areas. The NDVI values were obtained according to the ground reference information collected during the field survey. The monthly average LAI values were derived from the NDVI values using geospatial techniques (Fig. 8.3). The yield data were compared in a scatter plot through regression analysis using both the NDVI and the LAI.

## 8.3 Results

### 8.3.1 Reclassification

The raster layers of criteria were reclassified in accordance with the suitability levels into highly suitable, moderately suitable, marginally suitable, and not suitable categories. In the classification of LULC, the user accuracy (UA) was 100% for

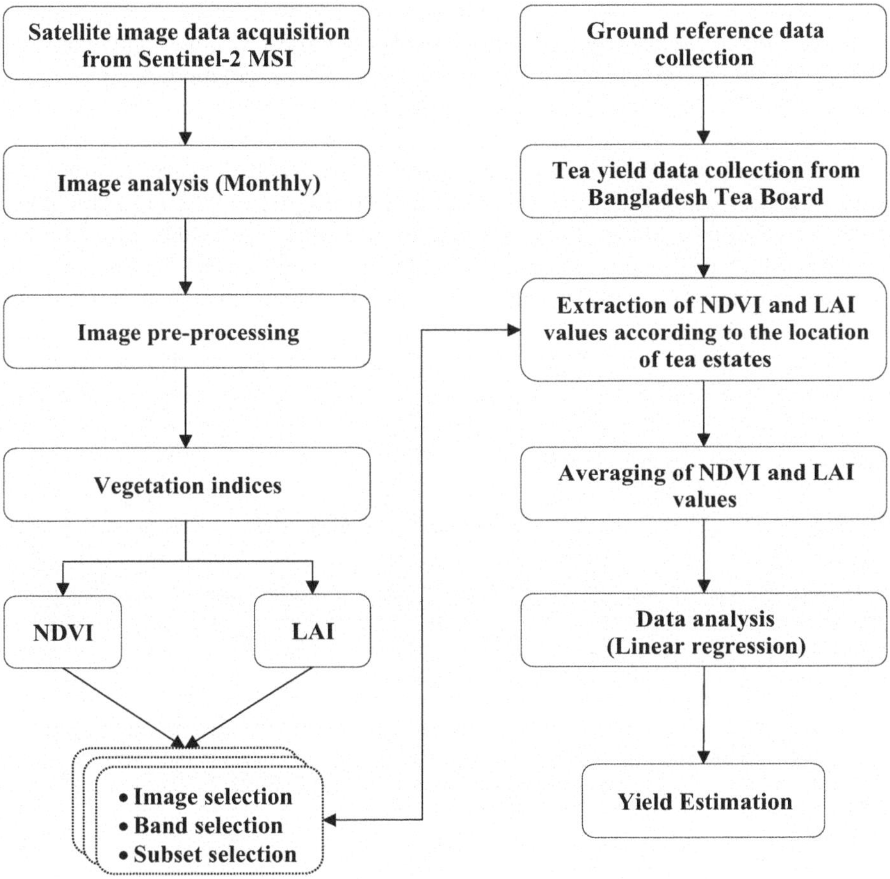


**Fig. 8.3** Yield estimation procedure from phenological datasets extracted from Sentinel 2 MSI

forests, tea estates, high agricultural lands, and wetlands; 85.71% for water bodies and settlements; and 71.43% for rivers. On the other hand, producer accuracy (PA) was 100% for tea estates, water bodies, settlements, high agricultural lands, and rivers and 87.5 and 72.73% for forests and wetlands, respectively (Table 8.7). The overall accuracy (OA) was 92%, and the Kappa coefficient was 90.65%.

In the reclassification of multicriteria, for LULC, 16.41% of lands (201,818 ha) were highly suitable, 10% (123,008 ha) were moderately suitable, 11.87% (145,964 ha) were marginally suitable, and 61.72% (759,050 ha) were not suitable. Considering the NDVI, 2.31% of the lands (28,362 ha) were highly suitable, 26.51% (325,998 ha) were moderately suitable, 67.77% (833,430 ha) were marginally suitable, and 3.42% (42,051 ha) were not suitable. According to elevation, 35.05% of the areas (431,045 ha) were highly suitable, 28.75% (353,533 ha) were moderately suitable, 22.28% (274,016 ha) were marginally suitable, and 13.92% (171,246 ha) were not suitable.

In the reclassification of precipitation, 38.18% of lands (469,553 ha) were in the highly suitable category, 46.54% (572,386 ha) were moderately suitable, and 15.28% (187,901 ha) were marginally suitable. In the reclassification of temperature, it was noted that 100% (1,229,840 ha) of lands were in the highly suitable category. In the case of slope, 14.73% of lands (181,103 ha) were highly suitable, 85.12% (1,046,828 ha) of lands (the majority of the area) were moderately suitable, and only 0.16% (1909 ha) of lands were marginally suitable. In the reclassification of soil texture, 71.36% (877,555 ha) of lands were highly suitable, 27.22% (334,729 ha) of lands were moderately suitable, and 1.43% (17,556 ha) of lands were marginally suitable. In the reclassification of soil pH, 13.99% (172,008 ha) of lands were highly suitable, 81.05% (996,795 ha) were moderately suitable, and 4.96% (61,037 ha) were marginally suitable. According to the classification of drainage, moderately well-drained to well-drained lands, which accounted for 13.36% (164,353 ha), were highly suitable; imperfectly drained lands, which accounted for 9.60% (118,083 ha), were moderately suitable; poorly drained lands, which accounted for 66.30% (815,421 ha), were marginally suitable; and very

**Table 8.7** Accuracy assessment for LULC

| Components | Forests | Tea estates | Water bodies | Settlements | High agril. land | Rivers | Wet lands | Total (user) | % Accuracy |
|---|---|---|---|---|---|---|---|---|---|
| Forests | 7 | 0 | 0 | 0 | 0 | 0 | 0 | 7 | 100 |
| Tea estates | 0 | 7 | 0 | 0 | 0 | 0 | 0 | 7 | 100 |
| Water bodies | 0 | 0 | 6 | 0 | 0 | 0 | 1 | 7 | 85.71 |
| Settlements | 1 | 0 | 0 | 6 | 0 | 0 | 0 | 7 | 85.71 |
| High agril. land | 0 | 0 | 0 | 0 | 7 | 0 | 0 | 7 | 100 |
| Rivers | 0 | 0 | 0 | 0 | 0 | 5 | 2 | 7 | 71.43 |
| Wetlands | 0 | 0 | 0 | 0 | 0 | 0 | 8 | 8 | 100 |
| Total (producer) | 8 | 7 | 6 | 6 | 7 | 5 | 11 | 50 | |
| % Accuracy | 87.5 | 100 | 100 | 100 | 100 | 100 | 100 | | |

poorly drained lands, estimated as 10.73% (131,983 ha), were in the not suitable category. Brown hill soils belong to the highly suitable category, accounting for 13.24% (162,782 ha), whereas gray piedmont soils, accounting for 11.84% (145,659 ha), belonged to the moderately suitable category. Noncalcareous alluvium, brown floodplain soils, dark gray floodplain soils, gray floodplain soils, acid basin clays, and deep red-brown terrace soils belonged to the marginally suitable category, accounting for 73.72% (906,642 ha), and peat soils, water bodies, and urban areas, accounting for 1.20% (14,757 ha), were in the not suitable category.

According to the reclassification of distance from roads, it was observed that 13.97% (171,759 ha) of lands were in the highly suitable category, 14.51% (148,404 ha) were moderately suitable, 21.81% (268,246 ha) were marginally suitable, and 49.72% (611,431 ha) were not suitable. The reclassification of distance from rivers showed that 6.23% (76,601 ha) of the total lands were highly suitable, 11.68% (143,618 ha) were moderately suitable, 18.03% (221,779 ha) were marginally suitable, and 64.06% (787,842 ha) were not suitable for tea cultivation in the Sylhet division of Bangladesh (Table 8.2) (Fig. 8.4).

### 8.3.2 *AHP Weights*

In the AHP analysis, the comparison of the criteria scale matrices was accomplished according to the experts' opinions, where judgments for ranking the criteria influenced the suitability classes of lands. Twelve criteria for the land suitability evaluation of tea were determined according to the baseline survey and a review of the literature. The AHP for the selected criteria was supported by a pairwise matrix, and the weights were determined from the normalized matrices based on expert knowledge to prioritize the criterion layer in the weighted overlay (Rashid 2019). The results of the AHP demonstrated that precipitation (23%) was highly influential, followed by drainage (19%), temperature (15%), soil pH (10%), and slope (8%). There were similar influences of elevation, soil texture and soil type (5%), LULC, and distance from rivers (3%), with the least influence of the NDVI and distance from roads (2%) (Table 8.8).

### 8.3.3 *Land Suitability*

The suitability map was developed using weighted overlay spatial analysis according to the AHP weights (Fig. 8.5). The result of the weighted overlay showed that 3.37% of the total lands (41,460 ha) were highly suitable, 9.01% (110,767 ha) were moderately suitable, 49.87% (613,367 ha) were marginally suitable, and 37.75% (464,246 ha) were not suitable (Table 8.9). It was also observed that among the 135 tea estates in the Sylhet division, 31 were located in highly suitable areas, 79 in moderately suitable areas, 24 in marginally suitable areas, and only one in a not suitable area (Fig. 8.6).

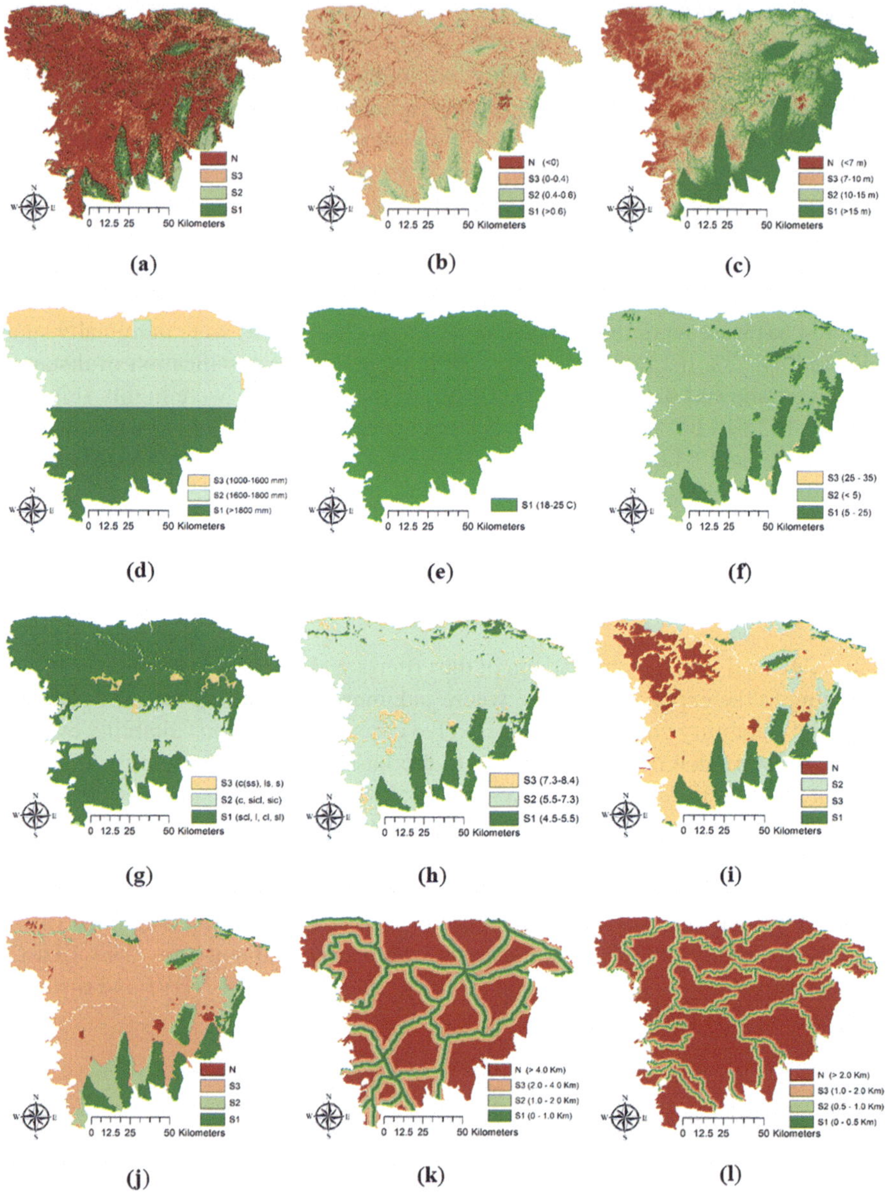


**Fig. 8.4** Reclassification of criteria: (**a**) NDVI, (**b**) LULC, (**c**) Elevation, (**d**) Precipitation, (**e**) Temperature, (**f**) Slope, (**g**) Soil texture, (**h**) Soil pH, (**i**) Drainage, (**j**) Soil type, (**k**) Distance from roads, and (**l**) Distance from rivers

**Table 8.8** AHP weights for the assessment of the relative importance of the criteria

| Criteria | Expert A (30 years) | Expert B (10 years) | Expert C (12 years) | Expert D (12 years) | Expert E (8 years) | Expert F (10 years) | Expert G (12 years) | Expert H (15 years) | Expert I (12 years) | Expert J (8 years) | Average | Weight |
|---|---|---|---|---|---|---|---|---|---|---|---|---|
| LULC | 0.024 | 0.023 | 0.022 | 0.033 | 0.037 | 0.038 | 0.030 | 0.018 | 0.019 | 0.020 | 0.026 | 3 |
| NDVI | 0.024 | 0.018 | 0.022 | 0.021 | 0.020 | 0.017 | 0.034 | 0.017 | 0.019 | 0.021 | 0.021 | 2 |
| Elevation | 0.035 | 0.040 | 0.051 | 0.088 | 0.058 | 0.068 | 0.054 | 0.040 | 0.063 | 0.047 | 0.054 | 5 |
| Precipitation | 0.214 | 0.244 | 0.248 | 0.200 | 0.249 | 0.258 | 0.202 | 0.217 | 0.232 | 0.205 | 0.227 | 23 |
| Temperature | 0.133 | 0.177 | 0.133 | 0.145 | 0.135 | 0.181 | 0.155 | 0.150 | 0.142 | 0.129 | 0.148 | 15 |
| Slope | 0.095 | 0.082 | 0.090 | 0.078 | 0.077 | 0.066 | 0.076 | 0.086 | 0.071 | 0.100 | 0.082 | 8 |
| Soil texture | 0.066 | 0.045 | 0.048 | 0.048 | 0.049 | 0.038 | 0.055 | 0.048 | 0.052 | 0.048 | 0.050 | 5 |
| Soil pH | 0.104 | 0.105 | 0.085 | 0.091 | 0.108 | 0.095 | 0.114 | 0.100 | 0.093 | 0.089 | 0.098 | 10 |
| Drainage | 0.186 | 0.182 | 0.191 | 0.200 | 0.178 | 0.157 | 0.189 | 0.219 | 0.207 | 0.226 | 0.194 | 19 |
| Soil type | 0.071 | 0.046 | 0.065 | 0.053 | 0.051 | 0.041 | 0.050 | 0.039 | 0.062 | 0.050 | 0.053 | 5 |
| Distance from roads | 0.024 | 0.019 | 0.023 | 0.023 | 0.020 | 0.020 | 0.020 | 0.033 | 0.020 | 0.020 | 0.022 | 2 |
| Distance from rivers | 0.024 | 0.019 | 0.023 | 0.023 | 0.020 | 0.020 | 0.020 | 0.033 | 0.020 | 0.046 | 0.025 | 3 |

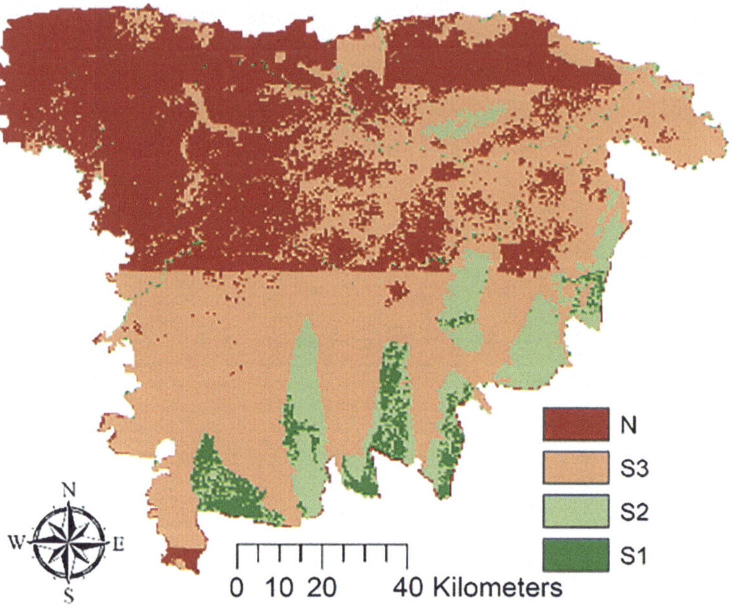


**Fig. 8.5** Suitability classes for tea estates

**Table 8.9** Areas in the land suitability classes of tea

| Suitability Level | Pixel counts | Area (%) | Area (ha) |
|---|---|---|---|
| S1 (highly suitable) | 1535 | 3.37 | 41,460 |
| S2 (moderately suitable) | 4101 | 9.01 | 110,767 |
| S3 (marginally suitable) | 22,709 | 49.87 | 613,367 |
| N (not suitable) | 17,188 | 37.75 | 4,642,451 |

### *8.3.4 Validation of Yield*

Yield estimation and validation were performed using both NDVI and LAI values with the observed yield data. The trendline obtained from the scatter plot showed the effects of the NDVI and the LAI on yield. According to the regression analysis between the NDVI and yield, the coefficients of determination were 0.69, 0.66, and 0.67, and between the LAI and yield, they were 0.68, 0.65, and 0.63 for 2017, 2018, and 2019, respectively (Fig. 8.7). The predicted yield maps were developed from the obtained linear equation using geospatial techniques. The red color in the map indicates the restricted area, and the light green to deep green color shows the tea-producing area (Fig. 8.8).

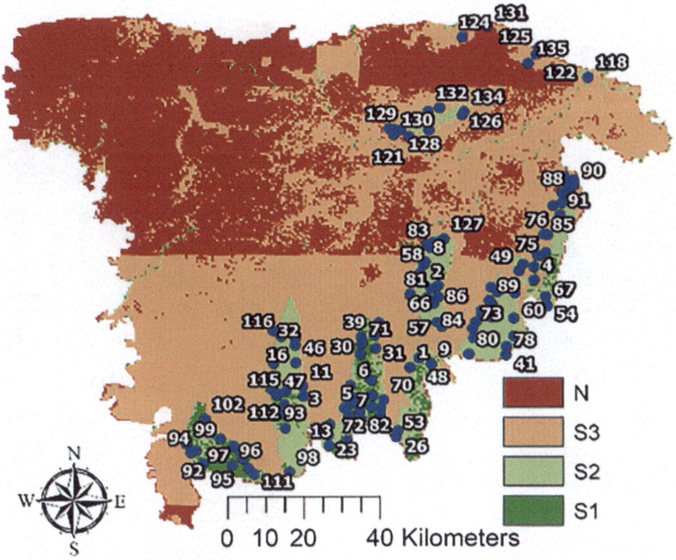


**Fig. 8.6** Validation of tea estates in different land suitability classes

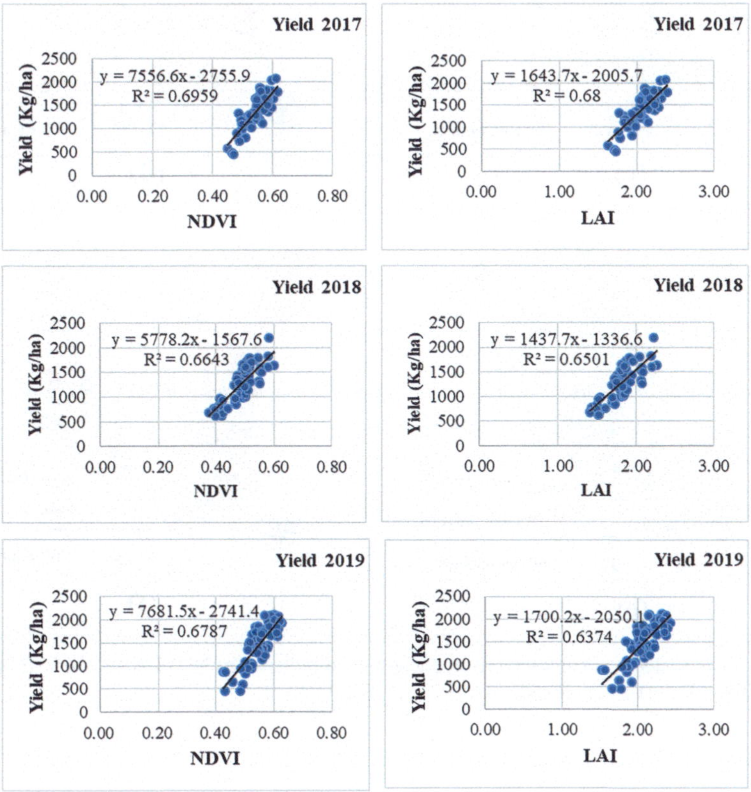


**Fig. 8.7** Regression analysis for yield prediction of tea using phenological indices and ground reference time series yield information

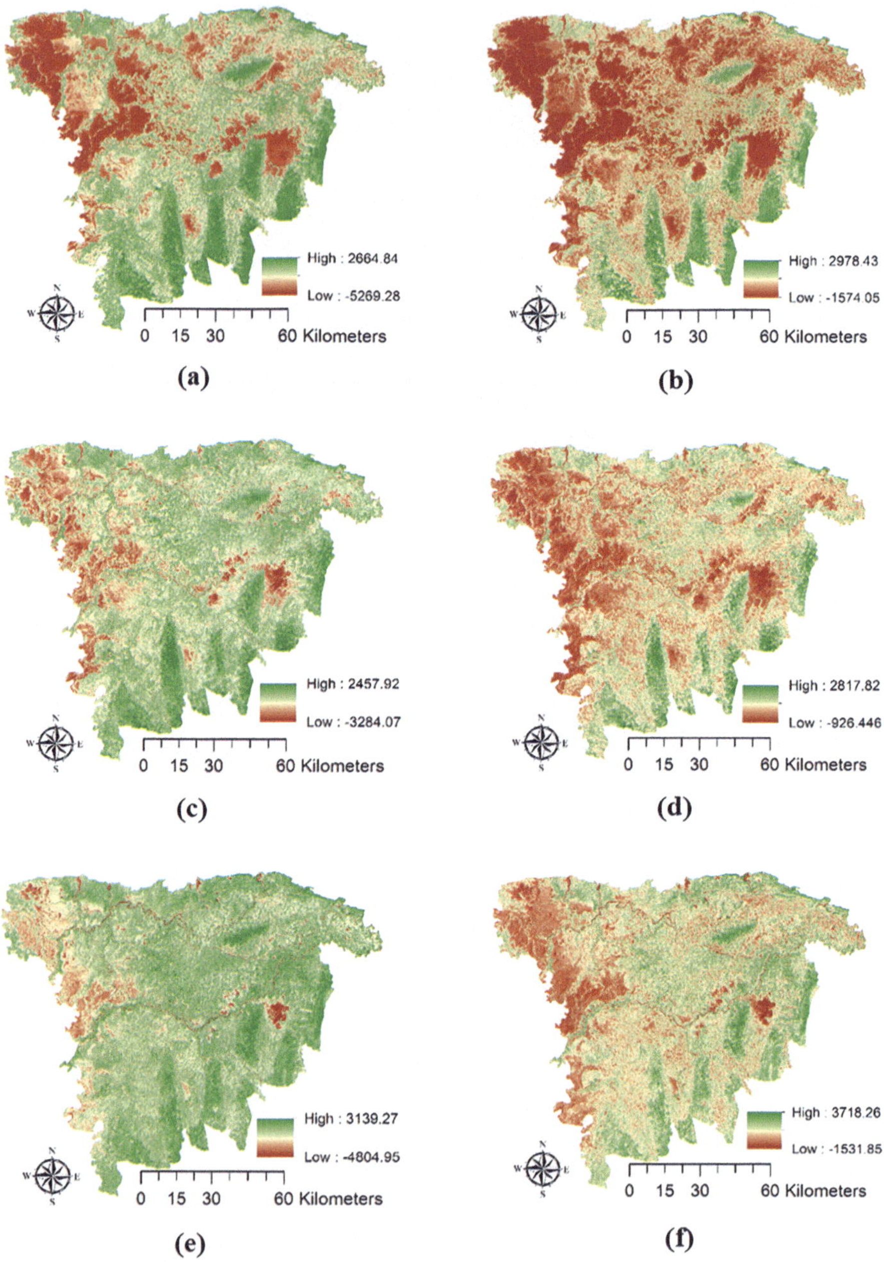


**Fig. 8.8** Yield maps of tea obtained from phenological indices: (**a**) NDVI 2017, (**b**) LAI 2017, (**c**) NDVI 2018, (**d**) LAI 2018, (**e**) NDVI 2019, and (**f**) LAI 2019

## 8.4 Discussion

Incorporation of Sentinel-2 MSI and geospatial datasets was significant in this study to assess environmental limitations as well as to evaluate land suitability for tea production (Hinton 2007). LULC classification and NDVI measurements were performed from Sentinel-2 datasets that serve as a high-resolution remote sensing data. Edaphic, climatic, and topographic factors are critical and important for sustainable tea production (Amarathunga et al. 2008). The SRTM digital elevation datasets, prepared by NASA, and used to generate the elevation layer of the study area, are significantly important in the digital mapping of terrain due to its accessibility of high-quality elevation data. Edaphic and climatic parameters for this study were selected according to the reference of the previous studies. Distance from roads is an important criterion with respect to transportation, and distance from rivers, with the advantages of irrigation facilities and transportation, was considered another criterion in this research (Appendix 1). An important step in the land suitability evaluation is to determine the weight of each criterion that affects suitability assessment (Jayasinghe et al. 2019). Multiple factors affect the land suitability evaluation because the criteria are of unequal importance (Elsheikh et al. 2013). In this study, a multicriteria decision-making process was used that integrates AHP with biophysical and remote sensing parameters. This study represents the application of AHP along with the weighted overlay model for land suitability evaluation of tea production resulting in a value of consistency ratio (CR) less than 0.1 (Purnamasari et al. 2019).

Most of the lands suitable for tea cultivation were in the southern and eastern parts of the Sylhet division. This result might be due to the suitable drainage system, slope, soil type, soil pH, and elevation in this area along with the most important factors, such as precipitation and temperature. On the other hand, around one-third portion of lands, mostly located in northwestern part of the Sylhet division, were not suitable due to the presence of wetlands that are not arable for tea cultivation along with other adverse edaphic factors. This research finds that drainage is an influencing factor after precipitation. One of the novel points of this research is the validation of yield using vegetative and biophysical indices based on time series NDVI and LAI datasets.

Previous research had the limitation of obtaining inappropriate validation results due to inadequate ground reference information. Validation of the results was accomplished in this research by physical verification with GPS identification of tea estate locations and corresponding time series yield data from tea estates. In previous studies on the land suitability evaluation of tea, only a few edaphic and climatic parameters were used to determine the areas in different suitability classes (Nguyen et al. 2020; Jayasinghe et al. 2019; Gahlod et al. 2017; Bo et al. 2012; Chanhda et al. 2010). However, this research integrated the use of geospatial and remote sensing data with AHP to locate tea estates in different suitability classes. The limitation of this research was disregarding the influence of shade trees on tea estates. A new method needs to be incorporated in future studies to remove the shade of trees from high-resolution remote sensing data.

## 8.5 Conclusions

This study launched a method of determining suitable lands for tea cultivation in Bangladesh utilizing GIS, satellite remote sensing, and AHP. Among the criteria used, precipitation had the greatest influence (23%), followed by drainage (19%), temperature (15%), and other factors. The weighted overlay using the AHP demonstrated that only 41,460 hectares (3.37%) of land were highly suitable, followed by 110,767 hectares (9.01%) of moderately suitable land. The majority of the area (613,367 hectares), which accounted for 49.87%, was marginally suitable, and a considerable portion of lands (464,246 hectares), estimated as 37.75%, were not suitable for tea cultivation. Among the 135 tea estates, 58% were in moderately suitable areas, 23% were in highly suitable areas, 18% were in marginally suitable areas, and less than 1% were in not suitable areas. The results of the land suitability evaluation for tea in Bangladesh would be very significant in the decision-making process to boost production as well as for the sustainable management of agricultural lands. Thus, land suitability evaluation is essential for understanding the future land use and production trend of tea for the growth of the tea industry in Bangladesh.

**Acknowledgments** The authors express thanks to the open-access journal *Remote Sensing* published by Multidisciplinary Digital Publishing Institute (MDPI) for their policy to support the authors for reusing the published article. In this regard, we would like to extend our gratitude to *Remote Sensing Journal* for publishing this article (Animesh Chandra Das, Ryozo Noguchi, Tofael Ahamed. Integrating an Expert System, GIS, and Satellite Remote Sensing to Evaluate Land Suitability for Sustainable Tea Production in Bangladesh. *Remote Sensing*. *12*(24):4136. https://doi.org/10.3390/rs12244136, 2020). Some minor modifications have been made to this book chapter. Furthermore, we would like to thank the University of Tsukuba for supporting this research to develop the multicriteria land suitability evaluation and yield forecasting of tea in Bangladesh. We also express our sincere thanks to the European Space Agency (ESA), the Bangladesh Agricultural Research Council (BARC), and the Bangladesh Tea Board (BTB) for remote sensing, geospatial, geographic, and statistical data. Finally, we express our gratitude to international experts and field enumerators involved in this research.

**Funding** This research received no external funding.

**Conflicts of Interest** The authors declare no conflict of interest.

## Appendix 1. Criteria for the Land Suitability Evaluation of Tea

| Criteria | Suitability class | Sub-criteria | References |
|---|---|---|---|
| LULC | S1 | Tea estates | Su et al. (2016), Su et al. (2017) |
| | S2 | Forest | Su et al. (2016), Su et al. (2017), Hajiboland (2017) |
| | S3 | High agricultural land | Su et al. (2017) |
| | N | Settlements, water bodies, rivers, and wetlands | Pezeshki (2001), Purnamasari et al. (2019) |
| NDVI | S1 | >0.6 | Choudhary et al. (2019) |
| | S2 | 0.4–0.6 | Choudhary et al. (2019) |
| | S3 | 0.4 | Choudhary et al. (2019) |
| | N | <0 | Choudhary et al. (2019) |
| Elevation | S1 | >15 m | Hajiboland (2017), Jayasinghe et al. (2019) |
| | S2 | 10–15 m | Hajiboland (2017), Jayasinghe et al. (2019) |
| | S3 | 7–10 m | Hajiboland (2017), Jayasinghe et al. (2019) |
| | N | <7 m | Hajiboland (2017), Jayasinghe et al. (2019) |
| Precipitation | S1 | >1800 mm | Gahlod et al. (2017) |
| | S2 | 1600–1800 mm | Gahlod et al. (2017) |
| | S3 | 1000–1600 mm | Gahlod et al. (2017) |
| Temperature | S1 | 18–25 °C | Gahlod et al. (2017) |
| Slope | S1 | 5–25° | Jayasinghe et al. (2019) |
| | S2 | <5° | Jayasinghe et al. (2019) |
| | S3 | >25° | Jayasinghe et al. (2019) |
| Soil texture | S1 | scl, l, cl, sl | Gahlod et al. (2017) |
| | S2 | c, sicl, sic | Gahlod et al. (2017) |
| | S3 | c(ss), ls, s | Gahlod et al. (2017) |
| Soil pH | S1 | 4.5–5.5 | Jayasinghe et al. (2019), Sultana et al. (2014), Hussain et al. (2012), Natesan (1999) |
| | S2 | 5.5–7.3 | Jayasinghe et al. (2019), Sultana et al. (2014), Hussain et al. (2012), Natesan (1999) |
| | S3 | 7.3–8.4 | Jayasinghe et al. (2019), Sultana et al. (2014), Hussain et al. (2012), Natesan (1999) |

| Criteria | Suitability class | Sub-criteria | References |
|---|---|---|---|
| Drainage | S1 | Moderately well drained to well drained | Nguyen et al. (2020), Sys et al. (1993) |
| | S2 | Imperfectly drained | Nguyen et al. (2020), Sys et al. (1993) |
| | S3 | Poorly drained | Nguyen et al. (2020), Sys et al. (1993) |
| | N | Very poorly drained | Sys et al. (1993) |
| Soil type | S1 | Brown hill soils | Egashira et al. (2007) |
| | S2 | Gray piedmont soils | Prokop & Płoskonka (2014), Akhtaruzzaman et al. (2014) |
| | S3 | Non-calcareous alluvium, Brown flood plain soils, Dark gray flood plain soils, gray flood plain soils, acid basin clays, deep-red brown terrace soils | Islam et al. (2017), Egashira et al. (2007), Egashira et al. (1998), Bhuiya (1987) |
| | N | Peat, water bodies, urban | Islam et al. (2017) |
| Distance from roads | S1 | 0–1.0 km | Pramanik (2016) |
| | S2 | 1.0–2.0 km | Pramanik (2016) |
| | S3 | 2.0–4.0 km | Pramanik (2016) |
| | N | >4.0 km | Pramanik (2016) |
| Distance from rivers | S1 | 0–0.5 km | Pramanik (2016) |
| | S2 | 0.5–1.0 km | Pramanik (2016) |
| | S3 | 1.0–2.0 km | Pramanik (2016) |
| | N | >2.0 km | Pramanik (2016) |

# References

Ahamed TN, Rao KG, Murthy JSR (2000) GIS-based fuzzy membership model for crop-land suitability analysis. Agric Syst 63(2):75–95. https://doi.org/10.1016/S0308-521X(99)00036-0

Ahamed T, Noguchi R, Takigawa T, Tian L (2016) Bioproduction engineering: automation and precision agronomics for sustainable agricultural systems. Nova Science Publishers, Inc., New York. https://novapublishers.com/shop/bioproduction-engineering-automation-and-precision-agronomics-for-sustainable-agricultural-systems-second-edition/. Last Accessed 15 Oct 2023

Ahmed S, Stepp JR, Orians C, Griffin T, Matyas C, Robbat A, Buckley S (2014a) Effects of extreme climate events on tea (Camellia sinensis) functional quality validate indigenous farmer knowledge and sensory preferences in tropical China. PLoS One 9(10):e109126

Ahmed T, De Costa WAJM, Wijeratne MA (2014b) Effects of different rates of Gibberellic Acid (GA3) on shoot growth dynamics and yield of mature clonal tea (Camellia sinensis L.). Tea J Bangladesh 43:1–9. https://teaboard.portal.gov.bd/sites/default/files/files/teaboard.portal.gov.bd/publications/c1e35269_c436_4254_aeba_fcb2aa912f72/Tea%20J.%20Bangladesh%20 2014.pdf. Last Accessed 15 Oct 2023

Akhtaruzzaman M, Haque ME, Osman KT (2014) Morphological, physical and chemical characteristics of hill forest soils at Chittagong University, Bangladesh. Open J Soil Sci 4:26–35. https://doi.org/10.4236/ojss.2014.41004

Aldababseh A, Temimi M, Maghelal P, Branch O, Wulfmeyer V (2018) Multi-criteria evaluation of irrigated agriculture suitability to achieve food security in an arid environment. Sustainability 10(3):803. https://doi.org/10.3390/su10030803

Amarathunga SLD, Panabokke CR, Pathiranage SRW, Amarasinghe L, Wijeratne MA (2008) Land suitability classification and mapping of tea lands in Ratnapura district. Sri Lanka J Tea Sci 73:1–10

Bandyopadhyay S, Jaiswal RK, Hegde VS, Jayaraman V (2009) Assessment of land suitability potentials for agriculture using a remote sensing and GIS based approach. Int J Remote Sens 30(4):879–895. https://doi.org/10.1080/01431160802395235

Bangladesh Tea Board (2020) Statistical handbook on Bangladesh tea industry 2019. Project Development Unit, Bangladesh Tea Board, Srimangal-3210, Moulvibazar, Sylhet, Bangladesh, vol VIII, pp 1–146. https://teaboard.portal.gov.bd/sites/default/files/files/teaboard.portal.gov.bd/monthly_report/fc0fdf97_6559_4f2c_ad17_d595843dae52/2020-06-16-18-11-37cad04cafc8b1c3ce3b2d90a02e018a.pdf. Last Accessed 15 Oct 2023

Bhagat RM, Baruah RD, Safique S (2010) Climate and tea [Camellia sinensis (L.) O. Kuntze] production with special reference to north eastern India: a review. J Environ Res Dev 4(4):1017–1028. https://www.cabdirect.org/cabdirect/abstract/2011322704

Bhattacharya A, Saini U, Joshi R, Kaur D, Pal AK, Kumar N, Gulati A, Mohanpuria P, Yadav SK, Kumar S, Ahuja PS (2014) *Osmotin*-expressing transgenic tea plants have improved stress tolerance and are of higher quality. Transgenic Res 23:211–223. https://doi.org/10.1007/s11248-013-9740-5

Bhuiya ZH (1987) Organic matter status and organic recycling in Bangladesh soils. Resour Conserv 13(2–4):117–124. https://doi.org/10.1016/0166-3097(87)90055-1

Bo LI, Zhang F, Zhang LW, Huang JF, Zhi-Feng JIN, Gupta DK (2012) Comprehensive suitability evaluation of tea crops using GIS and a modified land ecological suitability evaluation model. Pedosphere 22(1):122–130. https://doi.org/10.1016/S1002-0160(11)60198-7

Bozdağ A, Yavuz F, Günay AS (2016) AHP and GIS based land suitability analysis for Cihanbeyli (Turkey) County. Environ Earth Sci 75(9):813. https://doi.org/10.1007/s12665-016-5558-9

Chanhda H, Wu CF, Ye YM, Ayumi Y (2010) GIS based land suitability assessment along Laos-China border. J For Res 21(3):343–349. https://doi.org/10.1007/s11676-010-0080-5

Choudhary K, Shi W, Boori MS, Corgne S (2019) Agriculture phenology monitoring using NDVI time series based on remote sensing satellites: a case study of Guangdong, China. Opt Mem Neural Netw 28:204–214. https://doi.org/10.3103/S1060992X19030093

Clevers JGPW, Büker C, Van Leeuwen HJC, Bouman BAM (1994) A framework for monitoring crop growth by combining directional and spectral remote sensing information. Remote Sens Environ 50(2):161–170. https://doi.org/10.1016/0034-4257(94)90042-6

Collette BB, Nauen CE (1983) FAO species catalogue. Volume 2. Scombrids of the world. An annotated and illustrated catalogue of tunas, mackerels, bonitos and related species known to date (No. 125)

Cooper R, Morre DJ, Morre DM (2005) Medicinal benefits of green tea: part II. Review of anti-cancer properties. J Altern Complement Med 11:639–652. https://doi.org/10.1089/acm.2005.11.639

Deka A, Deka PC, Mondal TK (2006) Tea. In: Parthasarathy VA, Chattopadhyay PK, Bose TK (eds) Plantation crops-I. Naya Udyog, Calcutta, pp 1–148

Egashira K, Hagimine M, Moslehuddin AZM (1998) Fixed ammonium in some Bangladesh soils. Soil Sci Plant Nutr 44(2):269–272. https://doi.org/10.1080/00380768.1998.10414449

Egashira K, Aramaki K, Yoshimasa M, Takeda A, Yamasaki SI (2007) Variation of rare earth element of soils in the process of soil formation. Clay Sci 13(4–5):117–124. https://doi.org/10.11362/jcssjclayscience1960.13.117

Elsheikh R, Shariff ARBM, Amiri F, Ahmad NB, Balasundram SK, Soom MAM (2013) Agriculture Land Suitability Evaluator (ALSE): a decision and planning support tool for tropical and subtropical crops. Comput Electron Agric 93:98–110. https://doi.org/10.1016/j.compag.2013.02.003

Food and Agriculture Organization (1976) A framework for land evaluation, 1st edn. FAO, Rome

Foy CD, Chaney RL, White MC (1978) The physiology of metal toxicity in plants. Annu Rev Plant Physiol 29:511–526. https://doi.org/10.1146/annurev.pp.29.060178.002455

Gahlod NS, Binjola S, Ravi R, Arya VS (2017) Land-site suitability evaluation for tea, cardamom and rubber using Geo-spatial technology in Wayanad district, Kerala. J Appl Nat Sci 9(3):1440–1447. https://doi.org/10.31018/jans.v9i3.1381

Gunathilaka RD, Smart JC, Fleming CM (2017) The impact of changing climate on perennial crops: the case of tea production in Sri Lanka. Clim Chang 140(3–4):577–592. https://doi.org/10.1007/s10584-016-1882-z

Habibie MI, Noguchi R, Shusuke M, Ahamed T (2019) Land suitability analysis for maize production in Indonesia using satellite remote sensing and GIS-based multicriteria decision support system. GeoJournal 86:777–807. https://doi.org/10.1007/s10708-019-10091-5

Hajiboland R (2017) Environmental and nutritional requirements for tea cultivation. Folia Hortic 29(2):199–220. https://doi.org/10.1515/fhort-2017-0019

Hinton JC (2007) GIS and remote sensing integration for environmental applications. Int J Geogr Inf Syst 10:877–890. https://doi.org/10.1080/02693799608902114

Hussain SG, Chowdhury MKA, Chowdhury MAH (2012) Land suitability assessment and crop zoning of Bangladesh. Bangladesh Agricultural Research Council (BARC), Farmgate, Dhaka

International Tea Committee (2017) Annual bulletin of statistics. International Tea Committee Ltd, 1st Floor, Millbank Tower, 21–24 Millbank, London SW1P4QP, UK

Islam MN, Al-Amin M (2019) Life behind leaves: capability, poverty and social vulnerability of tea garden workers in Bangladesh. Labor Hist 60(5):571–587. https://doi.org/10.1080/0023656X.2019.1623868

Islam MA, Hasan MA, Farukh MA (2017) Application of GIS in general soil mapping of Bangladesh. J Geogr Inf Syst 9(5):604–621. https://doi.org/10.4236/jgis.2017.95038

Jayasinghe SL, Kumar L, Sandamali J (2019) Assessment of potential land suitability for tea (*Camellia sinensis* (L.) O. Kuntze) in Sri Lanka using a GIS-based multi-criteria approach. Agriculture 9(7):1–26. https://doi.org/10.3390/agriculture9070148

Kamruzzaman M, Shaw R (2018) Flood and sustainable agriculture in the Haor basin of Bangladesh: a review paper. Univ J Agric Res 6(1):40–49. https://doi.org/10.13189/ujar.2018.060106

Kamruzzaman M, Parveen S, Das AC (2015) Livelihood improvement of tea garden workers: a scenario of marginalized women group in Bangladesh. Asian J Agric Ext Econ Sociol 7:1–7. https://doi.org/10.9734/AJAEES/2015/18690

Khormali F, Ayoubi S, Kananro Foomani F, Fatemi A (2012) Tea yield and soil properties as affected by slope position and aspect in Lahijan area, Iran. Int J Plant Prod 1(1):99–111. https://doi.org/10.22069/IJPP.2012.529

Lambert JD, Yang CS (2003) Mechanisms of cancer prevention by tea constituents. J Nutr 133:3262–3267. https://doi.org/10.1093/jn/133.10.3262S

McDowell RW, Snelder T, Harris S, Lilburne L, Larned ST, Scarsbrook M, Curtis A, Holgate B, Phillips J, Taylor K (2018) The land use suitability concept: introduction and an application of the concept to inform sustainable productivity within environmental constraints. Ecol Indic 91:212–219. https://doi.org/10.1016/j.ecolind.2018.03.067

Mukhopadhyay M, Mondal TK (2017) Cultivation, improvement, and environmental impacts of tea. In: Oxford research encyclopedia of environmental science. https://doi.org/10.1093/acrefore/9780199389414.013.373

Natesan S (1999) Tea soils. In: Jain NK (ed) Global advances in tea science. Aravali Books International Pty Ltd, New Delhi, pp 519–532

Nguyen H, Nguyen T, Hoang N, Bui D, Vu H, Van T (2020) The application of LSE software: a new approach for land suitability evaluation in agriculture. Comput Electron Agric 173:1–16. https://doi.org/10.1016/j.compag.2020.105440

Pezeshki SR (2001) Wetland plant responses to soil flooding. Environ Exp Bot 46(3):299–312. https://doi.org/10.1016/S0098-8472(01)00107-1

Pramanik MK (2016) Site suitability analysis for agricultural land use of Darjeeling district using AHP and GIS techniques. Model Earth Syst Environ 2(2):56. https://doi.org/10.1007/s40808-016-0116-8

Prokop P, Ploskonka D (2014) Natural and human impact on the land use and soil properties of the Sikkim Himalayas piedmont in India. J Environ Manag 138:15–23. https://doi.org/10.1016/j.jenvman.2014.01.034

Purnamasari RA, Ahamed T, Noguchi R (2019) Land suitability assessment for cassava production in Indonesia using GIS, remote sensing and multi-criteria analysis. Asia Pac J Reg Sci 3:1–32. https://doi.org/10.1007/s41685-018-0079-z

Rama Rao N, Kapoor M, Sharma N, Venkateswarlu K (2007) Yield prediction and waterlogging assessment for tea plantation land using satellite image-based techniques. Int J Remote Sens 28(7):1561–1576. https://doi.org/10.1080/01431160600904980

Rashid MFA (2019) Capabilities of a GIS-based multi-criteria decision analysis approach in modelling migration. GeoJournal 84(2):483–496. https://doi.org/10.1007/s10708-018-9872-5

Rouse JW, Haas RH, Schell JA, Deering DW (1973) Monitoring vegetation systems in the Great Plains with ERTS. In: Proceedings of the third earth resources technology satellite—1 symposium; NASA SP-351, pp 309–317. https://ntrs.nasa.gov/citations/19740022614

Saaty TL (1989) Group decision making and the AHP. In: Golden BL, Wasil EA, Harker PT (eds) The analytic hierarchy process. Springer, Berlin, Heidelberg

Saaty TL (1990) How to make a decision: the analytic hierarchy process. Eur J Oper Res 48(1):9–26. https://doi.org/10.1016/0377-2217(90)90057-I

Su S, Zhou X, Wan C, Li Y, Kong W (2016) Land use changes to cash crop plantations: crop types, multilevel determinants and policy implications. Land Use Policy 50:379–389. https://doi.org/10.1016/j.landusepol.2015.10.003

Su S, Wan C, Li J, Jin X, Pi J, Zhang Q, Weng M (2017) Economic benefit and ecological cost of enlarging tea cultivation in subtropical China: characterizing the trade-off for policy implications. Land Use Policy 66:183–195. https://doi.org/10.1016/j.landusepol.2017.04.044

Sultana J, Siddique MNA, Kamaruzzaman M, Halim MA (2014) Conventional to ecological: tea plantation soil management in Panchagarh District of Bangladesh. J Sci Technol Environ Inform 01(01):27–37. https://doi.org/10.18801/jstei.010114.03

Sys C, Van Ranst E, Debaveye J, Beernaert F (1993) Land evaluation—part III—crop requirements. Agricultural Publications-No7. General Administration for Development Cooperation, Brussels. http://hdl.handle.net/1854/LU-233235

Tewari S, Kulhavy J, Rock BN, Hadas P (2003) Remote monitoring of forest response to changed soil moisture regime due to river regulation. J For Sci 49(9):429–438. https://doi.org/10.17221/4716-jfs

Tripathi NK, Rajapakse SS, Honda K (2004) Tea yield modeling based on satellite derived LAI. Geocarto Int 19(3):51–54. https://doi.org/10.1080/10106040408542317

Tuvshinbayar D, Erdenetuya B, Erkhembayar E, Batbileg B, Sarangerel J (2017) Some results of crop stress monitoring by remote sensing in northern Mongolia. Mong J Agric Sci 21(02):59–63. https://doi.org/10.5564/mjas.v21i02.906

Vitousek PM (1994) Beyond global warming: ecology and global change. Ecology 75(7):1861–1876. https://doi.org/10.2307/1941591

Wang F, Brent Hall G, Subaryono (1990) Fuzzy information representation and processing in conventional GIS software: database design and application. Int J Geogr Inf Syst 4(3):261–283. https://doi.org/10.1080/02693799008941546

Zhang WL, Xie H, Li T, Wang DD (2017) Effects of different soil texture and management pattern on soil properties and tea quality of tea garden. Hubei Agric Sci 10:7. http://en.cnki.com.cn/Article_en/CJFDTotal-HBNY201710007.htm

# Chapter 9
# Land Suitability Assessment for Arabica Coffee Plantations in Central Aceh Indonesia Using Fuzzy Membership Theory

**Ivana Joy Pauline Pangaribuan, Raka Thoriq Araaf, Tofael Ahamed, Arzyana Sunkar, and Syafitri Hidayati**

**Abstract** Arabica coffee is an important commodity for the people of Central Aceh, both in terms of economic and cultural value. This plant requires land suitability criteria to support productivity, but the existing global climate change conditions could lead to persistent changes in ecosystems. Geographical information system (GIS) learning has offered solutions by providing land suitability assessment to mitigate potential climate change problems. The aim of this study is to assess the land suitability for arabica coffee plantations in Central Aceh, Indonesia. The parameters used are land use and land cover (LULC), elevation, slope, temperature, and rainfall, which are then combined with fuzzy membership analysis, and the output is provided in the form of a land suitability map for arabica coffee plantations. Fuzzy systems are very helpful for further development and application in land suitability assessment by generating different values of each parameter, ranging from 0 to 1. The results show that the land suitability in Central Aceh can be summarized as follows: 43.77% comprises S1 (highly suitable), 37.51% comprises S2 (moderately suitable), 18.41% comprises S3 (marginally suitable), and 0.31% comprises N (not suitable). The analysis shows that the northern area of

I. J. P. Pangaribuan
Tropical Biodiversity Conservation, Faculty of Forestry and Environment, IPB University, Bogor, West Java, Indonesia

Graduate School of Science and Technology, University of Tsukuba, Tsukuba, Ibaraki, Japan

R. T. Araaf
Graduate School of Science and Technology, University of Tsukuba, Tsukuba, Ibaraki, Japan

T. Ahamed (✉)
Institute of Life and Environmental Sciences, University of Tsukuba, Tsukuba, Ibaraki, Japan
e-mail: tofael.ahamed.gp@u.tsukuba.ac.jp

A. Sunkar · S. Hidayati
Department of Forest Resources Conservation and Ecotourism, Faculty of Forestry and Environment, IPB University, Bogor, Indonesia



T. Ahamed (ed.), *Remote Sensing Application II*, New Frontiers in Regional Science: Asian Perspectives 77, https://doi.org/10.1007/978-981-97-1188-8_9

Central Aceh Province is included in land suitability class S1 or the high suitability class, which can be developed for the expansion of arabica coffee plantations in Central Aceh. Therefore, by determining the suitability state, farmers can better understand which areas should be prioritized for developing coffee plantations.

**Keywords** Arabica coffee · Land suitability assessment · GIS · Central Aceh

## 9.1 Introduction

Climate change has become a global issue resulting in changes in ecosystems, which will continue. The existing trend shows that the global surface temperature has increased by 0.99 °C from the previous period (1985–1900) (IPCC 2021) and is predicted to continue to increase until the end of the twenty-first century (by 1.8–4 °C) (UNFCCC 2007). Specifically, climate change affects agricultural aspects in terms of changes in temperature, rainfall, and extreme weather that could lead to reduced crop yields, increased pest infestations, and disease pressures. The agricultural and plantation sectors are enabled through the selection of crops suitable for the climate of a specific region and the application of appropriate farming methods. Therefore, agriculture is a bioindustry that depends on the climate and exhibits distinctive regional features. Evidence of the impact of climate change in Mongolia has been reported by Uuld and Magda (2019), showing that there has been a decrease in wheat production due to extreme climates in the form of prolonged droughts and high temperatures, causing damage to crops.

Notably, the coffee plant is very sensitive to the surrounding climatic conditions (Bacon et al. 2017). Global coffee production and sustainability could be threatened by current and ongoing climate changes (DaMatta et al. 2016). Coffee grown in tropical areas is highly vulnerable to the impacts of climate change, thus affecting productivity and quality. According to Gay et al. (2006) and Lin (2007), increasing temperature and rainfall levels could reduce the quality of flowering, fruiting, and seeds. Changes in rainfall patterns could also affect production, as coffee plants require certain amounts of water at different growth stages.

Indonesia is one of the fourth largest coffee-producing countries globally, with 1.241 million ha of coffee land consisting of robusta coffee (75.4%) and arabica coffee (24.6%) plantations. The demand for Indonesian arabica coffee continues to increase over time. Coffee is cultivated in almost all parts of Indonesia, one of which is Central Aceh Province, which is included in the fifth largest national coffee-producing provinces, especially after a regional coffee variety was established, namely, Gayo coffee (Kupi Gayo). Gayo coffee has been produced in the Gayo Highlands, Central Aceh Province, since 1926 and has become a source of economic income for the local community (Damayanti and Setadi 2019). The Gayo tribe is one of the tribes on the island of Sumatra, Indonesia, and is one of the oldest tribes in Indonesia. The Gayo community and coffee have become cultural and social units.

The first step in adapting to climate change is to estimate the land area suitable for coffee plantations. Land evaluation and assessment can be conducted to classify the sustainability of agricultural land, determine land management, and predict land productivity over time. This contributes to making effective strategic land use decisions in biophysics and ecosystems while minimizing the effects of climate change on agricultural productivity (Yitbarek et al. 2013). In previous studies, fuzzy systems have been applied in land assessment (Purnamasari et al. 2019; Tsiko and Haile 2011; Suh and Jeffrey 2016). Fuzzy systems can enable information processing, where the use of classical set theory and binary logic is impossible or difficult (Czabanski et al. 2017). The fuzzy system converts data with varying values, randomness levels, and uncertainty degrees into a membership function value to be used as a representative of the land suitability class (Zhang et al. 2015). Function values are standardized into a range between 0 and 1 (Suh and Jeffrey 2016). With a fuzzy system, the inaccuracy of the weighting process for each parameter can be minimized, so a fuzzy system is a suitable tool for decision-making purposes (24.33) and for assessing land suitability (Kurtener et al. 2008). By conducting a land suitability assessment, we can provide an early ideal frame to determine the level of land suitability for arabica coffee plantations and obtain a land suitability map for arabica coffee plantations in Central Aceh. Thus, the objective of this research is to conduct land suitability assessment for arabica coffee plantations in Central Aceh using fuzzy expert systems.

### *9.1.1 Gayo Coffee for the Community*

Coffee is the main source of life for the Gayo people. Among the Gayo people, coffee is traditionally offered when welcoming guests. For the Gayo people, coffee has economic, historical, scientific, and sociocultural value. Coffee has been cultivated from generation to generation. In addition to economic needs, coffee cultivation indirectly affects environmental sustainability. The community achieves cultivation in an agroforestry system (Hartoyo et al. 2023), which indicates that coffee plants are combined with layered plant canopies and the soil can be protected from direct impact by raindrops, which reduces the risk of erosion (Gemilang et al. 2021) and provides a water supply, especially since the community coffee plantations are located in the highlands. Gayo coffee received a Fair Trade Certified classification from the International Fair-Trade Organization on May 27, 2010, and a Geographical Indication (GI) certificate from the Indonesian Ministry of Law and Human Rights. However, Syakir and Surmaini (2017) issued a statement that climate change could reduce global coffee production by up to 50% by 2050. Arabica coffee plantations are very important for the local community, but coffee production must be accompanied by proper management to ensure productivity.

## 9.2 Methodology

### *9.2.1 Study Area*

The study area is located in Indonesia (Fig. 9.1a), Central Aceh Province (Fig. 9.1b and Fig. 9.1c). Geographically, this area is located between 4°10′33″ and 5°57′50″ north latitude and 95°15′40″–97°20′25″ north longitude. Central Aceh is an area with an altitude ranging from 200 to 2600 masl (Fig. 9.1). This area exhibits an average temperature of 23–28 °C. Considering the altitude and temperature, Central Aceh is suitable for coffee plantations. In 2021 (BPS 2021), as many as 61,945 residents of Central Aceh worked as farmers, with the main commodity of Arabica coffee. The area of arabica coffee plantations in Central Aceh has reached 49,891 ha and continues to increase every year (BPS 2021). This shows the importance of coffee plantations for the local people of Central Aceh.

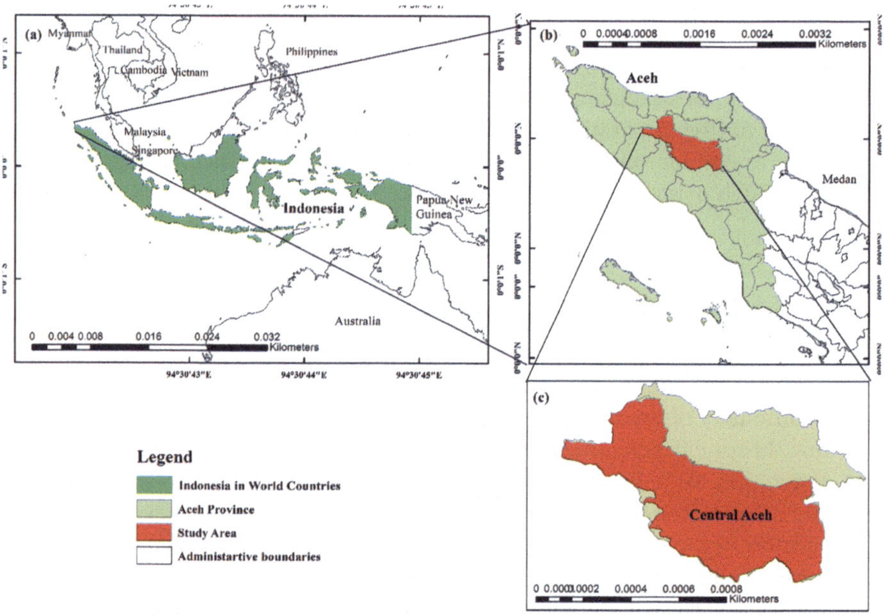


Fig. 9.1 Study area, Central Aceh Province, Indonesia

## 9.2.2 *Data*

### 9.2.2.1 Datasets

In this study, parameters and criteria were utilized that can be adjusted according to the land suitability assessment needs. The five most suitable parameters were selected based on a literature study for assessing land suitability. The parameters are climatic conditions, rainfall, temperature, physical conditions, land use and land cover (LULC), elevation, and slope. We used a Landsat image of the current study location to analyze the spatial difference between the existing land use and land suitability patterns. Data were collected from the Landsat 8 OLI for 2020 (Table 9.1).

**Table 9.1** Selected parameters and data sources in this study

| Criteria | Source | Landsat image |
|---|---|---|
| LULC | ESRI, livingatlas.arcgis.com/landcoverexplorer/ | 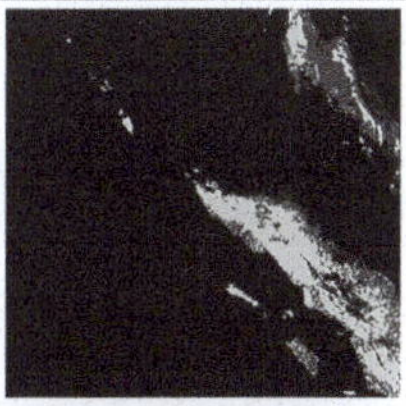 |
| Elevation | Indonesian Geospatial Digital Elevation Model (DEM) National | 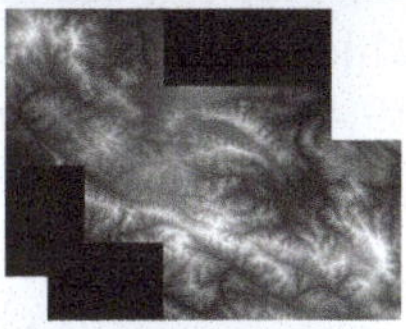 |
| Slope | Indonesian Geospatial—National Batimetri | 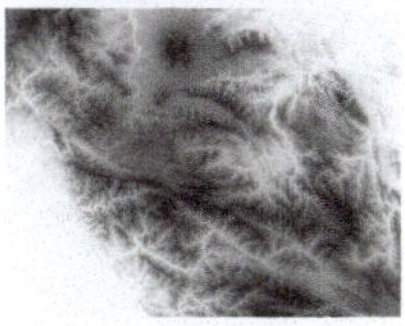 |
| Temperature 2020 | USGS, Modis MOD 11A2V6 2020 | 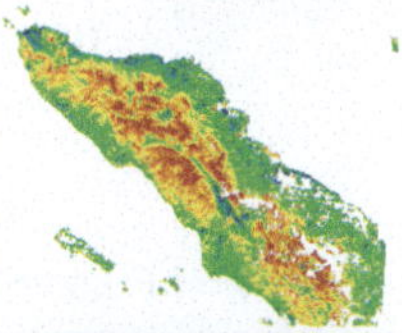 |
| Rainfall 2020 | Climate Hazards Group InfraRed Precipitation with Station data (CHIRPS) | 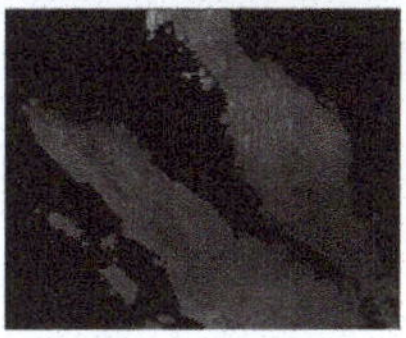 |

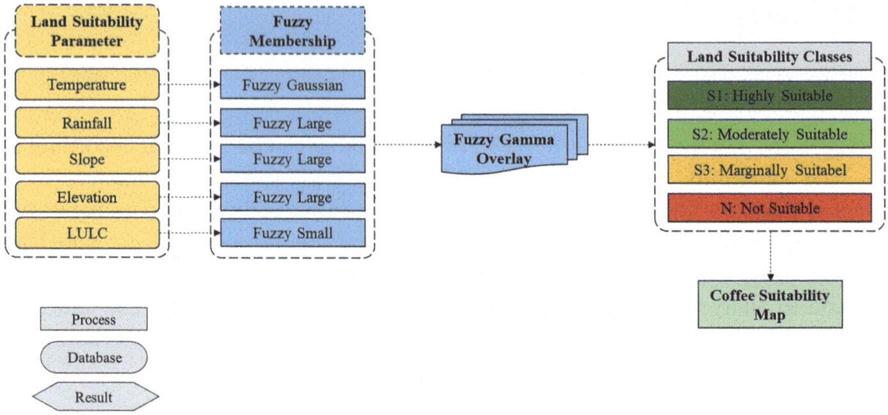


**Fig. 9.2** Research framework for developing land suitability assessment for arabica coffee plantations

### 9.2.2.2 Research Framework

The research framework, as shown in Fig. 9.2, covers four stages:

(a) Identify the most suitable arabica coffee land suitability parameters based on the literature, (b) analyze the temperature, rainfall, slope, elevation, and land use and land cover (LULC) in ArcGIS, (c) reclassify the fuzzy membership analysis results (FM Gaussian (9.1); FM large (9.2); and FM small (9.3)) to obtain the land suitability class, which can be prioritized and weighted using the fuzzy gamma overlay technique (9.4), and (d) generate a arabica coffee land suitability map based on the land suitability parameter.

$$\mu(x) = e^{\left(-f1*(X-f2)^2\right)} \tag{9.1}$$

$$\mu(x) = \frac{1}{1+\left(\frac{x}{f2}\right)^{-f1}} \tag{9.2}$$

$$\mu(x) = \frac{1}{1+\left(\frac{x}{f2}\right)^{f1}} \tag{9.3}$$

$$\mu(x) = (\text{FuzzySum})^{\gamma} \times (\text{FuzzyProduct})^{1-\gamma} \tag{9.4}$$

Notably, $f_1$ is the spread, and $f_2$ is the midpoint.

### 9.2.3 *Parameter Classification*

Parameters must be determined to measure the suitability of agricultural land; in this case, the determined parameters can provide reasonable and suitable land standards for arabica coffee commodities. In previous research, Estrada et al. (2017) used eight parameters, Abdisa (2020) employed four parameters, and Rahmawaty et al. (2021) adopted up to 13 parameters. In this study, a total of five parameters were selected based on the literature (Ritung et al. 2011) and the suitability of the study area. The parameters selected are related to climatic and physical conditions (Table 9.2). The parameters were mapped and classified by the FAO land suitability classification, as follows: Class S1 (highly suitable), Class S2 (moderately suitable), Class S3 (marginally suitable), and N (not suitable).

#### 9.2.3.1 Temperature

Temperature influences physiological activity and plant metabolic levels the most (Silva and Mazafera 2008). Every plant has a different temperature tolerance; for coffee plants, temperature affects the vegetative growth period (Drinnan and Menzel 1995). Low temperatures reduce the vegetative growth period, decrease the flower initiation period, and promote the emergence of floral onset. High temperatures can affect the balance of respiration and photosynthesis (DaMatta and Ramalho 2006). The appropriate temperature for coffee plants ranges from 25 to 28 °C (Table 9.2). Temperature datasets were obtained from the USGS with the MODIS MOD11A2 V6 dataset for land surface temperature (LST) analysis. The LST was calculated using Landsat 8 OLI data, which were selected during the period of low cloud coverage in 2020.

**Table 9.2** Arabica coffee growing conditions

| Parameter | S1 | S2 | S3 | N | Reference |
|---|---|---|---|---|---|
| Temperature | 25–28 | 22–25<br>28–32 | 20–22<br>32–35 | <20<br>>35 | Ritung et al. (2011) |
| Rainfall | 2000–3000 | 1750–2000 | 1500–1750 | <1250<br>>4000 | Ritung et al. (2011) |
| Slope | 0–8 | 8–16 | 16–30 | >30 | Ritung et al. (2011) |
| Elevation | 1100–1500 | 1500–1750 | 1750–2000 | <1100<br>>2200 | FAO (1984) |
| LULC | Agricultural Land | Forest Area | Grazing Land | Settlement and Water Body | Abdisa (2020) |

#### 9.2.3.2 Rainfalltriad

Rainfall refers to the total amount of water reaching the ground surface during a certain period. Due to the different topographic conditions across Indonesia, rainfall values vary according to the altitude of the area (Triatmodjo 2008). Excessive rainfall conditions can affect flower and fruit conditions, the potential for disease, excessive fermentation, and coffee bean defects (Taniwaki et al. 2014). We collected rainfall data for 2020 from the Climate Hazards Group InfraRed Precipitation with Station data (CHIRPS). For assessment, the data were reanalyzed and reclassified in ArcMap to obtain the annual average temperature and rainfall.

#### 9.2.3.3 Slope

The slope of land is related to the availability of nutrients in soil, and a steep slope could inhibit the storage of nutrients in soil; the steeper the land, the less nutrients can be stored (Arvi et al. 2019). Increasing the nutrient levels can support the growth of arabica coffee. The appropriate slope for arabica coffee is less than 8% (Table 9.2). Data on the slope in the study area were obtained from the Indonesia Geospatial website, National Bathymetry.

#### 9.2.3.4 Elevation

As plants that are sensitive to environmental conditions, coffee plants have certain height criteria to produce quality varieties, chiefly affecting the taste and quality of arabica coffee (Qadry et al. 2017). Land elevation data or digital elevation model (DEM) data were obtained through the Indonesian National Geoportal (https://tanahair.indonesia.go.id/demnas/#/).

#### 9.2.3.5 Land Use/Land Cover

Land use refers to the use of land by humans, especially land used for economic activities. Moreover, land cover refers to human social, cultural, and economic activities that influence the outline and form of the environment. LULC analysis was conducted to determine the land use and utilization in the study area that could affect arabica coffee growth. The LULC datasets in this study were collected from the ESRI (livingatlas.arcgis.com/landcoverexplorer/).

## 9.2.4 *Data Analysis*

### 9.2.4.1 Fuzzy Membership and Fuzzy Gamma Overlay

One GIS method that can be used for land suitability assessment is fuzzy logic modeling. This method involves the application of the fuzzy membership and fuzzy overlay tools in ArcGIS. Fuzzy membership (FM) can address the problem of inaccuracy in geographic database management, spatial data analysis, and spatial decision support system development. The values of the different parameters are transformed and standardized through fuzzification into a common scale from 0 to 1 for determining their membership of the set, so the values can be compared and layered among the parameters. A value close to 0 indicates that the data are not a member of the set because they do not fulfill the variable requirement, while a value close to 1 indicates that the data are a member of the set because they satisfy the requirement (Adhikari and Li 2013).

There are seven fuzzy membership functions in ArcGIS, and in this study, the most suitable functions are used after all the FM calculations are completed (Table 9.3), such as Gaussian, large, and small. The results for each parameter are reclassified with the natural breaks (Jenks) tool, thereby categorizing each parameter into four classes: S1 (highly suitable), S2 (moderately suitable), S3 (marginally suitable), and N (not suitable). After the data are fuzzified, in each layer, each criterion is combined using the fuzzy overlay function to determine the relationships and interactions among the criteria. In this research, the fuzzy gamma overlay function is employed.

**Table 9.3** Fuzzy suitability membership function

| | | FM function | | | |
|---|---|---|---|---|---|
| No | Parameter | Midpoint | Spread | FM Type | Equation |
| 1 | Temperature | 11.42 | 0.1 | Gaussian | $\mu(x) = e^{\left(-0.1*(X-11.42)^2\right)}$ |
| 2 | Rainfall | 2584.01 | 5 | Large | $\mu(x) = \frac{1}{1+\left(\frac{x}{2584.01}\right)^{-5}}$ |
| 3 | Slope | 1507 | 5 | Large | $\mu(x) = \frac{1}{1+\left(\frac{x}{1507}\right)^{-5}}$ |
| 4 | Elevation | 2584.01 | 5 | Large | $\mu(x) = \frac{1}{1+\left(\frac{x}{2584.01}\right)^{-5}}$ |
| 5 | LULC | 9.5 | 0.1 | Small | $\mu(x) = \frac{1}{1+\left(\frac{x}{9.5}\right)^{0.1}}$ |

#### 9.2.4.2 Data Reclassification

Land suitability classes capture the level of suitability. In this study, the suitability classes refer to the FAO classification (1983), which can be categorized into 4 groups:

1. **S1,** highly suitable: Land with no significant limitations or only minor limitations to sustain a given land use type without significant reduction in productivity or benefits and does not require major inputs above the acceptable level.
2. **S2,** moderately suitable: Land with moderately severe limitations overall for sustained application of the given land use type; the limitations can reduce productivity or benefits and increase the needed inputs to the extent that the overall advantage to be gained from the use of land, although still attractive, is appreciable less than that expected from Class S1 land.
3. **S3,** marginally suitable: Land with severe limitations overall for sustained application of the given land use type; the limitations can reduce productivity or benefits or increase the needed inputs to the extent that any expenditure can only be marginally justified.
4. **N,** not suitable: Not suitable, as the range of inputs needed is unjustifiable.

## 9.3 Results and Discussion

### *9.3.1 Arabica Coffee Land Suitability Criteria*

Arabica coffee is a type of coffee known for its more delicate and complex flavor relative to robusta coffee. This crop requires special attention in the selection of the planting area, including surrounding environmental conditions and agricultural factors, to ensure optimal plant growth and harvesting results. Arabica coffee cultivation requires specific land suitability criteria such as rainfall, ambient temperature, rainfall, elevation, slope, and land cover. Each criterion has a certain threshold for ensuring optimal arabica coffee growth, which can be assessed by land suitability analysis. Land suitability assessment can help optimize arabica coffee growth, improve the crop quality, mitigate risks, improve the resource efficiency, promote sustainable agriculture practices, and address environmental changes such as climate change.

Arabica coffee plantations are sensitive to temperature conditions. An inappropriate ambient temperature could cause plant stress and reduce the growth rate, especially during flower development. The temperature in Aceh Tengah ranges from 12.91 to 36.49 °C (Fig. 9.3a), and the highland area of Aceh Tengah provides suitable conditions for arabica coffee growth due to the lower temperature and higher elevations. Elevation and temperature are two interrelated criteria, where the higher the elevation value is, the lower the temperature. Arabica coffee grown at elevations less than 900 masl tends to experience coffee leaf rust, while an elevation above

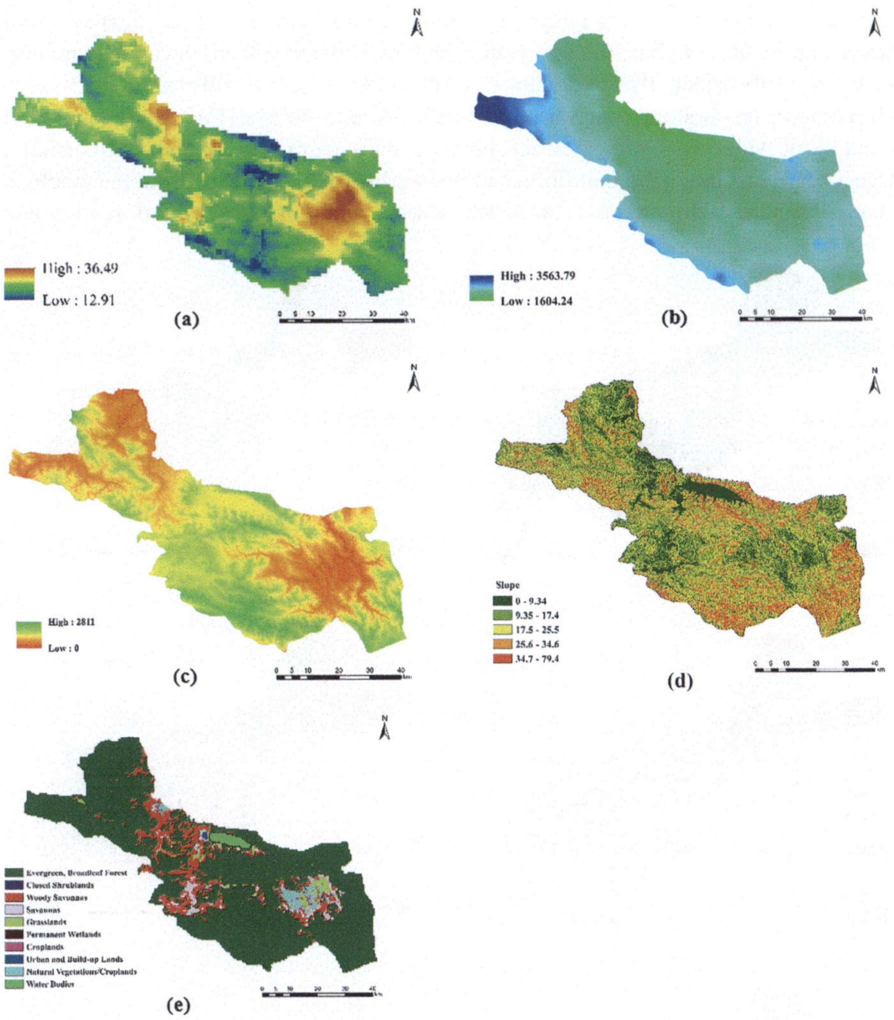


**Fig 9.3** Spatial distribution of land suitability for each criterion: (**a**) Temperature, (**b**) rainfall, (**c**) elevation, (**d**) slope, and (**e**) LULC

1700 m could affect the reproductive ability of arabica coffee plants (Elyanti et al. 2012). In addition, the elevation affects rainfall levels.

As a tropical country, rainfall is the dominant climatic factor in Indonesia, and its value varies in each region. Central Aceh exhibits the lowest rainfall value of 1604.24 mm/year and the highest rainfall value of 3563.79 mm/year (Fig. 9.3b). Rainfall can provide a natural source of water for arabica coffee, and the presence of water is necessary for nutrient transportation, cell formation, and metabolic processes. However, excessive rainfall can cause waterlogging that could damage the roots, while a lack of rainfall could cause drought and plant stress, hindering production and reducing the coffee fruit quality. The slope determines various landscape conditions, such as the soil water content, erosion potential, runoff, and surface and subsurface flow velocities. Central Aceh, as a hilly area (the Gayo Highlands), has a slope that varies between 0% and 79.5% (Fig. 9.3d), while the ideal slope value for arabica coffee plants ranges from only 0–8%. Furthermore, LULC explains how land conditions are viewed from a remote sensing perspective. Aceh Tengah, with an area of 4318.39 km², encompasses varied land cover

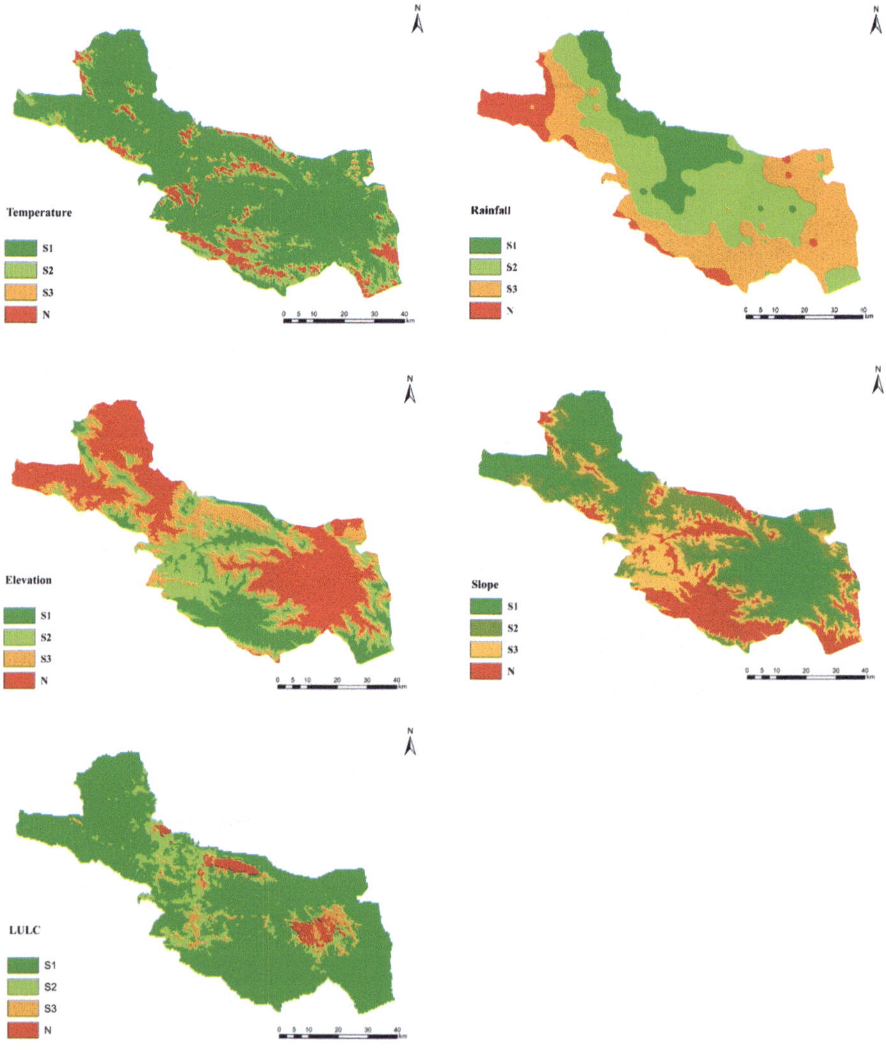


**Fig 9.4** Reclassification of the criteria used in land suitability assessment: Temperature, rainfall, elevation, slope, and LULC

conditions. Spatial analysis shows that there are 10 land cover groups, with the main group of green areas (Fig. 9.3e).

Furthermore, the five criteria (temperature, rainfall, elevation, slope, and LULC) are fuzzified according to their membership functions. This fuzzification process aims to convert all data values into a range of 0–1 to generalize, integrate, and solve the different magnitudes of the data in each layer. Fuzzy sets have been applied in land suitability analysis in previous studies to determine the suitability of land for various types of crops, such as avocado (Selim et al. 2018) and grapes (Arab and Ahamed 2022). In this research, Gaussian, large, and small fuzzy membership functions are employed (Table 9.3). After each criterion is fuzzified, the land suitability class is divided into four groups (S1, S2, S3, and N) (Fig. 9.4).

### *9.3.2 Land Suitability for Arabica Coffee Plantations in Central Aceh*

This study reveals that most of the highly suitable areas occur in the northern part of Aceh Tengah Province. These areas include Ketol Subdistrict, Bintang Subdistrict, Kute Panang Subdistrict, Kebayakan Subdistrict, LUT Tawar Subdistrict, Bebesen Subdistrict, Bies Subdistrict, and Silih Narah Subdistrict.

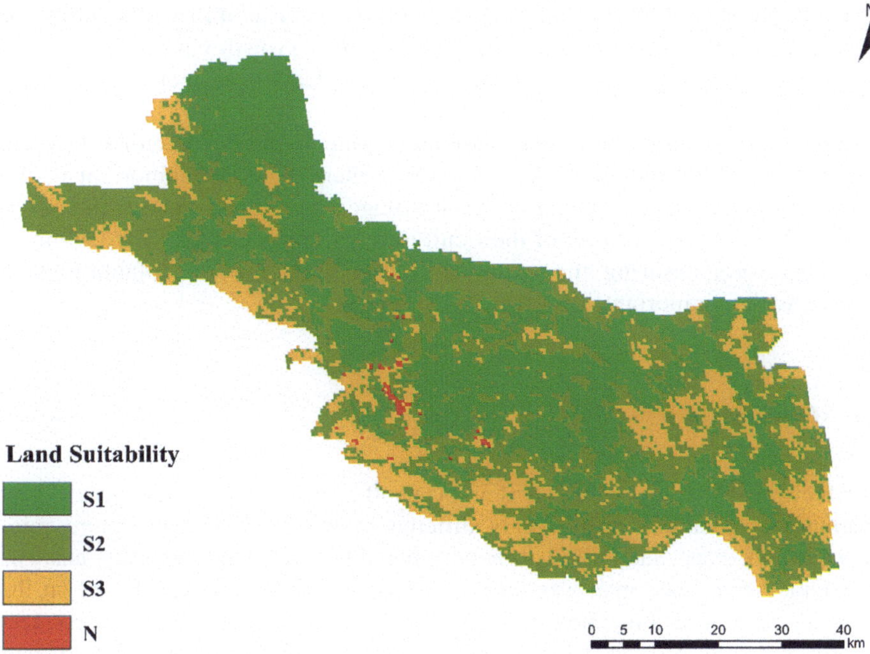


**Fig 9.5** Land suitability map of arabica coffee plantations in Central Aceh

**Table 9.4** Suitable areas for arabica coffee plantations in Central Aceh

| Suitability classes | Area (ha) | Area (%) |
|---|---|---|
| S1—highly suitable | 195,122.02 | 43.77 |
| S2—moderately suitable | 167,203.44 | 37.51 |
| S3—marginally suitable | 82,090.56 | 18.41 |
| N—not suitable | 1371.42 | 0.31 |
| Total | 445,787.44 | 100 |

According to Gayo et al. (2018), the Silih Narah and Bies subdistricts are among the subdistricts with the third largest amount of arabica coffee land in Central Aceh after the Pegasing subdistrict. Unfortunately, the Pegasing subdistrict occurs in the south of Central Aceh, indicated as a marginally suitable area in Fig. 9.5. This may suggest that most coffee cultivation areas in Central Aceh are of the S3 suitability class rather than the S1 suitability class (Table 9.4).

One potential area for arabica coffee cultivation can be found in the Lut Tawar Subdistrict, specifically Nosar village (divided into three villages: Mude Nosar, Bamil Nosar, and Bale Nosar). The main occupation of the people of Nosar village is farming, including rice, onions, fruit trees, and coffee. The geographical conditions of Nosar village, which is located between Lake Lut Tawar and Mount Bur Kelieten, facilitate the adoption of different types of planting methods depending on the altitude of the area. On the lakeside, the community grows rice and onions using a crop rotation system to ensure food and economic availability every month. On the mountainside, the community grows coffee and fruit trees. The plantation system is applied in agroforestry. This is implemented by the community to ensure the availability of shading for arabica coffee plants, which supports arabica coffee growth via sun supply and pest management (Beer et al. 1998). The shade plants used include *Leucaena leucocephala* and *Artocarpus heterophyllus*.

Most of the community owns arabica coffee plantations, either through inheritance or land purchase. Arabica coffee plantations are important to the people of Central Aceh, especially Nosar village, not only because of their economic value but also because of their cultural significance to the Gayo community. Therefore, ensuring the sustainability of arabica coffee plantations in Central Aceh is important.

## 9.4 Conclusion

To achieve sustainable production of arabica coffee, it is crucial to identify the best location for plantations. In this study, different criteria, such as temperature, rainfall, elevation, slope, and LULC, were considered in coffee land suitability analysis of Central Aceh. Each map was then subjected to overlay analysis based on the fuzzy function. The utilization of fuzzy functions in land suitability assessment research serves to resolve data uncertainty and imprecision. In this study, it was found that the area with the most coffee plantations (Pegasing District) is included in land suitability class S3, which indicates that the coffee plantation land only provides marginally suitable environmental conditions to support arabica coffee growth. Land suitability analysis showed that the northern area of Central Aceh Province is categorized in land suitability class S1 or the high suitability class, which can be developed for the expansion of arabica coffee plantations in Central Aceh. The results obtained with the fuzzy set method are very helpful for further development and application in land suitability assessment. Land assessment could help farmers create sustainable coffee plantations including production sustainability. Therefore, by obtaining the suitability state, farmers can better understand which areas should be prioritized for developing coffee plantations. Ultimately, this could lead to the optimal use of inputs and the achievement of sustainable production.

**Conflict of Interest** The author declares no conflict of interest.

## References

Abdisa T (2020) Land suitability analysis for Arabica coffee (Coffea arabica) production using geospatial and multi-criteria evaluaton in Jarfega Jarte Distric, Westernd Ethiophia.

Adhikari B, Li J (2013) Modelling ambiguity in urban planning. Ann GIS 19(3):143–152

Arab ST, Ahamed T (2022) Land suitability analysis for potential vineyards extension in Afghanistan at Regional Scale using remote sensing datasets. Remote Sens 14:4450

Arvi D, Syakur S, Karim A (2019) Hubungan ketinggian tempat dan kelerengan terhadap produksi kopi Arabika Gayo 1 di Kabupaten Gayo Lues. Jurnal Ilmiah Mahasiswa Pertanian 4(4):596–602

Bacon CM, Sundstrom WA, Stewart IT, Beezer D (2017) Vulnerability to cumulative hazards: coping with the coffee leaf rust outbreak, drought, and food insecurity in Nucaragua. World Dev 93:136–152

Beer J, Muschler R, Kass DCL, Somarriba EJ (1998) Shade management in coffee and cacao plantations. Agrofor Syst 38(1):139–164

BPS (2021) Kabupaten Aceh Tengah dalam Angka. Badan Pusat Statistik Aceh Tengah, Aceh Tengah

Czabanski R, Jezewski M, Leski J. (2017). Introduction to Fuzzy Systems. P. Prokopowicz et al. (eds.), Theory and Applications of Ordered Fuzzy Numbers,Studies in Fuzziness and Soft Computing 356

DaMatta FM, Ramalho JDC (2006) Impacts of drought and temperature stress on coffee physiology and production: a review. Braz J Plant Physiol 18(1):55–81

DaMatta FM et al (2016) Sustained enhancement of photosynthesis in coffee trees grown under free-air CO2 enrichment conditions: disentangling the contributions of stomatal, mesophyll, and biochemical limitations. J Exp Bot 167:341–352

Damayanti T, Setadi H (2019) The influence of certification of Gayo Coffee Geographical indication against value addition of coffee in Gayo Highlands, Aceh. IOP Conf Ser Earth Environ Sci 338:1–10

Da Silva EA, Mazzafera P (2008) Influence of temperature and water on coffee culture. The Americas journal of plant science and biotechnology 2(2):32–41

Drinnan JE, Menzel CM (1995) Tempeprature affects vegetative growth and flowering of coffee (coffea arabica L). Journal of Horticultural Science 70:25–34

Elyanti, Karim A, Basri A (2012) Analisis indikasi gografis kopi arabika gayo ditinjau dari rencana tata ruang wilayah kabupaten. Jurnal Afrista 16(2):46–61

Estrada LL, Rasche L, Schneider UA (2017) Modeling land suitability for Coffea arabica L. in Central America. Environ Model Softw 95:196–209

FAO (1984) Guidelines: Land evaluation for forestry. forestry paper 48. FAO: Rome.

Gay C, Estrada F, Conde C, Eakin H, Villers L (2006) Potential impacts of climate change on agriculture: a case study of coffee production in Veracruz, Mexico. Clim Change 79:259–288

Gayo MUC, Rusdi M, Fazlina YD (2018) Distribusi spasial lahan kopi eksisting berdasrkan ketinggian dan arahan fungsi kawasan di Kabupaten ceh Tengah. Jurnal ilmiah mahasiswa pertanian Unsyiah 3(4):1–7

Gemilang A, Nurjanti H, Bastaman AF (2021) Zonasi Agrowisata kopi sumedang sebagai upaya konservasi lahan danau Sunyanyuri dan perkebunan kopi rakyat. Jurnal ilmu pertanian 03(2):68–77

Hartoyo APP, Hutagulung L, Kulsum F, Sunkar A, Herliyana EN, Hidayati S (2023) Species composition, structure, and management in Gayo coffee-based agroforestry system: the case of Mude Nosar Village, Central Aceh Regency. J Nat Resour Environ Manag 13(1):37–48

IPCC (Intergovernmental panel on climate change). Climate change 2021, The physical science basis, summary for policymakers. IPCC, Switzerland.

Lin BB (2007) Agroforestry management as an adaptive strategy against potentialmicroclimate extremes in coffee agriculture. Agric For Meteorol 144:85–94

Kurtener D, Torbert HA, Kruege E (2008) Evaluation of agricultural land suitability: Application of fuzzy indicators. Computational Science and Its Applications - Iccsa 2008, Pt 1. PRO 5072:475–490

Purnamasari RA, Noguchi R, Ahamed T (2019) Land suitability assessments for yield prediction of cassava systems and remote sensing. Comput Electron Agric 166

Qadry NA, Radiansyah, Abubakar Y (2017) Pengarh Ketinggian tempat tumbuh dan varietas terhadap mutu fisik dan fisiko kimia kopi arabika gayo. Jurnal ilmiah mahasiswa pertanian Unsyiah 2(1):279–287

Rahmawaty R, Ginting YA, Batubara RR, Carenina G, Ginting CF, Ginting RK, Angelita L, Rauf A (2021) Land suitability assesment for coffea arabica on the land overgrowed by Uncaria gambir, IOP Conf. Series: Earth and Environmental. Science 886:012121

Ritung S, Nugroho K, Mulyani A, Suryani E (2011) Technical guidelines for land evaluation for agricultural commodities (Revised Edition) Indones. Cent. Agric. L. Resour. Res. Dev. Agency Agric. Res. Dev. Bogor 168

Selim S, Koc-San D, Selim C, San BT (2018) Site selection for avocado cultivation using GIS and multi-criteria decision analyses: Case study of Antalya. Turkey Comput Electron Agric 154(2):450–459

Suh J, Brownson JRS (2016) Solar Farm Suitability Using Geographic Information System Fuzzy Sets and Analytic Hierarchy Processes: Case Study of Ulleung Island Korea Energies 9(8):648. https://doi.org/10.3390/en9080648

Syakir M, Surmaini E (2017) Perubahan iklim dalam konteks sistem produksi dan pengembangan kopi di Indonesia. Jurnal litbang pertanian 36(2):77–90. https://doi.org/10.21082/jp3.v36n2.2017

Taniwaki MH, Teixeira AA, Teixeira ARR, Copetti MV, Iamanaka BT (2014) Ochratoxigenic fungi and ochratoxin A in defective coffee beans. Food Res Int 61:161–166. https://doi.org/10.1016/j.foodres.2013.12.032

Triatmodjo (2008) Hidrologi Terapan. Beta Offset. BR. Yogyakarta, Yogyakarta

Tsiko RG dan Haile TS (2011) Integrating Geographical Information Systems, Fuzzy Logic and Analytical Hierarchy Process in Modelling Optimum Sites for Locating Water Reservoirs. A Case Study of the Debub District in Eritrea. WaterSA 3, 254-290.

UNFCCC (2007) Climate change: impacts, vulnerabilities and adaptation in developing countries. Bonn Germany

Uuld A, Magda R (2019) Import demand analysis for wheat in Mongolia; economics & working capital: London, UK, 2019; Available: http://eworkcapital.com/import-demand-analysis-for-wheat-in-mongolia/. Accessed 21 Jan 2021

Yitbarek T, Kibret K, Gebrekidan H, Beyene S (2013) Physical Land Suitability evaluation for rainfed production of cotton maize, upland rice and sorghum in Abobo Area, Western Ethiopia. Am J Res Commun 1(10):296–318

Zhang Z, Lin, C, Chen B. 2015. New stability and stabilization conditions for T-S fuzzy systems with time delay. Fuzzy Sets and Systems. 263:8

# Chapter 10
# Rice Land Suitability and Insurance Premiums: A GIS-Based Multicriteria Analysis Approach for Sustainable Rice Production in Bangladesh

**Md Monjurul Islam, Tofael Ahamed, and Ryozo Noguchi**

**Abstract** Rice is indeed one of the most important staple crops globally, and its production plays a crucial role in ensuring food security for a significant portion of the world's population. The purpose of this research is to develop a land suitability model for rice production based on suitability levels and to propose insurance premiums to obtain maximum returns based on the harvest index and subsidy dependence factor for the marginal and moderately suitable lands in the northern part of Bangladesh. A multicriteria analysis was undertaken and a rice land suitability map was developed using geographical information system and analytical hierarchy process. The analysis identified that 22.74% of the area was highly suitable, while 14.86% was marginally suitable, and 28.54% was moderately suitable for rice production. However, 32.67% of the area, which was occupied by water bodies, rivers, forests, and settlements, is permanently not suitable; 1.19% is presently not suitable. If agriculture at risk, it can adversely impact livelihoods and cause food insecurity. To motivate low-quality landowners to produce rice, there is no alternative but to provide protection through crop insurance. We suggest producing rice up to marginally suitable lands to obtain support from insurance. The minimum coverage is marginal coverage (70%) to cover the production costs, while the maximum coverage is high coverage (90%) to enable a maximum return. This new crop insurance model, based on land suitability can be a coherent support for owners of different quality land to increase production.

---

M. M. Islam
Department of Rural Development, Bangabandhu Sheikh Mujibur Rahman Agricultural University, Gazipur, Bangladesh

T. Ahamed (✉)
Institute of Life and Environmental Sciences, University of Tsukuba, Tsukuba, Ibaraki, Japan
e-mail: tofael.ahamed.gp@u.tsukuba.ac.jp

R. Noguchi
Graduate School of Agriculture, Kyoto University, Kyoto, Japan



T. Ahamed (ed.), *Remote Sensing Application II*, New Frontiers in Regional Science: Asian Perspectives 77, https://doi.org/10.1007/978-981-97-1188-8_10

**Keywords** Land suitability · Agricultural crop insurance · AHP · GIS

## 10.1 Introduction

Bangladesh is one of the most vulnerable countries in the world considering climate change and natural disasters. It is ranked 5th among the countries most exposed to natural disasters, including floods, cyclones, and droughts. The production losses of major cereal crops due to natural disasters over the past 29 years have been equivalent to an average of 6.4% of the national crop production every year (Salman et al. 2010). The Bangladesh Climate Change Strategy and Action Plan 2009 has identified "insurance" as an effective disaster management tool (Habiba and Shaw 2009). A key challenge for the poor farmers of Bangladesh is the management of crop damage risk. One potential strategy to cope with the distressing effects of crop damage is to introduce a crop insurance scheme. The continuing growth of the population and consumption mean that the universal demand for food will increase for at least another 40 years (Godfray et al. 2010). Agriculture in Bangladesh is already under pressure, both from large and increasing demands for food and from obstacles related to the degradation of agricultural land (Ahmed and Ryosuke 2000). Therefore, an increase in production with less land has become a global challenge (Chen and Lu 2018). In this regard, a land suitability analysis (LSA) could be the most appropriate approach to evaluate and fulfill the required criteria for these facilities.

The risk of production, or risk of return, concerns events of chance origin, related to nature, to which the producers are exposed. Farmers throughout developing countries face multiple risks in agricultural production (Clarke et al. 2015). Two major risks are of concern to the agricultural sector: price risk caused by potential volatility in prices and production risk resulting from uncertainty in the levels of production that primary producers can achieve from their current activities (Iturrioz 2018; Assouto et al. 2020). It is estimated that during a major disaster year, occurring once every 100 years, 23% of the national paddy and wheat production is lost due to adverse weather. Due to the scarcity of affordable and suitable risk-management tools, and exposure to adverse shocks, low-income households may be forced to reduce food consumption, which threatens their economic and human development prospects (Holzmann 2002). Agricultural production can be increased if the uncertainty of nature and the risks associated with it can be managed in a better way. Crop insurance can play a vital role as an alternative ex ante risk coping instrument to enable poor farmers in developing countries to cope with weather-related production risks. They may significantly contribute to poverty alleviation (Hazell 2001; Mechler et al. 2006). The National Adaptation Program of Action, prepared by the Ministry of Environment and Forest (MoEF), advises exploring options for spreading natural disaster risks by investigating in the possible of crop insurance markets to cope with the increased risk of crop damage (NAPA 2005).

The history of crop insurance has not been successful in developing countries based on standard marketable criteria, and the experience in Bangladesh is also not good (ADB 2019). In developing countries where the poorest parts of the population often find themselves in a spiral of recurrent damages due to natural calamities, disaster insurance schemes fail to earn sufficient premium incomes to cover payouts as well as administrative costs (Hazell 1992; Anderson 2001; Quiggin et al. 1993; Spaulding et al. 2003; Eugene and Mahul 2004). Data availability is the pivotal issue in designing and implementing a successful crop insurance program in developing countries (Bryla and Syroka 2007). A large body of theoretical and pragmatic literature advises that uninsured agricultural risk causes farmers to refrain from advancing in agricultural production. Recent experimental evidence in India has shown that once farmers have access to some form of insurance against agricultural risk, they are more motivated to increase their investment in higher-return, higher-risk investments (Cole et al. 2017). The government of India is very active in promoting agricultural insurance under the "Pradhan Mantri Fasal Bima Yojana" through large subsidies, while in Bangladesh, insurance is not a prominent feature of agricultural development strategies or policies (Berber and Patrick 2018). When considering options to reduce the financial drain on farmers and local communities, and cut down on post-disaster government expenses, crop insurance is a valuable climate adaptation tool. It offers farmers the capacity to plan and put aside some money for the long term against the rainy days of life when harvests are suddenly damaged or destroyed (Chatterjee 2019).

Land Suitability Analysis (LSA) using multicriteria analysis is a systematic approach to evaluate the suitability of land for specific purposes, such as agriculture, forestry, or urban development. It involves the integration of multiple criteria or factors to assess the suitability of land parcels or areas (Muhsin et al. 2018; Anusha et al. 2023). The LSA is a tool that is used to identify the most suitable places for locating future land uses (Collins et al. 2001). It identifies the most appropriate spatial patterns for future land uses according to specific requirements, preferences, or predictors of some activity (Malczewski 2004). The LSA is a spatial multicriteria decision analysis (MCDA) process in which several elements are evaluated to conform to the necessities of stakeholders and ecological issues. Multicriteria analysis using a geographical information system (GIS) has the advantages of supporting the decision-making process in a systematic manner and reflecting a transparent decision using thematic maps (Ferretti and Pomarico 2013). This multidisciplinary approach has diverse applications in spatial analyses for land use (Hassan and Nazem 2016; Hossain et al. 2009) and is found to be a prevailing integrated method for complex land use planning (Abdullahi et al. 2015; Talukder et al. 2017). A variety of MCDA methods can be used in a GIS setting, such as weighted linear combinations (WLC), Boolean overlays, ordered weighted averaging, analytical hierarchy process (AHP), and multiple-objective land allocation (Rikalovic et al. 2014). Bangladesh has a mixed land use pattern; hence, experts' opinions in LSA studies are very important (Muhsin et al. 2018; Mugiyo et al. 2021). The AHP is also a multicriteria decision-making tool that gathers the spatial analysis functions of GIS in a land suitability analysis to identify the suitable sites (Chandio et al.

2013; Habibie et al. 2021). The AHP is the most widely used MCDA method that has the flexibility to include experts' decisions in a systematic manner that decomposes the problem into a hierarchy of subproblems to evaluate the problem more easily and subjectively (Saaty 1980). The subjective evaluations are converted into numerical values that are ranked according to a numerical scale (Bhushan and Rai 2004). Thus, this study uses AHP as a multicriteria decision analysis tool with spatial solutions in a geographic information system and conducts a land suitability analysis to understand the risk of production in the moderately and marginally suitable lands.

Agricultural insurance can play a crucial role in mitigating the risks faced by farmers in Bangladesh. The country is highly susceptible to natural disasters such as floods, cyclones, and droughts, which can severely impact agricultural production (FAO 2021). Agricultural insurance solutions are an important tool for farmers to manage risks (Dalhaus et al. 2018). Designing a cost-effective insurance product that helps farmers to manage risks and launching persistent insurance markets for small-holder farmers are challenging propositions in Bangladesh (Platteau et al. 2017). Farms are scattered and small in size, and the quality of land is also different. The lower quality landowners suffer more when production is lower, prices fall, or crops are damaged by natural calamities in comparison to the other farmers (Halewood and Surya 2012). There are significant numbers of crop insurance products globally, but none of them coincide directly with the Bangladesh agriculture context, and no insurance products consider the quality or suitability of land for a specific crop. Almost all crop insurance considers the weather and, at times, natural calamities. Considering the limitations of data, knowledge, the awareness of the farmers, costs, and experience, this research attempts to propose a new insurance premium model using land suitability, harvest index, and the subsidy dependence factor through community involvement. To the best of our knowledge, no research has been undertaken on crop insurance premiums based on land suitability, harvest index, and subsidy dependence factors.

Therefore, the objectives of this study were to first identify the suitable areas for rice production in poverty and risk prone areas, and second, to propose a crop insurance premium model based on land suitability, harvest index, and subsidy dependence factors.

## 10.2 Methodology

### *10.2.1 Study Area and Data Collection*

The study was conducted in the Rangpur, Lalmonirhat, and Kurigram districts of Rangpur Division, which are in the Teesta and Dharla river plains and are known as poverty-prone risk areas (Fig. 10.1). Approximately 47.23% of the population lives below the upper poverty line, while the national average is 24.3% (HIES 2016). This region has an area of 5879.60 km$^2$ and is located in the northern part of Bangladesh between 25°18′ and 26°33′N latitudes and 88°56′ and 89°54′E longitudes.

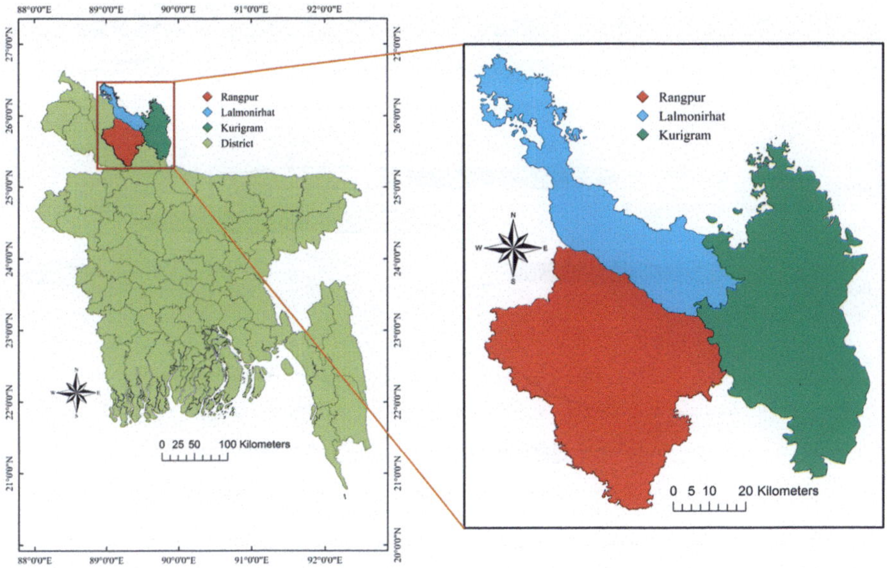


**Fig. 10.1** Northern part of Bangladesh, covering three districts: Rangpur, Lalmonirhat, and Kurigram

The area consists of 19 administrative units in three districts, with Rangpur, Lalmonirhat, and Kurigram having an overall population of 6.3 million (BBS 2014). The elevations range from 5 to 30 m above sea level. Agriculture is the main source of income for the population. Based on weather data, the minimum and maximum mean annual temperatures vary between 8.47 and 36.3 °C. The annual average rainfall recorded is 1932–2931 mm, with high humidity at 41–77% (BBS 2017).

This study attempted to develop an agricultural insurance premium model considering suitable lands for rice cultivation in the northern part of Bangladesh. Basically, the area is rice intensive and most of the farmers cultivate three types of rice (Aus, Aman, and Boro) all year round (Fig. 10.2). Major production influencing factors were used for rice land suitability analysis in the geospatial platform. Subsequently, considering the importance of the factors, the availability of data, the literature review, and expert opinions, the study identified eight criteria (slope, elevation, topsoil texture, land type, soil pH, temperature, rainfall, and flood prone) and one constraint (land use) for the analysis. Land use was identified with settlements, rivers, forests, and water bodies as constraints. Different statistical data were obtained from the Bangladesh Bureau of Statistics (BBS). Rainfall and temperature data were collected from the regional meteorological stations, and data for agricultural production were collected from the Department of Agricultural Extension (DAE). The base maps of different factors were collected from the Local Government and Engineering Division (LGED) and the Bangladesh Country Almanac (BCA) to carry out this research work.

The research was conducted in two phases. In the first phase, a land suitability analysis was conducted to identify the areas suitable for rice production (Fig. 10.3).

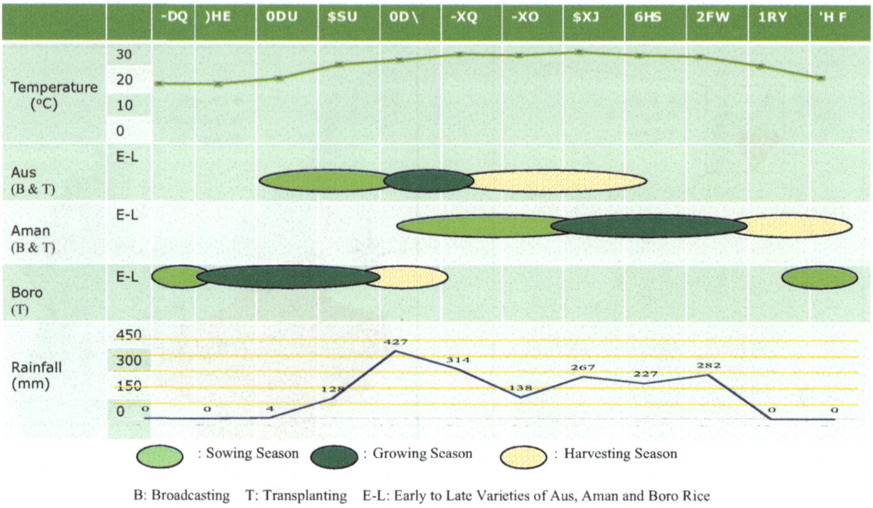


**Fig. 10.2** Rice growing season of Bangladesh with the monthly average temperature (°C) and rainfall (mm)

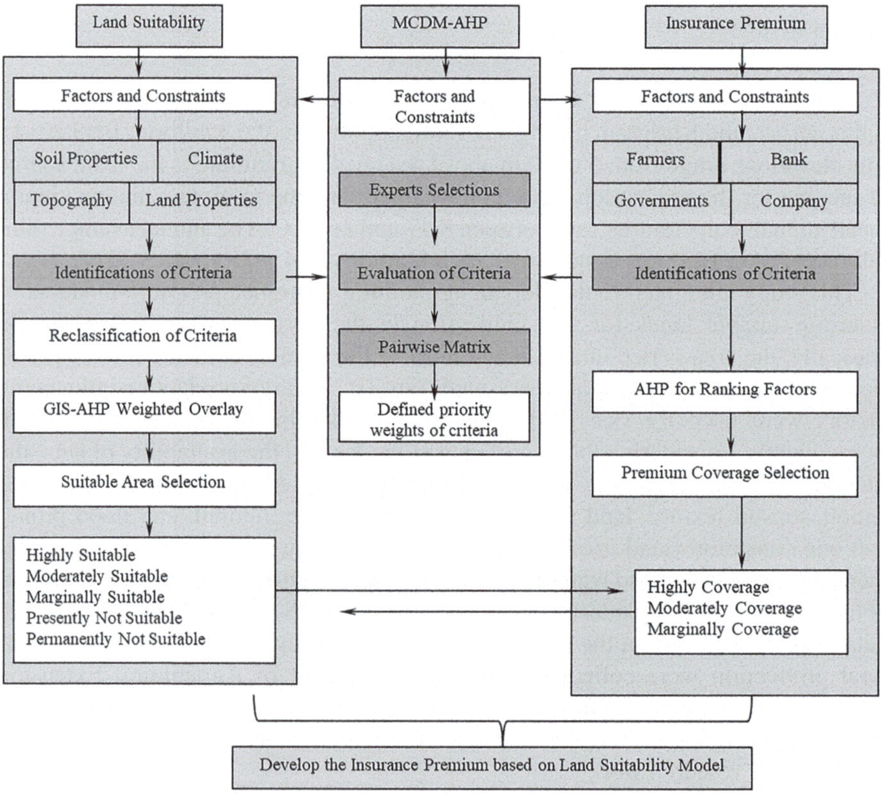


**Fig. 10.3** Conceptual framework for land suitability and insurance premiums

A land suitability map was prepared in ArcGIS 10.3 based on the rice production requirement criteria. An MCDA technique was adopted to carry out the suitability analysis. AHP and MCDA were used with eight factors and one constraint to incorporate the preferences of the experts and assign weights to the factors. On the other hand, GIS was applied to manage the thematic spatial dataset of the factors and constraints' layer maps using spatial analysis tools. A decision rule was applied to integrate GIS and AHP to develop a suitability map. In the second phase, the insurance premium model was developed based on the land suitability classifications. The following sections discuss land suitability using multicriteria analysis and insurance premiums.

## 10.2.2 Land Suitability with Multicriteria Analysis

This research applied MCDA in several steps in the spatial environment. The following nine factors were selected to conduct the MCDA: (1) slope; (2) elevation; (3) land type; (4) topsoil texture; (5) soil pH; (6) flood prone; (7) rainfall; (8) temperature; and (9) land use (Fig. 10.4). Only the land use factor was considered as a restriction on the study. A factor is a criterion that enhances or detracts from suitable alternatives for the activity under consideration, and a constraint serves to limit any alternative (Euro and Nijkamp 1998).

### 10.2.2.1 Slope

In the context of land suitability analysis, slope is typically measured as the ratio of vertical rise to horizontal distance and is expressed as a percentage, degree, or gradient. The thickness of the soil layer decreases with the increase of the slope (Ashford et al. 1997). The original Shuttle Radar Topography Mission (SRTM) and digital elevation models (DEM) were used to produce the slope and elevation for the study area. ArcGIS was used to rectify the topographical maps using the Universal Transverse Mercator (UTM) projection and WGS 84 datum. The datasets prepared in this process were compared to the SRTM dataset to identify the accuracy of both datasets. The slope was calculated from the maximum rate of change between each cell and its neighbors. Every cell in the output raster had a slope value. A lower slope value indicates a flatter terrain and a higher slope value indicates a steeper terrain. Steeper slopes pose challenges for land use due to erosion risks, limited water retention capacity, and difficulties in mechanized farming or construction activities. On the other hand, gentle slopes are generally more suitable for cultivation, as they allow for better soil conservation, water infiltration, and ease of management. Ninety-five (95) percent of the study area has less than 5% slope (Table 10.1); almost all the study areas are suitable for rice production in the context of slope.

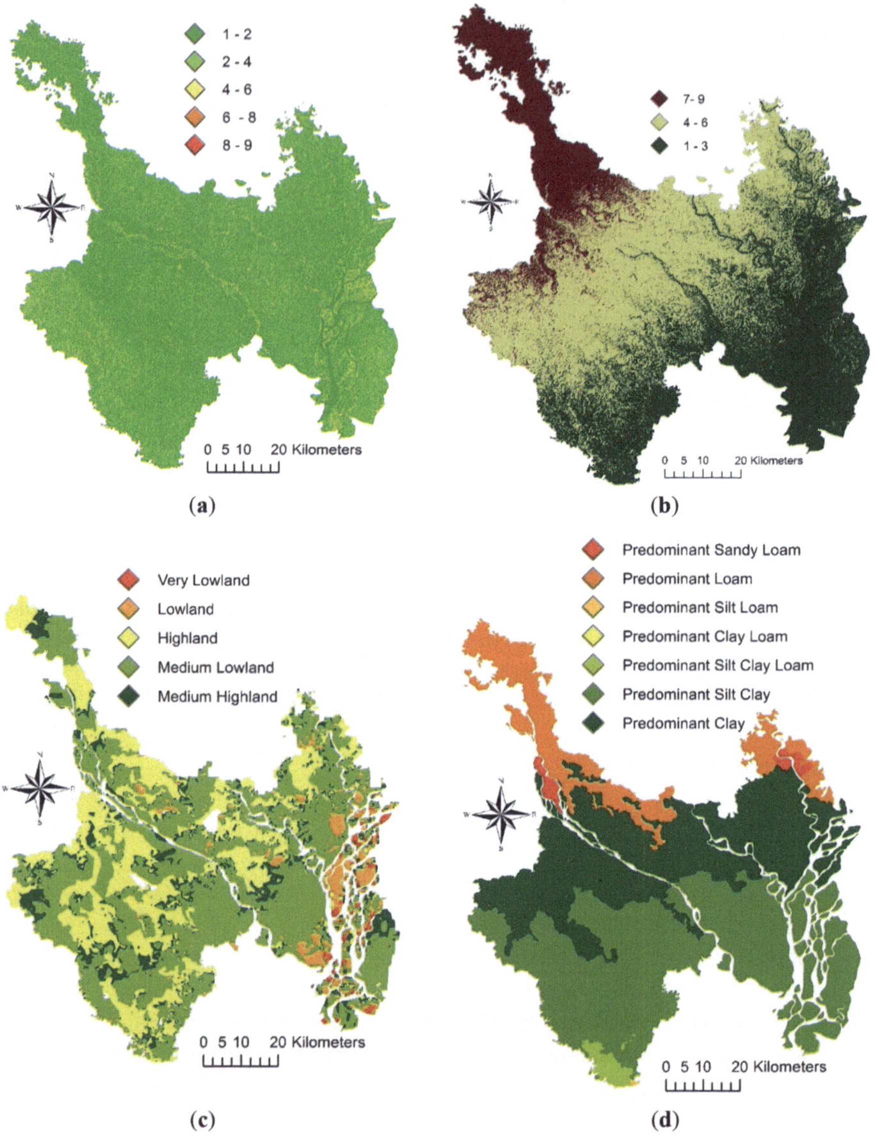


**Fig. 10.4** Criteria layers: (**a**) Slope; (**b**) Elevation; (**c**) Land type; (**d**) Topsoil texture; (**e**) Soil pH; (**f**) Flood prone; (**g**) Rainfall; (**h**) Temperature; and (**i**) Land use for the northern part of Bangladesh

#### 10.2.2.2 Elevation

Elevation is a crucial factor in land suitability analysis and refers to the height or vertical distance of a location above sea level (Gesch 2018). By integrating elevation data into land suitability analysis, stakeholders can make informed decisions regarding land

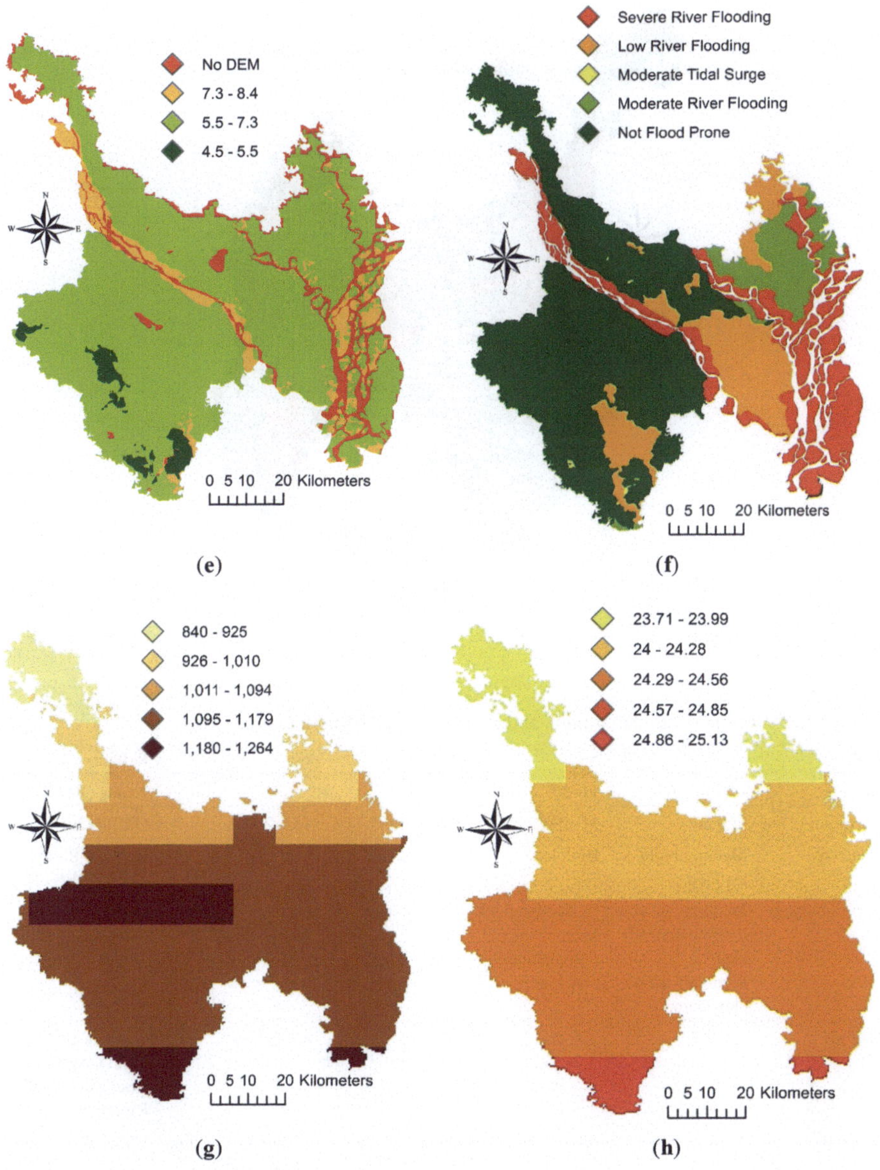


**Fig. 10.4** (continued)

management, conservation efforts, infrastructure development, and other land-use planning activities. Rice can grow in extremely diverse elevations, including fields that are leveled, gently rolling, or steep, at altitudes of up to 2000 m (GRiSP 2013). However, when the elevation is less than 375 m, rice production is better (Hussain et al. 2012). Most of the study areas are plain lands and the elevation is less than 70 m. Considering elevation, all the lands of the region are suitable for rice production (Table 10.1).

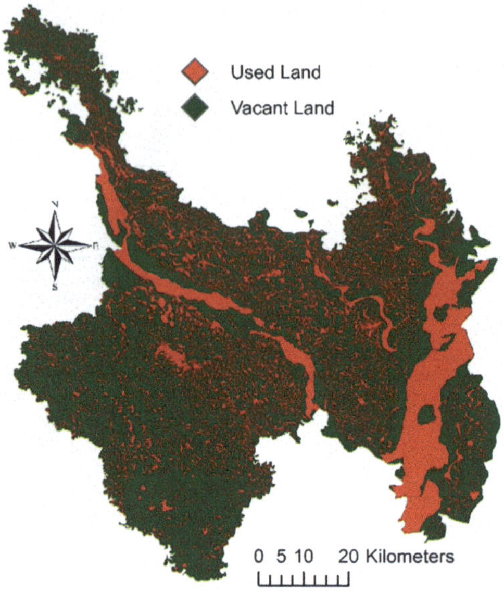


(i)

**Fig. 10.4** (continued)

**Table 10.1** Criteria used, data range, and suitability class for rice

| Criteria | S1 | S2 | S3 | N1 | Reference |
|---|---|---|---|---|---|
| Slope | <10° | 11–30° | 31–50° | | Atalay (2006) |
| Elevation | <50 m | 51–100 m | 101–150 m | 151–200 m | GRiSP (2013) |
| Land type | Medium high/ lowland | High land | Lowland | Very lowland | Paul and Rashid (2016), BBS (2016) |
| Top soil texture | Predominant clay Predominant silt clay Predominant silt clay loam | Predominant silt loam Predominant clay loam | Predominant loam | Predominant sandy loam | Dou et al. (2016), USDA, Asai et al. (2009) |
| Soil pH | 5.5–7.3 | 7.3–8.4 | 4.5–5.5 | <4 and >8 | Samanta et al. (2011), Ayehu and Besufekad (2015); Kihoro et al. 2013 |
| Flood prone | Not flood prone | Moderate tidal surge | Moderate river flooding | Low river flooding, severe river flooding | BBS (2016) |
| Temperature | 20–30 | 10–20 | 30–40 | | Samanta et al. (2011), Kihoro et al. (2013) |
| Rainfall | 1000–2000 mm | 2000–3500 mm | More than 3500 mm | | GRiSP (2013) |
| Land use | Settlements, rivers, water bodies, forests | | | | Restrictions |

#### 10.2.2.3 Topsoil Texture

Soil texture influences various soil properties, including water-holding capacity, nutrient retention, drainage characteristics, and workability. These properties, in turn, affect the suitability of soil for different crops, land uses, and management practices. Rice grain yield is significantly affected by topsoil (bulk density, water holding capacity, and permeability) (Asai et al. 2009). The rice grain yield in clay soil was 46% higher than in sandy loam soil averaged across cultivars and water areas (Dou et al. 2016). The effective soil depth is defined as the thickness of soil above a layer restricting root growth (e.g., consolidated rock or cemented materials, such as gravel). Most annual crops have a rooting depth of approximately 50 cm, while for tree crops, the rooting system can reach beyond 150 cm. However, most crops produce good yields in soils with an effective soil depth of approximately 100 cm, and this value has been used as an upper limit (Billings et al. 2018). There were seven categories of topsoil format available in the study area: predominant clay, predominant silt clay, predominant silt clay loam, predominant silt loam, predominant clay loam, predominant silt loam, and predominant sandy loam (Table 10.1). These categories were converted into a land suitability class according to their characteristics for rice cultivation based on the United States Department of Agriculture (USDA) soil texture suitability rating for rice and are observed as sands, loamy sands, sandy loams, sandy clay loam, and silt (Table 10.2).

#### 10.2.2.4 Land Type

Land type refers to the classification or categorization of land based on various characteristics, such as land cover, land use, soil type, and topography (Chughtai et al. 2021). It provides a broad overview of the dominant features and

**Table 10.2** USDA topsoil texture suitability rating for rice

| Textural classes | Description (USDA) | Rating |
|---|---|---|
| Sands | More than 85% sand, 0–10% clay, and 0–15% silt | 3 |
| Loamy sands | 70–91% sand, 0–15% clay, and 0–30% silt | 3 |
| Sandy loams | More than 43% sand, less than 7% clay, and less than 50% silt | 3 |
| Sandy clay loam | More than 45% sand, 20–35% clay, and less than 28% silt | 3 |
| Sandy clay | 45% or more sand and 35% or more clay | 2 |
| Loam | Less than 52% sand, 7–27% clay, and 28–50% silt | 2 |
| Silt loam | Less than 12% clay and 50–80% silt | 2 |
| Clay loam | 20–45% sand and 27–40% clay | 2 |
| Silt | Less than 12% clay and 80% or more silt | 2 |
| Silty clay loam | Less than 20% sand and 27–40% clay | 1 |
| Silty clay | 40% or more clay and 40% or more silt | 1 |
| Clay | Less than 45% sand, 40% or more clay, and less than 40% silt | 1 |

characteristics of a particular area or parcel of land. Good water retention capacity and a high amount of clay with organic matter are considered ideal for rice cultivation. Highlands are used to grow broadcast Aman and Aus rice, medium highlands are suitable for broadcast Aus and Aman, and medium lowlands are used to grow transplanted Aman and mixed broadcast Aus and transplanted Aman. Broadcast Aman is typically grown in lowland areas (Paul and Rashid 2016). Based on seasonal flooding, the government of Bangladesh (GoB) has divided the land into five categories: highland, medium highland, medium lowland, lowland, and very lowland (BBS 2016). Most of the area is medium lowland (54%), followed by highland (28%), medium highland (12%), and lowland and very lowland (<6%). Medium highland and lowland are considered to be highly suitable, highland to be moderately suitable, lowland to be marginally suitable, and very lowland is considered to be presently not suitable for rice cultivation (Table 10.1).

#### 10.2.2.5 Soil pH

Soil pH is a measure of the acidity or alkalinity of the soil, indicating the concentration of hydrogen ions (H+) in the soil solution (Hong et al. 2018). It is an essential soil property that influences various biological, chemical, and physical processes within the soil and has a significant impact on plant growth and nutrient availability. The pH scale ranges from 0 to 14, with 7 considered neutral. pH values below 7 indicate acidic soil, while values above 7 indicate alkaline (basic) soil. The optimal pH range for most plants is slightly acidic to neutral, typically between 6 and 7.5. However, specific plants may have more specific pH requirements depending on their adaptation to different soil conditions. The pH is an important factor in the quality assessment for paddy soils (Guo et al. 2018). Slightly acidic soils having a pH value of 6 to 7 are better for rice cultivation. However, rice can grow in a wide range of pH varying from values of 4–8 (Samanta et al. 2011; Ayehu and Besufekad 2015; Kihoro et al. 2013). Considering the availability of the data format, the study ranked pH 5.5–7.3 as highly suitable (79%), pH 7.3–8.4 as moderately suitability (8%), and pH 4.5–5.5 as marginally suitable (<3%) (Table 10.1).

#### 10.2.2.6 Temperature

Temperature is a significant factor in land suitability analysis as it directly influences the growth and development of plants, as well as other organisms and yield of rice (Shahzad et al. 2021). Temperature patterns and ranges vary across different regions, seasons, and elevations, and understanding the temperature characteristics of an area is essential for determining its suitability for specific land uses. Being a tropical and subtropical crop, rice is normally grown at a fairly high temperature–high rainfall regime ranging from 20 to 40 °C. The capacity of soil is the amount of water held in the soil between tensions corresponding to field capacity and the permanent wilting point (Samanta et al. 2011; Kihoro et al. 2013). Although the research area had three administrative districts, the temperature was almost similar in all of the areas. The average temperature was 23.5 °C (Table 10.1).

#### 10.2.2.7 Rainfall

Rainfall is a critical factor in land suitability analysis as it directly impacts various aspects of land use, agriculture, ecosystem functioning, and water resource management (Wu et al. 2020). It refers to the amount and distribution of precipitation, including rain, snow, and other forms of atmospheric moisture, that falls onto the land surface over a specific period. Rainfall, in the perilous stages of paddy growth, increases crop yield through the rapid dissolution of nutrients for uptake by plants (Amin et al. 2015). Three districts in the study area receive 1250–2000 mm of annual rainfall. The annual average rainfall was recorded between 1932 and 2931 mm (Table 10.1).

#### 10.2.2.8 Flood Prone

Flood-prone areas are those that are susceptible to temporary or permanent inundation during periods of high-water flow, typically caused by heavy rainfall, river overflow, storm surge, or other hydrological events. Bangladesh ranks as the sixth most flood-prone country in the world (UNDP 2004). Flood-prone environments include deep water areas submerged under more than 100 cm of water from 10 days to a few months, and areas that are affected by flash floods of longer than 10 days (BBS 2016; Chauhan et al. 2017). Due to the availability of the data on flood prone areas, the area was categorized into five types: not flood prone (54%), low river flooding (16.6%), moderate river flooding (8.6%), moderate tidal surge (0.13%), and severe river flooding (20%). According to the depth and duration of water stagnation, not flood prone is considered as highly suitable, low river flooding is considered as moderately suitable, moderate river flooding is considered as marginally suitable, and moderate tidal surge and severe river flooding are considered as not suitable for rice farming (MoEF 2008) (Table 10.1).

#### 10.2.2.9 Land Use

Land use refers to the human activities and purposes for which land is utilized. It encompasses the various ways in which land is managed, developed, and allocated for specific functions or activities (Xie et al. 2020). Land use decisions play a vital role in shaping the physical, economic, and social characteristics of a region or area. Land use is an important factor in identifying the suitable sites for any crop production. In this study, we considered rivers, forests, water bodies, and settlements as land use and they are considered as restrictions in the analysis. Approximately 33%

of the total land was occupied by settlements, rivers, and forests. As such, only 67% of the land was considered suitable for rice production (Table 10.1).

### *10.2.3 Preferences of Criteria in Decision Analysis*

In decision analysis, the preferences of criteria refer to the relative importance or weight assigned to different criteria used to evaluate alternative courses of action. These preferences are subjective and can vary depending on the decision maker's goals, values, and the context of the decision (Belton and Stewart 2002). The purpose of weighting is to express the importance or preference of each criterion related to other criteria on crop yield and growth rate. A literature review of various sources and interviews with local experts, agronomists, and researchers in this field was selected to identify the ideal requirements for suitable rice growing areas. Factors were identified for selecting suitable sites for rice production. The suitability levels for each of the factors were defined; these levels were used as a base to construct the maps (one for each factor). The suitability levels were: highly suitable-S1, moderately suitable-S2, marginally suitable-S3, and not suitable-N, based on the structure of FAO land suitability classification (FAO 1976).

It's important to note that the preferences for criteria may not be universal and can differ across individuals, organizations, and specific decision contexts. Decision analysis techniques, such as multicriteria decision analysis (MCDA), can help quantify and incorporate these preferences into a structured decision-making process (Adem Esmail and Geneletti 2018). The procedure of the MCDA included several stages. First, the relevant factors and constraints were established (Eastman et al. 1998). Using the abovementioned factors, a pairwise comparison matrix was constructed. Although there are a variety of existing procedures for the development of weights, a pairwise comparison matrix in the context of a decision-making process known as the AHP was used (Saaty 1980). The comparison concerns the relative importance of the two factors involved in determining the suitability of the stated objective (Eastman et al. 1998). In a land suitability analysis, a map represents each evaluation criterion with ordinal values (such as S1, S2, S3, and N), indicating the degree of suitability with respect to a sub-criterion and based on the crop requirements (Mustafa et al. 2011).

### *10.2.4 Analytical Hierarchy Process (AHP)*

The Analytical Hierarchy Process (AHP) is a multicriteria decision-making method developed by Thomas L. Saaty. It provides a structured framework for evaluating and comparing alternatives based on multiple criteria and their relative importance or preferences. AHP helps decision-makers systematically analyze complex decision problems and make informed choices (Saaty et al. 2022). The AHP is one of the most widely accepted procedures and is considered as the most reliable multicriteria decision-making technique (Engineering et al. 1995; Odu 2019). The process was applied to a set of factors to establish a hierarchical structure by providing a weight for each factor in the complete decision-making process (Kiker et al. 2005; Naveed et al. 2020). Therefore, many decision-making methods attempt to determine the relative importance, or weight, of the alternatives in terms of each factor involved in each decision-making problem (Alshurideh et al. 2022). The AHP provides a structural ground for quantifying the strong comparison of design factors in a pairwise technique, and thus decreases the complexity of the decision-making process (Saaty 1990; Khan et al. 2022). Weights were used to determine the priorities of factors (elevation, slope, land type, soil pH, topsoil textures, flood prone, temperature, and rainfall) and to identify the suitability of different land uses for rice production. The resultant AHP weights were used to determine the priority of each factor for weighted overlay applications using GIS.

In the first stage of the analysis, we organized the factors of the decision model into a hierarchy for land suitability and an insurance premium (Fig. 10.2). The second stage involved scoring the factors via pairwise comparisons and scoring scales of relative importance (Table 10.3). Questionnaires were used to gather expert opinions on the relative importance of the factors considered (Appendix 1). The comparative results (for each factor pair) were described as integer values of 1 (equal value) to 9 (extremely different), where a higher number denotes that the chosen factor was more important than the other factor to which it was compared. A rating of 9 indicates that in relation to the column factor, the row factor is more important. On the other hand, a rating of 1/9 indicates that relative to the column factor, the row factor is less important (Mustafa et al. 2011). In cases where the column and row factors are equally important, they have a rating value of 1. For example, when comparing elevation and slope angle factors, a score of 1 indicates that both were equally relevant to evaluate suitability, and a score of 9 indicates that elevation is more important than the slope angle. All scores were assembled in a pairwise comparison matrix with diagonal and reciprocal scores located in the lower left-hand triangle. Reciprocal values (1/3, 1/5, 1/7, and 1/9) were used where the row factor was found to be less important than the column factor (Table 10.3).

**Table 10.3** AHP pairwise comparison scale (Saaty 1990)

| Intensity of importance | Explanation |
|---|---|
| 1 | Equally important |
| 3 | Moderately important to one activity over another |
| 5 | Strongly important to one activity over another |
| 7 | Very strong important to one activity over another |
| 9 | Extremely important to one activity over another |
| 2, 4, 6, 8 | When compromise is needed intermediate values |
| Reciprocals | Values for inverse comparison |

Third, we calculated the matrix and ensured the consistency of the pairwise comparison factors. The AHP also provided the measurements to calculate the normalized values of each factor and to determine the normalized principal eigenvalue and priority vectors. The pairwise matrix was calculated and is given by the following expression:

$$\begin{bmatrix} C_{11} & C_{12} & ..\,.C_{1n} \\ C_{21} & C_{22} & ..\,.C_{2n} \\ . & . & .\,. \\ . & . & .\,. \\ C_{n1} & C_{n2} & .C_{nn} \end{bmatrix} \tag{10.1}$$

The sum of each column of the pairwise matrix was denoted as follows:

$$C_{ij} = \sum_{i=1}^{n} C_{ij} \tag{10.2}$$

We then divided each element of the matrix by its column total to generate a normalized pairwise matrix as follows:

$$X_{ij} = \frac{C_{ij}}{\sum_{i=1}^{n} C_{ij}} = \begin{bmatrix} X_{11} & X_{12} & ..\,.X_{1n} \\ X_{21} & X_{22} & ..\,.X_{2n} \\ . & . & .\,. \\ . & . & .\,. \\ X_{n1} & X_{n2} & .X_{nn} \end{bmatrix} \tag{10.3}$$

Finally, we divided the sum of the normalized matrix column by the number of factors used ($n$) to generate the weighted matrix of priority factors as follows:

$$W_{ij} = \frac{\sum_{j=1}^{n} X_{ij}}{n} = \begin{bmatrix} W_{11} \\ W_{12} \\ . \\ . \\ W_{1n} \end{bmatrix} \tag{10.4}$$

The initial consistency vectors were derived by multiplying the pairwise matrix by the vector of weights as follows:

$$\begin{bmatrix} C_{11} & C_{12} & ..\,.C_{1n} \\ C_{21} & C_{22} & ..\,.C_{2n} \\ . & . & .. \\ . & . & .. \\ C_{n1} & C_{n2} & .C_{nn} \end{bmatrix} \mathrm{x} \begin{bmatrix} W_{11} \\ W_{12} \\ . \\ . \\ W_{1n} \end{bmatrix} = \begin{bmatrix} C_{11}W_{11}+ & C_{12}W_{11}+ & \ldots + C_{13}W_{11} \\ C_{21}W_{12}+ & C_{22}W_{12}+ & \ldots + C_{23}W_{12} \\ . & . & . \; . \\ . & . & . \; . \\ C_{n1}W_{1n} & C_{n1}W_{1n} & .C_{n1}W_{1n} \end{bmatrix} = \begin{bmatrix} V_{11} \\ V_{12} \\ . \\ . \\ V_{1n} \end{bmatrix} \tag{10.5}$$

The principal eigenvector ($\lambda_{max}$) was then calculated by averaging the values of the consistency vector as follows:

$$\lambda_{max} = \sum_{i}^{n} CV_{ij} \tag{10.6}$$

Eigenvalues were calculated by averaging the rows of each matrix. Eigenvalues were also referred to as relative weights. The largest eigenvalue was equal to the number of factors, and when $\lambda_{max} = n$, judgments were consistent. Normalized eigenvalues were generated as weights of priority factors. The principal value suggests that the eight criteria were consistent, as the calculation results reveal a maximum value of 8.34 (Table 10.4). The judgments were also checked to determine the consistency index (*CI*), which was calculated as follows:

$$CI = \frac{\lambda_{max} - n}{n-1} \tag{10.7}$$

Here, $n$ is the total number of criteria. Saaty introduced the consistency ratio (*CR*) and compared it to the *CI* and random index (*RI*) values, which is the calculated value for matrices of different sizes (Saaty and Vargas 2012) (Table 10.5). The consistency ratio was calculated as follows:

$$CR = \frac{CI}{RI} \tag{10.8}$$

A lower *CR* ratio indicates more consistency. If the value of *CR* is 0.10 or more, then the weight values of the matrix indicate inconsistencies and the method (AHP) may not provide a meaningful result (Saaty 1980).

**Table 10.4** The judgements of experts in prioritizing the weights of the different factors for the land suitability analysis (A–J, experts from different fields)

| Factors | A | B | C | D | E | F | G | H | I | J | Average |
|---|---|---|---|---|---|---|---|---|---|---|---|
| Slope | 0.110 | 0.022 | 0.091 | 0.127 | 0.039 | 0.065 | 0.069 | 0.033 | 0.126 | 0.037 | 0.072 |
| Elevation | 0.175 | 0.040 | 0.052 | 0.179 | 0.041 | 0.049 | 0.050 | 0.038 | 0.053 | 0.038 | 0.071 |
| Land type | 0.147 | 0.399 | 0.278 | 0.127 | 0.084 | 0.281 | 0.215 | 0.268 | 0.221 | 0.306 | 0.233 |
| Topsoil texture | 0.108 | 0.037 | 0.259 | 0.127 | 0.040 | 0.065 | 0.116 | 0.070 | 0.192 | 0.077 | 0.109 |
| Soil pH | 0.053 | 0.029 | 0.023 | 0.055 | 0.062 | 0.034 | 0.021 | 0.035 | 0.039 | 0.039 | 0.039 |
| Flood prone | 0.103 | 0.092 | 0.028 | 0.133 | 0.073 | 0.094 | 0.084 | 0.096 | 0.045 | 0.103 | 0.085 |
| Temperature | 0.130 | 0.234 | 0.131 | 0.107 | 0.358 | 0.221 | 0.138 | 0.308 | 0.076 | 0.188 | 0.189 |
| rainfall | 0.173 | 0.148 | 0.139 | 0.145 | 0.303 | 0.191 | 0.306 | 0.152 | 0.246 | 0.212 | 0.202 |
| *CR* | 0.086 | 0.096 | 0.098 | 0.035 | 0.094 | 0.090 | 0.096 | 0.055 | 0.097 | 0.072 | |

**Table 10.5** Random inconsistency indices for $n = 10$ (Saaty 1980)

| $n$ | 1 | 2 | 3 | 4 | 5 | 6 | 7 | 8 | 9 | 10 |
|---|---|---|---|---|---|---|---|---|---|---|
| Random consistency index (RI) | 0 | 0 | 0.58 | 0.9 | 1.12 | 1.24 | 1.32 | 1.41 | 1.45 | 1.49 |

### *10.2.5 Weighted Overlay Analysis for Land Suitability*

Weighted overlay analysis is a spatial analysis technique used in GIS to assess land suitability by combining and weighting multiple criteria or factors. It allows decision-makers to integrate and analyze various data layers representing different criteria to generate a composite suitability map that ranks areas based on their overall suitability for a specific land use or activity. The rice crop suitability map was recognized by the weighted sum overlay using the spatial analyst tools in ArcGIS 10.3 with both vector and raster layers (Fig. 10.5). Initially, nine thematic vector layers of slope, elevation, soil pH, topsoil texture, land type, temperature, rainfall, flood prone, and land use (rivers, water bodies, settlements, and forests) were taken in a base geographical coordinate system, WGS 1984 (Table 10.6). These vector layers were projected in WGS 1984 UTM Zone 45N to obtain the same geographic extent. The thematic layers were then converted to raster layers to conduct the spatial analysis. The AHP weights were obtained from the experts' opinions using a designed questionnaire for the priority weights of the factors. The constraints were stated as "restrictions." To aggregate the weights of the factors and constraints, a decision rule, WLC, was applied to determine suitable locations in the GIS environments.

### *10.2.6 Crop Insurance*

Crop insurance is a risk management tool that provides financial protection to farmers against the loss of crops due to various perils, such as adverse weather conditions, pests, diseases, and other specified causes (Vyas et al. 2021). It helps farmers mitigate

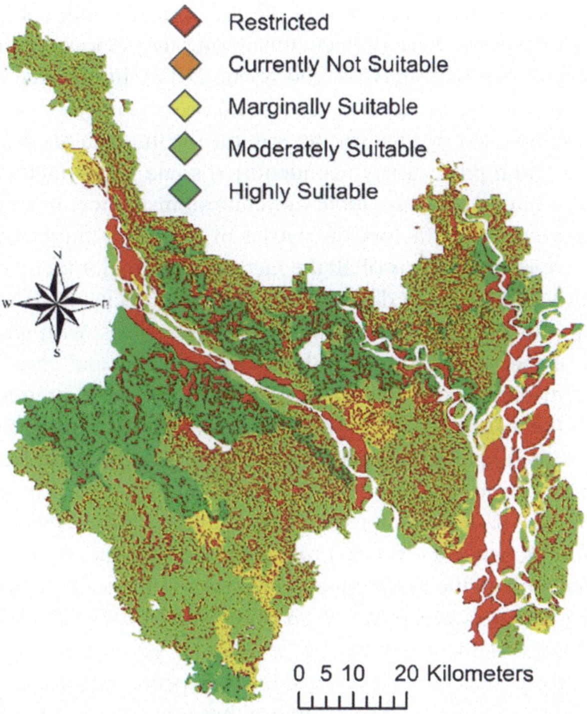


**Fig. 10.5** Rice land suitability map for the northern part of Bangladesh

**Table 10.6** List of data and original data sources used for land suitability assessment for rice production

| No | Data | Description | Source |
|---|---|---|---|
| 1 | Land use map | Scale at 1:50,000 | 2015, LGED, Bangladesh |
| 2 | Slope map | Derived from 30-m resolution | 2017, DEM STRM |
| 3 | Elevation map | Derived from 30-m resolution | 2017, DEM STRM |
| 4 | Land type | Scale 1:50,000 | 2015, BCA, Bangladesh |
| 5 | Topsoil texture map | Scale 1:50,000 | 2015, BCA, Bangladesh |
| 6 | Soil pH map | Scale 1:50,000 | 2015, BCA, Bangladesh |
| 7 | Flood prone map | Scale 1:50,000 | 2015, BCA, Bangladesh |
| 8 | Temperature map | Scale 1:50,000 | 2015, BCA, Bangladesh |
| 9 | Rainfall map | Scale 1:50,000 | 2015, BCA, Bangladesh |

the financial impact of crop failures and stabilize their income, ensuring their ability to recover and continue farming operations (Alam et al. 2020). Farmers face a variety of substantial risks in the production and marketing phases, yet agricultural insurance has rarely been available outside of subsidized government offerings, which typically reach only a small fraction of small-scale farmers (Kramer et al. 2022). A large number of newly developed products were piloted in developing countries including Bangladesh; however, none were perfect considering the socioeconomic characteristics of the farmers in Bangladesh (Cole and Xiong 2017). In this study, the insurance premium mainly considers the quality of the land or suitability of the land. An AHP survey was first conducted to identify the criteria for the insurance premium. After literature reviews and a pilot survey, we identified some major factors that could be highly influential to take decisions about agricultural insurance. In different perspectives, we tried to identify the factors and causes of the agricultural insurance market. Subsequent to the careful analysis of all the factors, eight major factors were selected, which were highly influential in different stakeholder's decisions. The major factors identified were product price, government subsidy, awareness, willingness of the government, belief of the benefits, trust in the company, rules and regulations, and the collection and distribution of money. To fix the crop insurance premium, the factors were ranked by the experts and the influence in the crop insurance program was obtained; most of the experts mentioned product price as the main factor for participating in a crop insurance program (Table 10.7). Finally, the study proposed an insurance premium model based on land suitability with the consideration of the harvest index and subsidy dependence factor to increase rice production domestically. The product price depends on the production, which was influenced by the harvest index (HI). The HI can be calculated from the crop information of past years based on the biological yield, and yield of dry matter or yield of grain and straw (Donald and Hamblin 1976; Ahamed et al. 2016); the subsidy dependence factor was considered because it is difficult to find any crop insurance that was successful without a government subsidy.

**Table 10.7** The judgements of experts in prioritizing the weights of the different factors for introducing crop insurance premium model (A–J, experts from different fields)

| Factors | A | B | C | D | E | F | G | H | I | J | Average |
|---|---|---|---|---|---|---|---|---|---|---|---|
| Product price | 0.268 | 0.341 | 0.278 | 0.279 | 0.084 | 0.281 | 0.215 | 0.399 | 0.221 | 0.306 | 0.267 |
| Government subsidy | 0.070 | 0.223 | 0.259 | 0.112 | 0.040 | 0.065 | 0.116 | 0.037 | 0.192 | 0.077 | 0.119 |
| Awareness | 0.033 | 0.165 | 0.091 | 0.065 | 0.039 | 0.065 | 0.069 | 0.022 | 0.126 | 0.037 | 0.071 |
| Willingness of the government | 0.035 | 0.022 | 0.023 | 0.029 | 0.062 | 0.034 | 0.021 | 0.029 | 0.039 | 0.039 | 0.033 |
| Believe of benefits | 0.038 | 0.049 | 0.052 | 0.065 | 0.041 | 0.049 | 0.050 | 0.040 | 0.053 | 0.038 | 0.048 |
| Trust to the company | 0.096 | 0.027 | 0.028 | 0.028 | 0.073 | 0.094 | 0.084 | 0.092 | 0.045 | 0.103 | 0.067 |
| Rules and regulations | 0.152 | 0.091 | 0.139 | 0.239 | 0.303 | 0.191 | 0.306 | 0.148 | 0.246 | 0.212 | 0.203 |
| Collection and distribution of money | 0.308 | 0.082 | 0.131 | 0.183 | 0.358 | 0.221 | 0.138 | 0.234 | 0.076 | 0.188 | 0.192 |
| *CR* | 0.055 | 0.097 | 0.098 | 0.097 | 0.094 | 0.090 | 0.096 | 0.096 | 0.097 | 0.072 | |

#### 10.2.6.1 Product Price

Product price is a key factor in consumer decision-making and business profitability. The price of a product is determined by various factors, including production costs, competition, market demand, and pricing strategies (Rajasa et al. 2023). Most of the farmers in Bangladesh have limited incomes and consider insurance premiums as another cost of production. During the 1980s and 1990s, agricultural insurance was exclusively provided through the state-owned insurance company, Shadharan Bima Corporation (SBC). SBC offered an individual grower multiple peril crop insurance product as well as livestock mortality and aquaculture insurance. Insurance demand is moderately price elastic (Hill et al. 2017), while crop insurance demand varies across the household head's primary occupation, land ownership, and the size of the farm land (Akter et al. 2009).

#### 10.2.6.2 Awareness

Awareness refers to having knowledge or understanding about a particular subject, issue, event, or concept. It involves being conscious and informed about something, often through the acquisition of information or experiences (May and Perry 2022). Awareness can relate to a wide range of topics, including social issues, personal well-being, current events, environmental concerns, and more (Wray-Lake et al. 2019). Farmers had limited awareness of agricultural insurance. In most cases, no field-level practice of crop insurance is the main cause for this. The lack of knowledge and awareness in the design, rating, and implementation of agricultural insurance need to be considered.

#### 10.2.6.3 Belief of the Benefits

The belief in the benefits of a product or service is an essential factor that influences consumers' purchasing decisions and their willingness to pay a certain price. When consumers perceive that a product offers substantial advantages or value, they are more likely to have a positive attitude toward it and be willing to invest in it (Zhang et al. 2020). Most of the farmers had some bad experiences, not directly owing to the insurance company, but some other microfinance organizations. Due to the bad experiences, farmers do not have sufficient self-confidence for a payment for any loss.

#### 10.2.6.4 Collection and Distribution of Money

The collection and distribution of money for crop insurance typically involves multiple parties, including farmers, insurance companies, and government agencies. There are some fixed costs and high administrative costs for any operation, as rural infrastructure is not well developed and most of the farms are small in size. The sizes of many farms were less than one hectare, which causes the costs of insurance delivery, underwriting, and claims administration to be prohibitively high (Mottaleb and Mohanty 2015). The main constraints and opportunities for crop and livestock insurance in developing countries are considered, such as insurance product types, hazards, vulnerability, and rural institutions that can support organization and distribution (Dick and Wang 2010).

#### 10.2.6.5 Trust in the Company

Trust in a company is crucial for its success and reputation. It is built over time through various factors and actions (Julianelli et al. 2020). Building trust takes time and effort, and it requires consistent actions aligned with the values and commitments of the company. Trust is a valuable asset that can positively impact a company's relationships with customers, employees, partners, and the broader community. There is no clear policy framework for agricultural insurance in Bangladesh, including a lack of clarity on the role of government in supporting agricultural insurance through the private insurance sector. The commercial insurance companies are not willing to take a lead in developing crop insurance products.

#### 10.2.6.6 Rules and Regulations

Rules and regulations are a set of guidelines and directives established by governing bodies, organizations, or authorities to govern the conduct and operations of individuals, businesses, and society as a whole. These rules and regulations serve various purposes, including ensuring public safety, promoting fairness, protecting rights, and maintaining order (Gupta et al. 2022). Establishing a business and continuing it in an appropriate manner needs some appropriate rules and regulations. How a company can work, what are the terms and conditions, and how a farmer participates in a crop insurance policy and gains benefit should be clear and understandable. While a premium subsidy is the most common intervention, other enabling measures are important, such as the legal and regulatory framework, reinsurance, technical and administrative assistance, and linkages to government extension services in agriculture, animal health, or meteorology (Dick and Wang 2010).

#### 10.2.6.7 Willingness of the Government

The willingness of a government refers to its readiness, motivation, and commitment to taking action on various matters, such as implementing policies, addressing issues, and engaging with stakeholders. The willingness of a government can vary depending on several factors, including political priorities, public sentiment, economic conditions, and the specific issue at hand. In developing countries, most of the decisions are made through government intervention. Agricultural insurance is normally undertaken as a market-based activity by private or state sector insurance companies, often with support measures from the government. In high and some middle-income countries, such as the United States, Canada, Spain, Italy, Japan, and South Korea, governments have significantly intervened, created sophisticated market structures, and accepted significant annual budgets for agricultural insurance (Dick and Wang 2010).

#### 10.2.6.8 Government Subsidy

Government subsidies are financial assistance provided by the government to individuals, businesses, or organizations to support specific activities, industries, or causes. These subsidies are typically aimed at achieving various policy objectives and can take different forms (Liu et al. 2022). Private commercial insurance companies in Bangladesh have limited financial capacity, and are, in general, reluctant to take a lead in investing in agricultural insurance staff and the design of products and policies, systems, and procedures, as this is a high-risk class of insurance. Thus, introducing crop insurance in a vast area requires a good budget from the government for a subsidy. The estimated discrepancies between an expected indemnity and expected premium appear too large to be financed by government subsidy on a continuous basis (Akter et al. 2009).

### *10.2.7 Insurance Coverage and Premium Rate*

Insurance coverage refers to the extent of protection provided by an insurance policy. It outlines the specific risks and perils for which the policyholder is insured. The coverage details can vary depending on the type of insurance, such as life insurance, health insurance, auto insurance, property insurance, and crop insurance (Fang et al. 2021). Agricultural insurance is generally undertaken as a market-based activity by private or state sector insurance companies, often with support measures from the government (Dick and Wang 2010). The demand for crop insurance is said to be moderately price elastic (Hill et al. 2017; Sinha and Tripathi 2016). Government-subsidized crop insurance programs have been established and developed for decades in developed countries, particularly in the United States and Canada (Wang et al. 2016). In many developing countries including Bangladesh, crop insurance was introduced approximately three decades ago. Some of these countries are continuing to offer such insurance, while in some others, it has stopped functioning owing to heavy losses. Insurance in the United States is provided by 17 private companies. They work in agreement with the USDA Risk Management Agency and approximately 45% of the field crops' production value are insured (23% in the European Union (EU)). The average premium rate is close to 9%, which is much higher than in Europe (4%), mainly because they offer a wider coverage—revenue or yield insurances versus mainly single-peril or combined-risk insurances. The premium subsidies amount to \$1900 million (58% of the total premiums). The United States government also provides funds for the administrative costs of insurance companies and provides reinsurance. Thus, the total support provided to insurance amounts to 72% of the total premiums (in the EU, approximately 500 M€ = 32% support) (European Commission 2006).

The premium rate is the amount of money the policyholder pays to the insurance company for coverage. It is typically paid on a regular basis, such as monthly, quarterly, or annually. In this research, the insurance premium mainly considers the quality of land or suitability of the land to cover the crop revenue of farmers. An AHP survey was conducted to first identify the criteria involved for insurance premiums (Appendix 2). To determine the premium rate, different low interest rates were considered that could have the ability or "willingness to pay." The range was 2–6% of the insured revenue, which was considered as the willingness to pay the premium for their crops. The premium rate varies due to the land quality and risks (FAO 1976). The expected revenue was calculated with the help of the regional average production of time series data and the government floor price.

The insurance coverage was characterized as high coverage (90%), moderate coverage (80%), and marginal coverage (70%) considering the average production costs of the production of rice (Iqbal 2005). The lowest insurance coverage was considered marginal coverage (70%) to overcome the production costs, while the supreme coverage was estimated as high coverage (90%) to ensure the production efficiency of the factors of productions. Insurance coverage was the same for different land suitability, harvest index, and subsidy dependence factors; however, the premiums were different. Thus, it could benefit land owners with moderately suitable or marginally suitable lands. No insurance was covered for the presently not suitable areas due to the physical and economic viability. A farmer could buy a product of insurance from the three categories and could pay the premium in 12 installments (every month). The future value of annuity ($V$) was calculated using regional yield and the price of rice varieties as follows (Chen and Tse 2017):

$$V = 12P\frac{\left\{1+\left(\frac{i}{12}\right)\right\}^{12n}-1}{i} \tag{10.9}$$

where $P$ is the premium for different coverages and can be expressed by combining the harvest index and subsidy dependence index. The future expected revenue from rice production can be expressed as follows:

$$V_j = \sum_{i=1}^{n} PPV_i \times PRT \tag{10.10}$$

where $V_j$ is the cumulative return from the yield for different coverages ($j$) of three seasonal of rice practices in one calendar year. The $PPV$ is the average production capacity for seasonal rice (ton/ha; *i* is the seasonal variety: Aus, Aman, and Boro; and *n* is the number of years) and PRT is the average price of rice ($/ton). Therefore, the premium for different coverages (*j*) is based on the expected revenue, production function, and subsidy which can be expressed as follows:

$$P_j = \frac{V_j \times i}{12\left\{1+\left(\frac{i}{12}\right)\right\}^{12n} - 1} \times HI \times SDF \times C_j \times C_f \tag{10.11}$$

where $P_j$ is the premium for the specific coverage (*j*) of three seasonal rice practices in one calendar year and HI is the harvest index and ratio between the observed yield and biological yield. The *SDF* is the subsidy dependence factor and was assumed to be 0.50 as a welfare consideration. The parameter $C_f$ is the conversion factor for the pixel resolution for Landsat 8, OLI. The government of Bangladesh provides agricultural subsidies not only in the form of cash but also on price of agricultural inputs, such as seeds, fertilizer, irrigation, and fuel. Among production inputs, the cost of irrigation holds the highest share (nearly 20%) as a single cost item followed by fertilizer (approximately 18%) because of the price increase of these items at both the national and international markets (Akram 2017). Electricity in agriculture enjoyed a 20% subsidy for a long time; however, it was discontinued later to offset the impact of a rise in oil prices. Presently, the policy is proposed for at least a 25% subsidy on electricity and fuel needed to continue to meet future challenges, particularly, the *Boro* crop for which the irrigation is powered 40% by electricity and 60% by fossil fuel (Byron 2017). If the price of electricity and fuel is within the capacity of farmers, they will not lose interest in *Boro* cultivation. The subsidy expenditure as a total subsidy is presently offered at 17.30%, the proposed budget (2017–2018) by the government is 22.13%, and electricity or fuel for irrigation is proposed to increase from 20% to 25% (Islam 2009). The parameter $C_j$ is the coverage rate (highly, moderately, and marginally) and $C_f$ is the conversion factor to scale down the pixel level (30 m × 30 m, Landsat 8, OLI).

## 10.3 Results

### *10.3.1 Land Suitability*

The study found that highly suitable land (S1) covers an area of 1337.02 km$^2$ (22.74%) of the total land. Hence, a significant part of the study area was highly suitable for rice cultivation (Table 10.8). The S1 (highly suitable) areas were characterized using a slope level of 0–10%, elevation less than 50 m, soil pH level between 5.5 and 7.3, the topsoil texture class that included predominant clay, predominant silt clay, and predominant clay loam, the land type of medium high and medium lowland, a rainfall range between 1000 and 2000 mm, and temperatures

**Table 10.8** Land suitability analysis for rice production sites

| Classes | Area (km$^2$) | % |
|---|---|---|
| Highly suitable | 1337.02 | 22.74 |
| Moderately suitable | 1678.04 | 28.54 |
| Marginally suitable | 873.71 | 14.86 |
| Currently not suitable | 69.97 | 1.19 |
| Permanently not suitable | 1920.87 | 32.67 |
| Total | 5879.60 | 100 |

between 20 and 30 °C. The result showed that 1678.04 km$^2$ (28.54%) of the area were moderately suitable (S2) for rice production, and 873.71 km$^2$ (14.86%) of the area were marginally suitable (S3) for the cultivation of rice with limitations of land type, soil texture, and the soil pH. This class was used less for rice production as its limitations and profitably corrected the suitability class to a moderate level. Currently, not suitable areas (*N*) had also covered a relatively insignificant portion of the study area of 69.97 km$^2$ (1.19%). Generally, not suitable areas (*N*) were in very lowland areas with a sandy loam texture. In addition, the total lands that were permanently not suitable for rice production were 1920.87 km$^2$ (32.67%), occupied by settlements, rivers, forests, and water bodies. The results show that the majority (up to 67%) of the total land in the study area had varying degrees of suitability for rice production (Fig. 10.5). The highly and moderately suitable areas, taken altogether, which have, according to the analysis, covered 51% of the total land were commendable for rice production. Overall, these results indicate that the study area has enormous potential for rice production. Therefore, economic levels of agricultural production can be achieved by cultivating rice crops in highly (S1) and moderately (S2) suitable areas and practicing the diversification of marginally (S3) suitable areas to crops other than rice.

### *10.3.2 Insurance Premium and Coverage*

The insurance coverage was categorized as high coverage (90%), moderate coverage (80%), and marginal coverage (70%) (Table 10.9). Insurance coverage was the same for different land suitability, but the premiums were different. Thus, it could benefit landowners with moderately suitable or marginally suitable lands. Presently, not suitable areas are not recommended for insurance coverage because of cost-effectiveness.

A farmer could buy a product of insurance from the three categories and could pay the premium in 12 installments of three rice crops (Aus, Aman, and Boro) during crop plantation, in the middle of cultivation, and after harvest. The average weights for each factor were derived using individual judgments. Considering the fact that farmers want to produce rice with secured and minimum premiums, the experts' opinions were listed to recognize the criteria to introduce an insurance premium according to land suitability.

**Table 10.9** The premium rate, premium per hectare per month, and premium per pixel per month based on the proposed coverage

| | Marginally coverage (70% yield coverage) | | | Moderately coverage (80% yield coverage) | | | Highly coverage (90% yield coverage) | | |
|---|---|---|---|---|---|---|---|---|---|
| | Percentage (%) | Per hectare (USD) | Per pixel (USD) | Percentage (%) | Per hectare (USD) | Per pixel (USD) | Percentage (%) | Per hectare (USD) | Per pixel (USD) |
| Highly suitable | 2 | 238.98 | 8.07 | 3 | 271.87 | 9.18 | 4 | 304.45 | 10.28 |
| Moderately suitable | 3 | 237.89 | 8.03 | 4 | 270.62 | 9.13 | 5 | 303.05 | 10.23 |
| Marginally suitable | 4 | 236.80 | 8.00 | 5 | 269.38 | 9.09 | 6 | 301.66 | 10.18 |

Product price was the key factor (29%) to participate in crop insurance, which was highly involved with the production factor as mentioned in the harvest index. Farmers had inadequate awareness of agricultural insurance. There were 17% of the experts who mentioned that there was a low level of awareness and knowledge among Bangladeshi farmers on the role of agricultural insurance. Although the villagers had no understanding of crop insurance, when they understood the theme, they were rather worried (9%) about the benefits. In the case of collection and distribution of money, the collection of the premium and distribution of the insurer's money had a large involvement in the cost issue (3%). Trust in the company was also a challenge to prove (5%). In the rules and regulations criteria, the lack of a national framework (4.5%) for agricultural insurance and limited agricultural insurance provisions were mentioned by the experts. The willingness of the government was the second (18%) most important criterion for the establishment of the crop insurance market. In the case of the government subsidy, private commercial insurance companies in Bangladesh have limited financial capacity and are generally reluctant to take a lead in investing in agricultural insurance. Therefore, introducing crop insurance in the poverty and risk-prone areas requires the government to provide a subsidy. In the survey, we found that 16% of the experts' opinions consider a government subsidy as an important factor for farmer participation in crop insurance.

## 10.4 Discussion

In this research, the suitability mapping of rice production in poverty-prone and risk-prone regions of Bangladesh is a valuable initiative aimed at increasing rice production and ensuring food security in areas that face significant challenges. From the expert analysis information, a key challenge for poor farmers of the study area was the management of the activities to reduce crop damage risk. One potential strategy for coping with the devastating effects of crop damage is to identify the suitability areas and introduce a crop insurance scheme on a cooperative basis. Agricultural insurance along with the cooperation of the community could help farmers to stabilize farm income and investment and guard against disastrous effects of losses due to natural hazards, low production, or low market prices. Crop insurance not only stabilizes farm incomes but also helps farmers to initiate production activity after a bad agricultural year, shock of crop losses, or low prices by providing farmers with a minimum amount of protection and helping farmers invest more in agriculture.

The socioeconomic identities of farmers, poor rural infrastructure, very low household incomes, large family sizes, low productivity, and insufficient training on agriculture made the regions more risk prone. To achieve sustainable livelihoods, a crop insurance policy needs to be introduced with a cooperative procedure, as most of the farmers do not deal with banks and nongovernmental organizations. In this research, a procedure to calculate premiums was introduced, based on land suitability areas, a harvest index, and a subsidy dependence factor. Considering the above factors, the Government of Bangladesh needs to take initiatives by subsidizing this sector initially, which suggests a premium rate of 2–6% of the expected outcome based on land suitability, harvest index, and subsidy dependence factor. Considering land suitability, there are three types of insurance coverage: high coverage (90%), moderate coverage (80%), and marginal coverage (70%). The study suggests producing rice up to marginally suitable land; a low harvest index could gain support through insurance. The minimum coverage was marginal coverage (70%), which covered the production costs, while maximum coverage was designed using high coverage (90%) to utilize the maximum benefit against the risk-prone marginal areas. The expected outcome can be calculated from the time-series production data. The insurance premium introduced in India has offered a monumental and path-breaking bounty of crop insurance at a nominal premium—as low as 1.5% and 2% of the sum assured—which is unprecedented, to boost the farm sector and derisk the vast farming community. Under this new scheme, the Government of India has substantially reduced the premium to be paid by the farmers—1.5% on Rabi crops (dry season), 2% on Kharif crops (wet season), and 5% on commercial/horticultural crops. To reduce delays and leakages, the payment of compensation was undertaken through direct transfers into the bank accounts of farmers (Rathore and Rao 2017). Further research needs to be undertaken considering community acceptance to justify the premium as proposed in the research considering land suitability, harvest index, and subsidy in the risk-prone areas of Bangladesh.

## 10.5 Conclusions

Land suitability analysis for crop production is indeed crucial in addressing the challenge of meeting the growing demand for food with limited land resources. It involves assessing the suitability of land for specific crops based on various factors such as soil quality, climate conditions, water availability, and topography. The study conducted an LSA analysis on the northern part of Bangladesh and found that approximately 33% of the total land, which was occupied by water bodies, rivers, forests, and settlements, was permanently not suitable, and 1.19% was presently not suitable. However, 22.74% of the study area was highly suitable, 28.54% was moderately suitable, and 14.86% was marginally suitable for rice production. Considering the demand for rice by the vast population, the regions need to secure rice production up to the marginally suitable lands. In the marginally suitable lands, the risks are higher than the moderately suitable areas, and the costs of production are also higher; the production could be lower than the moderately suitable lands. To motivate owners of low-quality land to produce rice, protection through crop insurance support must be provided. This new crop insurance model, "crop insurance based on land suitability," can be a very effective tool to manage production risks and provide rational support to owners of land of different quality, which will increase production to ensure food security. The socioeconomic characteristics of farmers in Bangladesh are far different from those in developed countries, and if we merely follow the latter's products and policies, rice production may not be sustainable. This study suggests a premium rate of 2–6% of the expected outcome based on land suitability. The premium calculation that is based on costs and risks does not provide a suitable environment for the introduction of crop insurance. The strategy of the Government of Bangladesh to provide subsidies to the farmers at times of shocks needs to be changed to subsidies to pay insurance premiums with appropriate regulations.

The study suggests producing rice up to marginally suitable land, and this can be supported by insurance. The minimum coverage is marginal coverage (70%) to cover the production costs, while maximum coverage is ensured through high coverage (90%) to enable maximum efficiency. Based on these judgments, the highest priority was the product price (26.72%), followed by willingness of the government (20.27%), government subsidy (19.18%), awareness (11.92%), belief of the benefits (7.12%), rules and regulations (6.70), trust in the company (4.76%), and collection and distribution of money (3.34%). These results are helpful for policymakers and agricultural extension workers who assist the agriculture sectors to increase production and ensure the welfare of the farmers. The information provided, at a local level, could be used by the growers to select suitable land and can be relevant to other crops to increase production. The government could encourage private participation to provide varied products with premium rates for commercial farmers on an individual basis. It would be wise to integrate microinsurance with microfinance in order to manage the risk of default and enterprise losses in a more proactive manner.

**Acknowledgments** The authors would like to express their sincere gratitude to the University of Tsukuba, Japan for providing GIS facilities to carry out the research as well as the open access journal Sustainability published by MDPI for their policy to support the authors for reusing the published article (https://doi.org/10.3390/su10061759). The authors are grateful to the Department of Agricultural Extension (DAE), Ministry of Agriculture, Bangladesh, Bangladesh Metrological Department (BMD), and Local Government and Engineering Division (LGED) for their cordial support in data collection and base map preparation. They also express their sincere thanks to the United States Geological Survey (USGS) and the European Space Agency (ESA) for geographical and satellite data. Furthermore, they also express their gratitude to the Bangabandhu Fellowship Trust, Bangladesh for providing scholarship to pursue this research in University of Tsukuba, Japan.

**Conflicts of Interest** The authors declare no conflict of interest.

# Appendices

## *Appendix 1 "Land Suitability Analysis for Paddy in Northern Parts of Bangladesh: Introducing Geo-Spatial Insurance Premium" SURVEY*

With this questionnaire, we expect the willingness to fill out the questionnaire according to your opinion. The questionnaire aims to complement the research data in the context of the preparation of the thesis with the title: "Land Suitability Analysis for Paddy in Northern Areas of Bangladesh: Introducing Geo-Spatial Insurance Premium."

(***Your response will only be used for research purposes and analysis***.)
**Identity of Respondent**

| | | |
|---|---|---|
| 1 | Name | |
| 2 | Position and address | |

### Appendix 1.1. Section-1: Please Select Scale Number of Criteria for Land Suitability Mapping of Rice Production in the Northern Part of Bangladesh

***Instruction:*** *Please choose the scale number from 1 to 9.*
**Example:** Which is the most important criterion to produce rice?

| Criteria | Slope | Elevation | Land type | Topsoil texture | Soil pH | Flood prone | Temperature | Rainfall |
|---|---|---|---|---|---|---|---|---|
| Example | 6 | 5 | 9 | 7 | 7 | 4 | 3 | 3 |
| Your opinion | | | | | | | | |

If you choose a number on scale 9, it means that *Land type* is the most important, followed by *Topsoil texture, Soil pH*, and others. If you choose the same scale, it means equally important.

## Appendix 1.2. Section-2: Please Select Weight of Criteria for Land Suitability Mapping of Rice Production in the Northern Part of Bangladesh

***Instruction:*** *Please add a checklist (✓) on the column scale criterion (A) or at the column scale criterion (B) based on experts' opinion.*

**Definition of scale:**

| | |
|---|---|
| 1: the second equally important criterion | 3: criterion (A) is little more important than (B) |
| 5: criterion (A) is more important than (B) | 7: criterion (A) is much more important than (B) |
| 9: criterion (A) is absolutely more important than (B) | * and that goes for the reverse |

**Example:** In taking the decision to develop the crop insurance how important this criterion:

| | | Scale | | | | | | | | | Scale | | | | | | | | |
|---|---|---|---|---|---|---|---|---|---|---|---|---|---|---|---|---|---|---|---|
| | Criterion A | 9 | 8 | 7 | 6 | 5 | 4 | 3 | 2 | 1 | 2 | 3 | 4 | 5 | 6 | 7 | 8 | 9 | Criterion B |
| 1 | Land type | | | ✓ | | | | | | | | | | | | | | | Topsoil texture |

If you give a sign checklist (✓) on a scale of 7 in column A, then it means that criterion A in this example "*Land type*" is more important than criterion B in this example "*Topsoil texture.*"

**Question:** After deciding the most important criterion in Appendix 1.1, how important is the criterion if it compares one to another?

| | | Scale | | | | | | | | | Scale | | | | | | | | |
|---|---|---|---|---|---|---|---|---|---|---|---|---|---|---|---|---|---|---|---|
| | Criterion A | 9 | 8 | 7 | 6 | 5 | 4 | 3 | 2 | 1 | 2 | 3 | 4 | 5 | 6 | 7 | 8 | 9 | Criterion B |
| 1 | Slope | | | | | | | | | | | | | | | | | | Elevation |
| 2 | Slope | | | | | | | | | | | | | | | | | | Land type |
| 3 | Slope | | | | | | | | | | | | | | | | | | Topsoil texture |
| 4 | Slope | | | | | | | | | | | | | | | | | | Soil pH |
| 5 | Slope | | | | | | | | | | | | | | | | | | Flood prone |
| 6 | Slope | | | | | | | | | | | | | | | | | | Temperature |
| 7 | Slope | | | | | | | | | | | | | | | | | | Rainfall |
| 8 | Elevation | | | | | | | | | | | | | | | | | | Land type |
| 9 | Elevation | | | | | | | | | | | | | | | | | | Topsoil texture |
| 10 | Elevation | | | | | | | | | | | | | | | | | | Soil pH |
| 11 | Elevation | | | | | | | | | | | | | | | | | | Flood prone |
| 12 | Elevation | | | | | | | | | | | | | | | | | | Temperature |
| 13 | Elevation | | | | | | | | | | | | | | | | | | Rainfall |
| 14 | Land type | | | | | | | | | | | | | | | | | | Topsoil texture |
| 15 | Land type | | | | | | | | | | | | | | | | | | Soil pH |

| | Criterion A | Scale 9 | 8 | 7 | 6 | 5 | 4 | 3 | 2 | 1 | Scale 2 | 3 | 4 | 5 | 6 | 7 | 8 | 9 | Criterion B |
|---|---|---|---|---|---|---|---|---|---|---|---|---|---|---|---|---|---|---|---|
| 16 | Land type | | | | | | | | | | | | | | | | | | Flood prone |
| 17 | Land type | | | | | | | | | | | | | | | | | | Temperature |
| 18 | Land type | | | | | | | | | | | | | | | | | | Rainfall |
| 19 | Topsoil texture | | | | | | | | | | | | | | | | | | Soil pH |
| 20 | Top Soil Texture | | | | | | | | | | | | | | | | | | Flood prone |
| 21 | Top Soil Texture | | | | | | | | | | | | | | | | | | Temperature |
| 22 | Top Soil Texture | | | | | | | | | | | | | | | | | | Rainfall |
| 23 | Soil pH | | | | | | | | | | | | | | | | | | Flood prone |
| 24 | Soil pH | | | | | | | | | | | | | | | | | | Temperature |
| 25 | Soil pH | | | | | | | | | | | | | | | | | | Rainfall |
| 26 | Flood prone | | | | | | | | | | | | | | | | | | Temperature |
| 27 | Flood prone | | | | | | | | | | | | | | | | | | Rainfall |
| 28 | Temperature | | | | | | | | | | | | | | | | | | Rainfall |

Signature ......................................

Name ( )

## *Appendix 2 "Land Suitability Analysis for Paddy in Northern Parts of Bangladesh: Introducing Geo-Spatial Insurance Premium" SURVEY*

With this questionnaire, we expect the willingness to fill out the questionnaire according to your opinion. The questionnaire aims to complement the research data in the context of the preparation of the thesis with the title: "Land Suitability Analysis for Paddy in Northern Areas of Bangladesh: Introducing Geo-Spatial Insurance Premium."

(***Your response will only be used for research purposes and analysis***.)

**Identity of Respondent**

| | | |
|---|---|---|
| 1 | Name | |
| 2 | Position and address | |

### Appendix 2.1. Section-1: Please Select Scale Number of Criteria for Insurance Premiums Based on Land Suitability for Rice Production in the Northern Part of Bangladesh

***Instruction:*** *Please choose the scale number from 1 to 9.*

**Example:** Which is the most important criterion to participate in crop insurance for rice production?

| Criteria | Product price | Awareness | Believe of benefits | Collection and distribution of money | Trust to company | Rules and regulations | Willingness of govt. | Govt. subsidy |
|---|---|---|---|---|---|---|---|---|
| Example | 9 | 5 | 6 | 2 | 7 | 4 | 3 | 3 |
| Your opinion | | | | | | | | |

If you choose a number on scale 9, it means that *Product price* is the most important, followed by *Awareness*, *Believe of benefits*, and others. If you choose the same scale, it means equally important.

## Appendix 2.2. Section-2 Please Fill Out with Selecting Weight of Criteria for Insurance Premiums Based on Land Suitability for Rice Production in the Northern Part of Bangladesh

***Instruction:*** *Please add a checklist (✓) on the column scale criterion (A) or at the column scale criterion (B) based on experts' opinion.*

**Definition of scale:**

| | |
|---|---|
| 1: the second equally important criterion | 3: criterion (A) is little more important than (B) |
| 5: criterion (A) is more important than (B) | 7: criterion (A) is much more important than (B) |
| 9: criterion (A) is absolutely more important than (B) | * and that goes for the reverse |

**Example:** In taking the decision to develop the crop insurance, how important this criterion:

| | | Scale | | | | | | | | | Scale | | | | | | | | |
|---|---|---|---|---|---|---|---|---|---|---|---|---|---|---|---|---|---|---|---|
| | Criterion A | 9 | 8 | 7 | 6 | 5 | 4 | 3 | 2 | 1 | 2 | 3 | 4 | 5 | 6 | 7 | 8 | 9 | Criterion B |
| 1 | Product price | | | ✓ | | | | | | | | | | | | | | | Awareness |

If you give a sign checklist (✓) on a scale of 7 in column A, then it means that criterion A in this example "*Product price* " was more important than criterion B in this example " *Awareness*."

**Question:** After deciding the most important criterion in Appendix 2.1, how important the criterion if it compares one to another?

| | | Scale | | | | | | | | | Scale | | | | | | | | |
|---|---|---|---|---|---|---|---|---|---|---|---|---|---|---|---|---|---|---|---|
| | Criterion A | 9 | 8 | 7 | 6 | 5 | 4 | 3 | 2 | 1 | 2 | 3 | 4 | 5 | 6 | 7 | 8 | 9 | Criterion B |
| 1 | Product price | | | | | | | | | | | | | | | | | | Awareness |
| 2 | Product price | | | | | | | | | | | | | | | | | | Believe of benefits |

| | Criterion A | Scale | | | | | | | | | Scale | | | | | | | Criterion B |
|---|---|---|---|---|---|---|---|---|---|---|---|---|---|---|---|---|---|---|
| | | 9 | 8 | 7 | 6 | 5 | 4 | 3 | 2 | 1 | 2 | 3 | 4 | 5 | 6 | 7 | 8 | 9 | |
| 3 | Product price | | | | | | | | | | | | | | | | | | Collection and distribution of money |
| 4 | Product price | | | | | | | | | | | | | | | | | | Trust to the company |
| 5 | Product price | | | | | | | | | | | | | | | | | | Rules and regulations |
| 6 | Product price | | | | | | | | | | | | | | | | | | Willingness of government |
| 7 | Product price | | | | | | | | | | | | | | | | | | Govt. subsidy |
| 8 | Awareness | | | | | | | | | | | | | | | | | | Believe of benefits |
| 9 | Awareness | | | | | | | | | | | | | | | | | | Collection and distribution of money |
| 10 | Awareness | | | | | | | | | | | | | | | | | | Trust to the company |
| 11 | Awareness | | | | | | | | | | | | | | | | | | Rules and regulations |
| 12 | Awareness | | | | | | | | | | | | | | | | | | Willingness of government |
| 13 | Awareness | | | | | | | | | | | | | | | | | | Govt. subsidy |
| 14 | Believe of benefits | | | | | | | | | | | | | | | | | | Collection and distribution of money |
| 15 | Believe of benefits | | | | | | | | | | | | | | | | | | Trust to the company |
| 16 | Believe of benefits | | | | | | | | | | | | | | | | | | Rules and regulations |
| 17 | Believe of benefits | | | | | | | | | | | | | | | | | | Willingness of government |
| 18 | Believe of benefits | | | | | | | | | | | | | | | | | | Government Subsidy |
| 19 | Collection and distribution of money | | | | | | | | | | | | | | | | | | Trust to the company |
| 20 | Collection and distribution of money | | | | | | | | | | | | | | | | | | Rules and regulations |
| 21 | Collection and distribution of money | | | | | | | | | | | | | | | | | | Willingness of government |
| 22 | Collection and distribution of money | | | | | | | | | | | | | | | | | | Government subsidy |
| 23 | Trust to the company | | | | | | | | | | | | | | | | | | Rules and regulations |
| 24 | Trust to the company | | | | | | | | | | | | | | | | | | Willingness of government |
| 25 | Trust to the company | | | | | | | | | | | | | | | | | | Government subsidy |

| | | Scale | | | | | | | | | Scale | | | | | | | |
|---|---|---|---|---|---|---|---|---|---|---|---|---|---|---|---|---|---|---|
| | Criterion A | 9 | 8 | 7 | 6 | 5 | 4 | 3 | 2 | 1 | 2 | 3 | 4 | 5 | 6 | 7 | 8 | 9 | Criterion B |
| 26 | Rules and regulations | | | | | | | | | | | | | | | | | | Willingness of government |
| 27 | Rules and regulations | | | | | | | | | | | | | | | | | | Government subsidy |
| 28 | Willingness of government | | | | | | | | | | | | | | | | | | Government subsidy |

Signature ......................................

Name ( )

# References


Abdullahi S, Pradhan B, Mansor S, Shariff ARM (2015) GIS-based modeling for the spatial measurement and evaluation of mixed land use development for a compact city. GISci Remote Sens *52*:18–39

Asian Development Bank (ADB) (2019) Bangladesh: Pilot Project on Weather Index-Based Crop Insurance, 6 A 6 ADB Avenue, Mandaluyong City 1550, Metro Manila, Philippines, https://www.adb.org/sites/default/files/project-documents/46284/46284-001-icm-en.pdf

Adem Esmail B, Geneletti D (2018) Multi-criteria decision analysis for nature conservation: a review of 20 years of applications. Methods Ecol Evol 9(1):42–53

Ahamed T, Noguchi R, Takigawa T, Tian L (2016) Bioproduction engineering, automation and precision agronomics for sustainable agricultural systems, 2nd edn. Nova Science Publishers, New York, NY

Ahmed A, Ryosuke S (2000) Climate change and agricultural food production of Bangladesh: an impact assessment using GIS-based biophysical crop; Simulation Model. Center for Spatial Information Science, University of Tokyo, Komaba, Japan

Akram S (2017) Subsidy in agriculture, The Financial Express, December 08, 2017. Available: https://thefinancialexpress.com.bd/views/opinions/subsidy-in-agriculture-1512751118. Accessed 15 Jan 2018)

Akter S, Brouwer R, Choudhury S, Aziz S (2009) Is there a commercially viable market for crop insurance in rural Bangladesh? Mitig Adapt Strateg Glob Chang *14*:215–229

Alam AF, Begum H, Masud MM, Al-Amin AQ, Leal Filho W (2020) Agriculture insurance for disaster risk reduction: a case study of Malaysia. Int J Disaster Risk Reduction 47:101626

Alshurideh MT, Al Kurdi B, Alzoubi HM, Ghazal TM, Said RA, AlHamad AQ et al (2022) Fuzzy assisted human resource management for supply chain management issues. Ann Oper Res:1–19

Amin MR, Zhang J, Yang M (2015) Effects of climate change on the yield and cropping area of major food crops: a case of Bangladesh. Sustainability 7:898–915

Anderson J (2001) Risk management in rural development: a review rural strategy background paper 7. The World Bank, Washington, DC

Anusha BN, Babu KR, Kumar BP, Sree PP, Veeraswamy G, Swarnapriya C, Rajasekhar M (2023) Integrated studies for land suitability analysis towards sustainable agricultural development in semi-arid regions of AP, India. Geosyst Geoenviron 2(2):100131

Asai H, Samson BK, Stephan HM, Songyikhangsuthor K, Homma K, Kiyono Y, Horie T (2009) Biochar amendment techniques for upland rice production in Northern Laos: 1. Soil physical properties, leaf SPAD and grain yield. Field Crops Res 111:81–84

Ashford SA, Sitar N, Lysmer J, Deng N (1997) Topographic effects on the seismic response of steep slopes. Bull Seismol Soc Am *87*:701–709

Assouto AB, Houensou DA, Semedo G (2020) Price risk and farmers' decisions: a case study from Benin. Sci Afr 8:e00311

Atalay I (2006) Toprak Olusumu, Sınıflandırılması ve Cog˘rafyası. Meta Basım Matbaacılık, IzmirAyehu GT, Besufekad SA (2015) Land suitability analysis for rice production: a GIS based multi-criteria decision approach. Am J Geogr Inf Syst 4:95–104
Bangladesh Bureau of Statistics (BBS) (2014) Small area atlas of Bangladesh. Ministry of Planning, Dhaka
Bangladesh Bureau of Statistics (BBS) (2016) Yearbook of Agricultural Statistics-2015. Ministry of Planning, Dhaka
Bangladesh Bureau of Statistics (BBS) (2017) Statistical Pocket Book Bangladesh 2016. Ministry of Planning, Dhaka
Belton V, Stewart T (2002) Multiple criteria decision analysis: an integrated approach. Springer Science & Business Media
Berber K, Patrick W (2018) Research for agricultural insurance in South Asia: a regional dialogue. International Food Policy Research Institute, Washington, DC, pp 1–5
Bhushan N, Rai K (2004) Strategic decision making applying the analytic hierarchy process. Springer, CREAX Information Technologies Pvt. Ltd., Bangalore, pp 11–21
Billings SA, Hirmas D, Sullivan PL, Lehmeier CA, Bagchi S, Min K et al (2018) Loss of deep roots limits biogenic agents of soil development that are only partially restored by decades of forest regeneration. Elem Sci Anth 6:34
Bryla E, Syroka J (2007) Developing index-based insurance for agriculture in developing countries. Sustain Dev Innov Briefs 9:421–438
Byron RK (2017) Subsidy expense to go up 17.3 pc, The Daily Star, June 02, 2017. Available: https://www.thedailystar.net/Bangladesh-Budget-2017-18/frontpage/subsidy-expense-go-173pc-1414411. Accessed 15 Oct 2017
Chandio IA, Matori ANB, WanYusof KB, Talpur MAH, Balogun AL, Lawal DU (2013) GIS-based analytic hierarchy process as a multicriteria decision analysis instrument: a review. Arab J Geosci 6:3059–3066
Chatterjee, A.R. (2019) The state of crop insurance in Bangladesh, The Financial Express, Dhaka Bangladesh, Aug 31, 2019; Available: https://thefinancialexpress.com.bd/views/analysis/the-state-of-crop-insurance-in-bangladesh-1567174854. Accessed 26 May 2023
Chauhan BS, Jabran K, Mahajan G (2017) Rice production worldwide. Springer International Publishing, Basel, pp 255–277
Chen WS, Tse YK (2017) Financial mathematics for actuaries, 2nd edn. World Scientific Publishing Pte Ltd., Hackensack, NJ, pp 4–45. ISBN 978-981-3224-67-4
Chen Y, Lu CA (2018) Comparative analysis on food security in Bangladesh, India and Myanmar. Sustainability *10*:405. https://doi.org/10.3390/su10020405
Chughtai AH, Abbasi H, Karas IR (2021) A review on change detection method and accuracy assessment for land use land cover. Remote Sens Appl Soc Environ 22:100482
Clarke D, De Nicola F, Hill RV, Kumar N, Mehta PA (2015) Chat about insurance: experimental results from rural Bangladesh. Appl Econ Perspect Policy 37:477–501
Cole S, Giné X, Vickery J (2017) How does risk management influence production decisions? Evidence from a field experiment. Rev Financ Stud 30:1935–1970
Cole SA, Xiong W (2017) Agricultural insurance and economic development. Annu Rev Econ 9:235–262
Collins MG, Steiner FR, Rushman MJ (2001) Land-use suitability analysis in the united states: historical development and promising technological achievements. Environ Manag 28:611–621
Dalhaus T, Musshoff O, Finger R (2018) Phenology information contributes to reduce temporal basis risk in agricultural weather index insurance. Sci Rep 8:1–10
Dick WJA, Wang W (2010) Government interventions in agricultural insurance. Agric Agric Sci Procedia 1:4–12
Donald CM, Hamblin J (1976) The biological yield and harvest index of cereals as agronomic and plant breeding criteria. Adv Agron 28:361–405
Dou F, Soriano J, Tabien RE, Chen K (2016) Soil texture and cultivar effects on rice (*Oryza Sativa, L.*) grain yield, yield components and water productivity in three water regimes. PLoS One 11:e0150549. https://doi.org/10.1371/journal.pone.0150549

Eastman JR, Jiang H, Toledano J (1998) Multi-criteria and multi-objective decision making for land allocation using GIS. Multicriteria Anal Land Use Manag 9:227–251
Engineering I, Triantaphyllou E, Mann SH (1995) Using the analytic hierarchy process for decision making in engineering applications: some challenges. Int J Ind Eng 2:35–44
Eugene G, Mahul O (2004) Enabling productive but asset-poor Jamaican farmers to succeed: a risk financing framework; WB Policy Research Working Paper; vol. 3211. Available: http://siteresources.worldbank.org/CMUDLP/Resources/Gurenko_Farmer.pdf. Accessed 26 May 2018
Euro B, Nijkamp P (1998) Land-use management and the path towards sustainability, multicriteria analysis for land-use management. Springer, Dordrecht, pp 1–13
European Commission (2006) Agricultural insurance schemes. Summary report, Directorate General JRC, Joint Research Centre—ISPRA, Institute for the Protection and Security of the Citizen, Agriculture and Fisheries Unit. European Commission, Brussels, pp 1–16
Fang L, Hu R, Mao H, Chen S (2021) How crop insurance influences agricultural green total factor productivity: Evidence from Chinese farmers. J Clean Prod 321:128977
Food and Agriculture Organization (FAO) (1976) A framework for land evaluation; Food and Agriculture Organization of the United Nations, Soils Bulletin 32; FAO: Rome, Italy
Food and Agriculture Organization (FAO) (2021) The impact of disasters and crises on agriculture and food security. FAO, Rome. https://doi.org/10.4060/cb3673en
Ferretti V, Pomarico S (2013) An integrated approach for studying the land suitability for ecological corridors through spatial multicriteria evaluations. Environ Dev Sustain 15:859–885
Gesch DB (2018) Best practices for elevation-based assessments of sea-level rise and coastal flooding exposure. Front Earth Sci 6:230
Godfray HCJ, Beddington JR, Crute IR, Haddad L, Lawrence D, Muir JF, Pretty J, Robinson S, Thomas SM, Toulmin C (2010) Food security: the challenge of feeding 9 billion people. Science 327:812–818
Global Rice Science Partnership (GRiSP). (2013) Rice Almanac. GRiSP, Maisach
Guo X, Li H, Yu H, Li W, Ye Y, Biswas A (2018) Drivers of spatio-temporal changes in paddy soil pH in Jiangxi Province, China from 1980 to 2010. Sci Rep 8:2702. https://doi.org/10.1038/s41598-018-20873-5
Gupta MJ, Chaturvedi S, Prasad R, Ananthi N (2022) Principles and practice of management. AG PUBLISHING HOUSE (AGPH Books)
Habiba U, Shaw R (2009) Crop insurance as a risk management strategy in Bangladesh. In: Disaster risk reduction approaches in Bangladesh. Springer, Tokyo
Habibie MI, Noguchi R, Shusuke M, Ahamed T (2021) Land suitability analysis for maize production in Indonesia using satellite remote sensing and GIS-based multicriteria decision support system. GeoJournal 86:777–807
Halewood NJ, Surya P (2012) Mobilizing the agricultural value chain. Information and communications for development 2012: Maximizing mobile, 31–42.
Hassan MM, Nazem MNI (2016) Examination of land use/land cover changes, urban growth dynamics, and environmental sustainability in Chittagong City, Bangladesh. Environ Dev Sustain 18:697–716
Hazell P (2001) Potential role for insurance in managing catastrophic risk in developing countries. *Int. Food Policy Res. Inst.*, 1–12. Available: http://citeseerx.ist.psu.edu/viewdoc/summary?doi=10.1.1.562.3304. Accessed on 26 May 2018
Hazell PBR (1992) The appropriate role of agricultural insurance in developing countries. J Int Dev 4:567–581
HIES (2016) Preliminary report on household income and expenditure survey 2016. Bangladesh Bureau of Statistics (BBS), Statistics and Informatics Division (SID), Ministry of Planning, Dhaka
Hill RV, Kumar N, Magnan N, Makhija S, de Nicola F, Spielman DJ, Ward PS (2017) Insuring against droughts: evidence on agricultural intensification and index insurance demand from a randomized evaluation in rural Bangladesh. International Food Policy Research Institute, Washington, DC, p 40
Holzmann R (2002) Risk and vulnerability: the forward looking role of social protection in a globalizing world; Social Protection Discussion Paper No. SP 0109; The World Bank: Washington, DC, USA

Hong S, Piao S, Chen A, Liu Y, Liu L, Peng S et al (2018) Afforestation neutralizes soil pH. Nat Commun 9(1):520
Hossain MS, Chowdhury SR, Das NG, Sharifuzzaman SM, Sultana A (2009) Integration of GIS and multicriteria decision analysis for urban aquaculture development in Bangladesh. Landsc Urban Plan 90:119–133
Hussain W, Sohaib O, Ahmed A, Khan MQ (2012) Geographical information system based model of land suitability for good yield of rice in Prachuap Khiri Khan Province, Thailand. Sci Technol Dev *31*:1–9
Iqbal MT (2005) Cost requirements for cultivation of boro rice (Oriza Sativa) under different farming systems. J Agron 4:366–368
Islam N (2009) IFI Watch Bangladesh, Bangladesh Working Group on International Financial Institute and Trade Organizations, Agriculture Subsidies: Govt. Pledges vs. IFI's Aversion, January 2009. Available: http://www.unnayan.org/documents/International_Economic_Relations/IFI_Agriculture_Subsidies.pdf. Accessed 11 July 2017
Iturrioz R (2018) Agricultural Insurance. Available: http://siteresources.worldbank.org/FINANCIALSECTOR/Resources/Primer12_Agricultural_Insurance.pdf. Accessed 26 May 2018
Julianelli V, Caiado RGG, Scavarda LF, Cruz SPDMF (2020) Interplay between reverse logistics and circular economy: critical success factors-based taxonomy and framework. Resour Conserv Recycl 158:104784
Khan MA, Ahmad R, Khan HH (2022) Multi-criteria land suitability analysis for agriculture using AHP and remote sensing data of northern region India. In Geographic Information Systems and Applications in Coastal Studies. IntechOpen
Kihoro J, Bosco NJ, Murage H (2013) Suitability analysis for rice growing sites using a multicriteria evaluation and GIS approach in Great Mwea Region, Kenya. SpringerPlus *2*:265. https://doi.org/10.1186/2193-1801-2-265
Kiker GA, Bridges TS, Varghese A, Seager TP, Linkov I (2005) Application of multicriteria decision analysis in environmental decision making. Integr Environ Assess Manag *1*:95–108
Kramer B, Hazell P, Alderman H, Ceballos F, Kumar N, Timu AG (2022) Is agricultural insurance fulfilling its promise for the developing world? A review of recent evidence. Annu Rev Resour Economics 14:291–311
Liu C, Liu Y, Zhang D, Xie C (2022) The capital market responses to new energy vehicle (NEV) subsidies: an event study on China. Energy Econ 105:105677
Malczewski J (2004) GIS-based land-use suitability analysis: a critical overview. Prog Plan *62*:3–65
May T, Perry B (2022) Social research: issues, methods and process. McGraw-Hill Education (UK)
Mechler R, Linnerooth-bayer J, Peppiatt D (2006) Disaster insurance for the poor ? A review of microinsurance for natural disaster risks in developing countries. International Institute for Applied Systems Analysis, Laxenburg, pp 1–31
Ministry of Environment and Forests (MoEF) (2008) Bangladesh climate change strategy and action plan 2008. Government of the People's Republic of Bangladesh, Dhaka
Mottaleb KA, Mohanty S (2015) Farm size and profitability of rice farming under rising input costs. J Land Use Sci 10:243–255
Mugiyo H, Chimonyo VG, Sibanda M, Kunz R, Masemola CR, Modi AT, Mabhaudhi T (2021) Evaluation of land suitability methods with reference to neglected and underutilised crop species: a scoping review. Land 10(2):125
Muhsin N, Ahamed T, Noguchi R (2018) GIS-based multi-criteria analysis modeling used to locate suitable sites for industries in suburban areas in Bangladesh to ensure the sustainability of agricultural lands. Asia-Pac J Reg Sci:1–30. https://doi.org/10.1007/s41685-017-0046-0
Mustafa AA, Singh M, Sahoo RN, Ahmed N, Khanna M, Sarangi A (2011) Land suitability analysis for different crops: a multi criteria decision making approach using remote sensing and GIS. Water Technol 3:61–84
National Adaptation Programme of Action (NAPA) (2005) Ministry of Environment and Forest; Final Report 64, No. 642005; Government of the People's Republic of Bangladesh: Dhaka, Bangladesh

Naveed QN, Qureshi MRN, Tairan N, Mohammad A, Shaikh A, Alsayed AO et al (2020) Evaluating critical success factors in implementing E-learning system using multi-criteria decision-making. Plos One 15(5):e0231465

Odu GO (2019) Weighting methods for multi-criteria decision-making technique. J Appl Sci Environ Manag 23(8):1449–1457

Paul B, Rashid H (2016) Climatic hazards in Coastal Bangladesh: non-structural and structural solution. Butterworth-Heinemann, Oxford, UK, pp 121–152

Platteau JP, De Bock O, Gelade W (2017) The demand for microinsurance: a literature review. World Dev 94:139–156

Quiggin J, Kafwgiannis G, Stanton J (1993) Crop insurance and crop production: an empirical study of moral hazard and adverse selection. Aust J Agric Econ *37*:95–113

Rajasa EZ, Manap A, Ardana PDH, Yusuf M, Harizahayu H (2023) Literature review: analysis of factors influencing purchasing decisions, product quality and competitive pricing. Jurnal Ekonomi 12(01):451–455

Rathore V, Rao MJ (2017) The performance of PMFBY and other crop insurance models in India. Int J Adv Res Dev 2:2455–4030

Rikalovic A, Cosic I, Lazarevic D (2014) GIS based multi-criteria analysis for industrial site selection. Procedia Eng 69:1054–1063

Saaty TL (1980) The analytic hierarchy process. McGraw Hill, New York, NY

Saaty TL (1990) How to make a decision: the analytic hierarchy process. Eur J Oper Res 48:9–26

Saaty TL, Vargas LG (2012) Models, methods, concepts & applications of the analytic hierarchy process. Springer US, New York, NY

Saaty TL, Zoffer HJ, Vargas LG, Guiora A, Saaty TL, Zoffer HJ, ... Guiora A (2022) The analytic hierarchy process: beyond "getting to yes" in conflict resolution. overcoming the retributive nature of the Israeli-Palestinian Conflict, 17-29.

Salman SM, Mahul O, Bagazonzya HK (2010) Agricultural insurance in Bangladesh promoting access to small and marginal farmers. The World Bank, Washington, DC, p 146

Samanta S, Pal B, Pal DK (2011) Land suitability analysis for rice cultivation based on multi-criteria decision approach through GIS. Data Base:12–20. https://doi.org/10.5923/j.ajgis.20150403.02

Shahzad A, Ullah S, Dar AA, Sardar MF, Mehmood T, Tufail MA et al (2021) Nexus on climate change: agriculture and possible solution to cope future climate change stresses. Environ Sci Pollut Res 28:14211–14232

Sinha S, Tripathi NK (2016) Assessing the challenges in successful implementation and adoption of crop insurance in Thailand. Sustainability *8*:1–20

Spaulding A, Kanakasabai M, Hao J, Skees J (2003) Can weather derivative contracts help mitigating agricultural risk? Microeconomic policy implications for Romania. In Proceedings of the International Conference for Policy Model, Istanbul, Turkey, 3–5 July

Talukder B, Blay-Palmer A, Hipel KW, vanLoon GW (2017) Elimination method of multi-criteria decision analysis (MCDA): a simple methodological approach for assessing agricultural sustainability. Sustainability 9:1–17

United Nations Development Program (UNDP) (2004) Reducing disaster risk: a challenge for development—a global report. UNDP, New York, NY

Vyas S, Dalhaus T, Kropff M, Aggarwal P, Meuwissen MP (2021) Mapping global research on agricultural insurance. Environ Res Lett 16(10):103003

Wang M, Ye T, Shi P (2016) Factors affecting farmers' crop insurance participation in China. Can J Agric Econ 64:479–492

Wray-Lake L, DeHaan CR, Shubert J, Ryan RM (2019) Examining links from civic engagement to daily well-being from a self-determination theory perspective. J Posit Psychol 14(2):166–177

Wu GL, Liu YF, Cui Z, Liu Y, Shi ZH, Yin R, Kardol P (2020) Trade-off between vegetation type, soil erosion control and surface water in global semi-arid regions: a meta-analysis. J Appl Ecol 57(5):875–885

Xie H, Zhang Y, Zeng X, He Y (2020) Sustainable land use and management research: a scientometric review. Landsc Ecol 35:2381–2411

Zhang Y, Xiao C, Zhou G (2020) Willingness to pay a price premium for energy-saving appliances: role of perceived value and energy efficiency labeling. J Clean Prod 242:118555

# Chapter 11
# Drought Monitoring and Rice Yield Loss Assessment Using Various Drought Indices and the SPI by an Integrated Approach of the AHP in the Central Dry Zone of Myanmar

**Aung Myint Htun, Sara Tokhi Arab, and Tofael Ahamed**

**Abstract** Persistent water shortages in a given region can cause severe drought and agricultural yield fluctuations. Yield fluctuations can occur in areas where high temperatures are associated with low precipitation levels. Rice, among all agricultural products, is more susceptible to drought because it requires more water from the growing to the ripening stages. Consequently, water shortages due to drought have prompted significant yield losses in the central dry zone of Myanmar. Therefore, the main purpose of this research was to quantify the drought level in each drought-affected district by using physiological, meteorological, and moisture indices. The analytic hierarchy process (AHP)-determined weights were used to combine all indices into one composite drought index. Finally, the yield loss was calculated for the winter season using regression analysis. The results indicated that more than 50% of the center dry zone was affected by severe to abnormal conditions from 2019 to 2021. Furthermore, severe yield fluctuations were observed in the Sagaing and Mandalay districts in 2020. In this regard, the findings suggest that drought assessment on a regional scale may assist farmers and stakeholders in monitoring crop moisture levels and developing early warning systems by providing timely information on possible dangers throughout the rice growing season.

A. M. Htun
Graduate School of Life and Environmental Sciences, University of Tsukuba, Tsukuba, Ibaraki, Japan

S. T. Arab
Division of Intelligent Agriculture Machinery Research, Institute of Agriculture Machinery, National Agriculture and Food Research Organization, Tsukuba, Ibaraki, Japan

T. Ahamed (✉)
Institute of Life and Environmental Sciences, University of Tsukuba, Tsukuba, Ibaraki, Japan
e-mail: tofael.ahamed.gp@u.tsukuba.ac.jp



T. Ahamed (ed.), *Remote Sensing Application II*, New Frontiers in Regional Science: Asian Perspectives 77, https://doi.org/10.1007/978-981-97-1188-8_11

**Keywords** Composite drought index · Yield fluctuation · Central dry zone · Myanmar

## 11.1 Introduction

Drought is one of the most harmful natural hazards and can significantly impact the national economy, environment, and agriculture sector. It can reduce crop yields and increase the salinity of soils, which can negatively impact irrigation systems and crop yields in the agriculture sector (Mishra and Singh 2010; Gottschalck et al. 2005). Hence, drought monitoring is very important for detecting, measuring, and classifying risks in drought-affected regions. The regions with the highest drought risk encompass parts of Europe, the Middle East, North Africa, and Asia, including India, China, Iran, Afghanistan, and Myanmar. The central dry zone of Myanmar is one of the most drought-prone areas in the country (UN 2022). Residents suffer from uncertain livelihoods and limited agricultural productivity due to the regularly low rainfall in this region. Crops such as rainfed pulses, sesame, and pigeon peas are grown in the central dry zone. Yields are modest and unpredictable in this region due to dryness. Moreover, 22% of Myanmar's rice is grown in the central dry zone. Since 1990, the rainy season has been reduced due to the early arrival and late departure of the southwest monsoon, which has caused significant rainfall and floods. Due to the increasing temperatures, thin soils, extended dry season, more severe droughts, and poor crop management, agriculture is becoming increasingly vulnerable. Therefore, drought monitoring in this region is significant for the livelihoods of people and crop yields (Zin et al. 2019).

Different drought monitoring methods have been used based on ground monitoring and satellite remote sensing. The implementation of the ground-based method is costly and time consuming (Svoboda et al. 2002). However, satellite remote sensing imagery allows scientists to observe a number of atmospheric and land surface variables in real time that can be used to estimate soil moisture, precipitation, evaporation, and plant conditions. These data can be employed to assess and monitor drought characteristics, including intensity, timeframe, and geospatial extent (Zhang et al. 2017). Furthermore, the existing developments in the Google Earth Engine (GEE), a geospatial data monitoring platform that computes and displays unprocessed and processed satellite-based information, allow researchers to monitor drought in large areas quickly and sufficiently. The open-source GEE platform offers researchers free access to satellite datasets, high-speed parallel processing capabilities and artificial intelligence techniques, a library of application programming interfaces (APIs), and an innovation-driven environment with popular programming languages, such as JavaScript and Python (Gorelick et al. 2017; Tamiminia et al. 2020). These recent advancements in technology enable real-time drought monitoring with single and standardized indicators in large areas.

Monitoring drought characteristics with a single drought indicator is challenging due to the complexities of local drought expression across different timeframes

(Karamouz et al. 2009). The common indicators include the standardized precipitation index (SPI), standardized soil moisture index (SSMI), vegetation condition index (VCI), standard relative humidity index (SRHI), temperature condition index (TCI), and precipitation condition index (PCI) (Kogan 1995; Krishnaswamy et al. 2014; Hao and AghaKouchak 2013; Farahmand et al. 2015). However, recently, researchers have employed standardized indicators that can detect dry spells with persistent geographical and temporal resolutions utilizing historical records from vast geographical zones. These standardized indicators are a combination of the TCI, VCI, SPI, or other drought indices (Arab et al. 2022; Arab and Ahamed 2023; Liu et al. 2020a, b; Garba et al. 2023). Various mathematical methods have been used to combine single drought indices for standardizing drought indices. One of the well-known techniques is based on the AHP (Biswas et al. 2023). Most researchers have used a composite drought index to specify drought-affected and nonaffected areas. However, very few studies have focused on rice yield loss assessment due to drought (Bayissa et al. 2019; Daryanto et al. 2016).

Different parametric and nonparametric methods have been used for determining drought-related crop yield losses, including the autoregressive integrated moving average, regression, multiple linear regression exponential smoothing, neural network, and random forest models (Belal et al. 2014; Arab et al. 2022; Potopová et al. 2020). One of the most important methods is the use of a regression model to determine how each factor affects the measured rice yield loss. Therefore, the main objective of this study was to determine yield fluctuations using a composite drought index in the drought-prone areas of Myanmar.

## 11.2 Materials and Methods

### *11.2.1 Study Area*

The dry zone of Myanmar is situated in the country's central region, along the Ayeyarwady River. It occurs between latitudes 19°20′ and 22°50′ and longitudes 93°40′ and 96°30′. There are 54 townships in Sagaing, Mandalay, Magway Divisions, and Naypyitaw, which are divided into 15 districts. The dry zone covers a total area of 80,881 km$^2$ (Fig. 11.1). The dry zone borders the Rakhine Range to the west, the Bago Range to the south, and the Shan Plateau and the Eastern Range to the east. These ranges range in elevation from 500 to 1800 ft and include isolated low-lying ranges and hillocks in nearly all townships. Three mountain ranges run from north to south through central Myanmar. Rakhine and the Western Range, which run along the western coast, function as natural barriers, causing southwest monsoon rain clouds to shear upward and overpass the Ayeyarwady valley, resulting in a rain shadow effect. The region is traversed by the Chindwin, Ayeyarwady, Mu, Myitnge, Zawgyi, Yama, Yaw, Salin, Mone, and Mann Rivers. Tributaries from the west that drain into the Ayeyarwady River have mostly been sanded up and only flow in raging torrents

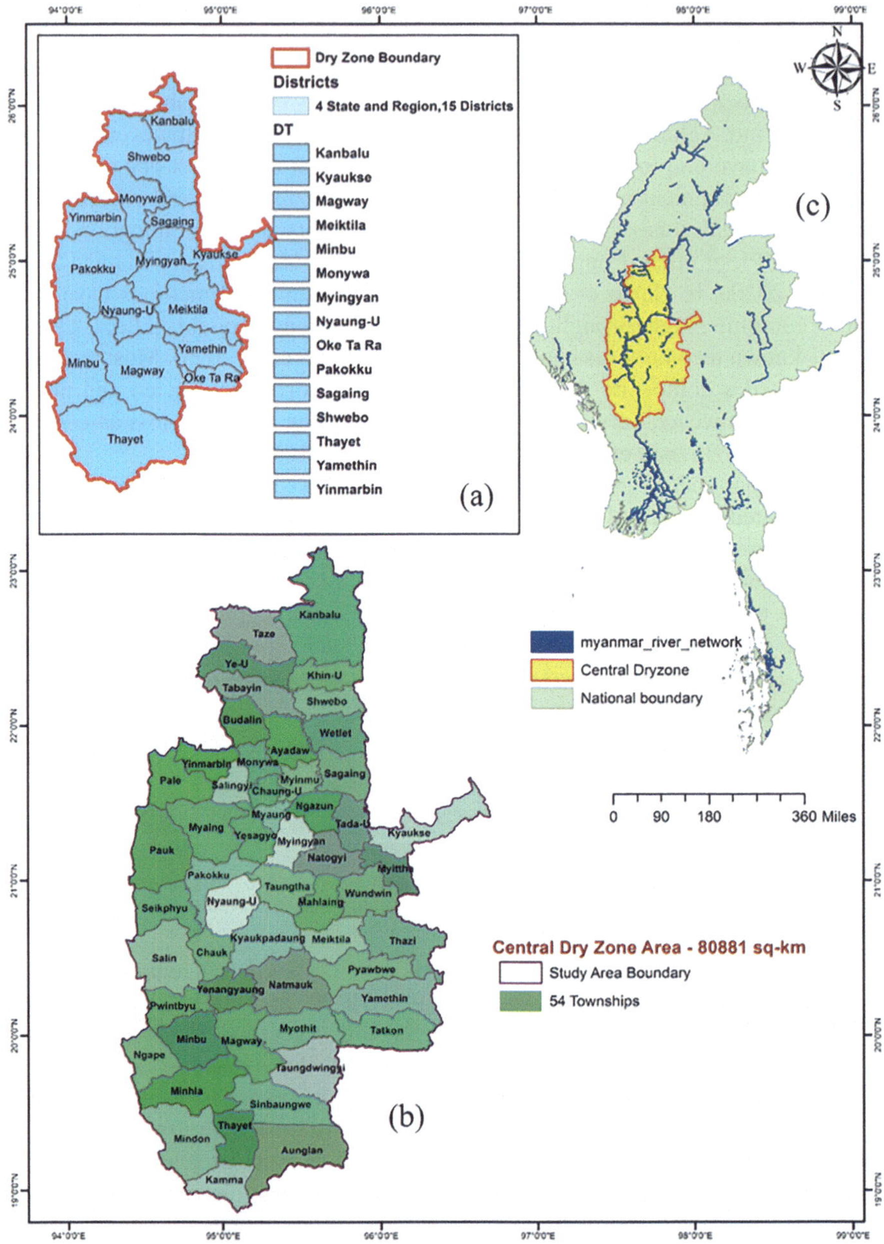


Fig. 11.1 Study area maps: (a) district boundary map, (b) township administrative boundary map, and (c) map of the central dry zone in Myanmar

during the monsoon season. The majority of the land is flat and undulating, facilitating extensive agriculture. Temperatures range from over 110 °F in the hottest months of March and April to 46 °F in January. The average maximum temperature is approximately 90 °F. The average annual humidity is approximately 63%. The humidity is approximately 42% in the hottest months of March and April and approximately 80% in the wettest month of September (Tun 2000). Compared to other regions of the country, this dry zone has lower precipitation levels, making it more susceptible to drought. Myanmar experienced droughts in 1972, 1979, 1982, 1983, 1986, and 1987. According to an examination of Myanmar's drought indices from 1951 to 2000, the drought trend is drastically increasing (https://www.ais.unwater.org/ais/pluginfile.php/597/mod_page/content/79/Myanmar_2.pdf).

### *11.2.2 Satellite Datasets*

The calibrated top-of-atmosphere (TOA) reflectance products of Landsat 8 OLI/TIRS Collection 2 were used on the Google Earth Engine platform. Two products are provided: the operational land imager (OLI) and the thermal infrared sensor (TIRS). Therefore, Level-1 (C1 Level-1) daytime image collections were utilized in this study. This product involves several data processing steps, such as radiometric calibration and atmospheric adjustment (Arab et al. 2022). The Landsat 8 OLI is an important source of remote sensing data because it has continuously recorded images of the Earth's surface (Wulder et al. 2019). In addition, long-term land cover change studies may be conducted at regional and global levels because of the availability of 40 years of Landsat data on the GEE platform (Liu et al. 2020a, b). Landsat 8 spectral and thermal bands (Table 11.1) were used in drought assessment by calculating vegetation indices and land surface temperatures. In addition, the Climate Hazards Group InfrRed Precipitation with Station (CHIRPS) rainfall dataset was used for SPI calculation over three seasons (summer, monsoon, and winter) in 2019–2021 (Fig. 11.2).

**Table 11.1** Land 8 operational land imager and thermal infrared sensor (OLI/TIRS) bands, spectral range (μm), and other details

| Name | Pixel size (m) | Wavelength (μm) | Description |
|---|---|---|---|
| B1 | 30 | 0.43–0.45 | Coastal aerosol |
| B2 | 30 | 0.45–0.51 | Blue |
| B3 | 30 | 0.53–0.59 | Green |
| B4 | 30 | 0.64–0.67 | Red |
| B5 | 30 | 0.85–0.88 | Near-infrared |
| B6 | 30 | 1.57–1.65 | Shortwave infrared 1 |
| B7 | 30 | 2.11–2.29 | Shortwave infrared 2 |
| B8 | 15 | 0.52–0.90 | Band 8 Panchromatic |
| B9 | 15 | 1.36–1.38 | Cirrus |
| B10 | 30 | 10.60–11.19 | Thermal infrared 1, resampled from 100 to 30 m |
| B11 | 30 | 11.50–12.51 | Thermal infrared 2, resampled from 100 to 30 m |

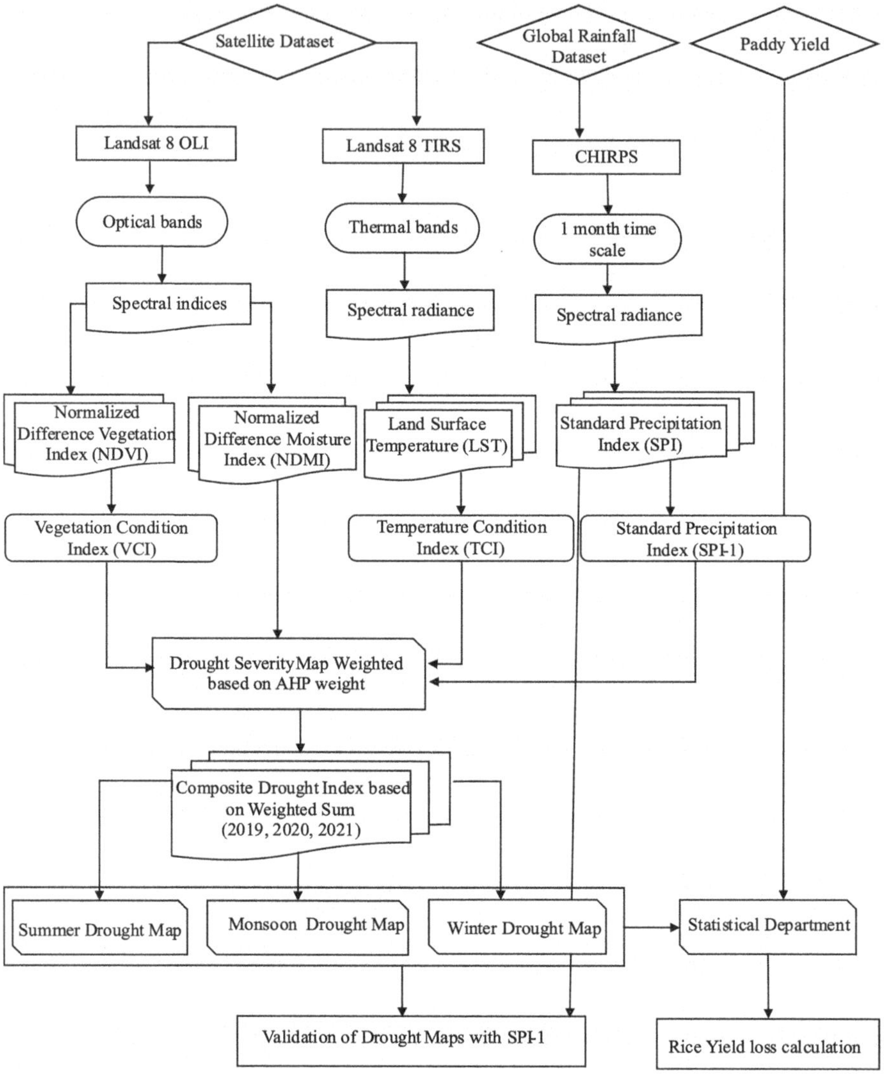


**Fig. 11.2** Research flowchart for drought monitoring in the central dry zone of Myanmar

## *11.2.3 Vegetation Indices*

The sensitivity of vegetation to climatic variability can be determined by the reconstructed phenology, which offers essential data for studying trends in ecological processes or climatology and may reveal climate effects on many scales, including regionally and globally (Habibie et al. 2020). The top-of-atmosphere (TOA) reflectance products of Landsat 8 OLI/TIRS Collection 2 were used to generate the normalized difference vegetation index (NDVI), normalized difference moisture index

(NDMI), and land surface temperature (LST) to determine the vegetation phenology. The code editor platform of the Google Earth Engine was utilized for the calculation and validation of the NDVI, normalized difference moisture index (NDMI), and LST over the three seasons of summer (March to May), monsoon (June to October), and winter (November to February) of 2019, 2020, and 2021.

#### 11.2.3.1 Normalized Difference Vegetation Index (NDVI)

The NDVI has been used in several studies on drought identification, assessment, and monitoring. There are a number of vegetation indices, but none of them have been assessed in a natural setting. Monitoring local vegetation is the main purpose of NDVI analysis. The NDVI should be paired with other elements for the best results (Mohan et al. 2021). Moreover, the NDVI can be used to display vertical spatial and temporal distributions as well as horizontal structures, such as vegetative cultivation and vegetation biomass. These vegetation features include plant traits, productivity, and impact of temperature on plants (Xue and Su 2017). The NDVI was calculated on the Google Earth Engine (GEE) platform as follows:

$$\mathrm{NDVI} = \frac{\mathrm{NIR} - \mathrm{RED}}{\mathrm{NIR} + \mathrm{RED}} \tag{11.1}$$

where NIR is the near-infrared band 5 of Landsat 8, and red is band 4.

#### 11.2.3.2 Land Surface Temperature (LST)

The LST helps to monitor agricultural droughts (Nivedha Deve et al. 2017). LST estimations utilizing thermal infrared bands, which have been shown to represent a solid measure for providing accurate data on ground moisture levels, have been developed into an important instrument to gauge the drought severity (Das et al. 2021; Vogt et al. 2000). Furthermore, understanding the impacts of climate change on both local and large scales strongly depends on climatic data produced by satellite imaging of land surface temperatures (LST) (Wongsai et al. 2017). The following equation can be used to explain temperature variations by analyzing the LST, which is affected by factors such as elevation, land cover, and NDVI (De Jesus and Santana 2017).

$$\mathrm{LST} = \frac{\mathrm{BT}}{1 + \left( \frac{\lambda \mathrm{BT}}{\mathrm{PV}} \times \ln \mathrm{LSE} \right)} \tag{11.2}$$

where BT is the satellite brightness temperature in °C, $\lambda$ is the top-of-atmosphere (TOA) spectral radiation at the sensor aperture, PV is the percentage of vegetation (PV), and LSE is the land surface emissivity.

#### 11.2.3.3 Normalized Difference Moistures Index (NDMI)

The NDMI and canopy water content are strongly correlated (Hardisky et al. 1983). The NDMI can be used to detect dryness and irrigation in agricultural plots, particularly in summer after weeks without rain (Mimić et al. 2022). This index originates from the NIR and SWIR bands and can be expressed as follows:

$$\mathrm{NDMI} = \frac{\mathrm{NIR} - \mathrm{SWIR}}{\mathrm{NIR} + \mathrm{SWIR}} \tag{11.3}$$

In Landsat 8 OLI images, NIR is the near-infrared band, and SWIR is the shortwave infrared band.

### 11.2.4 Drought Indices

#### 11.2.4.1 Vegetation Condition Index (VCI)

Kogan suggested the VCI in 1995. This index indicates the difference between the NDVI value in the current month and the lowest NDVI estimated from long-term data. A map was created by converting the NDVI values at each location and pixel from 0 to 1 (Kogan et al. 2019) (Figs. 11.3, 11.4, and 11.5). The VCI can be obtained as follows:

$$\mathrm{VCI} = \frac{\mathrm{NDVI}_{t,p} - \mathrm{NDVI}_{\mathrm{min}}}{\mathrm{NDVI}_{\mathrm{max}} - \mathrm{NDVI}_{\mathrm{min}}} \tag{11.4}$$

where $\mathrm{NDVI}_{t,p}$ is the current NDVI image of pixel $p$ at time $t$, and $\mathrm{NDVI}_{\mathrm{max}}$ and $\mathrm{NDVI}_{\mathrm{min}}$ are the absolute maximum and minimum NDVI values, respectively, for each pixel using the entire NDVI series (2019–2021).

#### 11.2.4.2 Temperature Condition Index (TCI)

Kogan (1990, 1995) also proposed the TCI, which can be calculated using the same method as that for the VCI. The more intense the drought is, the higher the temperature; nonetheless, the TCI is based on the normalization of the land surface temperature (LST) of each pixel using the maximum and minimum temperatures in the provided time series (Kogan et al. 2019). With the use of the thermal bands, LST values were derived from Landsat 8 TIRS datasets. Several data processing, geometric, and radiometric improvements, as well as a synthetic data dissemination process, were made in this product (Tariq et al. 2020) (Figs. 11.3, 11.4, and 11.5). The TCI can be expressed as follows:

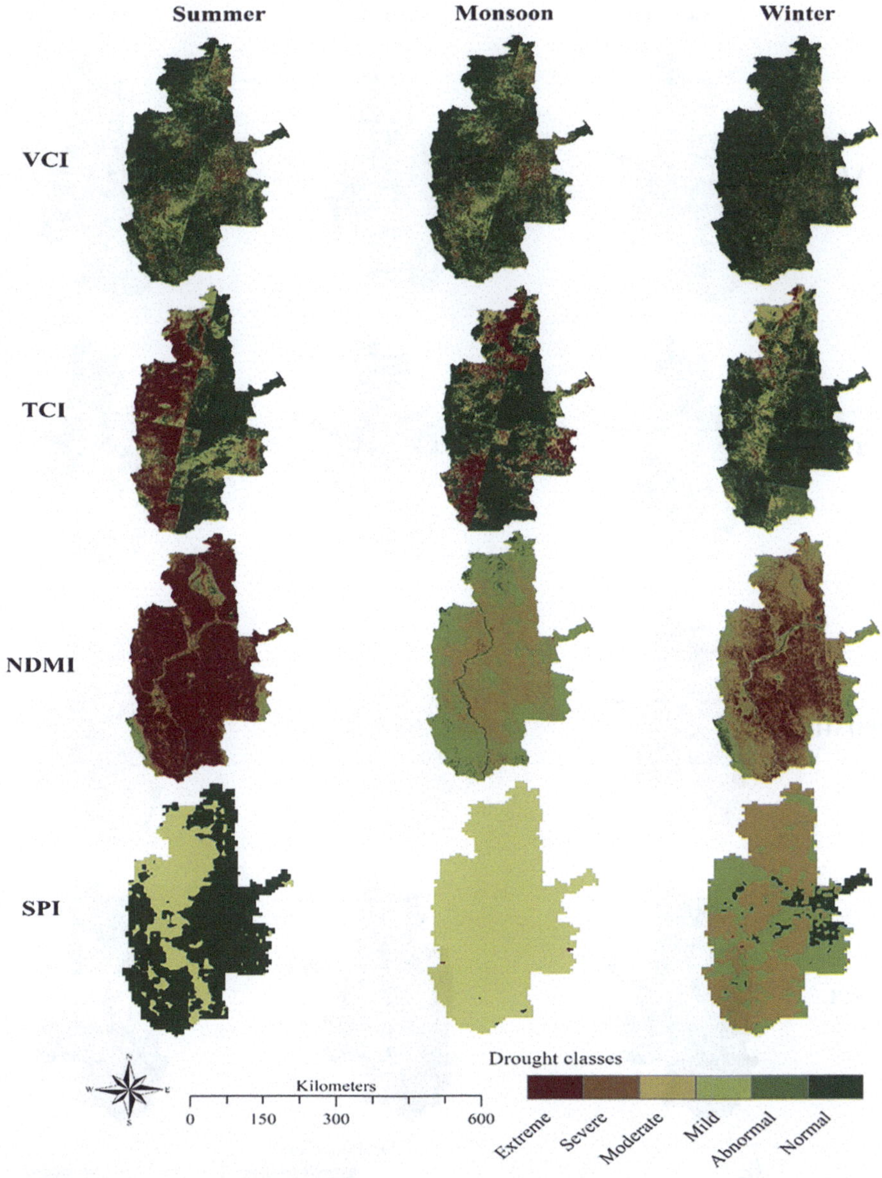


**Fig. 11.3** Drought severity classification based on the VCI, TCI, NDMI, and SPI for 2019 in the central dry zone of Myanmar

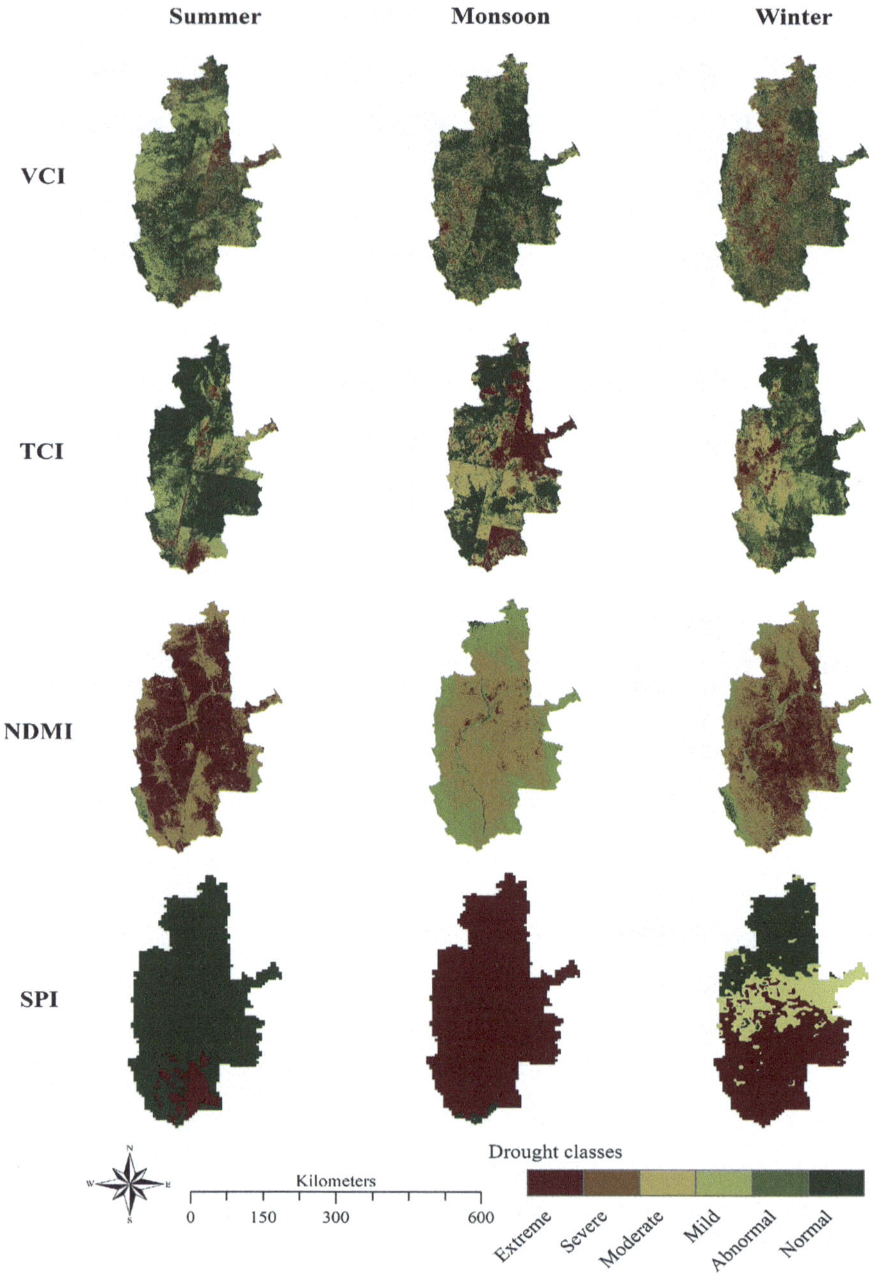


**Fig. 11.4** Drought severity classification based on the VCI, TCI, NDMI, and SPI for 2020 in the central dry zone of Myanmar

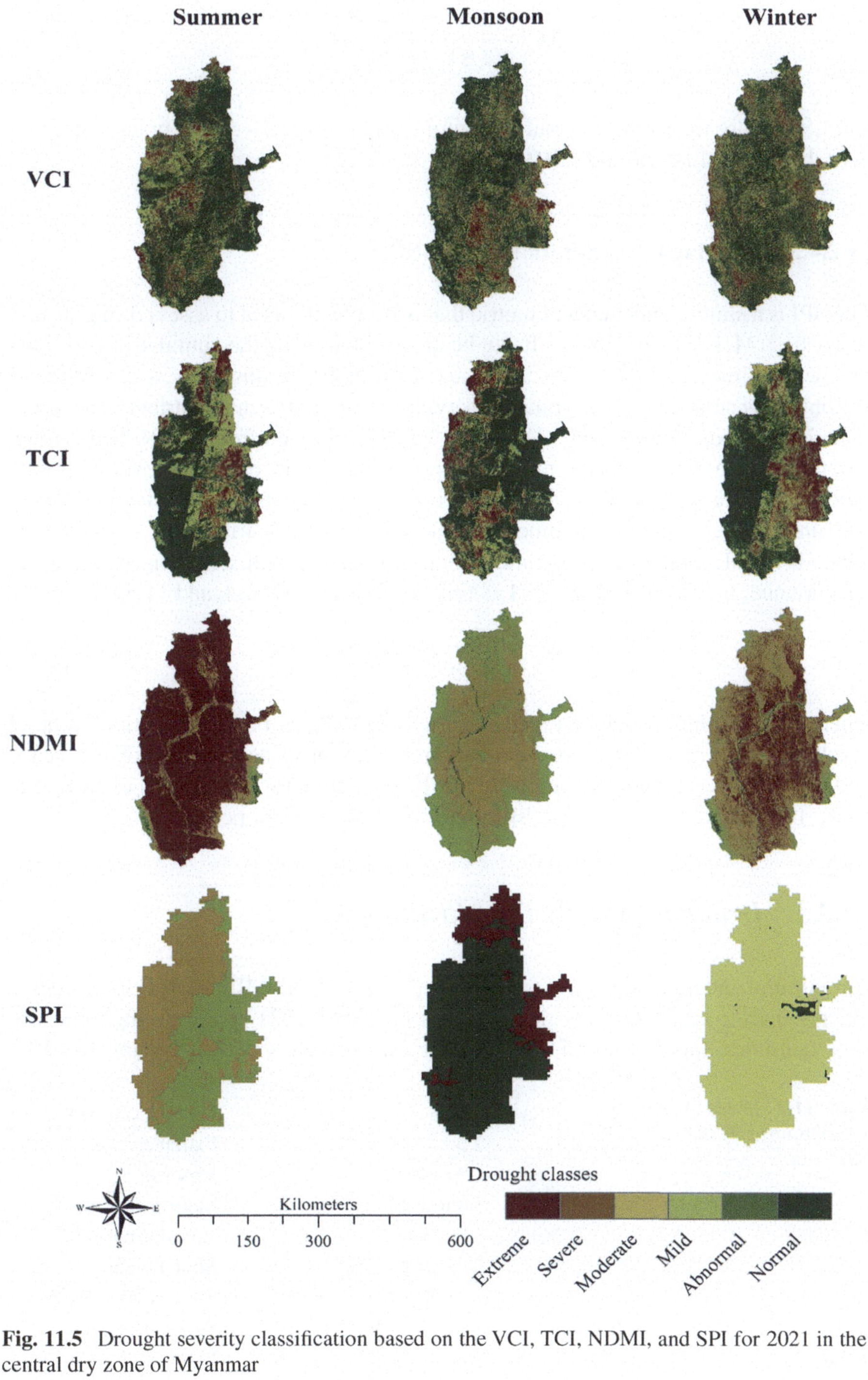


**Fig. 11.5** Drought severity classification based on the VCI, TCI, NDMI, and SPI for 2021 in the central dry zone of Myanmar

$$\mathrm{TCI} = \frac{\mathrm{LST}_{\max} - \mathrm{LST}_{t,p}}{\mathrm{LST}_{\max} - \mathrm{LST}_{\min}} \tag{11.5}$$

where $\mathrm{LST}_{t,p}$ is the actual land surface temperature of pixel $p$ at time $t$, and $\mathrm{LST}_{\min}$ and $\mathrm{LST}_{\max}$ are the minimum and maximum LSTs, respectively, of pixel $p$ at time $t$ over the entire study period (2019–2021).

#### 11.2.4.3 Standard Precipitation Index (SPI)

The SPI is a simple, independent metric that is frequently used to assess drought characteristics (Mckee et al. 1993b). It can be determined using the cumulative precipitation across timescales of 1, 3, 6, 9, 12, and 24 months. In this study, a timeframe of 1 month was considered. Precipitation is typically not uniformly distributed for accumulation periods shorter than 12 months. The SPI, derived from CHIRPS precipitation data for each pixel location of a composite period for each year over a specific time frame, is a $z$ score of the deviation from the mean, expressed in standard deviation units. Precipitation is assumed to follow a gamma distribution (Aksoy 2000), which is transformed into a standard normal distribution, with a zero mean and a single variance, for calculating the SPI as follows (Figs. 11.3, 11.4, and 11.5).

$$\mathrm{SPI}_{itk} = \frac{P_{itk} - P_{it}}{\sigma_{it}} \tag{11.6}$$

where $\mathrm{SPI}_{itk}$ is the $z$ value of pixel $i$ at $t$ in year $k$, $P_{itk}$ is the precipitation value of pixel $i$ at time $t$ in year $k$, $P_{it}$ is the mean precipitation value of pixel $i$ over $n$ years, and "$\sigma_{it}$" is the standard deviation of pixel $i$ over $n$ years. Drought is classified in Table 11.2 based on the average SPI value over the drought period.

### *11.2.5 Drought Severity Classification*

Agricultural and meteorological parameters (VCI, TCI, NDMI, and SPI) were used in this research to create drought risk maps for the 2019–2021 period during three seasons (summer, monsoon, and winter). SPI values were calculated using CHIRPS

**Table 11.2** States of drought as determined by the SPI value

| SPI | Drought category |
|---|---|
| 2 > | Extremely wet |
| 1.5 to 1.99 | Very wet |
| 1.0 to 1.49 | Moderately wet |
| 0 to 0.99 | Near normal |
| 0 to −0.99 | Mild drought |
| −1.0 to −1.49 | Moderately drought |
| −1.5 to −1.99 | Severe drought |
| ≤−2.0 | Extreme drought |

**Source:** World Meteorological Organization (2012)

rainfall data, while drought indices were calculated using Landsat 8 OLI and TIRS scenes (Figs. 11.3, 11.4, and 11.5). To evaluate the drought severity based on the aforementioned indicators during the near-optimal period, 3 years (2019–2021) were accounted for. The Kogan, Gao, and World Meteorological Organization drought classification techniques were utilized to categorize the obtained drought maps to determine the drought intensity and severity using the derived indices (VCI, TCI, NDMI, and SPI) on the GEE platform, examined in the ArcGIS environment (Table 11.3) (Gao 1996; McKee et al. 1993a).

### *11.2.6 Composite Drought Index Based on the AHP-Determined Weight*

The AHP (Saaty 1988) aims to determine how important the parts of each layer are relative to each other and to evaluate and rank each sample (Zhou 2015). These steps are based on the empirical knowledge of the experts making the decisions. In recent drought research, the AHP has been extensively and effectively applied (Wang et al. 2019; Wu et al. 2017). The primary AHP processes are constructing the assessment matrix, single-level sequencing, and consistency assessment. A qualitative issue may be quantified using the assessment matrix. The consistency ratio (CR) was utilized to measure the matrix deviation (AHP table). The evaluation matrix was consistent with a CR value below 0.1. In this study, after the calculation of all drought indices, the AHP technique was used for composing the drought indices based on the AHP-determined weights (Tables 11.4 and 11.5). The composite drought index can be calculated as follows:

$$\mathrm{CDI} = \mathrm{VCI} \times W_{\mathrm{VCI,AHP}} + \mathrm{TCI} \times W_{\mathrm{TCI,AHP}} + \mathrm{NDMI} \times W_{\mathrm{NDMI,AHP}} + \mathrm{SPI} \times W_{\mathrm{SPI,AHP}} \tag{11.7}$$

### *11.2.7 Statistical Dataset*

Rice yield and acreage were obtained from Myanmar's annual statistics book, which is accessible online for 2020 and 2019 (Statistical Book 2020, 2021) (https://www.csostat.gov.mm/PublicationAndRelease/StatisticalYearbook). The study area

**Table 11.3** Drought categorization from the VCI, NDMI, and TCI

| Drought category | VCI and TCI | SPI | NDMI |
|---|---|---|---|
| Extreme drought | 0–0.1 | ≤−2 | ≤0 |
| Severe drought | 0.1–0.2 | −1.5 to −1.99 | <0 to ≤0.2 |
| Moderate drought | 0.2–0.3 | −1.0 to −1.49 | <0.2 to ≤0.4 |
| Mild drought | 0.3–0.4 | 0 to −0.99 | – |
| Near normal | 0.4–0.5 | 0 to 0.99 | – |
| No drought | 0.5–1 | ≤2 | >0.4 |

**Table 11.4** Decision matrix based on the opinion of expert A

| | TCI | VCI | NDMI | SPI |
|---|---|---|---|---|
| TCI | 1 | 3 | 3 | 4 |
| VCI | 1/3 | 1 | 1 | 1 |
| NDMI | 1/3 | 1 | 1 | 1 |
| SPI | 1/4 | 1 | 1 | 1 |
| Weights | 0.52431 | 0.162195 | 0.162195 | 0.1513 |
| CR | 0.40% | | | |
| Eigen value = 4.010 | | | | |
| $n = 4$ | | | | |

**Table 11.5** AHP-determined weights for agricultural and meteorological droughts based on expert opinions

| Variables | A | B | C | D | E | Mean | Weight (%) |
|---|---|---|---|---|---|---|---|
| TCI | 0.52431 | 0.089073 | 0.616034 | 0.07466 | 0.409248 | 0.342665 | 34.2665 |
| VCI | 0.162195 | 0.195648 | 0.264563 | 0.351137 | 0.454052 | 0.285519 | 28.5519 |
| NDMI | 0.162195 | 0.042341 | 0.059701 | 0.075865 | 0.046717 | 0.077364 | 7.73638 |
| SPI | 0.1513 | 0.672938 | 0.059701 | 0.498337 | 0.089983 | 0.294452 | 29.44518 |
| A–E indicated the expert who worked in this sector respectively | | | | | | 1 | 99.99996 |

consists of 15 districts, but rice yield datasets were available only for 4 regions (Sagaing, Mandalay, Magway, and Naypyitaw). Therefore, in this study, we only relied on these datasets (CSO 2020/2021) (Table 11.6).

### *11.2.8 Evaluation of the Yield Fluctuations Due to Drought*

Yield fluctuations due to drought were considered in this research. In this regard, all agricultural fields were masked in the CDI drought map. Subsequently, the zonal mean of the four districts was used for regression analysis of 2019 and 2020. Since this research only focused on production losses that occurred as a result of severe weather or drought conditions, the effects of insects, pathogens, and floods were neglected in yield loss computation. The yield loss percentage can be computed as follows:

$$\text{Yield fluctuation} = \frac{Y_o - Y_p}{Y_o} \times 100 \tag{11.8}$$

where $Y_o$ is the observed yield obtained from the statistical department and $Y_p$ is the predicted yield derived from regression analysis.

**Table 11.6** Paddy yield data for the central dry zone regions

| Region | Year | Classification | Unit | Paddy |
|---|---|---|---|---|
| Sagaing | 2019 | Harvested | Acre | 2,155,076 |
| | | Production | Ton | 3,731,593 |
| | 2020 | Harvested | Acre | 1,772,169 |
| | | Production | Ton | 2,920,426 |
| Mandalay | 2019 | Harvested | Acre | 672,285 |
| | | Production | Ton | 1,158,397 |
| | 2020 | Harvested | Acre | 605,082 |
| | | Production | Ton | 1,035,520 |
| Magway | 2019 | Harvested | Acre | 706,661 |
| | | Production | Ton | 1,109,329 |
| | 2020 | Harvested | Acre | 661,256 |
| | | Production | Ton | 975,505 |
| Naypyitaw | 2019 | Harvested | Acre | 202,975 |
| | | Production | Ton | 361,677 |
| | 2020 | Harvested | Acre | 191,691 |
| | | Production | Ton | 340,210 |

**Source:** Statistical Yearbook 2020 and 2021, Central Statistics Organization (CSO)

## 11.3 Results

### *11.3.1 Drought Risk Assessment Using the Composite Drought Index*

In this study, drought risk maps were developed for the 2019–2021 period utilizing agricultural and meteorological parameters (VCI, TCI, NDMI, and SPI) (Figs. 11.3, 11.4, and 11.5). Landsat 8 OLI and TIRS scenes were considered for estimating drought indices, whereas CHIRPS rainfall data were employed calculating SPI values. This research focused on assessing drought during three seasons, namely, summer (March to May), monsoon (June to October), and winter (November to February), in the central dry zone of Myanmar. According to CHIRPS precipitation statistics, 2019–2021 exhibited a lower mean annual rainfall. Therefore, these 3 years were considered to predict the drought severity from the drought indicators over the optimal period (2019, 2020, and 2021). The Kogan and World Meteorological Organization drought classification techniques were adopted to classify the drought-affected regions by the intensity and severity using the various drought indices (VCI, NDMI, TCI, and SPI) (Fig. 11.2). Extreme drought occurs in a region when all parameter values are less than 10% (0.10); however, wet conditions occur when all parameter values are greater than 40% (0.40). To overlay the drought layers in all 3 years, the AHP weight technique was used. The highest weights, according to the AHP, were assigned to the TCI and SPI, at 34.2% and 29.4%, respectively. Finally, the drought composite index was used to determine the percentages of the drought-affected areas during the three seasons.

Based on the obtained results, the drought maps indicated that some parts of Shwebo, Monywa, and Thayet and all parts of Yinmarbin, Pakokku, and Minbu districts were moderately affected during the summer season and that Kanbalu, Shwebo, and Thayet were affected during the monsoon season of 2019. However, Sagaing, Kyaukse, and Myingyan were affected during the summer season; Kyaukse, Myingyan, Nyaung-U, and Minbu were affected during the monsoon season; Pakokku, Nyaung-U, Minbu, and Magway were affected during the winter season of 2020; Kanbalu, Shwebo, and Meiktila Yamethin were affected during the summer season; and Shwebo, Sagaing, Kyaukse, Myingyan, Meiktila, Yamethin, and Oke Ta Ra were affected during the winter season of 2021 (Figs. 11.6 and 11.7).

According to the findings based on the composite drought index, moderate drought impacts of 8% occurred during the summer and monsoon seasons of 2019 and 2020 and 10% during the winter season of 2021. However, the values of 38.2% in summer, 23.9% in the monsoon season of 2019, 22.2% in summer, and 40.4% in the monsoon season of 2020 were lower than the values of 25.1% in summer, 21.6% in the monsoon season, and 30.7% in the winter of 2021, which moderately affected all central dry zones. The agricultural drought-affected area results indicated that 9% moderate droughts and 35% mild droughts occurred in the summer of 2019, 13% moderate droughts and 43% mild droughts occurred in the monsoon season of 2020, and 10% moderate droughts and 31% mild droughts occurred in the winter of 2021. The pixel count, percentage, and area (square kilometers) in each agricultural risk zone are provided in Table 11.7.

### 11.3.2 Validation of the CDI with the SPI

The zonal average CDI value in all districts was validated with the SPI-1 from 2019 to 2020 during the three seasons (summer, monsoon, and winter). In the central dry zones of Myanmar, it is virtually always dry in summer, with a lack of rainfall. To reduce the impact of the occurrence of severe droughts in all years, the average values of the CDI and SPI-1 were considered. The findings showed a strong correlation between the zonal mean CDI value and the three-season mean SPI-1 value ($R^2 = 0.70$) (Fig. 11.8).

### 11.3.3 Rice Yield Loss Calculation

The rice yield (ton) per unit of area in 2019 and 2020 was used to calculate the yield loss with the composite drought index (CDI). In this regard, the effects of drought on rice production were explained by regression analysis on an annual basis in the four districts of the central dry zone in Myanmar. The regression analysis results indicated the highest $R^2$ values of 0.90 and 0.74 occurred during the winter season of 2019 and 2020, respectively, at the time of harvest (Fig. 11.9).

Subsequently, the yield loss was determined by subtracting the expected yield from the actual rice yield (ton) per unit of area. The yield fluctuation results

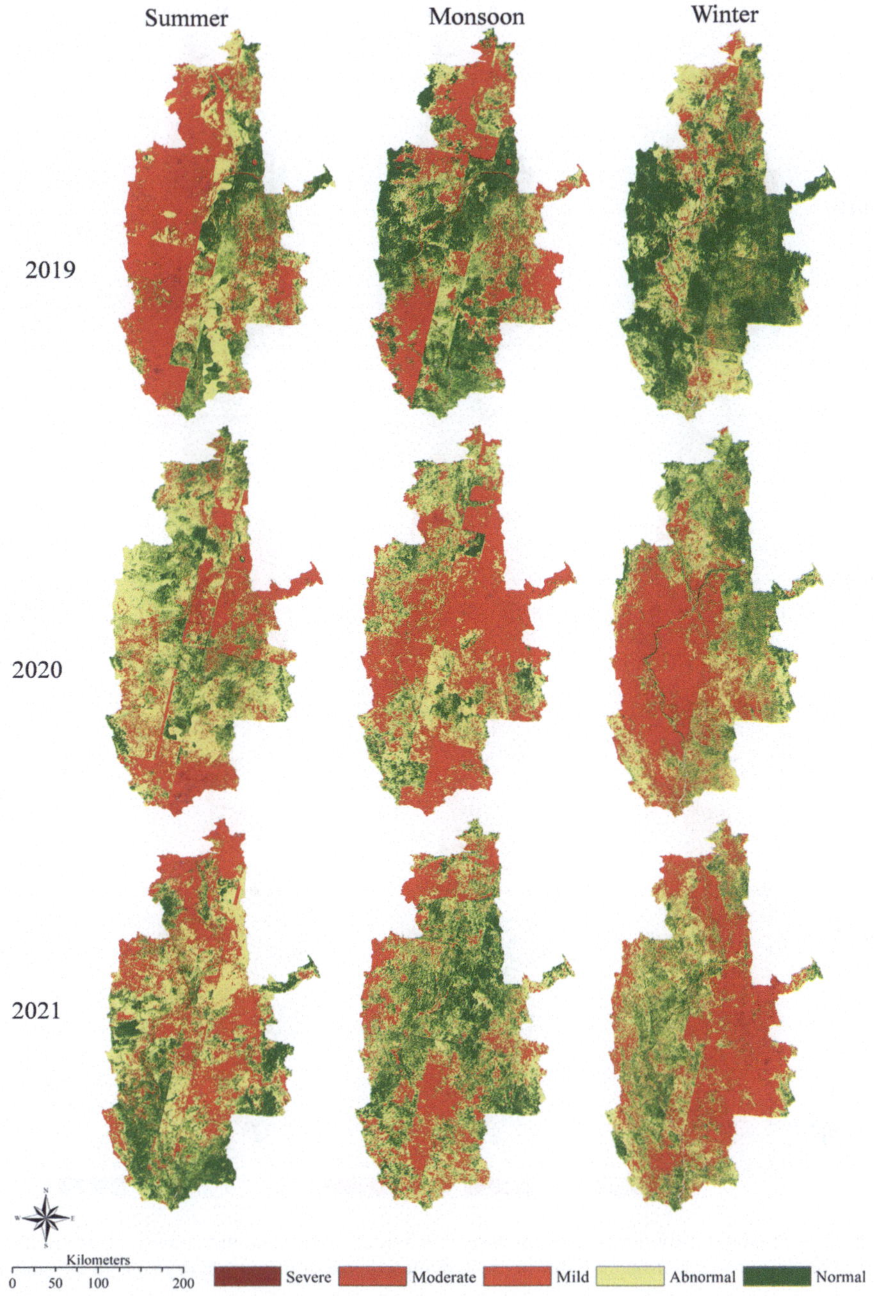


**Fig. 11.6** Spatial distribution of drought during the three seasons (summer, monsoon, and winter) in the dry zone of Myanmar based on the composite drought index

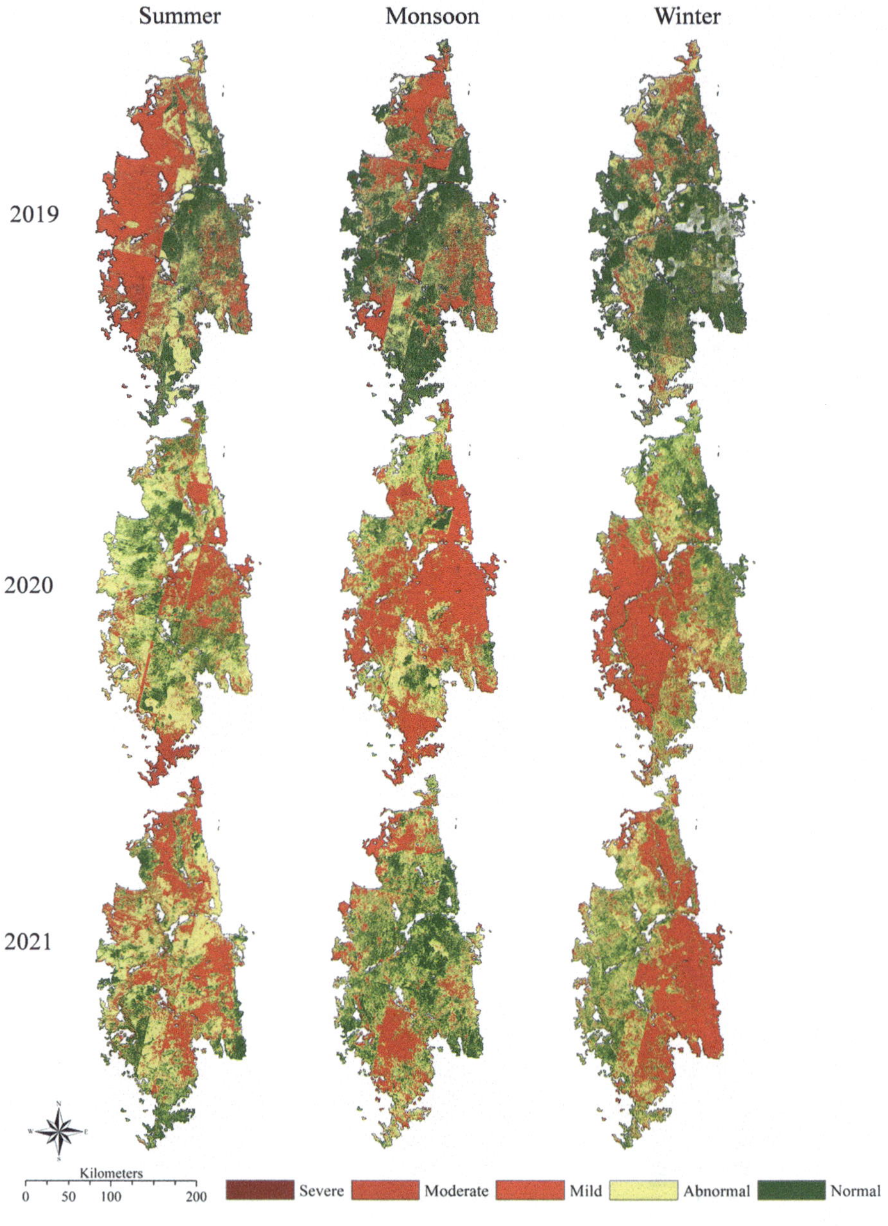


**Fig. 11.7** Spatial distribution of drought during the three seasons (summer, monsoon, and winter) based on the composite drought index in the agricultural area of the central dry zone of Myanmar

indicated that in all four districts, the yield loss was distinct from year to year. The yield loss mostly depends on the drought severity, paddy cultivars, or the environment. In some districts, such as Sagaing and Mandalay, there was a significant yield loss in 2020. However, certain districts, such as Magway and Naypyitaw, did not exhibit a major reduction in the rice output as a result of drought (Fig. 11.10).

**Table 11.7** Affected area in the dry zones during the summer, winter, and monsoon seasons

| | | | All central dry zone | | | | | Agricultural area | | | | |
|---|---|---|---|---|---|---|---|---|---|---|---|---|
| Year | Season | | Extreme | Severe | Moderate | Mild | Abnormal | Extreme | Severe | Moderate | Mild | Abnormal |
| 2019 | Summer | Pixel count | 0 | 439,112 | 6,636,011 | 30,833,767 | 28,023,842 | 0 | 300,431 | 4,865,326 | 19,753,244.0 | 20,332,830.0 |
| | | Area (km$^2$) | 0 | 418.5 | 6526.7 | 30,897.2 | 28,088.9 | 0 | 285.1 | 4782.8 | 19,753.4 | 20,363.7 |
| | | Area (%) | 0 | 0.5 | 8.1 | 38.2 | 34.8 | 0 | 0.5 | 8.5 | 35.2 | 36.3 |
| | Monsoon | Pixel count | 0 | 103,102 | 4,564,656 | 19,280,470 | 27,141,932 | 0 | 85,093 | 2,777,853 | 12,176,838.0 | 19,552,393.0 |
| | | Area (km$^2$) | 0 | 97.2 | 4492.0 | 19,280.8 | 27,113.1 | 0 | 80.6 | 2722.5 | 12,142.4 | 19,513.0 |
| | | Area (%) | 0 | 0.1 | 5.6 | 23.8 | 33.5 | 0 | 0.1 | 4.8 | 21.6 | 34.7 |
| | Winter | Pixel count | 0 | 32,494 | 703,393 | 7,197,139 | 28,814,198 | 0 | 27,043 | 550,106 | 5,754,410.0 | 20,039,770.0 |
| | | Area (km$^2$) | 0 | 30.5 | 665.7 | 6979.9 | 28,771.8 | 0 | 25.3 | 520.0 | 5590.4 | 19,950.0 |
| | | Area (%) | 0 | 0.04 | 0.9 | 9.4 | 39.0 | 0 | 0.0 | 1.0 | 10.6 | 37.9 |
| 2020 | Summer | Pixel count | 0 | 641,806 | 4,335,374 | 17,949,266 | 41,576,025 | 0 | 438,099 | 2,598,851 | 11,654,583.0 | 29,025,337.0 |
| | | Area (km$^2$) | 0 | 639.7 | 4273.7 | 17,856.0 | 41,854.3 | 0 | 438.8 | 2547.6 | 11,544.2 | 29,213.3 |
| | | Area (%) | 0 | 0.8 | 5.3 | 22.1 | 51.8 | 0 | 0.8 | 4.5 | 20.5 | 52.0 |
| | Monsoon | Pixel count | 0 | 172,551 | 6,290,909 | 32,638,049 | 31,605,403 | 0 | 164,439 | 4,535,861 | 24,174,773 | 20,927,354 |
| | | Area (km$^2$) | 0 | 160.7 | 6179.5 | 32,741.5 | 31,740.3 | 0 | 153.4 | 4442.7 | 24,242.4 | 21,004.2 |
| | | Area (%) | 0 | 0.2 | 7.6 | 40.5 | 39.2 | 0 | 0.3 | 7.9 | 43.1 | 37.4 |
| | Winter | Pixel count | 0 | 32,494 | 703,393 | 7,197,139 | 28,814,198 | 0 | 54,047 | 7,142,579 | 16,692,977.0 | 20,613,082.0 |
| | | Area (km$^2$) | 0 | 83.2 | 9016.8 | 23,024.2 | 31,810.0 | 0 | 51.6 | 7102.6 | 16,709.9 | 20,704.7 |
| | | Area (%) | 0 | 0.04 | 1.0 | 9.8 | 39.0 | 0 | 0.1 | 12.7 | 29.8 | 36.9 |

(continued)

**Table 11.7** (continued)

| | | | All central dry zone | | | | | Agricultural area | | | | |
|---|---|---|---|---|---|---|---|---|---|---|---|---|
| Year | Season | | Extreme | Severe | Moderate | Mild | Abnormal | Extreme | Severe | Moderate | Mild | Abnormal |
| 2021 | Summer | Pixel count | 0 | 22,609 | 3,702,545 | 20,290,812 | 35,869,812 | 0 | 8892 | 2,453,626 | 15,357,280.0 | 26,776,127.0 |
| | | Area (km$^2$) | 0 | 21.4 | 3601.6 | 20,166.7 | 36,012.1 | 0 | 8.4 | 2372.2 | 15,260.1 | 26,898.8 |
| | | Area (%) | 0 | 0.0 | 4.6 | 25.1 | 44.5 | 0 | 0.0 | 4.2 | 27.2 | 47.9 |
| | Monsoon | Pixel count | 0 | 4558 | 1,905,172 | 17,417,880 | 35,760,153 | 0 | 1013 | 1,398,144 | 10,966,320 | 23,741,558 |
| | | Area (km$^2$) | 0 | 4.0 | 1861.1 | 17,334.1 | 35,912.0 | 0 | 0.9 | 1369.0 | 10,889.7 | 23,806.2 |
| | | Area (%) | 0 | 0.0 | 2.4 | 21.6 | 44.2 | 0 | 0 | 2.4 | 19.4 | 42.4 |
| | Winter | Pixel count | 0 | 1,063,560 | 7,920,837 | 24,804,104 | 32,984,381 | 0 | 867,186 | 5,976,004 | 17,342,562 | 22,107,521 |
| | | Area (km$^2$) | 0 | 1017.1 | 7785.8 | 24,815.4 | 33,258.1 | 0 | 829.1 | 5850.8 | 17,368.2 | 22,264.2 |
| | | Area (%) | 0 | 1.3 | 9.8 | 30.7 | 40.8 | 0 | 1.5 | 10.4 | 30.9 | 39.7 |

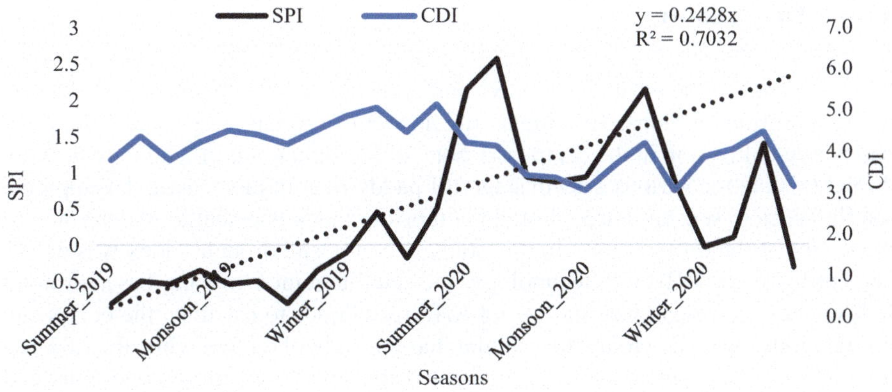


**Fig. 11.8** Validation of the CDI with the SPI during the three seasons of 2019 and 2020

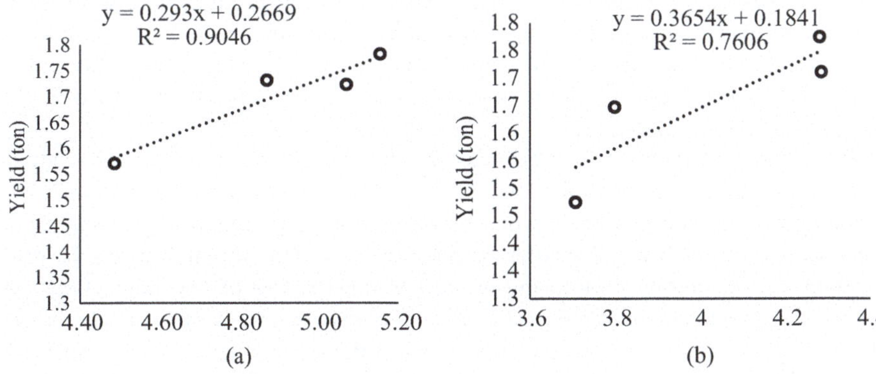


**Fig. 11.9** Spatial correlation between the paddy yield and DCI (drought composite index) during the winter season of (**a**) of 2019 and (**b**) 2020

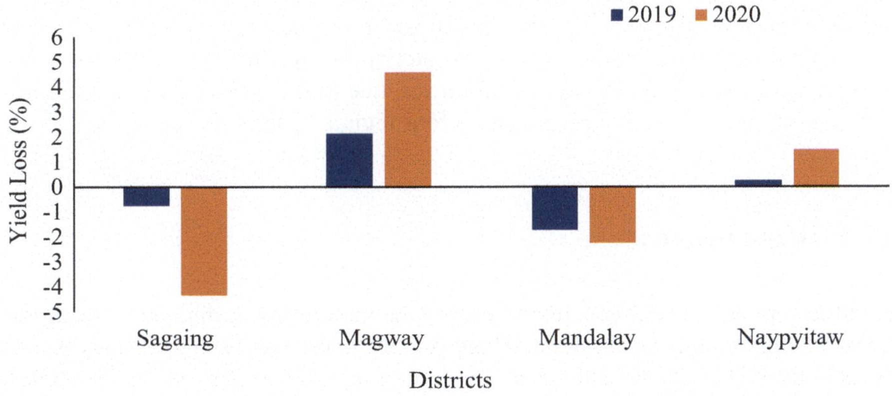


**Fig. 11.10** Percentage of the rice yield losses in the four districts (Sagaing, Magway, Mandalay, and Naypyitaw) of the central dry zone of Myanmar

## 11.4 Discussion

Drought is one of the most serious environmental factors that affects rice production in drought-prone regions. To mitigate the impact of drought on growers' lives and paddy rice cultivation in the central dry zone of Myanmar, it is crucial to reduce the drought risk at the various growth stages of paddy rice. In this regard, developing a satellite remote sensing-based drought composite index is essential to determining the extent of the drought risk. In this study, a composite drought index was developed utilizing the AHP, GIS technology, and satellite remote sensing-based drought indices. Several researchers have used AHP techniques to calculate the composite drought index and have found that rainfall has a weight of 44.5%, while the LST has a weight of 28%. Our results also indicated that the temperature conditions and standard precipitation index had the highest weights of 34% and 28%, respectively (Table 11.5) (Kumari and Kumar 2022; Kundu et al. 2021; Palchaudhuri and Biswas 2016). As indicated in Table 11.7, there was significant drought during all three seasons, with a percentage of drought coverage greater than 50%. Previous studies have indicated that nearly 80% of people suffer from food insecurity. In this study, it was also revealed that the farmers in this area are severely affected by drought (Boori et al. 2017; Dutta 2018).

The calculated paddy rice yield loss due to drought showed that some areas (Sagaing and Mandalay) exhibited a notable decline in yield due to drought in 2020. In Mangway and Naypyitaw, however, farmers may have found ways to manage water well or planted drought-tolerant rice varieties. This research offers helpful information for drought monitoring in the central dry area of Myanmar. Drought indices and meteorological indicators can be reliably used to identify the characteristics needed to define the drought conditions of the research area. The complexity resulting from the numerous parameters can be reduced via GIS technology. The weights obtained with the AHP mostly depend on the opinion of the researcher. Spatially diverse input data are necessary for the GIS model. The spatial variation in several factors, including evapotranspiration and relative humidity, was not considered in this investigation due to a lack of accurate data. However, the geographical variation in other components was considered. The quality of the results was not adversely affected on a large scale. Moreover, due to the lack of ground reference and statistical datasets, we only relied on four regional paddy rice datasets.

## 11.5 Conclusions

Satellite remote sensing can improve drought monitoring techniques over traditional drought monitoring systems. Therefore, the main purpose of this study was to develop the CDI and calculate the yield loss during drought periods. In this regard, to determine the risks of droughts and water shortages, this research was conducted in the central dry regions of Myanmar. Agricultural (VCI, TCI, and NDMI) and meteorological parameters (SPI) were used to create drought risk maps from 2019 to 2020. Moreover, to assess the level of dryness in the central dry zone, three seasons—summer (March to May), monsoon (June to October), and winter (November to February)—were accounted for. The results indicated that the central dry zone was mostly affected by moderate to mild droughts during the rice growing seasons of 2019–2020. Moreover, it was revealed that the drought length and severity during the rice-growing season affected rice productivity. Although short-term or low-intensity drought imposes little or no effect on the yield, based on our results, the effect was approximately 5% per unit of area. In some districts, drought did not affect the rice yield because the farmers may have adopted effective irrigation techniques during drought periods. In this study, we did not attempt to explain the yield variation based on environmental factors. This study highlights the importance of factors that are generally ignored, such as the status of yield fluctuations or other circumstances that could impact how farming techniques are implemented as well as the uncertainty in data. Additionally, this study focused on examining ongoing satellite-based drought monitoring to facilitate the forecasting of drought conditions and to allow stakeholders to plan for potential risks.

**Acknowledgements** The authors appreciate the Landsat 8 OLI and TIRS products of the United States Geological Survey and the Google Earth Engine developers, in addition to the CHIRPS precipitation dataset of the Climate Hazards Group and the yield datasets of the Myanmar Statistical Department. Moreover, the first author gratefully acknowledges the JICA for providing the scholarship for his master's degree study at the Graduate School of Life and Environmental Sciences at the University of Tsukuba.

## References

Aksoy H (2000) Use of gamma distribution in hydrological analysis. Turk J Eng Environ Sci 24(6):419–428. https://aj.tubitak.gov.tr/engineering/issues/muh-00-24-6/muh-24-6-7-9909-13.pdf. Last accessed 15 Feb 2023

Arab ST, Ahamed T (2023) Near-real-time drought monitoring and assessment for vineyard production on a regional scale with standard precipitation and vegetation indices using Landsat and CHIRPS datasets. Asia-Pac J Reg Sci 7:591. https://doi.org/10.1007/s41685-023-00286-7

Arab ST, Noguchi R, Ahamed T (2022) Yield loss assessment of grapes using composite drought index derived from Landsat OLI and TIRS datasets. Remote Sens Appl Soc Environ 26:100727. https://doi.org/10.1016/j.rsase.2022.100727

Bayissa YA, Tadesse T, Svoboda M, Wardlow B, Poulsen C, Swigart J, Van Andel SJ (2019) Developing a satellite-based combined drought indicator to monitor agricultural drought: a case study for Ethiopia. GIScience Remote Sens 56:718–748. https://doi.org/10.1080/15481603.2018.1552508

Belal AA, El-Ramady HR, Mohamed ES, Saleh AM (2014) Drought risk assessment using remote sensing and GIS techniques. Arab J Geosci 7:35–53. https://doi.org/10.1007/s12517-012-0707-2

Biswas T, Pal SC, Ruidas D, Islam ARMT, Saha A, Costache R, Shit M (2023) Modelling of groundwater potential zone in hard rock-dominated drought-prone region of eastern India using integrated geospatial approach. Environ Earth Sci 82(3):81. https://doi.org/10.1007/s12665-023-10768-8

Boori MS, Choudhary K, Paringer RA, Evers M (2017) Food vulnerability analysis in the central dry zone of Myanmar. Компьютерная оптика 41(4):552–558

Daryanto S, Wang L, Jacinthe PA (2016) Global synthesis of drought effects on maize and wheat production. PLoS One 11(5):e0156362. https://doi.org/10.1371/journal.pone.0156362

Das AC, Noguchi R, Ahamed T (2021) An assessment of drought stress in tea estates using optical and thermal remote sensing. Remote Sens 13(14):2730. https://doi.org/10.3390/rs13142730

De Jesus JB, Santana IDM (2017) Estimation of land surface temperature in caatinga area using Landsat 8 data. J Hyperspectr Remote Sens 7(3):150. www.periodicos.ufpe.br/revistas/jhrs

Dutta R (2018) Drought monitoring in the dry zone of Myanmar using MODIS derived NDVI and satellite derived CHIRPS precipitation data. Sustain Agric Res 7:46–55

Farahmand A, AghaKouchak A, Teixeira J (2015) A vantage from space can detect earlier drought onset: an approach using relative humidity. Sci Rep 5:8553. https://doi.org/10.1038/srep08553

Gao B-C (1996) NDWI—a normalized difference water index for remote sensing of vegetation liquid water from space. Remote Sens Environ 58:257–266. https://doi.org/10.1016/S0034-4257(96)00067-3

Garba I, Abdourahamane ZS, Mirzabaev A (2023) A drought dataset based on a composite index for the Sahelian climate zone of Niger. Data 8(2):28. https://doi.org/10.3390/data8020028

Gorelick N, Hancher M, Dixon M, Ilyushchenko S, Thau D, Moore R (2017) Google Earth Engine: planetary-scale geospatial analysis for everyone. Remote Sens Environ 202:18–27. https://doi.org/10.1016/j.rse.2017.06.031

Gottschalck J, Meng J, Rodell M, Houser P (2005) Analysis of multiple precipitation products and preliminary assessment of their impact on Global Land Data Assimilation System land surface states. J Hydrometeorol 6(5):573–598. https://doi.org/10.1175/JHM437.1

Habibie MI, Noguchi R, Matsushita S, Ahamed T (2020) Development of micro-level classifiers from land suitability analysis for drought-prone areas in Indonesia. Remote Sens Appl Soc Environ 20:100421. https://doi.org/https://doi.org/10.1016/j.rsase.2020.100421

Hao Z, AghaKouchak A (2013) Multivariate standardized drought index: a parametric multi-index model. Adv Water Res 57:12–18. https://doi.org/10.1016/j.advwatres.2013.03.009

Hardisky MA, Klemas V, Smart M (1983) The influence of soil salinity, growth form, and leaf moisture on the spectral radiance of. Spartina alterniflora 49:77–83. https://www.asprs.org/wp-content/uploads/pers/1983journal/jan/1983_jan_77-83.pdf

Karamouz M, Rasouli K, Nazif S (2009) Development of a hybrid index for drought prediction: case study. J Hydrol Eng 14(6):617–627. https://doi.org/10.1061/(ASCE)HE.1943-5584.0000022

Kogan FN (1990) Remote sensing of weather impacts on vegetation in non-homogeneous areas. Int J Remote Sens 11:1405–1419. https://doi.org/10.1080/01431169008955102

Kogan FN (1995) Application of vegetation index and brightness temperature for drought detection. Adv Space Res 15(11):91–100. https://doi.org/10.1016/0273-1177(95)00079-T

Kogan F, Guo W, Yang W (2019) Drought and food security prediction from NOAA new generation of operational satellites. Geomat Nat Haz Risk 10(1):651–666. https://doi.org/10.1080/19475705.2018.1541257

Krishnaswamy J, John R, Joseph S (2014) Consistent response of vegetation dynamics to recent climate change in tropical mountain regions. Glob Chang Biol 20(1):203–215. https://doi.org/10.1111/gcb.12362

Kumari M, Kumar D (2022) Dynamic drought risk assessment and analysis with multi-source drought indices and analytical hierarchy process. Int J Environ Sci Technol:1–18

Kundu A, Dutta D, Patel NR, Denis DM, Chattoraj KK (2021) Evaluation of socio-economic drought risk over Bundelkhand region of India using analytic hierarchy process (AHP) and geo-spatial techniques. J Indian Soc Remote Sens 49:1365–1377

Liu D, Chen N, Zhang X, Wang C, Du W (2020a) Annual large-scale urban land mapping based on Landsat time series in Google Earth Engine and OpenStreetMap data: a case study in the middle Yangtze River basin. ISPRS J Photogramm Remote Sens 159:337–351. https://doi.org/10.1016/j.isprsjprs.2019.11.021

Liu Q, Zhang S, Zhang H, Bai Y, Zhang J (2020b) Monitoring drought using composite drought indices based on remote sensing. Sci Total Environ 711:134585. https://doi.org/10.1016/j.scitotenv.2019.134585

McKee TB, Doesken NJ, Kleist J (1993a) The relationship of drought frequency and duration to time scales. In: Proceedings of the 8th conference on applied climatology, Anaheim, CA, USA, 17–22 January 1993, vol 17, pp 179–183. https://climate.colostate.edu/pdfs/relationshipofdroughtfrequency.pdf

McKee TB, Doesken NJ, Kleist J (1993b) The relationship of drought frequency and duration to time scales. In: Proceedings of the 8th conference on applied climatology. American Meteorological Society, Boston, pp 179–183

Mimić G, Živaljević B, Blagojević D, Pejak B, Brdar S (2022) Quantifying the effects of drought using the crop moisture stress as an indicator of maize and sunflower yield reduction in Serbia. Atmosphere 13(11):1880

Mishra AK, Singh VP (2010) A review of drought concepts. J Hydrol 391(1–2):202–216. https://doi.org/10.1016/J.JHYDROL.2010.07.012

Mohan K, Ramasamy A, Varghese J (2021) Drought severity assessment using automated land surface temperature retrieval technique. Arab J Geosci 14(22):2358. https://doi.org/10.1007/s12517-021-08672-1

Nivedha Deve S, Jasmineniketha M, Geetha P, Soman KP (2017) Agricultural drought analysis for Thuraiyur taluk of Tiruchirappalli District using NDVI and land surface temperature data. In: Proceedings of 2017 11th international conference on intelligent systems and control, ISCO 2017, pp 155–159. https://doi.org/10.1109/ISCO.2017.7855972

Palchaudhuri M, Biswas S (2016) Application of AHP with GIS in drought risk assessment for Puruliya district, India. Nat Hazards 84:1905–1920

Potopová V et al (2020) Statistical modelling of drought-related yield losses using soil moisture-vegetation remote sensing and multiscalar indices in the south-eastern Europe. Agric Water Manag 236:106168. https://doi.org/10.1016/j.agwat.2020.106168

Saaty TL (1988) What is the analytic hierarchy process? In: Mathematical models for decision support. Springer, Berlin, pp 109–121. https://doi.org/10.1007/978-3-642-83555-1_5

Statistical Yearbook (2020). Central statistics organization, Myanmar. https://www.csostat.gov.mm/PublicationAndRelease/StatisticalYearbook

Statistical Yearbook (2021) Central statistics organization, Myanmar. https://www.csostat.gov.mm/PublicationAndRelease/StatisticalYearbook

Statistical Yearbook (2022) Central statistics organization, Myanmar. https://www.csostat.gov.mm/PublicationAndRelease/StatisticalYearbook

Svoboda M, LeComte D, Hayes M, Heim R, Gleason K, Angel J et al (2002) The drought monitor. Bull Am Meteorol Soc 83(8):1181–1190

Tamiminia H, Salehi B, Mahdianpari M, Quackenbush L, Adeli S, Brisco B (2020) Google Earth Engine for geo-big data applications: a meta-analysis and systematic review. In: ISPRS journal of photogrammetry and remote sensing, vol 164. Elsevier B.V., Amsterdam, pp 152–170. https://doi.org/10.1016/j.isprsjprs.2020.04.001

Tariq A, Riaz I, Ahmad Z, Yang B, Amin M, Kausar R, ... & Rafiq M (2020) Land surface temperature relation with normalized satellite indices for the estimation of spatio-temporal trends in temperature among various land use land cover classes of an arid Potohar region using Landsat data. Environ Earth Sci 79:1–15

Tun T (2000) Greening the dry zone of Myanmar. https://www.osti.gov/etdeweb/biblio/20311679. Last accessed 14 Feb

United Nations (UN) Convention to Combat Desertification (2022) Drought in numbers 2022—restoration for readiness and resilience. Drought in Numbers (English Final) (unccd.int)

Vogt JV, Niemeyer S, Somma F, Beaudin I, Viau AA (2000) Drought monitoring from space. Drought and drought mitigation in Europe, 167–183. https://link.springer.com/chapter/10.1007/978-94-015-9472-1_13. Last accessed 14 Feb 2023

Wang S, Li R, Li X (2019) Inversion and distribution of soil moisture in belly of Maowusu sandy land based on comprehensive drought index. Trans Chin Soc Agr Eng 35:113–121. https://doi.org/10.11975/j.issn.1002-6819.2019.13.012

Wongsai N, Wongsai S, Huete AR (2017) Annual seasonality extraction using the cubic spline function and decadal trend in temporal daytime MODIS LST data. Remote Sens 9(12):1254. https://doi.org/10.3390/rs9121254

World Meteorological Organization, 2012: Standardized Precipitation Index User Guide (M. Svoboda, M. Hayes and D. Wood). (WMO-No. 1090), Geneva.

Wu L, Wang H, Liu H (2017) Characteristics and regionalization of drought risk distribution in Pastoral Areas of Inner Mongolia. J Arid Meteorol 35:1070–1076. https://doi.org/10.11755/j.issn.1006-7639(2017)-06-1070

Wulder MA, Loveland TR, Roy DP, Crawford CJ, Masek JG, Woodcock CE, Allen RG, Anderson MC, Belward AS, Cohen WB, Dwyer J, Erb A, Gao F, Griffiths P, Helder D, Hermosilla T, Hipple JD, Hostert P, Hughes MJ et al (2019) Current status of Landsat program, science, and applications. Remote Sens Environ 225:127–147. https://doi.org/10.1016/j.rse.2019.02.015

Xue J, Su B (2017) Significant remote sensing vegetation indices: a review of developments and applications. J Sens 2017:1353691. https://doi.org/10.1155/2017/1353691

Zhang X, Wei C, Obringer R, Li D, Chen N, Niyogi D (2017) Gauging the severity of the 2012 Midwestern US drought for agriculture. Remote Sens 9(8):767

Zhou B (2015) Research on evaluation model for grid partition scheme based on improved principal component analysis. North China Electric Power University, Beijing

Zin WYL, Teartisup P, Kerdseub P (2019) Evaluating traditional knowledge on climate change (TKCC): a case study in the central dry zone of Myanmar. Environ Nat Resour J 17(2):1–29

# Chapter 12
# Microclimate Characterization and Climate Risk Assessment of Agroforestry-Based Arabica Coffee Production

**Janet P. Pablo, Alexander W. Fagyan, Conrado C. Bao-Idang, Carlito P. Laurean, Louie C. Rimas, and Noreen A. Moreno**

**Abstract** The shade plants regulate the microclimate of the coffee plants when integrated with or under pine, alnus, mango, and chayote. Significant reduction was noted on the minimum and maximum air temperature, soil temperature, wind velocity, and light intensity, while no differences found in the relative humidity. The microclimate changes are favorable for the growth and development of the coffee trees. Likewise, the socioeconomic capacity and the agroforestry-based coffee growing practiced by farmers in Atok, Benguet, Philippines to adapt to the effect of the climatic hazards make them moderately adaptive to climate change. The coffee growers in the municipality of Tublay, Benguet, Philippines, are moderately vulnerable due to higher exposures to the climate hazards and few adaptation technologies practiced. Despite low wealth level that constrain them to readily spend for cost of technological adaptations, the coffee growers are planting cash crops to augment their income and increase their resilience. Further, agroforestry-based coffee planting also promotes favorable microclimate condition, biodiversity conservation, and soil protection.

**Keywords** Microclimate · Climate change · Sensitivity · Arabica coffee · Resilience · Vulnerability

---

J. P. Pablo (✉) · A. W. Fagyan · C. P. Laurean · L. C. Rimas · N. A. Moreno
College of Agriculture, Benguet State University, La Trinidad, Benguet, Philippines
e-mail: j.pablo@bsu.edu.ph

C. C. Bao-Idang
College of Forestry, Benguet State University, La Trinidad, Benguet, Philippines


T. Ahamed (ed.), *Remote Sensing Application II*, New Frontiers in Regional Science: Asian Perspectives 77, https://doi.org/10.1007/978-981-97-1188-8_12

## 12.1 Introduction

Coffee has existed for centuries in the provinces of the Cordillera Administrative Region (CAR) such as the Benguet Province in the Philippines. The coffee production areas in Benguet Province are usually upland sloping areas with varying degrees and types of degradation stemming from its rugged terrain. Three major species are identified in Benguet Province, *Arabica, Robusta,* and *Liberica*. Arabica is found mostly in elevations ranging from 900 to 1800 m above sea level or high mountain zone. *Robusta* and *Liberica* are found in low mountain zones, 600–1200 m above sea level (Tad-awan et al. 2010).

Most of the province's coffee trees are often grown as part of complex multistrata or agroforestry system based or intercropped with other taller plants. Coffee trees are planted under a naturally controlled shade provided by one or more tree species such as Alnus or pine trees serving as shade trees. Coffee can tolerate shade and share similar growth requirements with forest trees.

Growing of shade trees with the coffee plant has a great contribution to both the life span of the coffee trees and their yield (Alemu 2015). Coffee production with different levels of shade, delays, or slows down ripening of berry and reduced yield but yields a superior green coffee and cup quality (Pohlan and Janssens 2010). On the other hand, growing coffee under full sunlight with little or no forest canopy causes berries to ripen more rapidly and the bushes to produce higher yields. Consequently, full exposure to sunlight causes erratic yields in coffee trees caused by periodic overbearing and subsequent dieback (Alemu 2015).

The major physiological benefits that coffee receives from shade trees are an ideal microclimate for growth and production by reducing air and soil temperature extremes, reduction in wind speeds, and buffering of humidity and soil moisture availability. Aside from increasing biodiversity, shade trees reduce the quantity and quality of transmitted light and hence avoidance of overbearing for sustenance of coffee plantations (Beer et al. 1998).

While most coffee plants are naturally intolerant to direct sunlight and prefer a canopy of sun-filtering shade trees, the management of maintaining optimal shade levels is important. Alemu (2015) cited that the optimal shade level for coffee trees is around 40–50% of the full sunlight where the upper limit of acceptable shade for coffee (a $C_3$ plant) is considered to be between 40% and 70% and higher shade density had a negative effect on coffee yield (Beer et al. 1998). Beer et al. (1998) stated that the saturating light levels were as low as 300 μmol $m^{-2}$ $s^{-1}$ for shaded coffee and the level for unshaded plants was as high as 600 μmol $m^{-2}$ $s^{-1}$.

Several physiological benefits have been identified attributed to the shade trees providing an ideal microclimate for the growth and production of coffee trees. However, the microclimate data under the existing agroforestry system where coffee are grown in Benguet Province have not yet been documented to prove the claim that the cool climate of the province is suitable for the production of specialty coffee, the Arabica coffee. Likewise, coffee losses are documented and reported to affect the volume and area of production of the Arabica coffee due to the climate hazards.

The research then would like to characterize the microclimate conditions of coffee-based agroforestry system to determine the appropriate microclimate and shade level management for coffee under agroforestry system and to determine the resilience and adaptability of coffee to climate hazards.

## 12.2 Materials and Methods

### *12.2.1 Site Selection*

The research was conducted in the major coffee-producing areas in the municipalities of Tublay and Atok in Benguet Province in the Philippines where Arabica coffee is integrated with Alnus trees (*Alnus japonica*), mango fruit trees (*Mangifera indica*), and chayote (*Sechium edule*) as shade plants, respectively (Table 12.1). Geographic coordinates and elevations of the experimental sites were taken with the use of the Global Positioning System (GPS) which is then mapped through Quantum Geographic Information System (QGIS) (Fig. 12.1). The elevation of the coffee growing areas ranges from 1495 to 1648 masl considered as low-mountain zone (1000–1500 masl) and Mid-Mountain Zone (1500–2000 masl) under the landscape range of agriculture and forest ecological zones (AFEZ) of the Cordillera region. The soil series classification of the experimental area is Ambassador silt loam and considered as moderately suitable due to the steep slope (FAO 2005).

#### 12.2.1.1 Geographic Profile of the Sites

The municipality of Atok is geographically located in the central part of the Benguet, Philippines, with a total land area of 21,912.18 ha based on the cadastral survey conducted by the Department of Environment and Natural Resources (DENR). The

**Table 12.1** Location and profile of coffee and shade trees/plants present in the experimental sites

| Location | Coordinates | Elevation (masl) | Afez | Shade plants |
|---|---|---|---|---|
| Upper Sayet, Caliking, Atok | 16°31′9.23″ N 120°41′20.07″ E | 1648 | Mid-Mountain Zone | Alnus |
| Lower Sayet, Caliking, Atok | 16.31°18′0.19″ N 120°41′0.55″ E | 1534 | Mid-Mountain Zone | Mango |
| Caliking Proper, Caliking, Atok | 16°32′8.83″ N 120°41′11.62″ E | 1520 | Mid-Mountain Zone | Chayote |
| Coroz, Ambassador, Tublay) | 16°28′46.10″ N 120°39′37.01″ E | 1495 | Low-Mountain Zone | Alnus |

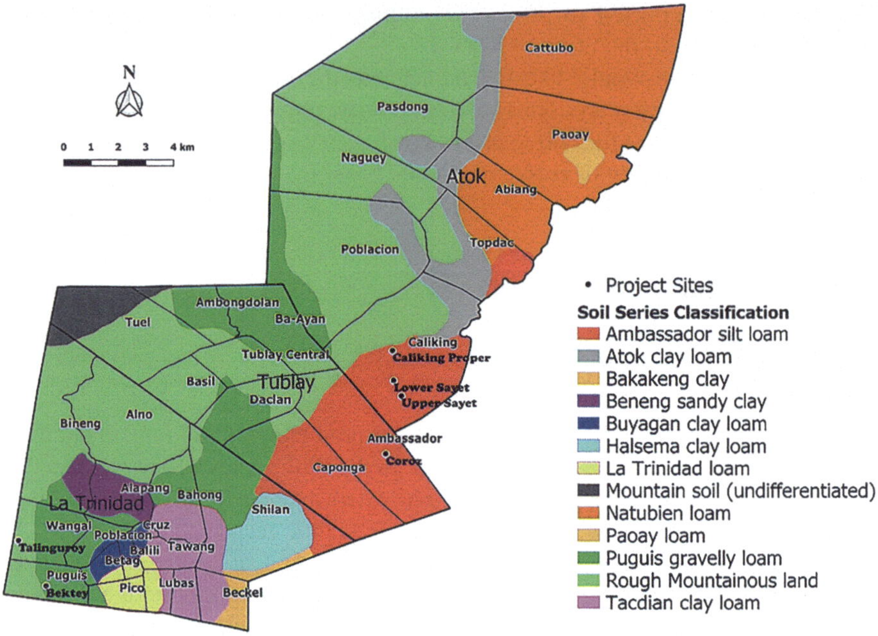


**Fig. 12.1** Map of Atok and Tublay showing the soil series classification of the experimental sites

experimental sites, Barangays Caliking and Topdac, are the top producers of Arabica coffee in the municipality.

Barangay Caliking has a total land area of 2920.91 ha, of which, 1.76 ha are residential and commercial areas, and 495.07 ha are devoted to agriculture. The terrain characteristics of the barangay are 65% hilly and the other 35% is mountainous. On the other hand, Barangay Topdac of Atok Municipality has a total land area of 2447.64 ha classified into five (5) according to its uses; agricultural land (787.39 ha), residential land (0.74 ha), commercial land (0.036 ha), educational land (3.27 ha), and grass/pastureland (1656.65 ha). The most dominant terrain in the barangay is hilly mountainous with a range of 30–50% gradient. It is followed by rolling-hilly with a range of 18–30% and the least is level to undulating with a range of 0–2%. Coffee trees are grown in areas devoted to agriculture and mostly on the hilly or mountainous especially on the slope.

Tublay is another municipality in Benguet Province and has a total land area of 10,255 ha. Barangay Ambassador, the project site, has a total land area of 3282 ha (32% of the total land area of the municipality). It is the largest barangay in the Municipality of Tublay with twenty-two (22) sitios including Ambassador and Daclan. Barangay Ambassador is dominantly hilly to mountainous with slopes ranging from 18% to 30%, while Barangay Daclan is characterized by hilly to mountainous areas having an approximate slope ranging from 18% to 30% and rough mountainous areas with an approximated slope ranging from 31% to 100%.

### 12.2.2 Microclimate Characterization of Coffee-Based AF System

Microclimate observations were recorded on the experimental sites of the major coffee-production areas in Tublay and Atok, wherein Arabica coffee is integrated with Alnus trees, mango fruit trees, and chayote as shade plants, respectively.

Above ground microclimatic parameters such as air temperature (°C), relative humidity (%), wind velocity (kph), and light intensity (klux) were measured using a thermohygrometer (Huato, Model HE710-TH), anemometer (Tenmars, Model TM-414), and light meter (Sekonic, Model L-308S). Soil temperature was measured using a soil thermometer (Extech Model 392052).

### 12.2.3 Vulnerability Risk Assessment and Biophysical Sensitivity of Arabica Coffee to Climate Change

Vulnerability to climate change of the Arabica coffee-based agroforestry system was assessed using the Vulnerability and Adaptive Capacity Assessment for different Agro ecosystems (VAST-Agro) tool, developed by the Agriculture Systems Cluster of the College of Agriculture, University of the Philippines-Los Baños. The VAST-Agro is a quick participatory method of assessing the vulnerability of a community's agricultural sector or particular crops in a community to the hazards of climate change based on their exposure to climate hazards, sensitivity to damages that the hazard may bring, and the community's capacity to cope with adversities (Garcia et al. 2011).

Two barangays in Atok and two barangays in Tublay were selected as sites to conduct the research. Among the criteria used for selection included the (1) top coffee producing barangay of the municipality, (2) willingness of the community (MAO and BLGU) to participate in the activity, and (3) farmers experience changes in the climate.

Secondary data gathering for the coffee profile from the municipal commodity profiles, results of national surveys, climate statistics, and agricultural data were taken from local weather stations and Municipal Agricultural Office.

Intergovernmental Panel on Climate Change (IPCC 2014) defines vulnerability to climate change as the degree to which geophysical, biological, and socio-economic factors are susceptible to, and unable to cope with adverse impact of climate change. The three key dimensions of vulnerability were (1) exposure, (2) sensitivity, and (3) adaptive capacity. These key dimensions were mapped in Quantum Geographic Information System through remote sensing. Remote sensing is mainly used for soil mapping; land type, vegetation, soil survey and land use elements that provides information about the physical characteristics of the area, in this case is the coffee-based agroforestry system in Atok and Tublay, Benguet. The climate risk vulnerability to estimate the exposure to hazard, sensitivity, and adaptive capacity of the coffee production under agroforestry system was assessed based on the procedures used by the Department of Agriculture (DA) and the International Center for Tropical Agriculture (CIAT) (Daipan et al. 2023).

On-site validation of the climate hazard exposure and technological adaptation assessment was through the conduct of community meetings and focus group discussions where a series of meetings/consultations and workshops was done with stakeholders to gather preliminary data and information in the study sites (Atok and Tublay, Benguet). Structured interviews were also used to gather data information from key informants.

(a) Climate statistics to include the occurrences of extreme events like tropical cyclones, drought, flooding, trends, and changes in climate and other CC-related disturbances.
(b) Agricultural data on land use, production areas, soil characteristics, crops and cropping system, seasonal calendar, management practices, crop losses, support services, government programs/projects, and other related data.
(c) Based on the data gathered, the frequency and impact of the climate events and their influence on agriculture and development were determined.

The assessment of the exposure of the project's sites in Atok and Tublay, Benguet was mapped using the biophysical indicators/climate-related pressures such as the following: tropical cyclone, flooding, landslide, drought, erosion, saltwater intrusion, sea level rise, and storm surge.

Assessment of the sensitivity to climate change hazards was done by getting the biophysical and socioeconomic characteristics of the community. Mapping of the assessment of the sensitivity of coffee was done through Maximum Entropy (Maxent) modeling in QGIS. The climatic data used were the following: annual mean temperature, mean diurnal range, isothermality, temperature seasonality, maximum temperature of the warmest month, minimum temperature of the coldest month, annual temperature range, mean temperatures of the wettest quarter, driest quarter, warmest quarter, coldest quarter, and annual precipitation, precipitation seasonality, and precipitation of the wettest and driest month, and precipitation of the wettest quarter, driest quarter, warmest quarter and coldest quarter, and finally the number of consecutive dry days (Daipan et al. 2023).

The adaptive capacity to climate change hazards incorporates information on the adaptive capacity indicators such as the following: physical capacity, cognitive ability and linguistic capacity, resource availability, degree of isolation, availability of support systems, economic capacity and income level and diversity of income sources and the technological ability of the community. These indicators go through preliminary statistical analysis, then to participatory expert ranking and weighing of relevant indicators across municipalities, then to secondary statistical analysis and finally mapping based on the adaptive capacity index (Daipan et al. 2023).

## 12.3 Results and Discussion

### *12.3.1 Results*

#### 12.3.1.1 Microclimate Characterization of Coffee-Based Agroforestry System

##### 12.3.1.1.1 Coffee Trees Under Alnus Shade (Tublay)

The coffee farms in Tublay have an elevation of 1495 masl, considered a Low-Mountain Zone supported for land uses such as agroforestry systems of production (Table 12.1). Coffee grown under Alnus tree shades in this zone showed significant differences by *t*-test at 5% probability for both the minimum and maximum air temperatures with 18.5–25.9 °C, respectively, with a temperature difference of 0.9 °C lower than the open field (Table 12.2). The soil temperature has a 2.8 °C difference lower under the Alnus shade trees (19.6 °C). The Alnus shade trees significantly reduce the wind velocity from 6.5 (open field) to 3.3 kph. Significant light intensity interception with an estimated 49% light attenuation values was recorded in coffee grown under Alnus trees . No significant differences in values of relative humidity under open field and under Alnus tree shade cover were recorded. The distance between the experiment set up for the open field and the shaded areas is about 75-100m.

Planting distance or spacing of shade trees affects wind velocity and the closer the spacing, the slower the wind speed. There is a greater relative frequency of high wind speeds in sparse shade trees as compared to dense shade trees with reduced speed. Moreover, the denser shade trees reduced the amount of light intercepted by the coffee. Due to this, more than one-third of the study area spaced nurse trees at an average of 8 m × 8 m almost one-third used an average of 6 m × 6 m. Those farmers using 6 m × 6 m or closer spacing reported that they do pruning, pollarding, and/

**Table 12.2** Averages and statistical analysis of microclimate variables for coffee grown under open field and Alnus shade in Coroz, Ambassador, Tublay

| Microclimate variables | Open field | Under Alnus | *P*-value (two-tailed) |
|---|---|---|---|
| Air temperature-min (°C) | 17.6 | 18.5 | 0.03* |
| Air temperature-max (°C) | 26.8 | 25.9 | 0.01* |
| Soil temperature (°C) | 22.4 | 19.6 | 0.00** |
| Relative humidity-min (%) | 66.1 | 69.3 | $0.37^{ns}$ |
| Relative humidity-max (%) | 88.2 | 86.6 | $0.63^{ns}$ |
| Wind velocity (kph) | 6.5 | 3.3 | 0.00** |
| Light intensity (klux) | 61.8 | 31.6 | 0.00** |

$^{ns}$ indicates not significant ($P > 0.05$); * indicates significant difference at $P < 0.05$; ** indicates significant difference at $P < 0.01$

or thinning of shade trees as the coffee trees grow bigger to provide enough space and allow penetration of sunlight to the coffee crops.

#### 12.3.1.1.2 Coffee Trees Under Alnus, Mango, and Chayote Shade Crops

The coffee growing areas in Atok municipality had an elevation of 1520–1648 masl in the landscape range of agriculture and forest ecological zone 2 (AFEZ 2) and are currently being used for agroforestry production and croplands.

Significant differences in the maximum temperature, soil temperature, wind velocity, and light intensity while no significant differences were recorded on the minimum and maximum relative humidity between the open field and under the different shade trees (Table 12.3). Under the Alnus shade trees, the minimum air temperature was 1.1 °C higher in the open field, while the maximum temperature was 1 °C lower on shaded coffee. Lower temperature slows down bean ripening processes and allows more time for bean filling (Muschler 2001), but according to Damatta (2004) and Winston et al. (2005) the optimum air temperature for Arabica coffee to produce greater bean quality ranges from 18 to 24 °C. Soil temperature was significantly lower in coffee under shade, with an average difference of 2.6 °C. Moreover, the differences in the values of wind velocity and light intensity were significant by *t*-test at 1% probability under open and shaded production. Soil temperature is an important environmental variable, but it is rarely reported as an indicator of climate change. A recent study of Getachew et al. (2022) found that soil temperatures were found to have a strong and significant negative relationship with some biochemical characteristics of coffee beans such as caffeine, trigonelline, and total chlorogenic acid. Wind velocity was reduced to 61% and light intensity was decreased to 49% in shaded production. Relative humidity, however, showed no significant differences in open and shaded coffee for both minimum and maximum values.

A similar trend with coffee under Alnus and chayote was found under mango shade. Minimum temperature does not significantly differ, while values for maximum temperature differ significantly. There was a high significant difference in the soil temperature of coffee under mango compared to open field production. Means for relative humidity do not differ significantly similar to the results of all the sites. Wind velocity and light intensity were also found to be highly significant.

Under chayote shade, there was no significant differences in the minimum temperature recorded, while maximum temperature showed a significant difference. Soil temperature, however, revealed a significant differences with 2.5 °C lower under the shaded area. Consistent with coffee under Alnus, relative humidity values under chayote do not differ significantly. Moreover, statistical differences between mean values of wind velocity and light intensity were significant by *t*-test at 1% of probability. Mean wind velocity recorded under chayote was observed to be as low as 1.5 kph that may be attributed to the geographical location of the farm and thick shading provided by the chayote plant.

**Table 12.3** Averages and statistical analysis of microclimate variables for coffee grown under open field and different shade crops in Caliking, Atok, Benguet, Philippines

| Microclimate variables | Optimum value | 1648 masl | | *P*-value | 1534 masl | | *P*-value | 1520 masl | | *P*-value |
|---|---|---|---|---|---|---|---|---|---|---|
| | | Open field | Under Alnus | | Open field | Under mango | | Open field | Under chayote | |
| Air temperature-min (°C) | 18.0 | 16.8 | 17.9 | 0.02* | 17.6 | 18.4 | $0.07^{ns}$ | 16.9 | 17.7 | $0.09^{ns}$ |
| Air temperature-max (°C) | 24.0 | 25.8 | 24.8 | 0.01* | 26.9 | 25.6 | 0.00* | 26.1 | 25.0 | 0.02* |
| Soil temperature (°C) | 14.5–21.0 | 21.5 | 18.9 | 0.00** | 22.5 | 19.9 | 0.00** | 21.5 | 19.0 | 0.00** |
| Relative humidity-min (%) | 70.0 | 64.5 | 67.7 | $0.27^{ns}$ | 63.9 | 67.2 | $0.24^{ns}$ | 66.4 | 69.9 | $0.20^{ns}$ |
| Relative humidity-max (%) | 80.0 | 86.4 | 88.5 | $0.44^{ns}$ | 87.4 | 85.7 | $0.56^{ns}$ | 88.5 | 86.9 | $0.49^{ns}$ |
| Wind velocity (kph) | – | 8.8 | 3.4 | 0.00** | 5.5 | 2.5 | 0.00** | 6.8 | 3.3 | 0.00** |
| Light intensity (klux) | 16.2–32.4 | 63.4 | 32.6 | 0.00** | 63.5 | 31.7 | 0.00** | 60.7 | 30.5 | 0.00** |

$^{ns}$ indicates not significant ($P > 0.05$); * indicates significant difference at $P < 0.05$; ** indicates significant difference at $P < 0.01$

### 12.3.1.2 Climate Change Resilience Assessment of Arabica Coffee Production

#### 12.3.1.2.1 Assessment of the Climate-Related Hazards

Climate hazards identified by coffee grower respondents from the two municipalities studied mentioned typhoons and strong winds, heavy rains, and the continuous monsoon that usually occur from June to November resulting in erosion and landslide that have an impact on coffee plants (Table 12.4).

For the past 10 years, farmers recalled four strong typhoons that eventually resulted in low yields during the harvest season. Mechanical damage to coffee trees and most of the coffee berries fell as a result of the strong typhoon. It was also during strong typhoons and continuous rain when rainfall amount was very high, where erosions and landslides occurred on coffee plantations, as most of the coffee trees were grown on slopes. Coffee-bearing trees planted especially on steep slopes were uprooted and washed out. Post-harvest processes are also difficult during unseasonal rain as farmers commonly dry the coffee beans under the sun. Farmers observed that monsoon rains experienced from June to October affected the fruit setting of coffee as berries do not fully develop and resulting to a high number of floaters during wet processing of berries.

Drought accompanied by a rise in temperature for the past 5 years was observed, and hailstorm that occurs from December to February was included as a hazard. Prolonged dry season was also observed to cause wilting and dieback of coffee trees. Two cases of hailstorms were also recalled for the past 10 years during the harvest season in Atok municipality, which caused ripe berries to fall off.

Other identified hazards were pests and diseases caused by weather extremes like high temperatures, monsoon rain, and altered rainfall patterns that usually coincide during the flowering stage of coffee.

Moreover, majority of the respondents agreed that there is an observed rise in the minimum temperature for the past 5 years resulting in an increased incidence of pests and diseases. Several pests recalled were coffee berry borer, stem borer, and coffee scale insect, while diseases mentioned were coffee rust and sooty molds. From 2013 to 2015, there was a significant decrease in yield due to the high

**Table 12.4** Frequency of occurrence of identified climate/climate-related hazards in Atok, and Tublay Benguet, Philippines

| Hazard | Atok | Tublay |
|---|---|---|
| Typhoon and strong winds | 4 times/year | 5 times/year |
| Erosion/landslide | Once a year | Once a year |
| Pest and diseases | 3 times in 10 years | 3 times in 10 years |
| Monsoon rain/persistent rain | Once a year | Once a year |
| Drought | Twice in 10 years | Twice in 10 years |
| Rise of minimum temperature | Twice in 10 years | 3 times in 10 years |
| Hailstorm | Twice in 10 years | – |

incidence of coffee rust and sooty molds. Coffee berry borer was the most destructive causing a great reduction in yield every year. Other pests and diseases mentioned were stem borer, coffee scale insect, and leaf rust. Farmers with infested cherries reported poor green bean quality that led to low-quality beans causing low buying price of beans in the market.

Chiefly, coffee plants have high climate risk vulnerability with a higher frequency of hazard exposure (Daipan et al. 2023; Pham et al. 2019). Despite the climate hazard assessment of the coffee growing areas, Figs. 12.2 and 12.3 showed a 0.00–0.20, with very low exposure to the identified hazards and no change in crop sensitivity index map (Fig. 12.3) when coffee is planted under shade trees. Shading is a likely management technique for adapting coffee to warmer climates (Pinto et al. 2008). A review conducted by Coltri et al. in 2019 shows that several authors suggested that shade trees protect crops from high temperatures (Barradas and Fanjul 1986; Morais et al. 2006), frosts (Caramori et al. 1986), strong winds (Pezzopane et al. 2011), and, in extended dry periods, increase soil moisture and air humidity (Beer et al. 1998), thus reducing coffee physiological stress, adapting the plants to high climate variability. Other studies suggest that low-to-moderate shade level (10–60%) is a good balance between ecological, climate, and yield gains (Atallah et al. 2018; Cerda et al. 2017; Jha et al. 2014; Staver et al. 2001). On the other hand, the full-sun system does not provide ecosystem service benefits, such as

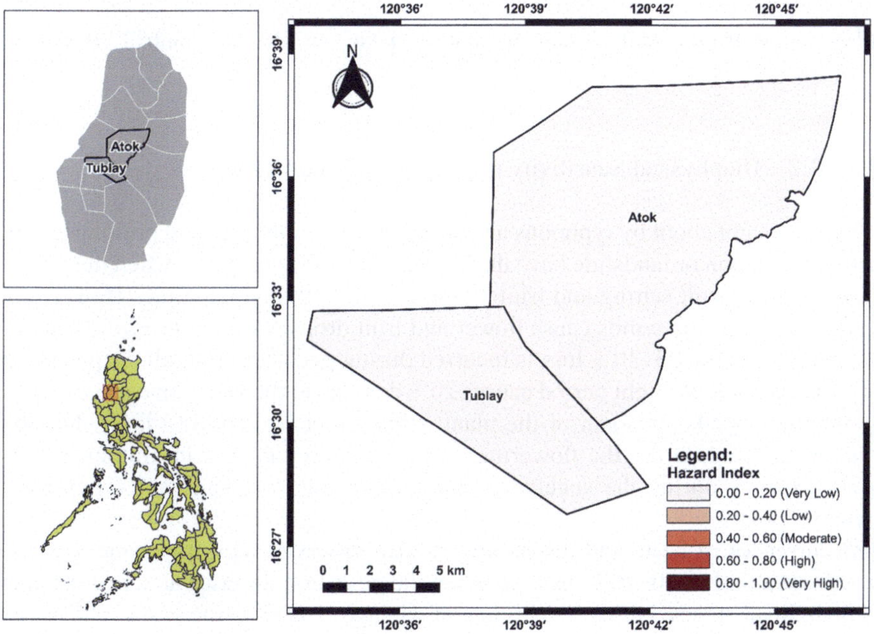


**Fig. 12.2** Climate hazard index in the municipalities of Atok and Tublay, Benguet Philippines based on the bioclimatic variables

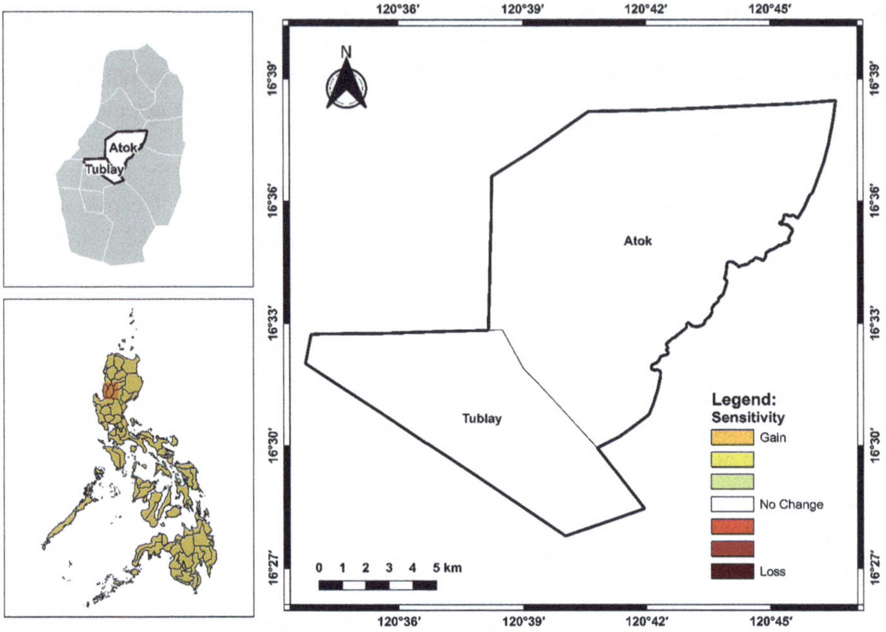


**Fig. 12.3** Crop sensitivity index map of Arabica coffee under agroforestry in the municipalities of Atok and Tublay, Philippines

carbon sequestration and climate regulation, pest, and disease control (Jha et al. 2014; Jose 2009).

#### 12.3.1.2.2 Biophysical Sensitivity to Climate-Related Hazards

Hazards brought about by typhoons and strong winds, monsoon, and prolonged rain leading to erosion or landslide have the highest sensitivity potential when the occurrence is during fruit setting and fruit maturation in both municipalities (Fig. 12.4). Typhoons and strong winds cause flower and fruit drops resulting in 12–30% damage and loss, while 25–30% loss is incurred during persistent rainfall. Exposure of the coffee plant to drought period causes 20% damage to the vegetative stage affecting the photosynthetic organ of the plant, while an estimated 10% loss when the drought occurred during the flowering stage. The observed rise in minimum temperature occurs during the vegetative and fruiting stages causing a 5% reduction in yield.

Occurrences of pests and diseases were also observed with the changes in the climate resulting in 30–40% damage when the pests and diseases attacked the coffee plant in all stages. Higher magnitude of damage caused by the pest and diseases was observed in Atok coffee growing areas. The overall biophysical sensitivity potential score is rated low, with 20% to 40% score considering the total agricultural

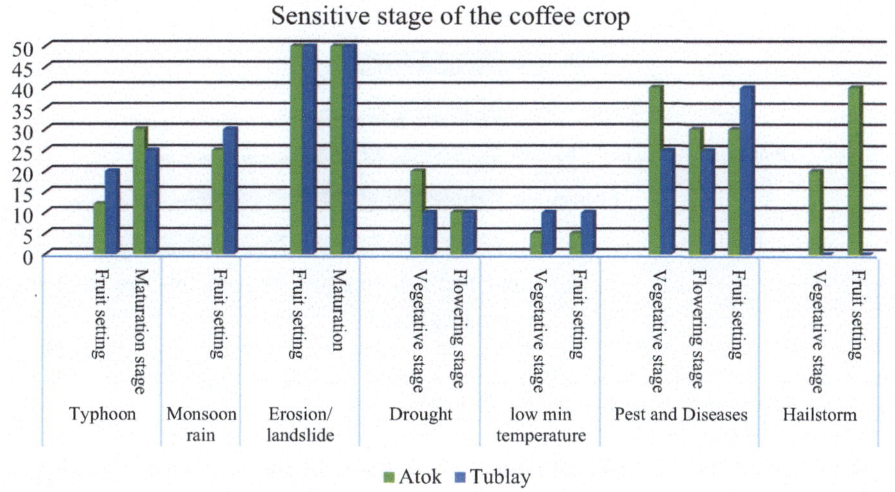


**Fig. 12.4** Sensitive stage or plant part of the coffee crop under different climate hazard exposure

area affected and the magnitude of damage to the coffee plants. An estimated 40% damage to the coffee trees is recorded when a hailstorm occurs during the fruit set.

#### 12.3.1.2.3 Sensitivity of Income Sources to Climate Hazards

The major sources of income of the respondents in both municipalities come from agriculture. The major crops grown in Atok are chayote, coffee, cabbage, and potatoes, while respondents in Tublay have more diverse crops to include spinach, zucchini, Pechay, beans, bell pepper, tomato, broccoli, and carrots and others are growing banana, lemon, and calamansi. Other sources of income include livestock poultry, off-farm, non-farm, and business such as restaurants and sari-sari stores and overseas Filipino workers (OFW).

At present, some portions of the forest areas of the barangay are already converted into agricultural areas and are being planted to vegetables and other root crops. The total agricultural land is being used as croplands planted with other commodities such as cut flowers, vegetables, and root crops (cabbage, Chinese cabbage, chayote, potatoes, and Pechay). Coffee and vegetable crops grown are sold in the market as a source of income.

#### 12.3.1.2.4 Adaptation Capacities

Adaptive capacity is assessed using socioeconomic and technological adaptations. The socioeconomic variables to indicate the ability of the farming community to adjust to climate change include the age of the respondents to represent the human

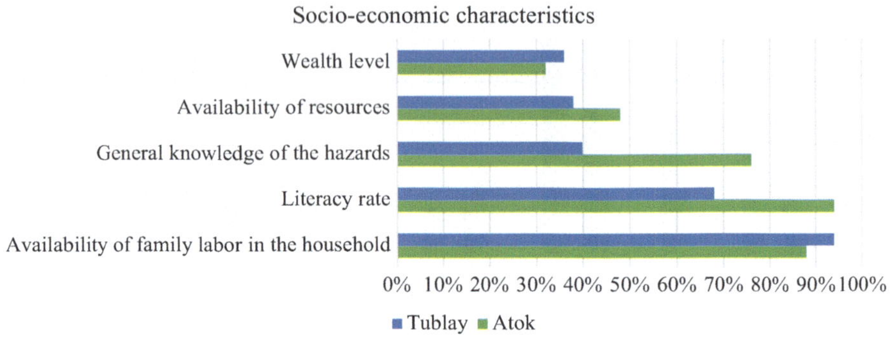


**Fig. 12.5** Socioeconomic characteristics of the adaptive capacity of respondents in Atok and Tublay municipalities of Benguet Province, Philippines

working capital and the available labor in the household, literacy rate, knowledge of the hazards, availability of resources, presence or early warning systems, the system of information dissemination, accessibility, and wealth level (Fig. 12.5).

Availability of family labor is high indicating enough manpower or family members that are capable to help in terms of hazards. Despite the high literacy rate and presence of early warning system, the general knowledge of the hazards in Tublay, Benguet, is moderate. The mode for disseminating information within the community about the hazards is the radio, television, text brigade, IEC materials, and social media. The municipality of Tublay has a good presence and accessibility of support systems from the government, non-government, private, and farmers associations. The ability of the farmers to spend for the cost of adaptation is low and, thus, needs assistance to increase their resilience.

Contributing to the adaptive capacity of the municipality of Atok is the high level of literacy rate and knowledge of climate-related hazards. The municipality has a community early warning system, access to support systems and the dissemination of hazard information is rated good and moderate. Interventions by the local government units and other government and private agencies for the provision of early warning systems, coffee postharvest equipment, availability of transportation, capacitation of the community, and regular information dissemination resulted in high adaptive capacity. However, for the ability of individual farmers to spend for adaptation cost, the rating is low where 20–40% of the total percent population can instantly pay for the expenses to change the damage.

#### 12.3.1.2.5 Technological Adaptations

Majority of the respondents from the interview and Focus Group Discussion done know only a few mitigation and adaptation strategies to combat the effects of the climate hazards for coffee trees. Coffee trees are being planted under shade trees like Alnus, pine, and mango and shade plants like chayote (Table 12.5).

**Table 12.5** Technological adaptations practiced by coffee growers in the production sites of Atok, Benguet, Philippines

| Hazard | Known adaptation |
|---|---|
| Typhoon and strong winds | • Planting coffee under shade trees (such as Alnus, fruit trees, chayote) to provide natural windbreaks and reduce wind damage |
| Erosion/landslide | • Planting coffee under shade trees and implementing agroforestry practices to enhance soil stability<br>• Tree planting near farms to reinforce the soil and minimize erosion<br>• Construction of drainage canals to manage water runoff and prevent soil erosion |
| Pest and diseases | • Manual removal of infested parts and regular monitoring for pest and disease presence<br>• Clearing the surrounding area to reduce pest and disease habitats<br>• Application of smoke under the coffee trees to deter pests<br>• Utilization of biopesticides to control pests and diseases |
| Monsoon rain/ persistent rain | • Planting coffee under shade trees to provide protection from excessive rainfall and reduce soil erosion |
| Drought | • Planting coffee under shade trees to mitigate the effects of direct sunlight and reduce water evaporation from the soil<br>• Application of organic mulches to conserve soil moisture |
| Rise of minimum temperature | • Planting of coffee under shade trees, planting of cash crop |
| Hailstorm | • Planting coffee under shade trees to provide a physical barrier against hail |

Accordingly, it is an effective way to lessen the effect of climate change as the coffee trees will not be greatly affected by the changes in temperature such as the hot temperature during the daytime and cold temperature during the nighttime, modifying the microclimate. Also, the shade trees serve as windbreaks for strong winds during typhoons experienced in June to November and as erosion/landslide control on the farm.

The low yield of coffee due to the rise of minimum temperature in the past years prompts farmers to plant cash crops in months with available rainwater for irrigation to have additional income. Several cash crops being planted include beans, pigeon pea, ginger, and sweet potato, while planting coffee alongside chayote is also very popular where chayote is an added cash crop.

Water tanks constructed by the Local Government Unit and National Irrigation Administration were also helpful in storing water for irrigation during the dry season.

Further, several mentioned adaptations being done by farmers to prevent or stop the attack of pests and diseases are the manual removal of damaged/infested plant parts, clearing of surrounding areas of the farm, and application of smoke under the coffee tree. Some farmers too in Ambassador, Tublay, apply wood vinegar *(Mokusako)* to plants infested with sooty molds and coffee rust and alcohol and vinegar mixture on stems attacked by a stem borer. The adaptive capacity map of the coffee growers in the Municipalities of Atok and Tublay, Benguet, Philippines, remains to be low to moderate (Fig. 12.6).

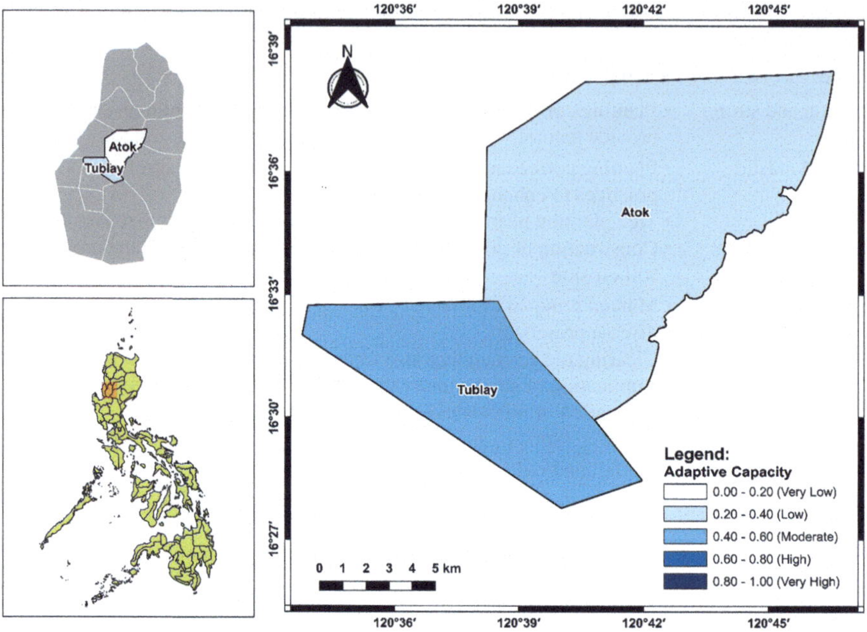


**Fig. 12.6** Adaptive capacity map of the coffee growers in the Municipalities of Atok and Tublay, Benguet, Philippines

**Table 12.6** Vulnerability index for Atok, Benguet

| | Typhoon and strong winds | Erosion/ landslide | Pest and diseases | Monsoon rain/ persistent rain | Drought | Rise of minimum temperature | Hailstorm |
|---|---|---|---|---|---|---|---|
| Qualitative description | Moderately vulnerable | Moderately vulnerable | Moderately adaptive | Moderately vulnerable | Highly adaptive | Highly adaptive | Moderately adaptive |
| *Overall Vulnerability Index* | *0.227* | | | | | | |
| *Over-all Qualitative Description* | *Moderately adaptive* | | | | | | |

#### 12.3.1.2.6 Vulnerability Assessment of the Coffee-Based Agroforestry in Atok, Benguet

The coffee production area in Atok, Benguet, has high exposures to typhoons and strong winds, monsoon, or prolonged rain that resulted in erosion and landslides (Table 12.6). Coffee trees are generally grown in sloping areas of about 30–100% degrees and flowering and fruiting stages are the most sensitive phenology of the coffee trees to most of the hazards as a result of the biological assessment. Social sensitivity indicates moderate to a high rating in most of the indicators used while low on the availability of resources and wealth level. Based on the social adaptive capacity assessment, the resilience of the farmers increases due to the available resources and interventions from collaborators and partners and local technologies adapted by the coffee growers such as growing coffee trees under shade trees or agrisilviculture system. The limited technology to adapt to hailstorms affected the total vulnerability score of the community.

##### 12.3.1.2.7 Vulnerability Assessment of the Coffee-Based Agroforestry in Tublay, Benguet

The production areas of coffee in the municipality of Tublay is highly exposed to typhoon and strong winds, monsoon or prolonged rains, and erosion or landslide due to the sloping condition of the locality (Table 12.7) and has a moderate to high climate risk vulnerability index (Fig. 12.7). The rise of minimum temperatures is attributed to the low elevation of the community. The biologically sensitive stage of the coffee crop is from vegetative to fruit setting stage with 40–80%. In most of the socioeconomic indicators, the community indicated a low wealth level and cannot readily spend for the cost of adaptations. However, the coffee growers are using shade trees and cultural management practices such as planting cash crops to augment their income from coffee and to increase their resilience that attributes to the moderate adaptive capacity of the farmers (Fig. 12.6).

### 12.3.2 Discussion

Arabica is normally grown in cooler, elevated areas of the tropics and sub-tropics at 3300 ft (1000 m) or more above sea level. Growth, yield, and good quality coffee are influenced by environmental factors such as elevation and temperature, rainfall

**Table 12.7** Vulnerability index for Tublay, Benguet

| | Typhoon and strong winds | Erosion/ landslide | Pest and diseases | Monsoon rain/ persistent rain | Drought | Rise of minimum temperature |
|---|---|---|---|---|---|---|
| Qualitative description | Moderately vulnerable | Moderately vulnerable | Moderately adaptive | Moderately vulnerable | Highly adaptive | Moderately adaptive |
| *Overall Vulnerability Index* | *0.160* | | | | | |
| *Over-all Qualitative Description* | *Moderately adaptive* | | | | | |

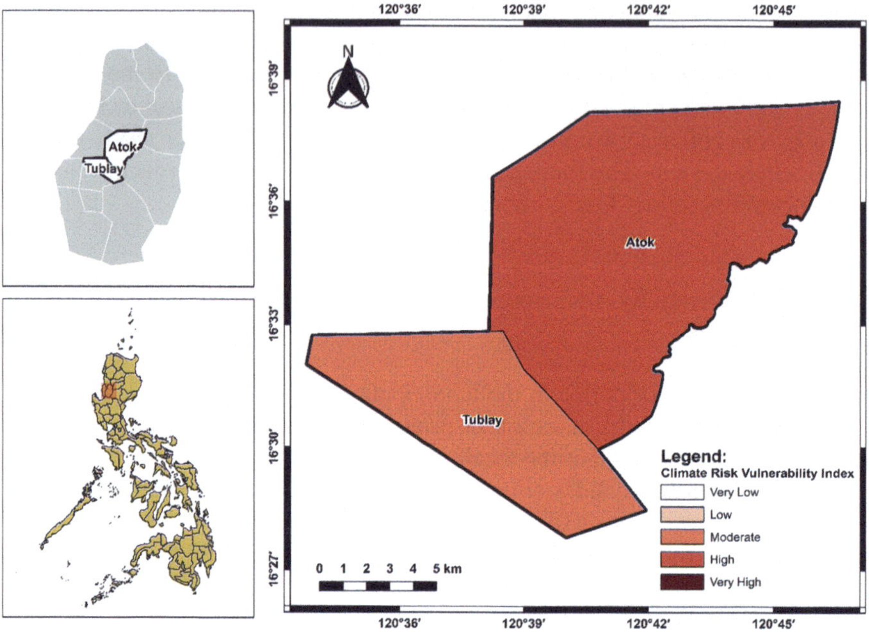


**Fig. 12.7** Climate risk vulnerability map of Arabica coffee production in Atok and Tublay, Benguet, Philippines

and water supply, soil, aspect and slope as well as the proper choice of variety. Also, coffee is a long-term crop with a lifespan of more than 10 years, and very much longer under good management.

The microclimate of coffee under Alnus has an average reduction in air temperature from 0.9 to 1.1 °C than in the open field; 2.4 °C lower soil temperature as high as 61% reduction in wind velocity, and 49% decrease in light intensity. Coffee under mango shade showed significant reduction in air and soil temperature, wind velocity, and light intensity. Chayote as a shade crop reduced the air and soil temperature, light intensity and the wind velocity.

The recorded measurements, however, for air temperature were beyond the optimum mean annual temperature range for Arabica coffee which is 18–21 °C (Alègre 1959). The reduced air temperature was mainly due to the reduced direct incidence of solar radiation on the coffee canopy (Campanha et al. 2005). Under the shade, however, reduced air temperature and light intensity increased the percentage relative humidity in the air around coffee plants (Bote and Struik 2010).

Thus, as concluded by Lin (2006), the reduced microclimate variability of shaded coffee sites suggests that shade trees mitigate microclimate variability. The reduction in microclimate variability will be useful in the future as climate variability increases because mitigating climate fluctuations will maintain crops at their optimal temperature and protect crops from reduced yield and associated risks (Semenov and Porter 1995).

The shade plants regulate the microclimate of the coffee plants when integrated with or under pine, Alnus, mango, and chayote. Significant reduction was noted in the minimum and maximum air temperature, soil temperature, wind velocity, and light intensity, while no differences in the relative humidity. Relative humidity is very variable which could be attributed to the cloudiness and aspects/exposure of the experimental site within the study area. Variations on the cloud cover in the experimental areas under open field and under shade trees resulted to equal or higher light penetration and not an effect of the shade trees. The microclimate changes are favorable for the growth and development of the coffee trees.

The major physiological benefits that coffee receives from shade trees is an ideal microclimate for growth and production, such as reduce wind speed and soil temperature extremes and buffering of humidity and soil moisture availability. Aside from increasing biodiversity, shade trees reduce the quantity and quality of transmitted light and hence avoid over-bearing for sustenance of coffee plantations (Beer et al. 1998). Conversely, the cool climate of the province is suitable for the production of coffee. Coffee is a plantation crop well adapted to different eco-physiological conditions of the tropics. *Coffea arabica* is adapted to cooler temperatures of the tropical highlands above 1000 m altitude along the equator and *C. canephora* to lowlands (Pohlan and Janssens 2010).

#### 12.3.2.1 Climate Change Resilience Assessment of Arabica Coffee Production

Arabica coffee prefers a cool temperature with an optimum daily temperature of 20–24 °C (Winston et al. 2005), but the optimum mean annual temperature range is 18–21 °C (DaMatta 2004). Higher temperatures of 30 °C cause plant stress leading to cessation of photosynthesis and may cause abortion of flowers and can provoke loss of physical and beverage quality. Mean temperatures of less than 15 °C limit plant growth and are considered suboptimal. As Arabica coffee is susceptible to frost damage, the use of shade trees will reduce the incidence. Ideal rainfall for Arabica coffee is greater than 47–60 in. (1200–1500 mm) per year, though a range between 800 and 4200 mm remains acceptable. Both the total amount and the distribution pattern are important. Rain should be uniformly distributed over 7–9 months of the year. Coffee requires adequate water during the growing and cropping period; however, it also requires a dry stress period followed by sufficient rain or irrigation to promote uniform flowering and a good fruit set. Water requirements can be reduced by the use of suitable, well-established shade trees and mulch (Winston et al. 2005; Pohlan and Janssens 2010).

## 12.4 Summary and Conclusion

The microclimate of coffee under Alnus has an average reduction in air temperature from 0.9 to 1.1 °C than in the open field; 2.4 °C lower soil temperature as high as 61% reduction in wind velocity and 49% decrease in light intensity. Coffee under mango shade showed significant reduction in air and soil temperature, wind velocity, and light intensity. Chayote as a shade crop reduced the air and soil temperature, light intensity and as low as 1.5 kph wind velocity was recorded. Overall, Alnus trees as shade plants for coffee trees provide a favorable microclimate environment to coffee trees as it decreases the maximum and minimum measurements during daytime and nighttime, respectively, resulting in a greater production potential for the coffee trees.

A review by Beer et al. (1998), as cited by Muleta et al. (2007), revealed that the upper limit of acceptable shade for cotrees (a C3 plant) is between 40% and 70%. As such, regular practice of pruning or thinning (for shade trees) and removal of leaves (for chayote) and with the aid of an instrument measuring light intensity is necessary to maintain the required amount of sunlight that will reach coffee crops from the canopy level.

The shade plants regulate the microclimate of the coffee plants when integrated with or under pine, Alnus, mango, and chayote. Significant reduction was noted on the minimum and maximum air temperature, soil temperature, wind velocity, and light intensity while no differences in the relative humidity. The microclimate changes are favorable for the growth and development of the coffee trees.

The Arabica coffee production in Atok (Caliking and Topdac) had high exposures to typhoons and strong winds, erosion/landslide, and monsoon or persistent rain and the area is more sensitive to the effects of typhoons and pests and diseases as these hazards have a big effect to yield. The adaptive capacity of the community is moderately high (30–31 out of 45), due to the availability of resources, support institutions, labor, and a high percentage of households implementing the adaptation technologies on the climate hazards mentioned. The potential impact of typhoons and strong winds, erosion/landslide, monsoon rain, and pest and diseases can be higher due to the geographical location of Atok and few adaptation technologies known to coffee growers. The community, therefore, is moderately vulnerable to typhoons and strong winds, erosion/landslide, and monsoon/persistent rain. Overall, the community is moderately adaptive to climate change.

Barangay Ambassador and Daclan in Tublay had high exposures to typhoons and strong winds, erosion/landslide, and monsoon/persistent rain. The area is more sensitive to the effects of typhoons and landslides and rise of minimum temperature since half of the coffee-growing areas is at low elevations which is expected to be hotter than the average temperature. The adaptive capacity of the community is moderately high, due to the availability of support systems both private and public, labor and resources. The potential impact of typhoons and strong winds, erosion/landslide, monsoon rain and rise of minimum temperature can be higher due to the topographical location and few adaptation technologies implemented. The community, therefore, is moderately vulnerable to typhoons and strong winds, erosion/landslide and monsoon/persistent rain. Overall, the community is moderately adaptive to climate change.

To attain high suitability and sustainability of the coffee production areas, appropriate climate assessment should be done. Among the agroforestry systems assessed, coffee grown under Alnus trees as shade plants have a more favorable microclimate environment that favors greater production potential and protection from climate hazard exposures.

**Acknowledgments** The authors would like to thank the Commission on Higher Education (CHED) for the fund support for this project, the cooperating agencies, and the farmer cooperators of the municipalities of Atok, Tublay and La Trinidad, Benguet, and the Benguet State University.

## References

Alègre C (1959) Climates et caféiers d'Arabie. Agron Trop 14:23–58

Alemu MM (2015) Effect of tree shade on coffee crop production. J Sustain Dev 8(9):66

Atallah SS, Gomez MI, Jaramillo J (2018) A bioeconomic model of ecosystem services provision: coffee berry borer and shade-grown coffee in Colombia. Ecol Econ 144:129–138

Barradas V, Fanjul L (1986) Microclimate characterization of shaded and open-grown coffee (Coffea Arabica L.) plantations in Mexico. Agric For Meteorol 38:101–112

Beer J, Muschler R, Kass D, Sommariba E (1998) Shade management in coffee and cacao plantations. Agrofor Syst 38:139–164

Bote A, Struik P (2010) Effects of shade on growth, production and quality of coffee (Coffea Arabica) in Ethiopia. J Hortic For 3(11):336–341

Campanha MM, Santos RHS, de Freitas GB, Martinez HEP, Garcia SLR, Finger FL (2005) Growth and yield of coffee plants in agroforestry and monoculture systems in Minas Gerais, Brazil. Agrofor Syst 63:75–82. https://doi.org/10.1023/B:AGFO.0000049435.22512.2d

Caramori PH, Ometto JC, Villa Nova NA, Costa JD (1986) Efeitos do vento sobre mudas de cafeeiro Mundo Novo e Catuaí Vermelho. Pesqui Agropecuaria Bras 21:1113–1118

Cerda R, Allinne C, Gary C, Tixier P, Harvey C, Krolczyk L, Mathiot C, Clément E, Aubertot J, Avelino J (2017) Effects of shade, altitude and management on multiple ecosystem services in coffee agroecosystems. Eur J Agron 82(Part B):308–319

Daipan L, Lumbres R, Lingaling M, Talkasen L, Godio Q, Mendoza G (2023) Climate risk vulnerability assessment on the production of heirloom rice (Oryza sativa L.) and Arabica coffee (Coffea arabica L.) in Mountain Province, Philippines. Sci Eng J 16(1):28–39

Damatta F (2004) Ecophysiological constraints on the production of shaded and unshaded coffee: a review. Field Crop Res 86:99–114

FAO (2005) The importance of soil organic matter. http://www.fao.org/3/a-a0100e.pdf. Accessed 5 Nov 2019

Garcia JNM, Wagan AM, Medina SM (2011) Vulnerability and adaptive capacity assessment in different agroecosystem (VAST-agro). Agricultural Systems Cluster, University of the Philippines, Los Banos

Getachew M, Tolassa K, Defrenne P, Verheyen K, Tack AJM, Hylander K, Ayalew B, Boeck P (2022) The relationship between elevation, soil temperatures, soil chemical characteristics, and green coffee bean quality and biochemistry in southwest Ethiopia. Agron Sustain Dev 42:61. https://doi.org/10.1007/s13593-022-00801-8

Intergovernmental Panel on Climate Change (IPCC) (2014) Climate change 2014: impacts, adaptation, and vulnerability. Part A: global and sectoral aspects. In: Field CB, Barros VR, Dokken

DJ, Mach KJ, Mastrandrea MD, Bilir TE, Chatterjee M, Ebi KL, Estrada YO, Genova RC, Girma B, Kissel ES, Levy AN, MacCracken S, Mastrandrea PR, White LL (eds) Contribution of working group II to the fifth assessment report of the intergovernmental panel on climate change. Cambridge University Press, Cambridge, pp 1–32

Jha S, Bacon CM, Philpott SM, Mendez VE, Läderach P, Rice RA (2014) Shade coffee: update on a disappearing refuge for biodiversity. Bioscience 64:416

Jose S (2009) Agroforestry for ecosystem services and environmental benefits: an overview. Agrofor Syst 76:1–10

Kiseve S (2012) Evaluation of legume cover crops intercropped with coffee. University of Nairobi, Nairobi

Lin B (2006) Agroforestry management as an adaptive strategy against potential microclimate extremes in coffee agriculture. Agric For Meteorol 144:85–94

López-Bravo D, Virginio-Filho E, Avelino J (2012) Shade is conducive to coffee rust as compared to full sun exposure under standardized fruit load conditions. Crop Prot 38:21–29

Morais H, Caramori P, Ribeiro A, Gomes J, Koguishi M (2006) Microclimatic characterization and productivity of coffee plants grown under shade of pigeon pea in Southern Brazil. Braz Agric Res 41:769–770

Muleta D, Assefa F, Nemomissa S, Granhall U (2007) Composition of coffee shade tree species and density of indigenous arbuscular mycorrhizal fungi (AMF) spores in Bonga natural coffee forest, Southwestern Ethiopia. For Ecol Manag 241:145–154. https://doi.org/10.1016/j.foreco.2007.01.021

Muschler R (2001) Shade improves coffee quality in a sub-optimal coffee-zone of Costa Rica. Agrofor Syst 51(2):131–139

Pezzopane JRM, de Souza PS, de Souza Rolim G, Gallo PB (2011) Microclimate in coffee plantation grown under grevillea trees shading. Acta Scient Agron 33(2):201–206

Pham Y, Reardon-Smith K, Mushtaq S, Cockfield G (2019) The impact of climate change and variability on coffee production: a systematic review. Clim Chang 2019(156):609–630. https://doi.org/10.1007/s10584-019-02538-y

Pinto HS, Assad E, Junior JZ, Marin FR (2008) A Nova Geografia da Produçao Agrícola no Brasil. Embrapa Informatica Agropecuaria, Unicamp

Pohlan HAJ, Janssens MJJ (2010) Growth and production of coffee in soils, plant growth and crop production. In: Encyclopedia of life support systems (EOLSS), developed under the auspices of the UNESCO. EOLSS Publishers, Oxford. http://www.eolss.net. Accessed 19 Nov 2019

Semenov MA, Porter JR (1995) Climate variability and the modeling of crop yields. Agric For Meteorol 73(3–4):265–283. https://doi.org/10.1016/0168-1923(94)05078-K

Staver C, Guharay F, Monterroso D (2001) Designing pest-suppressive multistrata perennial crop systems: shade-grown coffee in Central America. Agrofor Syst 53:151–170

Tad-Awan BA, Pablo JP, Macanes V, Basalong A (2010) Germplasm collection, diversity analysis, evaluation and gene banking of Coffea spp. In: Different agro-ecological zones of Benguet. BSU AIHR proceedings

Winston E, De Laak JO, Marsh T, Lempke H, Aung O, Nyunt T, Chapman K (2005) Arabica coffee manual for Myanmar. Food and Agriculture Organization of the United Nations, Rome

# Chapter 13
# Application of Satellite Remote Sensing Techniques to Identify Renewable Energy Sources for Agricultural Production and Energy Security in Lahendong, North Sulawesi, Indonesia: An Approach to Reduce Global Warming

**Mohammad Aftaf Muhajir, Kudang Boro Seminar, Leopold Oscar Nelwan, Harry Imantho, and Rizky Mulya Sampurno**

**Abstract** Addressing the energy problem and the environmental damage caused by global warming requires a new, sustainable, and eco-friendly energy source. The diminishing presence of fossil fuels and other nonrenewable energy sources underscores the increasing importance of harnessing and advancing renewable energy. The National Energy Policy for 2006–2025 in Presidential Regulation Number 5 of 2006 states that 17% of the country's energy supply must come from renewable sources. Geothermal, hydropower, biomass, solar, and marine energy are all innovative and renewable energy sources that are available in Indonesia. Based on studies, Lahendong is an area that has geothermal energy potential that can be utilized to meet energy needs. However, the existence of agricultural industry in a potential geothermal area also plays an importance role in providing sustenance for humans, as well as requiring significant energy and water inputs. This study discusses the

---

M. A. Muhajir
Information Technology for Natural Resources Management, Graduate School, IPB University, Bogor, Indonesia

K. B. Seminar (✉) · L. O. Nelwan
Department of Mechanical and Biosystem Engineering, Faculty of Agricultural Technology, IPB University, Bogor, Indonesia
e-mail: kseminar@apps.ipb.ac.id

H. Imantho
Earth Observatory and Change Section, SEAMEO BIOTROP, Bogor, Indonesia

R. M. Sampurno
Graduate School of Science and Technology, University of Tsukuba, Tsukuba, Ibaraki, Japan

Department of Agricultural and Biosystem Engineering, Faculty of Agroindustrial Technology, Universitas Padjadjaran, Jatinangor, Sumedang, Indonesia


T. Ahamed (ed.), *Remote Sensing Application II*, New Frontiers in Regional Science: Asian Perspectives 77, https://doi.org/10.1007/978-981-97-1188-8_13

utilization of remote sensing data to identify the geothermal area and the potential of land conversion for agricultural production and energy security. The methodology of the study involved computing radiative heat flux (RHF) that associated with land surface temperature (LST) and normalized difference vegetation index (NDVI). Subsequently, the NDVI classification was performed to determine the emissivity value. The process also entailed extracting lineaments from geological maps to assess fault and fracture density (FFD). Furthermore, we used a weighted overlay technique, such as the analytic hierarchy process (AHP), to determine potential geothermal areas. The inputs for this analysis included LST, FFD, NDVI, and the distance from surface manifestations. The results revealed the successful identification of 9 geothermal anomaly areas as surface manifestations, along with the recognition of 15.02 square kilometers of high-potential geothermal zones, estimating an energy output of 141.56 MW. The research also found potential agricultural uses for the study area, including heating greenhouse, agro-processing, and warming soil for outdoor root systems.

**Keywords** Agriculture · Energy estimation · Geothermal area · Renewable energy · Lahendong

## 13.1 Introduction

Uncontrolled population increases have put energy resources under tremendous pressure, and the world is reaching the threshold of an energy crisis (IEA 2022). To overcome the problem of providing the energy and minimizing the effect of environmental damage due to global warming, renewable and environmentally friendly alternative sources of energy are strongly needed. The National Energy Policy for 2006–2025 in Presidential Regulation Number 5 of 2006 stated that around 17% of the national energy supply must come from renewable sources. Renewable energy can be essential for the future. Some of the potential resources are available in Indonesia such as geothermal, biomass, marine energy, hydropower, and solar energy.

The islands in Indonesia are characterized by their location on one of the most active tectonic plates in the world. They are situated between the Indo-Australian, Pacific, Philippine, and Eurasian tectonic plates. This strategic place provides advantages as a country rich in geothermal energy. It is reported that 285 regional points are distributed along the volcanic arc, each with the potential to produce approximately 29,215 GW of electricity (Ministry of Energy and Mineral Resources 2012).

Geothermal systems usually can be found in the earth's crust in areas that have unusually high heat flows. The increased geothermal heat is often attributed to the existence of young igneous formation or hot rock found at lower crustal levels (Rybach 1981; DiPippio 2005). Then the warmth is propagated to the earth's surface by the convection of groundwater, forming a hydrothermal system. In this system, surface water circulates down to depth, where that water becomes heated before rising back to the surface through an underground network or closely spaced fracture or other permeable rock formations. Once the heat reaches the surface, they can manifest as distinctive geothermal phenomena like fumaroles, mud pots, geyser, and hot spring.

Remote sensing can be used to study geothermal features from temperature data. This approach allows for mapping and quantifying temperature anomalies associated with the characteristic of geothermal surfaces. Thermal infrared (TIR) remote sensing also has established as an effective tool for large-scale geothermal exploration, aiding in the initial identification of potential areas for more depth ground-based studies (Hodder 1970; Lee 1978). Furthermore, TIR remote sensing has found applications in the documentation and long-term observing of thermal characteristic related to protected and established geothermal structures (Seielstad and Queen 2009).

Lahendong, a part of North Sulawesi, has great potential for geothermal. However, until now, the area has had a serious problem due to the electricity supply. Therefore, the government-assigned Pertamina, a state-owned enterprises, to do exploration in advance about geothermal in that area which was expected can supply electricity (Polakitan 2018). Although fulfilling the electrical energy requirement, the agricultural industry which predominantly situated in rural and hilly regions like Lahendong, plays a crucial role demanding significant energy and water inputs that come from geothermal and hydrogeothermal.

Geothermal can not only be used as a source of energy but can also be utilized in agriculture, one of which is through the utilization of soil sulfur, nutrients, temperature, and hot water flow. Susanti et al. (2005) reported that in the hot spring area of West Sumatra, Indonesia, showed that there is a linear relationship between the structure and composition of vegetation and environmental factors such as temperature, pH, and sulfur content. Meanwhile, Marwan (2015) stated that soil composition such as resistivity is very closely related to plant vegetation, besides that surface conditions in the geothermal field dominated by relatively high conductivity zones (Marwan et al. 2017). Therefore, this study intends to determine geothermal area and estimate the potential energy using remote sensing data and identify the potential land replacements for agricultural production and energy security.

### *13.1.1 Geothermal System in Indonesia*

Based on the geological order, geothermal systems in Indonesia are divided into three categories: volcanic, volcanic-tectonic, and nonvolcanic. The volcanic is commonly located in the quaternary volcanic arc extending from Sumatra, Java, Bali and Nusa Tenggara, Maluku, and North Sulawesi. The formation of this geothermal system usually consists of intermediate volcanic rocks (andesite-basalt) to acidic rocks and generally has a reservoir of around 1.5 km with temperature (around 250–370 °C).

The volcanic–tectonic geothermal system, which is associated with the graben structure and the volcanic cone, is commonly found in the Sumatran region along the

Sumatra fault (Sesar Semangko). The nonvolcanic geothermal lacks a direct connection related to volcanism and is generally outside the quaternary volcanic belt.

The nonvolcanic environment in western part of Indonesia is predominantly prevalent in the eastern part of Sundaland (Sunda exposure). This is due to the dominated rocks that constitute the Asian continental crust, including metamorphic rocks and sediments. In eastern part of Indonesia, nonvolcanic environments are observed in regions such as the arms and legs of Sulawesi and the Maluku islands, extending to Irian. These areas are characterized by the prevalence of granitic rocks, metamorphic rocks, and marine sediments (Kasbani 2009).

### *13.1.2 Direct Use Geothermal for Agricultural Applications*

In contrast to geothermal force age, in which heat energy is changed over to power, direct-use applications use heat energy straightforwardly to achieve an expansive scope of purpose. The temperature scope of these applications is somewhere in the range of 10 and 150 °C. Given the pervasiveness of this temperature run in the shallow subsurface, these kinds of uses of geothermal energy can possibly be introduced anyplace that has adequate fluid accessible (Glassley 2014).

Indonesia, a country with abundant high-enthalpy geothermal resources, has been working to increase geothermal energy for power generation. Indonesia is also blessed with a vast of reserve of low-to-medium geothermal assets, including underground aquifers and natural geothermal wells, which can be useful for direct utilization. In addition to the geothermal resources, direct use applications additionally use energy from unexploited brine and little limit creation wells. Typically, the geothermal resources in Indonesia are found in hilly area with agricultural land, forests, spas, and other places that require heat for their processes or activities (Surana et al. 2010).

In agriculture, geothermal water provides essential warmth and moisture. Agricultural systems can transport heated water to cooler soil, enabling the cultivation of crops. Geothermal water can also be channeled into greenhouses to provide warmth and maintain moisture levels. This clearly shows that geothermal farming can work in areas with geothermal resources. However, lower temperature geothermal water can also be utilized for things like generating power or heating home.

Geothermal energy for farming in Indonesia started in 1999 when BPPT (Agency for the Assessment and Application of Technology) partnered with Pertamina to develop a pilot plant using geothermal directly in mushroom farming, specifically in geothermal field like Kamojang in West Java. In another geothermal field, Lahendong, which is allocated to a palm sugar estate. The nongovernmental organization Yayasan Masarang collaborated with Pertamina to introduced a large-scale facility for direct geothermal energy for palm sugar production, initially with a capacity of 1 ton with a capacity of 12 ton per day, utilizing flashed steam from the isolated hot water (saline solution). Currently, the facility is operating at a capacity of only 1 ton per day.

In 2003 and 2008, BPPT built pilot plants for the direct utilization of geothermal energy in Way Ratai (Lampung Province) for drying copra (coconut meat) and cocoa. They achieved this using a surface natural geothermal well and a man-made shallow geothermal well, respectively. Way Ratai is an underdeveloped geothermal field in Lampung Province, located in a coconut plantation area. Way Ratai boasts numerous naturally occurring shallow wells with temperatures ranging from 80 to 98 °C. In 2003–2004, BPPT implemented a pilot plant for using the natural geothermal well to dry copra in this field, with a capacity of 200 kg copra per batch.

Currently, the primary use of direct geothermal application is for balneology, bathing, and swimming pool. Some agricultural utilizations such as mushroom cultivation, palm sugar production, coconut meat (copra) and cocoa drying, are also being implemented, but their amounts are very limited compared to the potential all around the country. The prospects applications are extremely promising (Surana et al. 2010).

## 13.2 Methodology

### *13.2.1 Study Area*

The study area is Lahendong area, Tomohon Selatan subdistrict, Tomohon city, North Sulawesi Province, Indonesia. Lahendong is one of the urban villages which is in 124°48′E–124°50′20″E longitude and 1°15′45″N–1°17′15″N latitude. To the east, there is Lake Linow which is a volcanic lake that is a tourist attraction for the people of Tomohon and residents outside the city of Tomohon (Fig. 13.1).

### *13.2.2 Data*

#### 13.2.2.1 Primary Data

Remote sensing data used come from two satellite: Landsat 8, operational land imager (OLI) and thermal infrared (TIR) sensor and Terra satellite, ASTER sensor (Table 13.1). The inversion of LST image both day and night was used to quantitatively analysis to recognize the detection of the area with thermal abnormality. Therefore, influence of solar radiation on land surface can be eliminated. Then, all images were corrected both radiometric and geometric.

#### 13.2.2.2 Secondary Data

A topographic map at the scales of 1:50,000 acquired from Geospatial Information Agency of Indonesia or BIG was used to recognize roads, rivers, and settlements. Topographic map from BIG was used to obtain built-up area. Figure 13.2 depicts the area of Lahendong, where vegetation comprising forests, cropland, shrub, estates,

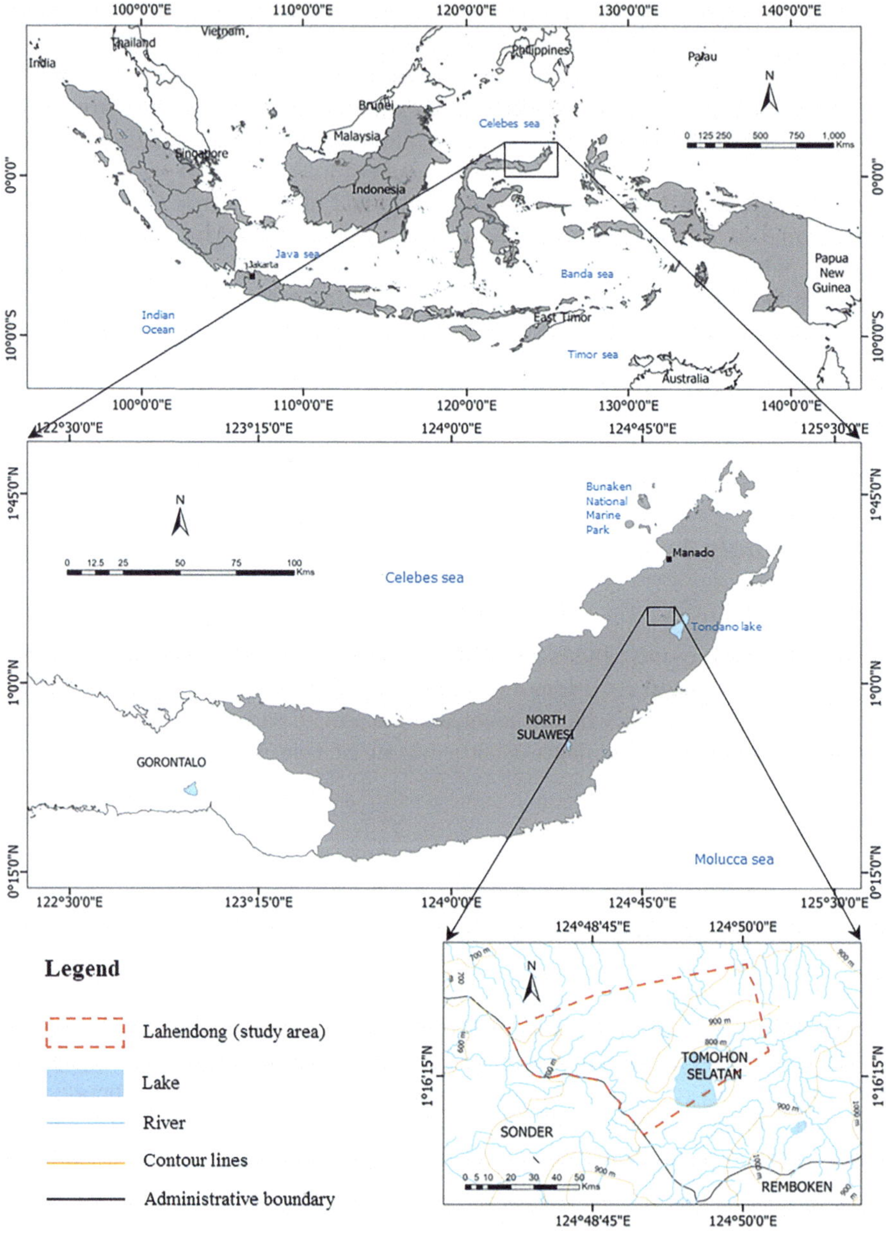


**Fig. 13.1** The study area of Lahendong, North Sulawesi, Indonesia

**Table 13.1** Landsat 8 and ASTER data used and the date of acquisition

| Platform | Acquisition date | Path | Row |
|---|---|---|---|
| Landsat 8 | 05/21/2015 | 112 | 59 |
| Landsat 8 | 06/08/2016 | 112 | 59 |
| Landsat 8 | 12/20/2017 | 112 | 59 |
| Landsat 8 | 09/02/2018 | 112 | 59 |
| ASTER TIR (Nighttime) | 09/22/2015 | 220 | 185 |
| ASTER TIR (Nighttime) | 03/09/2016 | 220 | 185 |
| ASTER TIR (Nighttime) | 11/14/2017 | 220 | 185 |
| ASTER TIR (Nighttime) | 12/03/2018 | 220 | 185 |
| ASTER TIR (Daytime) | 08/10/2015 | 220 | 185 |
| ASTER TIR (Daytime) | 06/09/2016 | 220 | 185 |
| ASTER TIR (Daytime) | 11/03/2017 | 220 | 185 |
| ASTER TIR (Daytime) | 08/02/2018 | 220 | 185 |

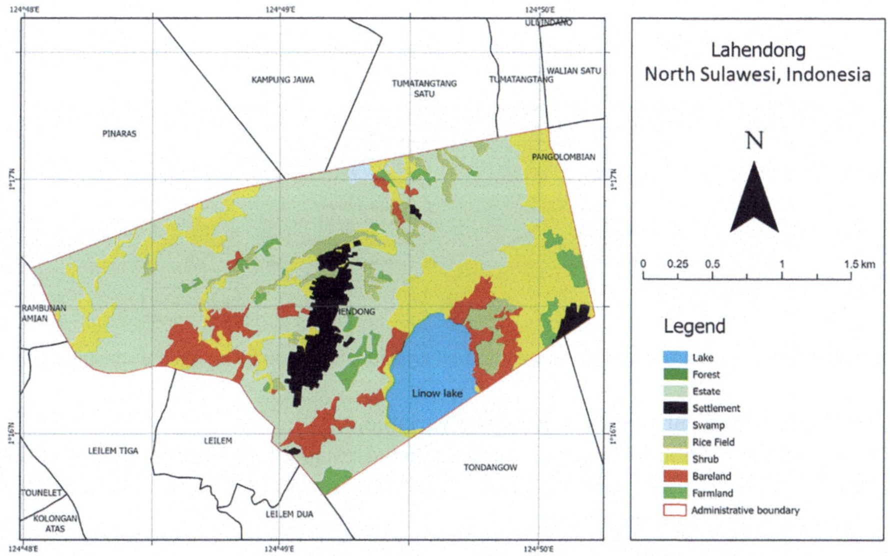


**Fig. 13.2** Topographic map of Lahendong

and rice fields cover nearly 70% of the total area. In addition, geologic map for analysis of geological profile taken from Siahaan (2005).

### *13.2.3 Method of Geothermal Area Identification*

Figure 13.3 is the framework for this study which covers five stages:

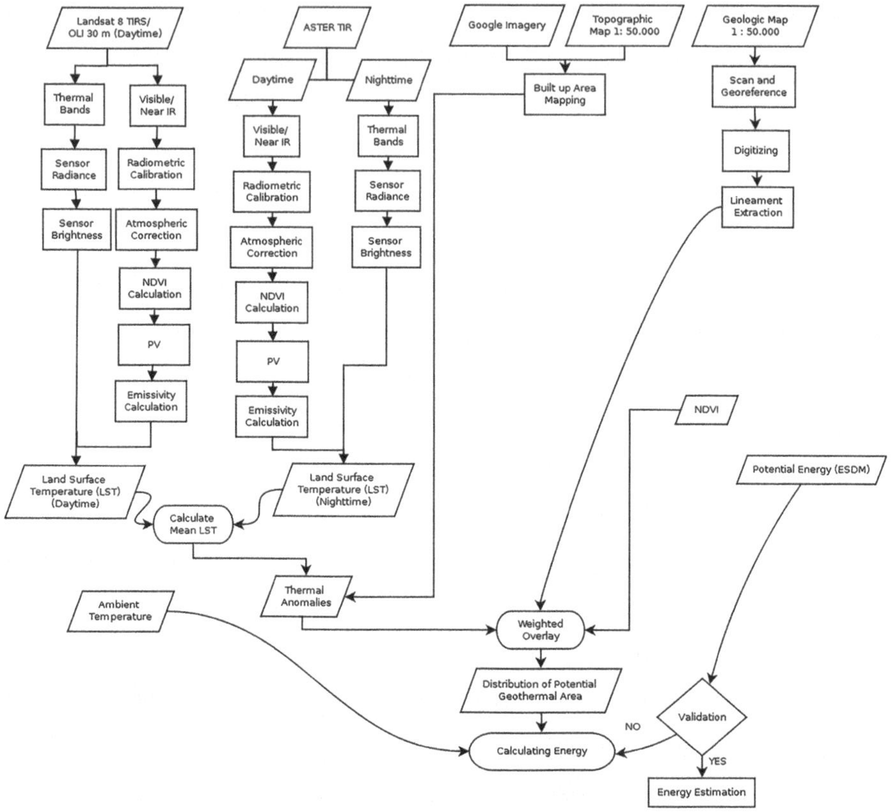


**Fig. 13.3** The research framework for geothermal area identification using remote sensing data

1. Collection and entry of data used as a variable.
2. Preparation of models based on empirical studies and the determination of common criteria for recognizing geothermal system symptoms in research area including image processing.
3. Weighting in determining geothermal prospect area.
4. Overlapping the map of the geothermal determination criteria with the weight.
5. Calculation of the potential of speculative geothermal resources for each research area (includes validation of model results with geothermal potential data from Ministry of Energy and Mineral Resources).

### *13.2.4 Identification of Geothermal Manifestation*

First, all remote sensing data were corrected (atmospheric, radiometric, and geometric). Second, LST and emissivity determined for Landsat data and ASTER data by an inversion of the Planck function. We used band 14 to get LST from ASTER because it has specific TIR bands that need a single-channel (SC) algorithm to acquire. Both day and night LST were analyzed statistically to explore the method of highlighting the anomaly of geothermal. Geothermal was then detected by using the difference of temperature background and temperature anomalies through a specific threshold. Finally, the anomalies of geothermal were analyzed by using medium-resolution preprocessed imagery.

### 13.2.5 *Image Preprocessing*

Radiometric calibration and atmospheric correction of Landsat 8 image were conducted to red, near infrared (NIR), and thermal infrared (TIR) or band 4, band 5, and bands 10, respectively. The orthorectified or terrain-corrected imagery of Landsat 8 were provided by the United States Geological Survey (USGS). Then, radiometric calibration was used to get atmosphere radiance data from the digital number (DN) data that were recorded by the remote sensors (Eq. 13.1). The radiance multiplicative scaling factor ($M_L$) and radiance additive scaling factor ($A_L$) used to compute the spectral radiance were also acquired from reference (USGS 2015).

$$L_1 = M_L Q_{cal} + A_L \tag{13.1}$$

$$T = \frac{K_2}{\ln\left(\frac{K_1}{L_1} + 1\right)} \tag{13.2}$$

$L_\lambda$ is Top of Atmosphere (TOA) spectral radiance (W m$^{-2}$ sr$^{-1}$ μm$^{-1}$), $M_L$ is band-specific multiplicative rescaling factor, $A_L$ is band-specific additive rescaling factor, and $Q_{cal}$ is quantized and calibrated standard product pixel values. Both multiplication and additive rescaling factor came from metadata of Landsat image.

Equation (13.2) shows the formula to get temperature from TIR data (USGS 2015). TIR data in spectral radiance transformed to brightness temperature by using thermal constants. $T$ is satellite brightness temperature (K), $L_\lambda$ is TOA spectral radiance (W m$^{-2}$ sr$^{-1}$ μm$^{-1}$), $K_1$ and $K_2$ are band-specific thermal conversion constant derived from thermal infrared image metadata.

The surface temperatures from ASTER image are also calculated from spectral radiance that derived from conversion of DNs to radiance (Eq. 13.3) (Boori et al. 2015). $L_\lambda$ is the spectral radiance, DN is the TIR band digital numbers, and UCC is the issued unit conversion coefficients (0.005225 W m$^{-2}$ sr$^{-1}$ μm$^{-1}$). For the temperature (Eq. 13.4), the coefficient of $K_1$ and $K_2$, in Table 13.2 are associated with ASTER TIR bands.

$$L_\lambda = (\mathrm{DN} - 1) \times \mathrm{UCC} \tag{13.3}$$

**Table 13.2** $K_1$ and $K_2$ coefficients of the ASTER TIR bands

| Band | $K_1$ (W m$^{-2}$ sr$^{-1}$ µm$^{-1}$) | $K_2$ (Wm$^{-2}$sr$^{-1}$µm$^{-1}$) |
|---|---|---|
| Band 10 | 3047.47 | 1736.18 |
| Band 11 | 2480.93 | 1666.21 |
| Band 12 | 1930.80 | 1584.72 |
| Band 13 | 865.65 | 1349.82 |
| Band 14 | 649.60 | 1274.49 |

$$T = \frac{K_2}{\ln\left(\frac{K_1}{L_1} + 1\right)} \tag{13.4}$$

### 13.2.5.1 Landsat 8 Land Surface Emissivity and Temperature

The transformations of radiance to reflectance and atmospheric correction of Landsat 8 OLI have been done using the DOS1 atmospheric corrections. In this study, the land surface emissivity (LSE) and land surface temperature (LST) were estimated by employing normalized difference vegetation index threshold method (NDVI$^{THM}$). Therefore, atmospheric correction was not conducted to the TIR band. Basically, TIR bands obtained have 100 m resolution, but since 2020, they resampled to 30 by USGS using cubic convolution to fit the OLI spectral bands. The NDVI$^{THM}$ is considered as an approach that has been further modified and developed (Sobrino et al. 2001; Valor and Caselles 1996). Band 10 of Landsat-8 was considered due to instability in the calibration of band 11 (Barsi et al. 2014). Then, the LSE was derived from the relationship between the NDVI and vegetation and soil emissivity (Eq. 13.5).

$$\varepsilon = 0.004P_v + 0.986 \tag{13.5}$$

In Eq. (13.5), $\varepsilon$ represents emissivity, and $P_v$ denotes the proportion of vegetation, which is calculated using NDVI value. Then substituted into Eq. (13.6).

$$P_v = \left(\frac{\text{NDVI} - \text{NDVI}_{\min}}{\text{NDVI}_{\max} - \text{NDVI}_{\min}}\right)^2 \tag{13.6}$$

LST was calculated using the Planck inversion function (Stathopoulou and Cartalis 2007). Where, $\lambda$ represents the wavelength of emitted radiance, BT is sensor brightness temperature, and $p$ is equal to 14,380.

$$\text{LST} = \frac{\text{BT}}{1 + l\left(\frac{\text{BT}}{p}\right)\ln e} \tag{13.7}$$

#### 13.2.5.2 ASTER Land Surface Emissivity and Temperature

Land surface emissivity (LSE) was gained from NDVI approach. This algorithm has been employed to estimate LSE using various sensors of visible and near infrared (VNIR) data (Eq. 13.8). NIR and *R* are the reflectance of the NIR band and red band, respectively.

$$\text{NDVI} = \frac{\text{NIR} - R}{\text{NIR} + R} \tag{13.8}$$

Jiménez-Muñoz et al. (2006) algorithm was used to estimate LSE from NDVI. The algorithm has been tested with in situ data for ASTER TIR band 10–14 (Eqs. 13.9–13.13). $\varepsilon$ is the emissivity of TIR of each channel. Then $P_v$ represents proportion of vegetation or fractional vegetation cover, which is derived from NDVI (Eq. 13.14). $\text{NDVI}_V$ is NDVI of vegetation and $\text{NDVI}_S$ is NDVI of soil, where the value is 0.7 and 0.05, respectively (Xiong et al. 2019).

$$\varepsilon 10 = 0.946 + 0.044 P_v \tag{13.9}$$

$$\varepsilon 11 = 0.949 + 0.041 P_v \tag{13.10}$$

$$\varepsilon 12 = 0.941 + 0.049 P_v \tag{13.11}$$

$$\varepsilon 13 = 0.968 + 0.022 P_v \tag{13.12}$$

$$\varepsilon 14 = 0.970 + 0.020 P_v \tag{13.13}$$

$$P_v = \left( \frac{\text{NDVI} - \text{NDVI}_s}{\text{NDVI}_v - \text{NDVI}_s} \right)^2 \tag{13.14}$$

The Planck function was utilized for calculating the intensity of the thermal radiation. The function presents the quantity of thermal electromagnetic waves that may be radiated by a blackbody below equilibrium circumstances at a recognized heat. As the LSE of a neighborhood is recognized, the LST of a neighborhood is also estimated through the Planck function inversion. The brightness temperature recorded by a sensor in space is calculated with the assumption that the land surface behaves like a black body, which means the land surface has an emissivity of 1. The Planck function validates the emissivity correction of brightness temperature. The Planck function also has been utilized to estimate LST. The emissivity corrected land surface temperature of each band with effective wavelengths of TIR is shown in Table 13.3.

After the LST map was obtained, the next step was the analysis of geothermal anomalies. Geothermal anomalies indicate the emergence of surface manifestation. The surface manifestation can be hot springs, mud pools, hot water pools, ground warm, fumaroles, and others. According to Geological survey of Japan (Noorollahi 2005), the spatial distribution analysis of geothermal wells showed that approximately 97% of geothermal wells were located within 4000 m of hot springs.

**Table 13.3** The effective wavelengths ($\lambda$) of the TIR channels of the ASTER instrument

| Band | Effective wavelength ($\lambda$) in μm |
|---|---|
| Band 10 | 8.287 |
| Band 11 | 8.685 |
| Band 12 | 9.079 |
| Band 13 | 10.659 |
| Band 14 | 11.289 |

Therefore, that distance was then used as the threshold distance to select potential area for geothermal based on the hot spring location. Based on the analysis, buffer conducted as far as 4000 m from the emergence of surface manifestations.

### *13.2.6 Fault and Fracture Density Processing*

The obtained Fault Density and Fracture Density in km/km$^2$ units, from DEM modelling was transformed into Hillshade Image through 8 different inclination directions of light, East (0°), Northeast (45°), North (90°), Northwest (135°), West (180°), Southwest (225°), South (270°), and Southeast (315°). Eight directions of light were then formed into two different images using overlay analysis. From two overlay display results, the best image associated with the direction of the main fault after the analysis using Roset Diagram was determined. After obtaining the image, geological structure can be identified by drawing the lineament, which is assumed to be discernible through variations in contrast tones. These are the ones that control the high-low density.

After that, the total length of each radius of density 1 km$^2$ in km/km$^2$ (Soengkono 1999) was measured. The purpose of this analysis was to obtain lineament density. Subsequently, a map of lineament density was created, assuming that high densities are associated with the center of the geothermal fluid movement or the permeable zone. In addition to calculating the density of the structure, the density of the intersecting structural points was also calculated. Areas with high structural intersection densities are assumed to be rock-exposed zones due to deformation or voltage forces (Nugroho et al. 2003).

### *13.2.7 Weighted Overlay*

In this study, the weight of each criterion was calculated using Analytical Hierarchy Process (AHP); the problem was organized into two parts: (1) the goal, which represents the objective of the study; (2) four criteria, which are the major factors influencing the goal, are LST, surface manifestation, fault and fracture density (FFD), and NDVI. Structuring the decision problem as a hierarchy is fundamental to the process of the AHP. The structure is shown in Fig. 13.4.

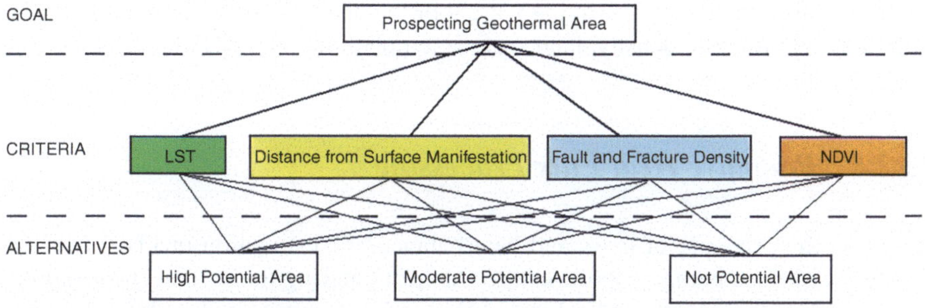


**Fig. 13.4** The goal and four criteria as major factors controlling the goal

#### 13.2.7.1 Create a Qualitative Pairwise Comparison Matrix of Alternatives Using Expert Judgments

This matrix displays how each criterion contributes; Experts are geologists/geoscientists who are experts on geothermal energy.

#### 13.2.7.2 Finding the Main Eigenvalue for Each Matrix

This can result in the weight of each criterion being compared. The weight reflects the relatively favorable criteria. By computing the weight or the Eigen value, the overall ranking of every decision alternative can be obtained.

#### 13.2.7.3 Matrix Consistency Evaluation

This is performed by determining the consistency index, abbreviated as CI, as follow:

$$\mathrm{CI} = \left(\lambda_{\max} - n\right) / \left(n - 1\right) \tag{13.15}$$

In this context, $\lambda_{\max}$ represents the highest Eigen value of the judgement matrix and $n$ *stands for* the matrix's order of size.

In this study, the assessment of a geothermal prospect area was the main objective for using the decision hierarchy. The decision problem's hierarchical structure process consists of two steps. The initial step has been categorized into four factors: land surface temperature, fault and fracture density, distance from surface manifestation, and NDVI value. The next step contains 13 subcategories used to assess the geothermal possibility. Then, each subcategory received a suitability rating value. Following the creation of these factors, the necessary maps were generated in either vector or raster format. Shapefile maps were subsequently transformed into raster maps to enhance their utility for reclassifying subcategories based on the updated rating. To arrange each category and subcategory, some analysis spatial process was implemented such as overlay, convert, and reclassify. Then, the potential area for geothermal is obtained by overlying weight of the distance from surface manifestation, fault and fracture density, land surface temperature, and NDVI.

### *13.2.8 Radiative Heat Flux Processing*

Based on the principle of mass and heat balance, the estimation involves the heat lost to the surface being proportional to the heat entering the system (heat influx). Based on that, the amount of heat lost to the surface is often used as the initial estimation in determining the number of resources (Saptadji 2012). Radiative Heat Flux (RHF) was calculated to obtain total energy emitted from the geothermal potential area of the area calculated using Stefan-Boltzmann's legal principle with a combination of ambient temperature data from air temperature data obtained from temperature measurement station of Meteorology, Climatology and Geophysics Agency (BMKG) by entering formula of RHF.

## 13.3 Results and Discussion

### *13.3.1 Daytime and Nighttime LST*

#### 13.3.1.1 Daytime LST Acquired from Landsat 8

Regarding the TIR band of the Landsat 8 TIRS sensor, the constants $K_1$ and $K_2$ from the metadata file of data remain fixed values (Qin et al. 2001). $NDVI_{min}$ and $NDVI_{max}$ were obtained through calculations involving NDVI value. Then, the land surface emissivity ($\varepsilon$) is computed using Eq. (13.8), and the temperature at the surface (Fig. 13.5) is calculated by Eq. (13.4) integrated with the coefficients in Table 13.2.

In 2015, daytime temperatures ranged from 290.71 to 302.06 K, with a mean value of 296.02 K and a standard deviation of 1.35 K. In 2016, daytime temperature ranged from 294.21 to 306.53 K, with a mean value of 298.82 K and a standard deviation of 2.34. During 2017, daytime temperature spanned from 286.95 to 300.03 K, with an average of 293.62 K and a standard deviation of 1.30 K. Meanwhile in 2018, daytime temperature varied from 293.76 to 304.43 K, featuring an average of 297.48 K and a standard deviation of 1.63 (Fig. 13.6).

#### 13.3.1.2 Nighttime LST Acquired from ASTER TIR

For ASTER TIR, the LST for a particular area can be approximated by employing the Planck function inversion. The Emissivity ($\varepsilon$) value was derived from Jimenez-Munoz Algorithm (2006), where $\varepsilon = 0.970 + 0.020P_v$, $P_v$ represents the proportion

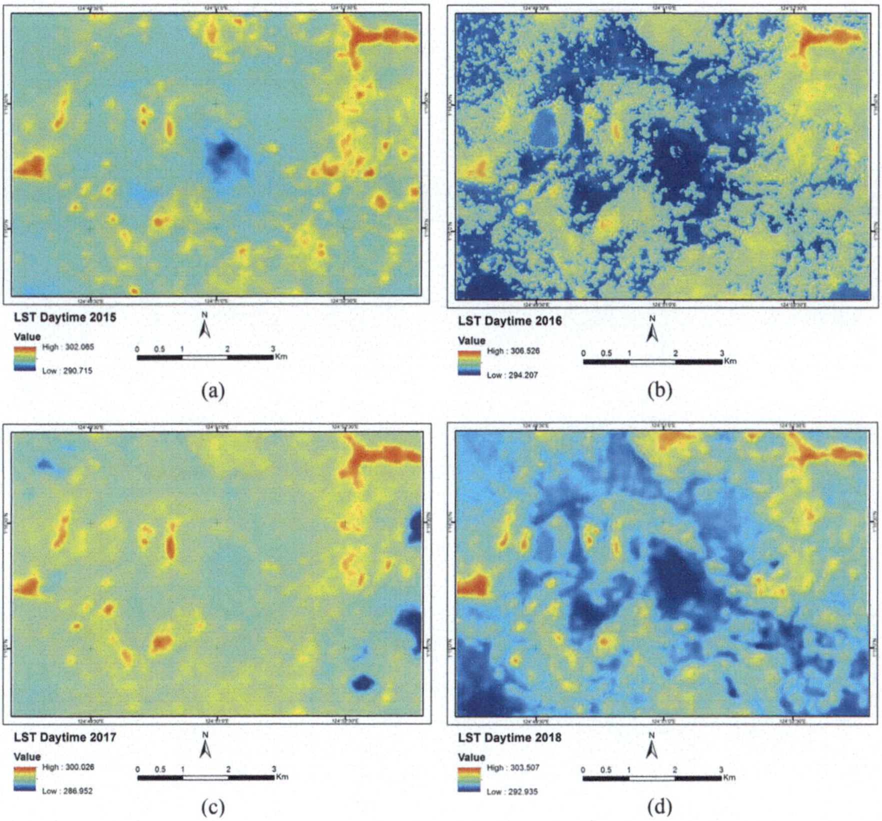


**Fig. 13.5** Daytime land surface temperature from Landsat 8 in band 10 for the years (**a**) 2015, (**b**) 2016, (**c**) 2017, and (**d**) 2018

of vegetation (sometimes known as fractional vegetation cover). The proportion of vegetation is calculated from NDVI. $NDVI_V$ and $NDVI_S$ are the NDVI vegetation and soil values, respectively, where $NDVI_V = 0.7$; $NDVI_S = 0.05$. These data are extracted from the NDVI histogram as outlined by Xiong et al. (2019). The outcomes of the inversion of nighttime LST within the study area for years 2015, 2016, 2017, and 2018 were obtained (Fig. 13.7).

During 2015, the nighttime ground temperatures ranged between 282.83 and 296.78 K, with a mean value of 290.16 K and a standard deviation of 1.63. In 2016, the nighttime ground temperature varied from 289.25 to 296.50 K, with an average of 291.18 K and a standard deviation of 0.79. In 2017, the nighttime ground temperatures spanned from 287.43 to 296.78 K, with a mean value of 290.16 K and a standard deviation of 0.96. Meanwhile in 2018, surface temperatures during the day were distributed between 288.18 and 295.10 K, featuring an average of 290.05 K and a standard deviation of 0.79 (Fig. 13.8).

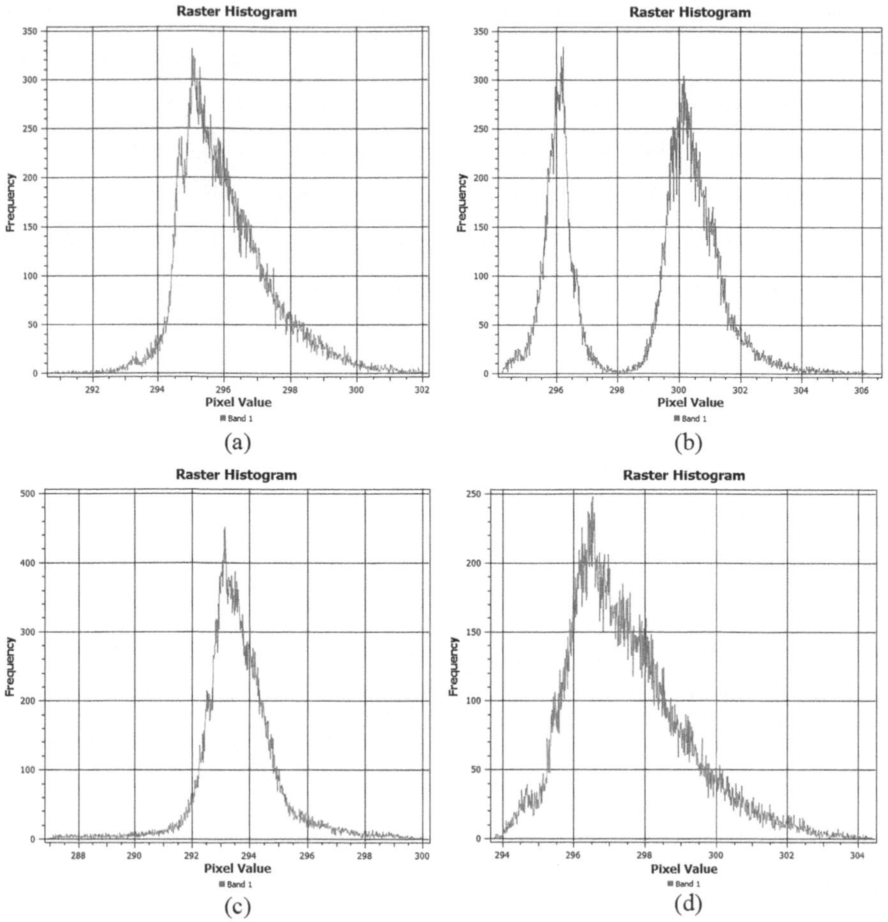


**Fig. 13.6** Histogram of daytime land surface temperature obtained from Landsat 8 band 10 for the years (**a**) 2015, (**b**) 2016, (**c**) 2017, and (**d**) 2018

## *13.3.2 Geothermal Anomalies and Surface Manifestation in Lahendong Area*

The research is conducted in urban and suburban regions, with predominant land cover consisting primarily of vegetation, bare soil, built-up area, and water. Through the temperature distribution for these four types of land cover (Tables 13.4 and 13.5), the water body daytime temperature is categorized as a relatively low temperature anomaly in each year. The water body night temperature is categorized as a high temperature anomaly. The vegetation daytime temperature displays a distinctive temperature pattern, showing relatively high values anomaly. When focusing on vegetation alone, temperature tends to be relatively low at night and high during the day.

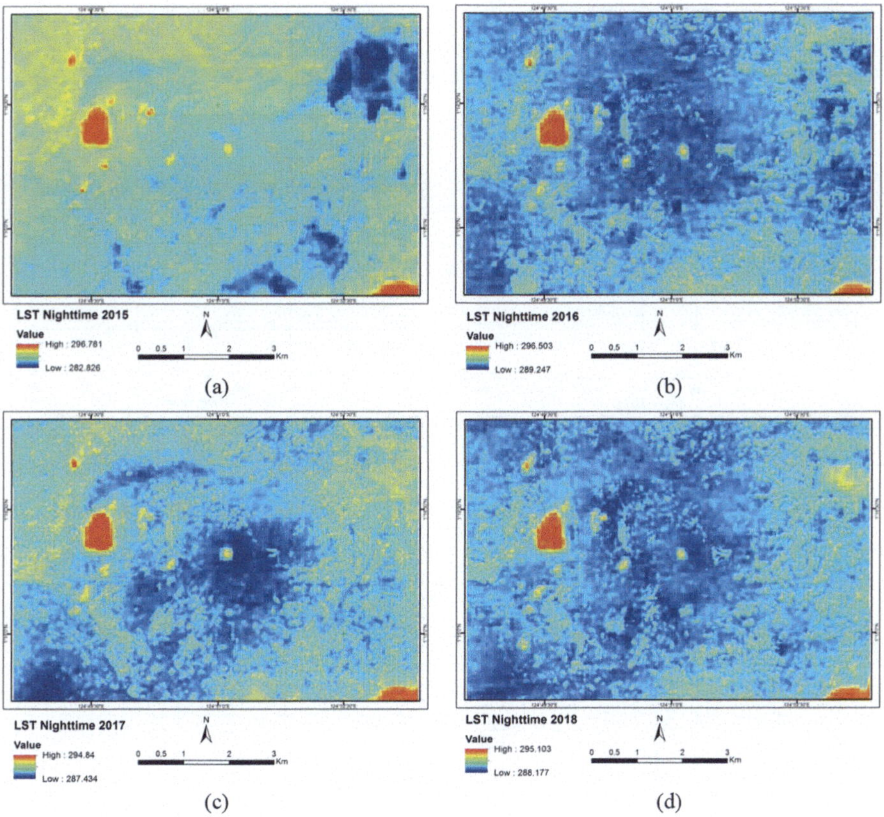


**Fig. 13.7** Nighttime land surface temperature from ASTER TIR in band 14 for the years (**a**) 2015, (**b**) 2016, (**c**) 2017, and (**d**) 2018

To emphasize the elevated temperature anomalies within the geothermal anomaly area, overlaying the analysis of daytime and nighttime LSTs effectively eliminates the comparatively cooler or thermal anomalies in land cover resulting from solar radiation (Wang et al. 2019). Further, geospatial analysis used to identify false anomalies in urban or villages area. Thematic maps such as built-up areas digitized from online imagery used to help identify urban or village areas. Geologic map used to validate the geothermal anomalous that emerge as surface manifestation in LSTs map. Within this study, the extraction of LST and the identification of anomalous regions involve detecting temperatures approximately 3–6 K higher than the surrounding background.

In a previous study, Utami et al. (2004) mapped the geological of the Lahendong geothermal field, including the altered ground and thermal manifestation inside (Fig. 13.9). The primary thermal phenomena are concentrated in the areas surrounding Linau Lake, particularly on its western and northern shores, as well as in the vicinity of Leilem, Lahendong, and Kasuratan Villages. All the thermal manifestations are steam-heated type, and mainly directed by the NE-SW trending faults. No

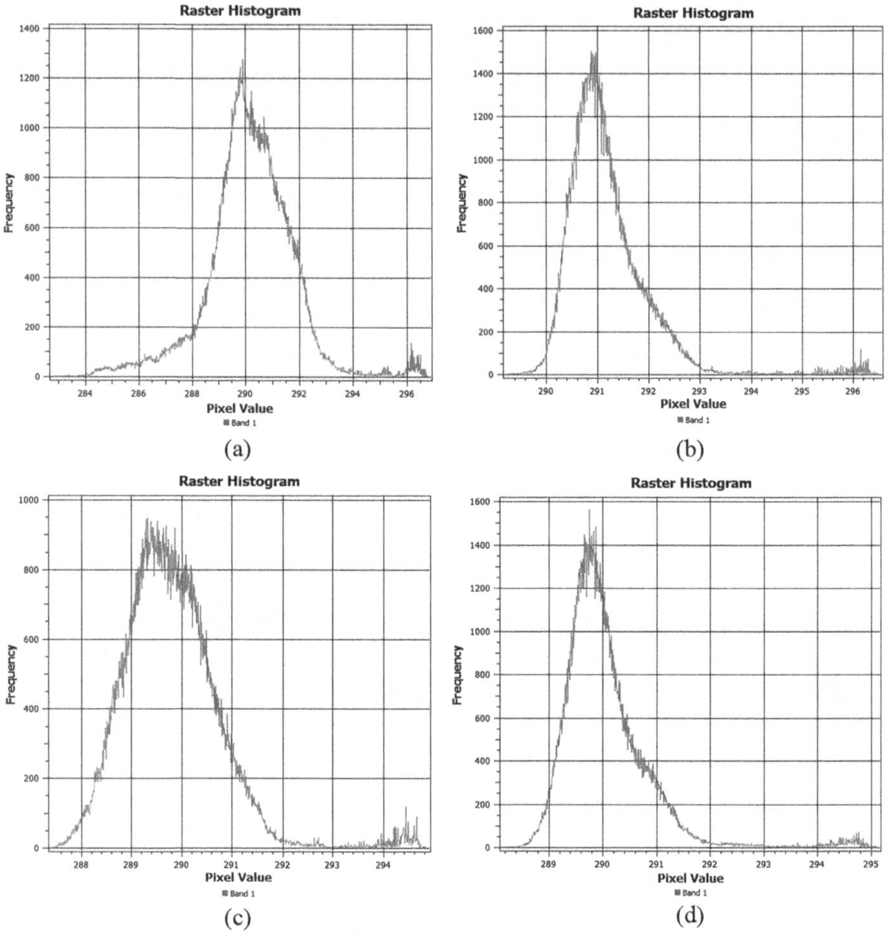


**Fig. 13.8** Histogram of nighttime land surface temperature obtained from ASTER TIR band 14 for the years (**a**) 2015, (**b**) 2016, (**c**) 2017, and (**d**) 2018

neutral pH water discharges within the Lahendong area. Typically, these manifestations consist of modified and vaporizing terrain, acid-sulfate hot springs, mud pots, mud pools (which may or may not include mud volcanoes), and the release of sulfide gas. The prevalent minerals found in the modified terrain include kaolin, residual silica, very fine-grained pyrite, aluminous salt, and, in certain areas, iron oxide. The most active thermal activity occurs within the Linau area, where there are some fumaroles as hot as 106 °C. The hottest steaming ground ($T$ = 60–98 °C at 45 cm depth) and, therefore, the hottest acid sulphate springs ($T$ = 80–90 °C, pH = 2–4) occur here.

According to the Ministry of Energy and Mineral Resources (2017), there are three manifestations that identified as fumarole, located near Lake Linau, Lailem Village, and Lahendong (Table 13.6).

**Table 13.4** Basic statistical information of daytime LST for different land cover types

| Landcover | 2015 | | | | | 2016 | | | | |
|---|---|---|---|---|---|---|---|---|---|---|
| | Min | Max | Range | Mean | Stdev | Min | Max | Range | Mean | Stdev |
| Barren land | 290.74 | 301.83 | 11.09 | 296.42 | 1.08 | 297.80 | 306.53 | 8.72 | 300.43 | 0.75 |
| Vegetation | 291.06 | 298.97 | 7.91 | 295.01 | 0.69 | 294.21 | 298.84 | 4.63 | 296.00 | 0.50 |
| Built-up area | 294.07 | 302.06 | 8.00 | 297.85 | 1.54 | 298.10 | 306.19 | 8.10 | 301.16 | 1.51 |
| Water | 290.71 | 298.60 | 7.89 | 295.09 | 0.90 | 294.66 | 299.59 | 4.92 | 296.69 | 0.56 |
| **Landcover** | **2017** | | | | | **2018** | | | | |
| | **Min** | **Max** | **Range** | **Mean** | **Stdev** | **Min** | **Max** | **Range** | **Mean** | **Stdev** |
| Barren land | 286.95 | 300.03 | 13.07 | 293.73 | 1.23 | 293.00 | 302.88 | 9.88 | 297.12 | 1.36 |
| Vegetation | 287.58 | 296.30 | 8.72 | 293.02 | 0.66 | 292.94 | 301.45 | 8.52 | 295.51 | 0.89 |
| Built-up area | 287.08 | 300.02 | 12.94 | 295.38 | 1.82 | 293.86 | 303.51 | 9.65 | 298.74 | 1.75 |
| Water | 292.07 | 296.67 | 4.60 | 293.75 | 0.86 | 293.42 | 300.46 | 7.04 | 295.43 | 0.98 |

**Table 13.5** Basic statistical information of nighttime LST for different land cover types

| Landcover | 2015 | | | | | 2016 | | | | |
|---|---|---|---|---|---|---|---|---|---|---|
| | Min | Max | Range | Mean | Stdev | Min | Max | Range | Mean | Stdev |
| Barren land | 282.83 | 296.39 | 13.56 | 289.94 | 1.67 | 289.25 | 296.21 | 6.96 | 291.22 | 0.62 |
| Vegetation | 284.78 | 295.43 | 10.65 | 290.38 | 1.04 | 289.30 | 294.77 | 5.47 | 290.82 | 0.41 |
| Built-up area | 283.71 | 296.42 | 12.71 | 289.80 | 1.90 | 289.44 | 296.39 | 6.96 | 291.74 | 0.70 |
| Water | 286.75 | 296.78 | 10.03 | 295.01 | 1.98 | 290.05 | 296.50 | 6.45 | 294.88 | 1.52 |
| **Landcover** | **2017** | | | | | **2018** | | | | |
| | **Min** | **Max** | **Range** | **Mean** | **Stdev** | **Min** | **Max** | **Range** | **Mean** | **Stdev** |
| Barren land | 287.53 | 294.69 | 7.17 | 289.95 | 0.76 | 289.30 | 295.10 | 5.80 | 293.68 | 1.41 |
| Vegetation | 287.43 | 294.48 | 7.04 | 289.28 | 0.68 | 288.18 | 294.60 | 6.43 | 290.44 | 0.69 |
| Built-up area | 287.45 | 294.69 | 7.23 | 290.43 | 0.78 | 288.21 | 293.80 | 5.59 | 289.73 | 0.43 |
| Water | 287.55 | 294.84 | 7.29 | 293.67 | 1.33 | 288.20 | 294.53 | 6.33 | 290.10 | 0.67 |

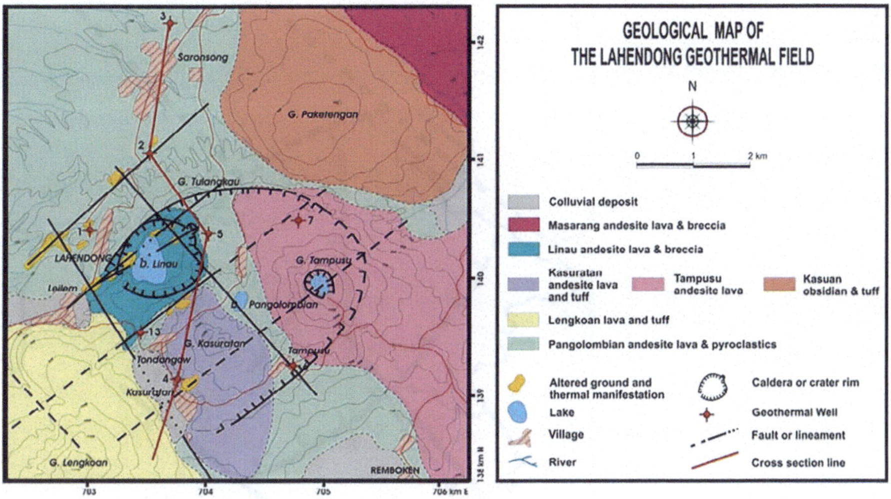


**Fig. 13.9** Geological map of the Lahendong geothermal field (Utami et al. 2004)

**Table 13.6** Surface manifestation of fumarole

| Location | Types | X | Y | Elevation (m) | $T$ (°C) | pH |
|---|---|---|---|---|---|---|
| Linau | Fumarole | 124°49′18.81″E | 1°16′08.65″N | 812 | 92 | 2 |
| Laliem | Fumarole | 124°49′11.21″E | 1°16′04.39″N | 779 | 90 | 2 |
| Lahendong | Fumarole, Mudpool | 124°48′45.32″E | 1°16′26.84″N | 700 | 89 | 2 |

Source: Ministry of Energy and Mineral Resources (2017)

Brehme et al. (2019) mapped the comparison of hot springs fault locations. The map denotes the presence of hot springs with the label "M" signifying surface manifestation. Additionally, there were other observed manifestations labeled as M1, M2, M3, M4, M10, M11, M13, M15, M16, and M17 (Fig. 13.11a–d). Meanwhile, in Lahendong, the characteristic of fault location with hot springs indicates that hot springs can be found under specific circumstances, including near the tips of fault, along fault lines, and at fault intersections. Furthermore, the assumption was made that concealed faulty lines may be the source of the unidentified hot springs (Fig. 13.10).

Nine geothermal anomalies were detected as a surface manifestation, which included fumaroles (M2, M11, and M17), springs and other indicators based on mean LST such as M1, M2, M3, M4, M11, M13, M16, and M17 (Brehme et al. 2019; Siahaan et al. 2005) (Fig. 13.11).

### 13.3.3 Fault and Fracture Density (FFD)

The Pangalombian structure is the edge of the caldera which is understood as an indication of favorable geothermal permeability in the area. The NE-SW fault structure serves as the primary volcanic axis, functioning as a normal and lateral fault. The E-W fault

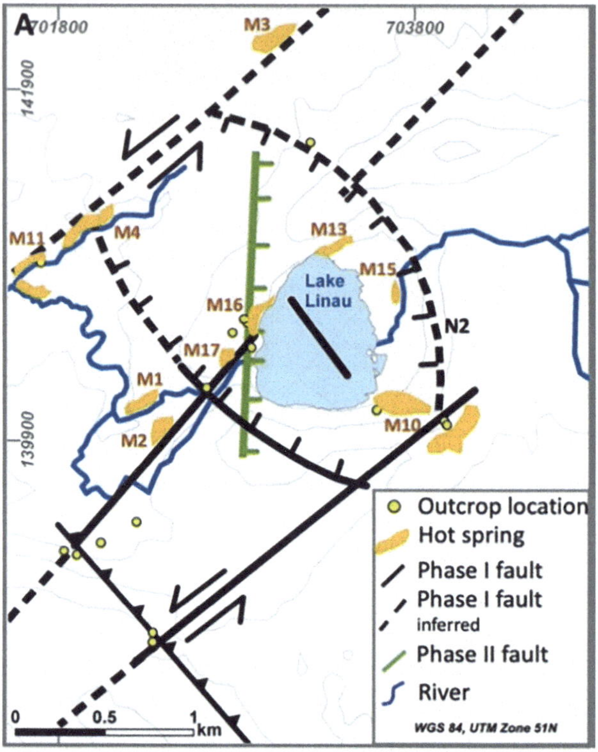


**Fig. 13.10** Fault set-up in Lahendong geothermal area (Brehme et al. 2019)

acts as a lateral and transcurrent fault, while the NW-SE fault and N-S faults are normal faults that are believed to support good permeability for geothermal system (Robert 1987).

The total value of line length density in satellite images of the Lahendong Geothermal field ranged from 0 to 5.46 km/km$^2$. In general, areas with high density values in the study area range from 3.49–5.46 km/km$^2$, indicating a high influx of fluid into a zone (Soengkono 1999). High density occurred around Lake Linow. Calculation of lineaments density was then displayed in the form of lineaments density contour map (Fig. 13.12a). The Lahendong geothermal area could be grouped into four density classes, namely high density (3.49–5.46 km/km$^2$) which was shown in red, medium density (1.92–3.49 km/km$^2$) which was shown in orange, low density (0.68–1.92 km/km$^2$) which was shown in green, and very low density (0–0.68 km/km$^2$) which was shown in dark green (Fig. 13.12b). Furthermore, the FFD map results were compiled with the location of surface manifestation. The location of manifestations of dominance hot spring was found in areas with high lineament density values (Fig. 13.12c).

### 13.3.4 Weighted Overlay Analysis

The analytical hierarchy process (AHP) (Saaty and Vargas 1991) was employed to establish weighting factors for subsequent overlay analysis. AHP is a popular decision-making tool that facilitates a pairwise comparison of pair-wise of criteria

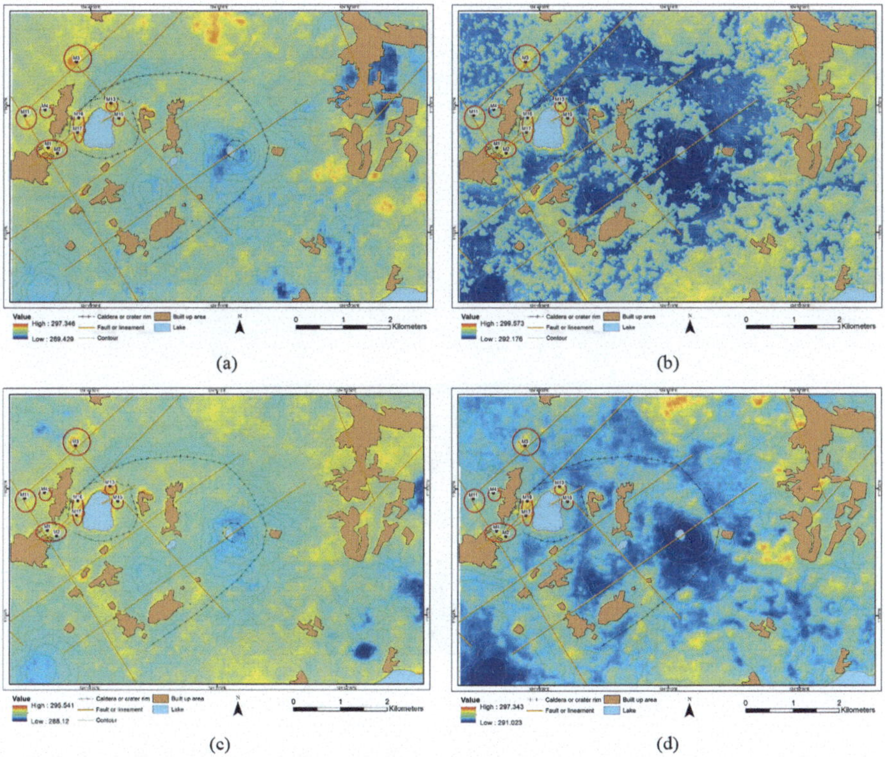

Fig. 13.11 Specific regions with unusual LST are marked by red circular and elliptical borders of the year (**a**) 2015, (**b**) 2016, (**c**) 2017, and (**d**) 2018

weights. Then, the decisions are decomposed to a hierarchical framework, and some criteria were assessed to achieve a goal.

To determine the importance of every layer of evidence, we compare them with each other, and then we assess their relative significance. Table 13.7 is used to giving score from 1 to 9. Pairwise comparisons, as explained by Boroushaki and Malczewski (2008), simplify things by comparing only two criteria at a time. Then the pairwise comparison conducted within a square matrix that represents preferences that lead to the derivation of eigenvalues and eigenvectors.

In the map of potential area for geothermal, we compare four evidence layers that indicated the potential geothermal area. The values from 1 to 9 were assigned considering the intensity of importance as shown in Table 13.8. For instance, the value of 5 was given for a judgment that FFD has strongly importance than NDVI. The system automatically computes 0.2, when comparing the importance of NDVI relative to FFD. We gave high importance values to the most straightforward indicators like surface distance. When comparing the same layer to itself, the preference values on the main diagonal of the matrix were always set to 1 because there is no preference between identical layers.

The square preference matrix was utilized to compute eigenvalues, with the count of eigenvalues matching the matrix's order. For example, in the temperature

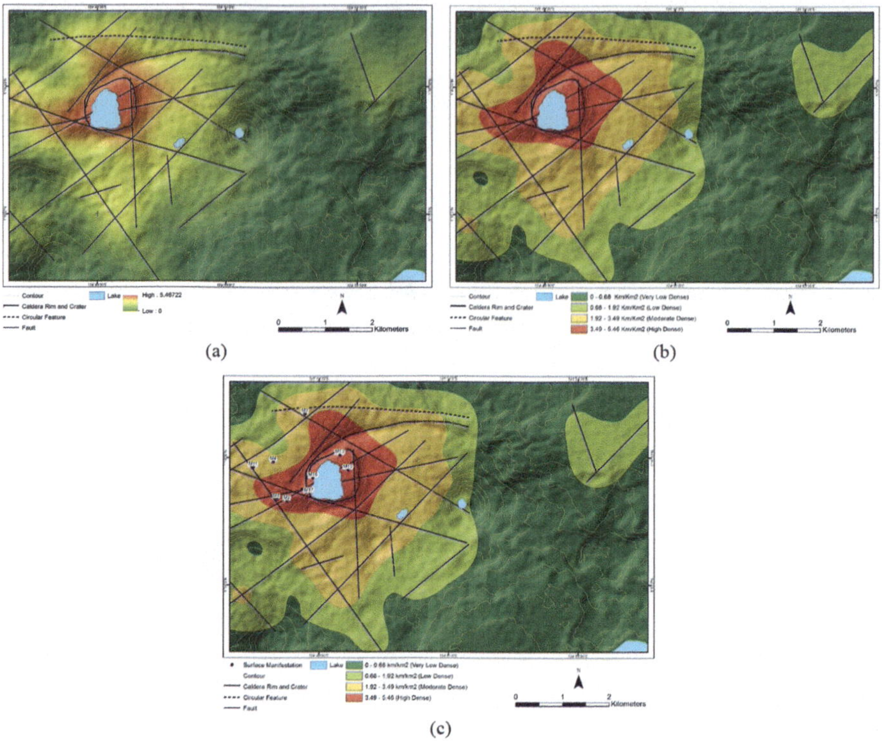


**Fig. 13.12** (**a**) Fracture density map of Lahendong geothermal area, (**b**) four classification of lineament density, (**c**) location of surface manifestation on lineament density

**Table 13.7** Pair-wise comparison score to determine geothermal potential area

| Degree of importance | Intensity of importance |
|---|---|
| 1 | Equal |
| 3 | Moderate |
| 5 | Strong |
| 7 | Very strong |
| 9 | Extreme strong |
| 2, 4, 6, 8 | Intermediate values |
| Reciprocals | Values for inverse comparison |

preference matrix, there were four eigen values. After calculating the largest eigenvalue, the corresponding eigen vector was found, consisting of four components, which align with the order of the matrix. The components in the largest eigen vector were then totaled, and each component was divided by this total to ensure a sum of 1 (normalization). These component values served as weights. Finally, each criterion that represents raster layer was multiplied by its respective weight and summed together for further analysis.

**Table 13.8** Preference matrix for geothermal potential area map

| | FFD | Surface manifestation | LST | NDVI |
|---|---|---|---|---|
| FFD | 1 | 0.25 | 2 | 5 |
| Surface manifestation | 4 | 1 | 5 | 7 |
| LST | 0.5 | 0.2 | 1 | 4 |
| NDVI | 0.2 | 0.14 | 0.25 | 1 |

**Table 13.9** Criteria weights for four evidence layers in geothermal potential area

| Layer | Calculated weight (%) |
|---|---|
| Distance from surface manifestation | 59.8 |
| Fault and fracture density | 21.4 |
| Land surface temperature | 13.5 |
| NDVI | 5.1 |
| Total | 100 |

Output criteria weights from the temperature preference matrix are shown in Table 13.9.

A numerical index was used to ensure the consistency of alternative prioritization in the pair-wise comparison matrix. An adjustment needs to be conducted once the ratio is greater than 0.1 (Saaty and Vargas 1991). The consistency ratio (CR) is a measure obtained by comparing the consistency index (CI) to an average consistency index, which is also called the random index (RI) (Eq. 13.16). The consistency index was determined using the preference matrix (Eq. 13.17), where $\lambda_{max}$ represents the largest eigenvalue from the preference matrix and $n$ is the matrix's size.

$$\mathrm{CR} = \frac{\mathrm{CI}}{\mathrm{RI}} \tag{13.16}$$

$$\mathrm{CI} = \frac{\lambda_{\max} - n}{n - 1} \tag{13.17}$$

The value RI was derived from expert knowledge and experience and depends on the number of criteria/the order of the matrix (Table 13.10). For the geothermal potential area, a consistency ratio of 0.062 was calculated from the matrix output.

The decision problem's hierarchical structure process consists of a pair of stages. The initial stage has been divided into four factors: land surface temperature, fault and fracture density, distance from surface manifestation, and normalized difference vegetation index (NDVI) value. The second stage comprises 13 subcategories employed to assess the potential of geothermal. Then, each subcategory was assigned a potentially rating value. After establishing these factors, the necessary maps for each layer were generated in either vector or raster. The vector maps were subsequently transformed into raster maps for greater utility in the reclassification of subcategories according to the new rating (Fig. 13.13).

**Table 13.10** Matrix of RI value

| Number of criteria | 2 | 3 | 4 | 5 | 6 | 7 | 8 | 9 | 10 |
|---|---|---|---|---|---|---|---|---|---|
| RI | 0 | 0.58 | 0.9 | 1.12 | 1.24 | 1.32 | 1.41 | 1.45 | 1.49 |

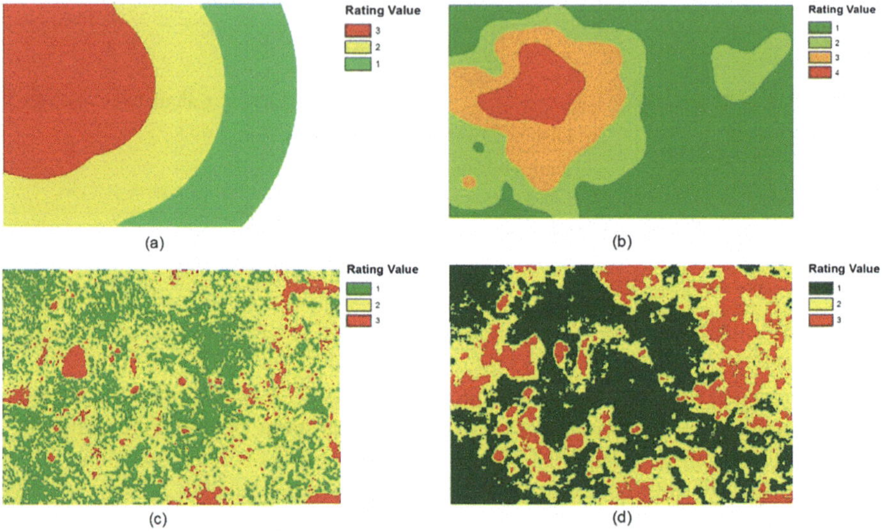


**Fig. 13.13** Map depicting the thematic layers following the application of weight; (**a**) distance from surface manifestation, (**b**) FFD, (**c**) NDVI, and (**d**) LST

#### 13.3.4.1 Weighting Criteria of Distance from Surface Manifestation

To identify suitable areas for geothermal prospecting, buffer distances of 2000, 4000, and 6000 meters were applied to all surface manifestations obtained from previous analyses to delineate suitable from unsuitable areas. The choice of the buffer distance was based on approximate maximum distance from which geothermal areas have been observed to impact the presence of geothermal energy.

According to Geological Survey of Japan (Noorollahi 2005), an examination of the spatial arrangement of hot springs and geothermal wells reveals that nearly 97% of geothermal wells were situated within a 4000 m radius of hot springs. Thus, this distance was employed as a reference for identifying prospective geothermal areas, guided by the presence and positioning of hot springs. Based on the analysis, a buffer was conducted up to 4000 m from the emergence of surface manifestations. Table 13.11 shows the criterion that was applied in determining the level of potential of an area based on the location from the center of the buffer zones.

Areas within buffers are determined as the most potential areas depending on whether they are in the first, second, or third buffered zones. Areas that are located outside the buffer zones are considered as nonpotential area for geothermal prospect.

**Table 13.11** Buffering criteria applied for geothermal area

| Buffer zone (m) | Reclassification | Rating | Description |
|---|---|---|---|
| 2000 | 1 | 3 | Most potential |
| 4000 | 2 | 2 | Moderately potential |
| 6000 | 3 | 1 | Least potential |

#### 13.3.4.2 Weighting Criteria of Fault and Fracture Density

The Lahendong geothermal area could be grouped into 4 density classes, namely high density (3.49–5.46 km/km$^2$), medium density (1.92–3.49 km/km$^2$), low density (0.68–1.92 km/km$^2$), and very low density (0–0.68 km/km$^2$). Areas with high density values in the study area range from 3.49 to 5.46 km/km$^2$, indicating a high influx of fluid entering a zone (Soengkono 1999). The denser the areas, the more permeable the zones are. It is indicated as pathway for thermal fluid to travel from the deeper reservoir to the shallower level. Table 13.12 shows the criterion that was applied in determining the level of potential of an area based on the fault and fracture density.

#### 13.3.4.3 Weighting Criteria of Normalized Difference Vegetation Index

NDVI was determined from a satellite image by analyzing the visible and near-infrared bands, which capture how vegetation interacts with sunlight. Healthy vegetation primarily absorbs visible light and reflects a significant amount of the near-infrared light. In contrast, less healthy or sparser vegetation reflects more visible light and less near-infrared light (Suryantini et al. 2013).

The value of NDVI threshold can recognize the type of vegetation in earth surface. For instance, NDVI values falling within the range of 0.501–1.0 signify abundant vegetation, such as trees found in urban areas. Value in the range of 0.2–0.5 indicates sparser vegetation, while those below 0.2 denotes areas of nonvegetation (Table 13.13).

#### 13.3.4.4 Weighting Criteria of Land Surface Temperature

LST is frequently employed for assessing temperature anomalies that come out on the earth's surface. These anomalies are often related to the emergence of geothermal manifestations when the location is associated with geological formations such as faults and fractures. In this study, land surface temperature classified into three classes: high, medium, and low (Table 13.14).

To organize every category and its corresponding subcategory (Table 13.15), some several analysis methods were implemented using ArcGIS software (i.e., overlay, convert, and reclassify raster). Output geothermal prospect area is carried out by calculating the weighted overlay of the distance from surface manifestation, FFD, NDVI, and LST (Fig. 13.13).

**Table 13.12** Reclassification and rating of fault and fracture density

| Density (km/km$^2$) | Reclassification | Rating | Description |
|---|---|---|---|
| 3.49–5.46 | 1 | 4 | Most potential |
| 1.92–3.49 | 2 | 3 | Moderately potential |
| 0.69–1.92 | 3 | 2 | Least potential |
| 0–0.68 | 4 | 1 | Not potential |

**Table 13.13** Reclassification and rating of NDVI

| NDVI value | Reclassification | Rating | Description |
|---|---|---|---|
| <0.2 | 1 | 3 | Most potential |
| 0.2–0.5 | 2 | 2 | Moderate potential |
| >0.5 | 3 | 1 | Least potential |

**Table 13.14** Reclassification and rating of LST

| LST value (K) | Reclassification | Rating | Description |
|---|---|---|---|
| <293.07 | 1 | 1 | Least potential |
| 293.07–294.32 | 2 | 2 | Moderate potential |
| >294.32 | 3 | 3 | Most potential |

**Table 13.15** The weights and scores assigned to the factors and subcriteria employed in geothermal area prospecting

| Factors | Weight (%) | Subcriteria | Rating |
|---|---|---|---|
| Distance from surface manifestation | 59.8 | 0–2000 m | 3 |
| | | 2000–4000 m | 2 |
| | | 4000–6000 m | 1 |
| Fault and fracture density | 21.4 | 3.49–5.46 km/km$^2$ | 4 |
| | | 1.92–3.49 km/km$^2$ | 3 |
| | | 0.69–1.92 km/km$^2$ | 2 |
| | | 0–0.68 km/km$^2$ | 1 |
| NDVI | 13.5 | <0.2 | 3 |
| | | 0.2–0.5 | 2 |
| | | >0.5 | 1 |
| LST | 5.1 | <293.07 K | 1 |
| | | 293.07–294.32 K | 2 |
| | | >294.32 K | 3 |

Furthermore, the calculation process is carried out by applying the appropriate weights to each factor, then the results obtained are three classes of areas that have the potential for geothermal areas as can be seen in Fig. 13.14. Red is an area that has high potential, yellow has medium potential, and green has no potential.

After being calculated from the AHP process and subtracted from the built-up area and lake area, the total area (km$^2$) of each criterion is shown in Table 13.16.

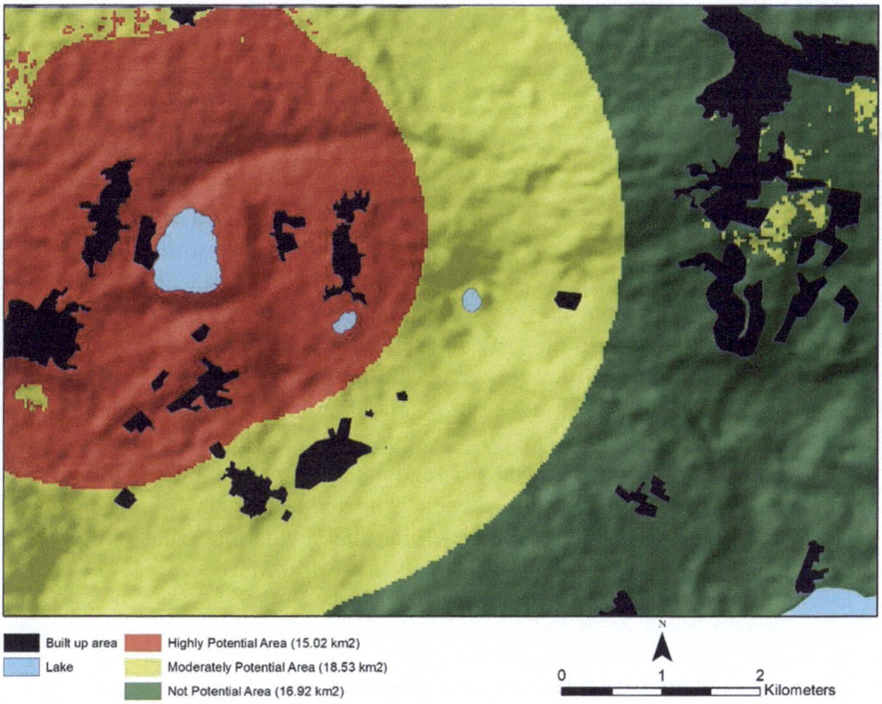


**Fig. 13.14** Potential geothermal area

**Table 13.16** Total area of each criterion of potential geothermal area

| Criteria | Area (km$^2$) |
|---|---|
| High potential area | 15.02 |
| Moderate potential area | 18.53 |
| Not potential area | 18.92 |

### *13.3.5 Calculating Radiative Heat Flux*

The overall heat loss or heat discharge rate (HDR) includes the combined value of conductive, convective, and radiative heat loss, excluding any influence from solar factors, in any volcano or geothermal field. Since satellite TIR data are restricted to examining radiative heat loss, the radiative heat flux (RHF) within the Lahendong thermal area is computed using Stefan-Boltzmann Law (Eq. 13.18) (Mia et al. 2012).

$$Q_r = \mathrm{ts}\varepsilon A\left(T_s^4 - T_a^4\right) \tag{13.18}$$

where:

$\tau$ = Atmospheric transmissivity, (obtained from https://atmcorr.gsfc.nasa.gov/, = 0.52)

$\sigma$ = Stefan–Boltzmann Constanta ($5.6704 \times 10^{-8}$ W/m$^2$ K)
$\varepsilon$ = Emissivity (0.97)
$A$ = Area (15,020,456.4 m$^2$)
$T_s$ = Land surface temperature, (max temperature in highly potential area = 296.8 K)
$T_a$ = Ambient temperature (is the cooler surrounding temperature, mean temperature area = 293.06 K)

Radiative heat flux (RHF) is a component of heat loss from volcanoes that is quantifiable with thermal infrared sensors on satellite. RHF is estimated by applied the Stefan-Boltzmann law of thermal heat loss using land surface temperature of Lahendong geothermal field derived from Landsat 8 TIR and ASTER. The total RHF value was about 141.56 Mega Watt Thermal (MWt) in high potential area. In order not to reduce the potential that exists in the geothermal prospect area, total potential area was added with moderate potential area. The resulting the value of RHF was approximately 316.27 MWt. Assuming that the efficiency of 15% for the Lahendong geothermal power plant, the estimated energy output is about 47.44 Mega Watt electricity (MWe). It has been validated with data from Ministry of Energy and Mineral Resources, calculated that the geothermal energy reserves in Lahendong are 80 MWe and the proven reserves are 140 MWe.

### *13.3.6 Potential Land for Agricultural Production and Energy Security*

According to the results of overlaying geothermal potential areas with topographic maps, all of the Lahendong areas are included in areas with high geothermal potential (Fig. 13.15). Apart from geothermal energy potential, agriculture in the area may undoubtedly profit from the availability of geothermal energy in both geothermal and hydro geothermal forms since the entire Lahendong area has been identified as being in a high geothermal potential area. Geothermal energy can be utilized for space heating, aquaculture, greenhouse heating, and crop drying in agriculture. Additionally, geothermal energy could find applications in various industrial operations such as concrete curing, pasteurizing milk, refrigerating food, drying organic materials (like seaweed and grass), desalinating water, processing wool, and preheating water in coal-fired power plants (Omenikolo et al. 2020).

Geothermal energy finds significant application in agriculture through its prominent use in greenhouse. The greenhouse buildings shield plants from extreme cold and heat by maintaining a consistent, year-round temperature. Then, farmers utilize geothermal energy to provide warmth to greenhouse structure, and hydrothermal within pipes is available to control humidity level if needed. This power source helps crops grow faster and be more reliable. Geothermal water allows farmers to lengthen the growing season, resulting in more harvests and higher earnings (Marsh 2022). Pests and unwelcome fungi around the growers' plants could be a problem.

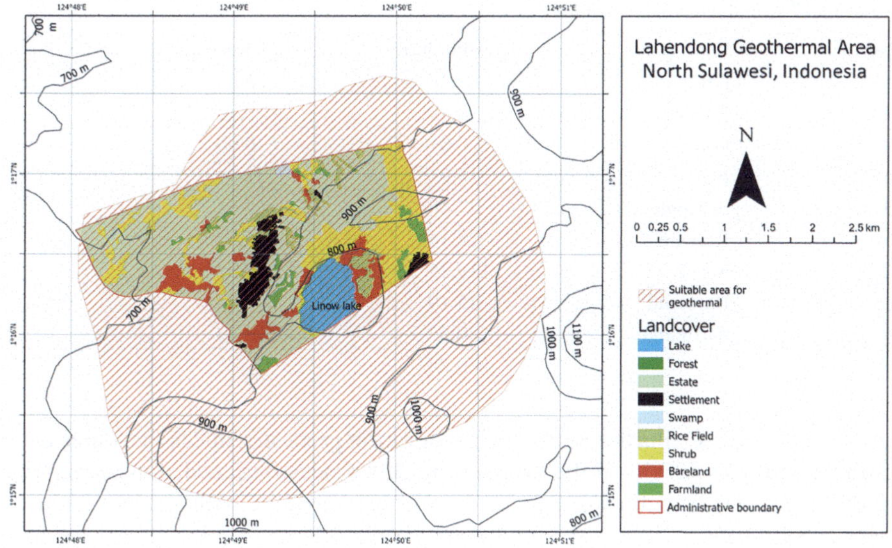


**Fig. 13.15** Land cover vs. high potential area for geothermal

These unwelcome invaders are eliminated when the soil is sterilized by the geothermal water, which is hot enough.

Since Indonesia has a tropical climate, the use of greenhouse is not common. In the open fields, farmers can also use geothermal water for their crops. Geothermal water can help regulate the soil's temperature, keeping warm and shielding the soil from the cold. Underground pipes can be put in place by farmers to carry this water for irrigation. Farmers can also mix water with geothermal energy. Although geothermal water has different chemical composition and salinity, by mixing with tap water, hydrothermal can be used for growing tomato, eggplant, and zucchini. Tomaszewska et al. (2022) employed ultrafiltration/reverse osmosis (UF/RO) technology to desalinate geothermal water and they applied this treated water to cultivate plant in a soil system. They utilized a mixture of 30% tap water and 70% treated geothermal water along with nutrients. The combined water has been reported to be safe and free of potential salt and ion toxicity issues.

Plantations or estate are the most common form of land cover, followed by shrubs (Table 13.17). It is possible to use geothermal water for irrigation. Even with its current size, open space can still be compared to municipalities. Open space and bushes can also be used as places for horticultural gardens or geothermal-based agricultural processing companies like sugar production. Chilies, spring onions, tomatoes, cabbage, eggplants, and chayote are horticulture products that come from South Tomohon, including Lahendong, according to BPS Tomohon Selatan.

Jamal et al. (2019) conducted research by directly utilizing geothermal steam for the palm sugar industry in Lahendong. Palm sap is heated using geothermal steam with a heating device called a juicer in the initial process or purification

**Table 13.17** Land cover in Lahendong

| No | Land cover | Area (ha) | Percentage (%) |
|---|---|---|---|
| 1 | Lake | 39.34 | 6.55 |
| 2 | Swamp | 1.89 | 0.32 |
| 3 | Forest | 0.01 | 0.00 |
| 4 | Estate | 318.91 | 53.13 |
| 5 | Settlement | 27.88 | 4.64 |
| 6 | Rice field | 27.16 | 4.52 |
| 7 | Shrub | 119.25 | 19.87 |
| 8 | Bareland | 45.19 | 7.53 |
| 9 | Farmland | 20.62 | 3.43 |
| | | 600.24 | |

process up to 70 °C. Furthermore, the evaporation process reduces the water content by heating to 120 °C in the evaporator. This process succeeded in converting 22,000 L per day of palm sap into 2.5 tons of crystal sugar with a steam requirement of ±4 tons/h.

The Lahendong area is highly suitable for geothermal business development, ensuring energy security. As an evidenced by the results of remote sensing analysis using Landsat and ASTER data, the Lahendong area is in a high potential zone. Figure 13.15 is an overlay of a geothermal high potential zone with a topographic map. Currently the Indonesian government through PT. Pertamina has operated a geothermal power plant (PLTP) to help meet electricity needs in the provinces of North Sulawesi and Gorontalo. It has a capacity of 120 MW per day derived from 6 units of geothermal power plants, capable of lighting 133,300 homes with a potential for reducing emissions of around 624,000 tons of $CO_2$ per year and potential for saving oil and gas foreign exchange reserves of 5676 barrels of oil equivalent per day (Pertamina 2023).

## 13.4 Conclusion

The study has shown that using thermal and infrared (TIR) remote sensing, both during the day and at night, is an efficient way to locate geothermal energy. This approach can help reduce the government's geothermal exploration cost. A total of 141.56 MW of energy was successfully extracted from 9 geothermal anomaly regions. The approach method can also be applied in other locations, provided that the local geological structure is considered.

Geospatial analysis in this study revealed the potential for establishing a geothermal-based industry throughout the entire Lahendong region. Agricultural areas can be developed on bare surface, and geothermal water can be utilized for irrigation and maintaining soil conditions, as well as for direct heat energy from the earth.

## References


Barsi JA, John RS, Nina GR, Brian LM, Robert GR (2014) Landsat-8 thermal infrared sensor (TIRS) vicarious radiometric calibration. Remote Sens 6(11):11607–11626

Boori MS, Vozenílek V, Balzter H, Choudhary K (2015) Land surface temperature with land cover classes in ASTER and Landsat data. J Geophys Remote Sens 4:138. https://doi.org/10.4172/2169-0049.1000138

Boroushaki S, Malczewski J (2008) Implementing an extension of the analytical hierarchy process using ordered weighted averaging operators with fuzzy quantifiers in ArcGIS. Comput Geosci 34(4):399–410

Brehme M, Giese R, Suherlina L, Kamah Y (2019) Geothermal sweetspots identified in a volcanic lake integrating bathymetry and fluid chemistry. Sci Rep 9(1):1–10

DiPippio R (2005) Geothermal power plants: principles, applications, and case studies. Elsevier, Kidlington

Glassley WE (2014) Geothermal energy: renewable energy and the environment. CRC Press, Boca Raton

Hodder DT (1970) Application of remote sensing to geothermal prospecting. Geothermics 1:368–380

IEA (2022) World energy outlook 2022. International Energy Agency, Paris

Jamal J, Marghariet L, Aristo R (2019) Pemanfaatan Uap PLTP Lahendong untuk Industri Pabrik Gula Aren. Jurnal Sinergi Jurusan Teknik Mesin 16(1):70. https://doi.org/10.31963/sinergi.v16i1.1204

Jiménez-Muñoz JC, Sobrino JA, Gillespie A, Sabol D, Gustafson WT (2006) Improved land surface emissivities over agricultural areas using ASTER NDVI. Remote Sens Environ 103(4):474–487

Kasbani K (2009) Tipe Sistem Panas Bumi di Indonesia dan Estimasi Potensi Enerinya. Buletin Sumber Daya Geologi 4(3):23–30

Lee K (1978) Analysis of thermal infrared imagery of the Black Rock Desert geothermal area. Q Colo Sch Mines (United States) 73(3):31–43

Marsh J (2022) Applications of geothermal energy for agriculture. Agritech tomorrow: online trade Magazine Agricultural Technology and Precision Farming, September 28, 2022. Accessed 3 Sep 2023

Marwan (2015) A preliminary study of the application of electromagnetic conductivity meter to soil properties of Paddy cultivation area Wue Village, Jantho, Aceh Besar District, Indonesia. Aceh Int J Sci Technol 4:14–19. ISSN 2088-9860

Marwan, Yanis M, Ismail N (2017) 2D electrical resistivity model of shallow structure of Seulawah Agam geothermal area based on VLF-R data. In: Joint convention, Malang

Mia MB, Bromley CJ, dan Fujimitsu Y (2012) Monitoring heat flux using Landsat TM/ETM+ thermal infrared data (a case study at Karapiti ('craters of the moon') thermal area, New Zealand). J Volcanol Geotherm Res 235–236:1–10

Ministry of ESDM [Kementerian Energi dan Sumber Daya Mineral Republik Indonesia] (2012) Rencana Umum Ketenagalistrikan Nasional 2012–2031. Kementerian Energi dan Sumber Daya Mineral, Jakarta

Ministry of Energy and Mineral Resources (2017) Potensi Panas Bumi Indonesia Jilid 1. Direktorat Panas Bumi, Kementerian Energi dan Sumber Daya Mineral, Jakarta

Noorollahi Y (2005) Application of GIS and remote sensing in exploration and environmental management of Namafjall geothermal area, N-Iceland. M.Sc. dissertation, University of Iceland, Reykjavik, 193 pp

Nugroho SD, Soetoto, Pri U (2003) Interpretasi Kontrol Struktur dan Komponen-Komponen Sistem Panas Bumi Gunung Ungaran, Jawa Tengah Berdasarkan Citra Landsat thematic mapper. In: Proceedings of joint convention Jakarta, the 32nd and 29th HAGI annual convention dan exhibition, Jakarta

Omenikolo AI, Ibe BO, Chile EN (2020) Application of geothermal energy in agriculture: a review. Int J Eng Appl Sci Technol 5(1):666–673. https://doi.org/10.33564/IJEAST.2020.v05i01.119

Pertamina (2023) Lahendong Pertamina Geotermal Energi. https://www.pge.pertamina.com/id/operasi-sendiri/lahendong. Accessed 20 Jul 2023

Polakitan KA (2018) Potensi Geothermal Area Lahendong Diperkirakan 350 Megawatt. Antara Sulut. https://manado.antaranews.com/berita/37102/potensi-geothermal-area-lahendong-diperkirakan-350-megawatt. Accessed 20 Jul 2023

Qin Z, Arnon K, Pedro B (2001) A mono-window algorithm for retrieving land surface temperature from Landsat TM data and its application to the Israel-Egypt border region. Int J Remote Sens 22(18):3719–3746

Robert D (1987) Geology model of Lahendong geothermal field: a guide for the development of this field. Report for Pertamina, Jakarta

Rybach L (1981) Geothermal systems, conductive heat flow, geothermal anomalies. In: Muffler LJP, Rybach L (eds) Geothermal systems: principles and case histories. Wiley, Chichester

Saaty TL, Vargas LG (1991) Prediction, projection, and forecasting: applications of the analytic hierarchy process in economics, finance, politics, games, and sports. Kluwer Academic Publishers, Norwell

Saptadji NM (2012) Sekilas Tentang Panas Bumi, Program Studi Magister Teknik Panas Bumi Fakultas Teknik Pertambangan dan Perminyakan ITB. Bandung

Seielstad C, Queen L (2009) Thermal remote monitoring of the Norris Geyser Basin, Yellowstone National Park. Final report for the National Park Service Cooperative Ecosystem Studies Unit, Agreement no H1200040001, 38 pp

Siahaan EE, Soemarinda S, Fauzi A, Silitonga T, Azimudin T, Raharjo IB (2005) Tectonism and volcanism study in the Minahasa compartment of the north arm of Sulawesi related to Lahendong geothermal field, Indonesia. In: Proceedings world geothermal congress 2005, Antalya, Turkey, 24–29 April 2005

Sobrino JA, Raissouni N, Li ZL (2001) A comparative study of land surface emissivity retrieval from NOAA data. Remote Sens Environ 75(2):256–266

Soengkono S (1999) Analysis of digital topographic data for exploration and assessment of geothermal system. In: Proceeding 21st New Zealand geothermal workshop

Stathopoulou M, Cartalis C (2007) Daytime urban heat islands from Landsat ETM+ and Corine land cover data: an application to major cities in Greece. Sol Energy 81(3):358–368

Surana T, Atmojo JP, Suyanto SA (2010) Development of geothermal energy direct use in Indonesia. In: GHC bulletin, pp 11–15

Suryantini S, Wibowo H, Rahman KR, Woldai T (2013) Application of normalized different vegetation index (NDVI) method to identify thermal anomaly area from remote sensing. In: 13th Indonesia international GEOTHERMAL convention and exhibition

Susanti R, Dayat E, Santri DJ (2005) Analisis Vegetasi Kawasan Mata Air Panas Gemurak Desa Penindaian, Kecamatan Semendo Darat Laut Kabupaten Muara Enim. Sumatera Selatan FORUM MIPA 9(1):11–19

Tomaszewska B, Mukti M, Czubernat M, Tyszer M, Bujakowski W, Kasztelewicz A, Kabay N, Baba A, Akkurt GG, Ökten HE, Meriç MK (2022) Geothermal resources for agriculture: growing tomato, eggplant, and zucchini in geothermal greenhouse using treated geothermal water for irrigation

USGS (2015) Landsat 8 (L8) data users, handbook version 1.0. EROS Sioux Falls, South Dakota

Utami P, Siahaan EE, Azimudin T, Browne PRL, Simmons SF (2004) Overview of the Lahendong geothermal field, North Sulawesi, Indonesia: a progress report. In: Proceedings of the 26th NZ geothermal workshop, pp 1–6

Valor E, Caselles V (1996) Mapping land surface emissivity from NDVI: application to European, African, and south American areas. Remote Sens Environ 57(3):167–184

Wang K, Jiang QG, Yu DH, Yang QL, Wang L, Han TC, Xu XY (2019) Detecting daytime and nighttime land surface temperature anomalies using thermal infrared remote sensing in Dandong geothermal prospect. Int J Appl Earth Obs Geoinf 80:196–205

Xiong Y, He J, Ma N, Ren X, Liu C (2019) Land surface temperature retrieval based on thermal infrared remotely sensed data of Aster. In: IOP conference series: earth and environmental science, vol 300, no 2, p 022027. IOP Publishing

# Chapter 14
# Physio-Climatic and Socio-Economic Suitability Assessment of Alternate Wetting and Drying (AWD) Irrigation Water Management in Bangladesh

**Md Rahedul Islam, Hyungjun Kim, and Wataru Takeuchi**

**Abstract** Alternate wetting and drying (AWD) irrigation system is a water-saving, methane emission-reducing, and cost-effective technique for rice cultivation but field level adoption rate is very slow in Bangladesh. To accelerate the AWD irrigation system, suitable areas and seasons are very important. The main objective of this study was to classify the suitability of AWD irrigated rice area of Bangladesh. The methodology combined physical parameters (slope, soil texture, moisture, drainage, permeability, pH) and climatic parameters (rainfall and temperature) to assess the physio-climatic suitability of AWD. In addition, irrigations pump density, farmers' wealth index poverty, arable land and farmers ratio, farm holding size and literacy rate were selected as socio-economic suitability parameters. The relevant data were collected from respective sources in statistical, geospatial and satellite imagery formats. The collected dataset was preprocessed, converted into the same formats and spatial resolution, interpolated, reclassified and scored according to the influences to assess the suitability of the AWD irrigated rice area estimation. The dataset used was from 2000 to 2020 and prepared the weighted overlay methods. The developed method has been applied with significance weights of different controlling parameters for assessing AWD irrigations suitable land and time for rice cultivated area. The result showed that physically the country was highly suitable (up to 36.76%) and moderately suitable (up to 62.46%) for all 3 different rice growing seasons. Physio-climatically, the boro rice season was highly suitable (83.36%), amon rice season was moderately suitable (55.2%) and Aus rice growing season was lowly suitable (78.92%) for AWD irrigation. On the other hand, socio-economically the country was moderately suitable (78.45%). Physio-climatically and

M. R. Islam (✉)
Department of Geography and Environment, Pabna University of Science and Technology, Pabna, Bangladesh
e-mail: rahe_ge@pust.ac.bd

H. Kim · W. Takeuchi
Institute of Industrial Science, University of Tokyo, Tokyo, Japan


T. Ahamed (ed.), *Remote Sensing Application II*, New Frontiers in Regional Science: Asian Perspectives 77, https://doi.org/10.1007/978-981-97-1188-8_14

socio-economically the country was suitable for AWD irrigation practices which could be a great potential tool to mitigate climate change challenges.

**Keywords** Physical parameters · Soil properties · Climatic parameters · Weighted overlay · Sensitivity

## 14.1 Introduction

Agriculture is the main livelihood in many Asian countries including Bangladesh. The country's agriculture contributes to 12.5% of gross domestic product, engaged nearly 40% of the labour force and in fiscal year 2020–2021, 76.5% of total population's livelihood is fully or partially dependent on it (BBS 2022). Rice is the main staple food and the main livelihood in Asia (Chang et al. 2022; Zhang et al. 2018a, b). Among Asian countries, Bangladesh is one of the top rice producers, consumer and exporter. The country uses 70% of the total agricultural land for rice cultivation and produces 56.8 million ton of rice (Hossain et al. 2013b). Rice is the staple food and 70% of total calorie intake comes from rice (Majumder et al. 2016). The food security of the country mainly depends on rice production. Bangladesh is a country with 160.5 million population and projected to be 255 million in 2050 (Mainuddin and Kirby 2015). Next 25 years, the country needs to increase rice production by 29% unless major dietary habit changes (Bishwajit et al. 2013). To increase rice production, the country intensively uses arable land, irrigation water, chemical fertilizer and pesticide. Intergovernmental Panel on Climate Change (IPCC) predicts that the country will possibly decline rice production 8% by 2050 (Cancelliere et al. 2007). The country's rice production as well as food security will face great challenges in future decades.

Among the challenges of the country's rice production, the main issues are irrigation, water scarcity and methane emission. Traditionally, rice is cultivated under fully flooded condition during the entire growing period, which required huge amount of water (9000–11,500 $m^3$/ha) (Hossain et al. 2013a, b). This huge amount of freshwater mostly (80%) comes from groundwater pumping (Alexandridis et al. 2009). Among the total groundwater withdrawal (35.87 million $km^3$), the 31.34 $km^3$ water is used for irrigation (FAO 2014; Thenkabail et al. 2010). The irrigation developments and yield improvement have been sustained by the construction of surface water use and shallow and deep tube-well irrigation pump also increased from 0.2 million to 10 million from 1985 to 2012 (Mainuddin and Kirby 2015). Due to excessive groundwater pumping, the groundwater table is decreasing 2–3 ft/year over the country. At the end of 2025, 15–20 million hectares of irrigated rice will suffer extreme water scarcity (Rahman and Parvin 2009). On the other hand, agriculture sector is the largest greenhouse gasses (GHG's) producer (39%) and rice accounts for 24% of total agriculture emission, 13% of country's total GHG's emission (*Climate Watch: Data for Climate Action—Bangladesh*n.d.). The total methane emission from rice paddy was 1120.43 Gg/year in 2017 and expected to increase in future ('FAOSTAT' 2015).

It's a great dilemma for the country to produce more rice for extended population with less irrigation water and greenhouse gas especially methane emission. Alternate Wetting and Drying (AWD) irrigation technique is one of the potential solutions for sustainable rice cultivation. AWD is a management practice in irrigated rice characterized by periodic drying and flooding of fields. Submergence of soil and organic residual material in rice paddies leads to anaerobic decomposition of organic matter that releases methane. Periodic drying events interrupt the duration of this process and reduce methane emissions up to a 70% compared to continuous flooding (Runkle et al. 2019). In addition, AWD can reduce water use by up to 30% without any production loss, thus also resulting in water conservation and reduced fuel consumption required for water pumping (Basak 2016; Carrijo et al. 2018b).

So far, a significant number of studies assessed biophysical suitability of different crops but a very few attempts on AWD irrigation suitability assessment for rice cultivation. Most of the study conducted on the socio-economic benefit (Alam et al. 2010; Carrijo et al. 2017; Education et al. 2016) and environmental advantages (Carrijo et al. 2018a; Minamikawa 2014; Runkle et al. 2019) of the AWD irrigation. Cultivated land quality assessment by using remote sensing based parameters on soil fertility, natural condition and constructive level has been used in microlevel land suitability assessment (Duan et al. 2022).

The land suitability for sustainable crop production has been evaluated with high resolution imageries on soil properties, topography and climate data using Liebig's law (Mesgaran et al. 2017). The study found about 50% of the existing croplands were identified being in low-quality lands which represent unsustainable practices. Nelson et al. (2015) evaluate the seasonally suitable rice area with the remote sensing and field data in the Philippines. The study used seasonal rice area, rainfall, potential evapotranspiration and soil percolation as the suitability determination factors and conducted intensive field study for sensitivity assessment. The dry season rice area found fully suitable for AWD irrigation and 60% of the wet season rice area are suitable. Sander et al. (2017) discuss the country scale AWD suitable rice area in the Philippines with remote sensing and climatic data. Based on water supply and loss of paddy rice field principles, the study estimated the water supply as effective precipitation and the field water loss calculated with soil percolation loss and evapotranspiration in the study area. The dry season rice area was fully suitable for AWD application, whereas the 30% of wet season rice area of the country was suitable for AWD irrigation implementation. Pearson et al. (2018) discuss on the socio-economic barrier to adoption AWD irrigation in Bangladesh. The study reviewed existing published article on AWD irrigation management in Bangladesh and explore the socio-economic barrier to widely adopt the technology in field level. They mentioned several barriers like as farm size, farm ownership, water rental system, yield decreasing and farmer education level as the barrier to adopt AWD irrigation in Bangladesh.

In Bangladesh, AWD irrigation system has been introduced in 2010 by International Rice Research Institute (IRRI). Although the national and international organizations are working on dissimilation of AWD irrigation system in the country, the AWD irrigation system has not been widely applied by farmers in the field level. The widespread uptake of AWD will need support from a range of different stakeholders, namely, irrigation authorities, extension services and local government units, to demonstrate the benefits and viability of AWD to farmers and to adapt the way in which water is provided, managed and valued. The physical environment as well as climatic parameter and socio-economic parameters have a great influence on AWD suitability but somehow there is huge research gap. In this study, we used physio-climatic as well as socio-economic parameters for AWD suitability analysis over Bangladesh. The overall objective of the study is to assess the AWD irrigation suitability in rice area in Bangladesh. To estimate the suitable AWD irrigated rice area, three specific objectives are used:

1. To assess the physically suitable rice area for AWD irrigation;
2. To estimate the physio-climatically AWD irrigated suitable rice area;
3. To assess the socio-economically suitable area for AWD irrigated rice.

## 14.2 Materials and Methods

The methodology consists of (a) AWD irrigation suitable parameters selection for rice cultivation; (b) data collections on the suitability factors for AWD irrigation; (c) data processing and analysis for different types of suitability assessment and (d) the research outcome comparison with collected field data on the study area. The overall research flowchart of this study is illustrated in Fig. 14.1.

### *14.2.1 Study Area*

Bangladesh is one of the largest rice-growing countries in South Asia. It extends from 20°44′00″ to 26°37′51″ N latitude and from 88°00′14″ to 92°40′08″ E longitude and covered 148,450 km$^2$ (Fig. 14.2a). Physically the country is almost at a low altitude, relatively flat except for the north-eastern and south-eastern hill tracts. Administratively, the country is divided into eight major divisions: Dhaka, Rajshahi, Chittagong, Rangpur, Sylhet, Barisal, Mymensingh and Khulna. Climatically, average annual rainfall varies from 1200 mm in the extreme west to over 4000 mm in the northeast and temperature 4 °C in January to 42 °C in April (BMD 2017). The country's mean annual lake evaporation is approximately 1040 mm, which is about 45% of the mean annual rainfall (Kirby et al. 2014). There is a huge seasonal variation in the weather parameter especially the temperature and rainfall (Fig. 14.2b). There are three major rice-growing seasons in the country. Boro rice season (December/January to March/April) is the second largest by area and accounts for 50% (4.84 million hectares) of total rice production of the country. It is the highest yield rice producing season, cultivated in summer with a fully irrigated system and high fertilizer inputs. Aus rice (April/

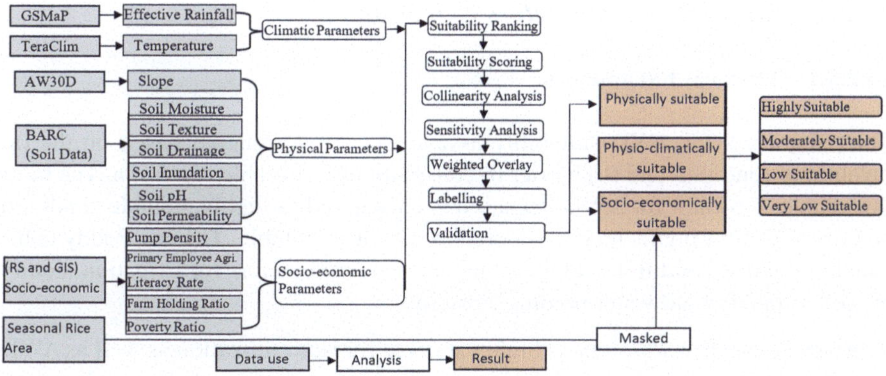


**Fig. 14.1** The overall research methodology includes data used, parameters selection, multicollinearity and sensitivity analysis of the physio-climatic and socio-economic parameters and outcome of this research

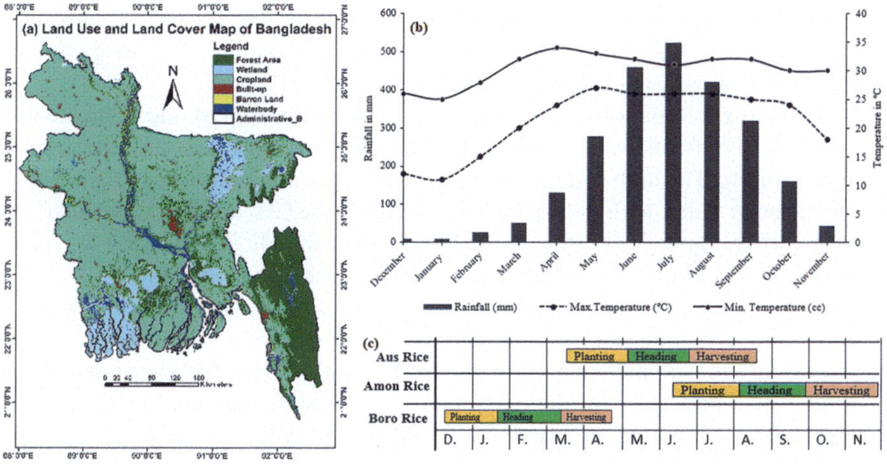


**Fig. 14.2** (**a**) Geographical location and generalized land use and land cover (LULC) map-2017 of the study area from MCD12Q1, (**b**) monthly average minimum and maximum temperature, and rainfall over Bangladesh (BMD (Bangladesh Meteorological Department) 2017), (**c**) rice paddy crop calendar in the study area

May to July) is a smaller rice area, accounts for 9% (1.04 million hectares) of total production and cultivated in monsoon season. The Aus is cultivated under almost rainfed condition but with great uncertainty due to the flood. Amon rice (July/August to November) is the largest (5.53 million hectares) rice-growing season and accounts for 41% of total rice production (Fig. 14.2c). This rice season is a very critical growing period due to the flood in the early growing stage and frequent drought in the later growth stage. The rice production in the country is seriously affected by climate change impacts.

## *14.2.2 Suitability Parameters Selections*

### 14.2.2.1 Physical Parameters

The parameters for AWD suitability assessment are considered based on the (a) favourable condition and (b) barrier or constraint to favourable condition. The most favourable condition for AWD irrigation is considered as highly suitable condition and the most limiting factors considered as very low suitable. The rice paddy cultivated in flooded condition and the slope are important factors for irrigation facility as well as paddy field water holding capacity.

1. **Slope:** Less than 2% slope is highly suitable for rice cultivation as well as AWD implementation (Rabia et al. 2013). The higher percentage of slope is less suitable for rice cultivation.
2. **Soil texture:** Soil texture is the particles size of the relevant content of the soil. The particles like sand, silt, clay amount in the soil can influence water and air holding capacity. The silty clay soil has less percolation rate and highly capable to hold water in the rice paddy and highly suitable for the rice paddy cultivation (Sander et al. 2017). The silty clay soil is highly suitable for AWD irrigation for higher irrigation water holding capacity.
3. **Soil moisture:** Soil moisture is a very important component of microclimate that both directly affects water availability and indirectly influences the temperature and humidity. The high soil moisture is good for rice cultivation (Sys et al. 1993). On the other hand, with increasing soil moisture, the $CH_4$ uptake decreased (Luo et al. 2013). The soil moisture from 200 to 300 mm is highly suitable for AWD irrigation and more or less than it, are comparatively lower suitable.
4. **Soil pH:** Soil pH is the indicator of soil alkalinity or acidity. The soil pH influences the soil nutrient content and methane emission. Methane emission increased with the methane emission (Wang et al. 2010). But the standard pH level is also important for the rice growth and yield maintenance. The pH level less than 4 and more than 8 is very low suitable for AWD irrigation.
5. **Soil drainage:** Soil drainage is a natural process which determines the water movement across, through and out of the soil as gravity pulls. It can determine the land suitability of a particular crop. The poor and high drainage on the soil are responsible for water logging and dry soil, which are not good for rice cultivation as well as AWD irrigation (Sys et al. 1991).
6. **Soil permeability:** Soil permeability is the quality of soil responsible for water and air transmission. It is directly influencing the soil percolation loss and seepage loss and affects the irrigation management. Rapid permeability rate response for fasterr field irrigation water loss (Fields 2009). The slow permeability is slow percolation rate and good for the AWD irrigation.
7. **Inundation:** Land inundation refers to the land area brought by water through natural causes such as river overflow, heavy rainfall or artificial causes like dam construction. Bangladesh is a low land area and inundation plays a very important role in rice cultivation. The very low inundated area is not suitable for rice

cultivation and as well as for AWD irrigation (Sys et al. 1993). Similarly, very high land is also not suitable for AWD irrigated rice cultivation.

#### 14.2.2.2 Climatic Parameters

1. **Precipitation:** Irrigated agriculture is responsible for 80% of freshwater withdrawal (Demarez et al. 2019) and rice is the major consumer of the irrigated water. The AWD irrigation technique is a control irrigation system and the irrigation schedule is managed as per the necessity of Crop Water Requirement (CWR). The irrigation scheduling and water cycle managed with the surface water level and soil moisture observation. As precipitation is a natural phenomenon and interrupt the AWD irrigation cycles. The high precipitation rate is low suitable for AWD irrigation application (Nelson et al. 2015).
2. **Temperature:** The land surface temperature is another important factor for rice cultivation (Neteler 2010) as well as AWD irrigation. The soil temperature both directly and indirectly influences the soil pH, evapotranspiration, and methane emission. Moreover, the low and very high temperatures are not good for rice production. The soil temperature more than 36 °C and less than 16 °C is not good for rice late tillering to panicle and gap filling stage (Fields 2009). The soil temperature more than 33 °C increased the methane emission rate and temperature between 26 and 29 °C is best suitable for AWD irrigated rice (Arunrat and Pumijumnong 2017).

#### 14.2.2.3 Socio-Economic Parameters

1. **Irrigation pump density:** Irrigation is the key component of AWD irrigation. The benefit of AWD irrigation in terms of water saving and production cost reduction fully depends on irrigation system. In Bangladesh, the irrigation facilities mostly owned by rich farmers, while the poor farmers rent water and pay a fixed amount to the pump owner, whatever the farmer apply AWD or Continuous flooded (CF) irrigation (Adnan 2007). Moreover, the government financial subsidies for irrigation are enjoyed by the pump owner. As a result, the AWD saving beneficiary is the pump owner. But the instalment of low-cost shallow pump owned by farmer helps to control the irrigation facilities. The pump density is considered as socio-economic parameter as the more irrigation pump means the higher suitable for AWD irrigation.
2. **Land holding size**: The average farm holding size in Bangladesh by household is 3.67 ha and it's differed from divisional variation in 2.76–4.92 ha (Mottaleb et al. 2016). Farm holding size is also strongly related to farm ownership. In Bangladesh, only 37% cultivate their own land, 34% are pure tenants and 29% cultivate their own plus rented land (Ahmed et al. 2013). The farm size and land ownership are one of the socio-economic barriers to adopt AWD irrigation as it decreases the farmer's technical efficiency (Majumder et al. 2016). In this study,

we consider the higher land ownership percentage is better for AWD irrigation application.

3. **Poverty:** Poverty means lack of resources or access to facilities. Most of the farmers are poor and their annual income is very low. It's very difficult for them to adapt or experiment with new facilities. Due to the poverty, farmers used microcredit loan from the NGO's during cultivation period (Pearson et al. 2018). The poor farmers have no land ownership and pump ownership, as a result the high poverty porn region has less potential to adopt AWD irrigation.
4. **Literacy rate:** Literacy rate or farmer level of education is an important factor to adopt AWD irrigation. The benefit of AWD irrigation depends on the proper application of AWD irrigation. Several studies reported the yield decreased due to the improper application of AWD (Pearson et al. 2018). Moreover, AWD not only has water saving benefits but also works for greenhouse gas emission reduction. To understand the proper implementation and advantage of AWD irrigation, farmers awareness in terms of education level was assessed on this study.
5. **Primary Employment and Population Density Ratio (PEA/PPP):** The small farm holder size, landless farmer and poverty are the major constraint for widely application of AWD irrigation. To overcome these problems, community level adaptation and co-operational farming are very effective for AWD adaptation (Rahman and Sander 2017). As a result, the primary employment in agriculture and the population density are set as indicators to large community level adaptation of AWD irrigation.

### *14.2.3 Dataset Used*

Several dataset have been used for this study. Moderate resolution imaging spectroradiometer (MODIS) MOD16A2 derived seasonal rice area map from earlier study, global satellite mapping of precipitation (GSMaP) data for precipitation estimation monthly temperature from Terra Climate, and ALOS-AW3D30m data for slope analysis has been used for this study. The remote sensing and associated data derived poverty dataset (Steele et al. 2017) and population density data (Stevens et al. 2015) were also used. Besides the remote sensing data, GIS-based shapefile data of soil properties like texture, soil drainage, soil inundation, soil pH, soil permeability, soil moisture data from Bangladesh Agriculture Research Council (BARC) have been used. The AWD irrigation application field location and water saving data from Rural Development Academy (Bogura) were used for model validation. The socio-economic data on irrigation pump density, poverty, literacy rate and other relevant data from BARC, Bangladesh Meteorological Department (BMD), Bangladesh Bureau of Statistics (BBS), Bangladesh Rice Research Institute (BRRI), Commonwealth Scientific and Industrial Research (CSIRO), Food and Agriculture Organization of United Nations Statistics (FAOSTAT), and relevant literature review from the journal and publications were also used. The data types, spatial and temporal resolution, purpose of the data are shown in Table 14.1.

### 14.2.4 Suitability Assessment

Land suitability assessment is highly needed for productive planning in climatic vulnerable countries (Hossen et al. 2021). This research used suitability assessment framework proposed by FAO (Sys et al. 1991). In this context, suitability can be summarized in a generic model of Eq. (14.1)

$$S = f\left(x\_1, x\_2, ?x?\_(3,?)\, x\_n\right) \tag{14.1}$$

where $S$ is defined as suitability index and $x_1$, $x_2$, …, $x_n$ are the factors affecting the overall suitability, such as soil, topology and climate parameters. To assess the bio-physical suitability of a given crop, geomorphological suitability, metrological suitability and socio-economic suitability are usually evaluated separately. In this study we considered 7 (seven) physical factors, 2 (two) climatic factors and 5 (five) socio-economic factors affecting suitability of AWD irrigated rice, considering geomorphological, metrological and socio-economic characteristics of environment based on detailed literatures review discussed earlier. The selected parameters data were collected from various sources and converted into same scale and formal. The best value of the parameters for AWD irrigation has been evaluated with FAO water

**Table 14.1** Remote sensing and geospatial dataset used this study from different sources, spatial and temporal resolution

| Dataset | Product | Spatial resolution | Temporal resolution | Purpose |
|---|---|---|---|---|
| Rice area (Islam et al. 2023) | Rice area | 1000 m | Seasonal (2001–2020) | Rice mapping |
| Poverty (World Bank 2020) | Wealth index | 1000 m | Annual (2005, 2015) | Poverty |
| Population (World Bank 2020) | Gridded | 1000 m | Annual (2005, 2015) | Population density |
| GSMaP (JAXA 2020) | Precipitation | 0.1° | Monthly (2001–2020) | Precipitation |
| Terr climate (Abatzoglou et al. 2018) | Temperature | 1 km | Monthly (2001–2020) | Temperature |
| ALOS | AW3D30 | 30 m | – | Slope |
| Soil properties (BARC (Bangladesh Agriculture Research Council) 1998) | Texture, moisture, drainage, pH, permeability, inundation | 1 km | Annual (1998) | Physical suitability |
| Pump density (BADC 2018) | Pump density | – | Annual (2005, 2015) | Irrigation facility |
| Literacy rate (World Bank 2020) | Education level | – | Annual (2001, 2011) | Awareness |
| Primary employment in agriculture (World Bank 2020) | PEA | 1 km | Annual (2005, 2015) | Adaptability |

requirement principles for rice crops, field observation, expert opinion and literature review. Based on the influences on AWD irrigation, the suitable value of a parameter is fixed as very low to very high suitability. The parameters are individually assigned scores from suitability value ranged from very low suitable (0) to highly suitable (9). The multicollinearity analysis had been conducted to assess the interrelationship among the parameters. The sensitivity analysis is used to determine the weighted values of each parameters towards the AWD suitability modelling (Seo et al. 2010). Finally, the parameters are grouped into three different group as physically suitable, climatic suitable and socio-economic suitable and three different types of suitability: (a) physically suitable, (b) physio-climatic suitable and (c) socio-economic suitable area are estimated.

#### 14.2.4.1 Physically AWD Suitability

The study area is located in the lower alluvial Ganges basin and physically almost flat landform with very few variations except the southeastern and northeastern hilly region. In this study, the ALOS Global Digital Surface Model, ALOS World 3D-30M model (AW3D30) dataset were used for the slope analysis (Takaku et al. 2016). The slope properties were converted into percentage and reclassified and rescaled according to the suitability. The soil properties, drainage, texture, pH, permeability, and inundation data produced by Bangladesh Agriculture Research Council (BARC) as GIS shapefile format with FAO soil standard (BARC 1998) in three different time frame on 2001, 2010 and 2020. The dataset reclassifies, resample and rasterized for the suitability analysis. Then, we assigned the weighted value for the physical parameter and analysed the sensitivity for the suitable weighted. Furthermore, we investigated the interrelationship among the physical parameter, analysed the multicollinearity of the physical parameters and used it for suitability model. Finally, estimated the physically suitable rice area for AWD irrigated rice area in 2001, 2010 and 2020. Table 14.2 shows the detail dataset with suitability classes value of the physical parameters.

#### 14.2.4.2 Physio-Climatic Suitability

The physical parameters are used as a single parameter with the climatic parameter to analyse the physio-climatic suitability. This study used TerraClimate dataset for all temperature related parameters with 2.5 arc minutes spatial resolution (Abatzoglou et al. 2018) and Global Satellite Mapping of Precipitation (GSMaP) with 10 $km^2$ spatial resolution for two precipitation related parameters (Kubota et al. 2006). All meteorological parameters were rescaled into 1000 m spatial resolution and monthly temporal resolution. The physical parameters are annual, but the climatic parameters are monthly averaged. To keep the harmony between the physical and climatic parameter, we prepared an annual average climatic dataset from 2001 to 2020. The monthly climatic data prepared the seasonal composite based on the seasonal rice calendar of the study area. Next, the physical and climatic

**Table 14.2** Physical parameters—slope and inundation; soil properties—texture, permeability, drainage, soil pH data and scored value according to the importance of suitability for AWD rice irrigation based on the FAO principle, field observation and estimation, sensitivity analysis and expert opinion

| Parameters and scored | Highly suitable (9) | Moderately suitable (7) | Low suitable (5) | Very low suitable (3) | Not suitable (1) |
|---|---|---|---|---|---|
| Soil texture | Silt clay | Fine sandy loam | Silt loam | Sandy loam | Fine sand |
| Soil permeability | Slow | Slow-moderate | Moderate | Mixed | Rapid |
| Soil moisture | 200–300 mm | 300–400 mm | 100–200 mm | >400 mm | <100 mm |
| Drainage | Imperfectly drain | Poor drained | M. well drain | Well drained | Very poor drain |
| Soil inundation | Medium lowland | Medium highland | Low land | Highland | Very highland |
| Slope | 0–2% | 2–4% | 4–8% | 8–25% | >25% |
| Soil pH | 5.5–7.3 | 7.3–7.8 | 7.8–8.4 | 4.0–5.0 | <4.0 and >8.4 |

**Table 14.3** Climatic parameters—effective precipitation and temperature data and scored according to the importance of suitability for AWD rice irrigation based on the FAO principle and sensitivity analysis

| Parameters/ scored | Highly suitable (9) | Moderately suitable (7) | Low suitable (5) | Very low suitable (3) | Not suitable (1) |
|---|---|---|---|---|---|
| Effective precipitation | 200–300 mm | 300–400 mm | 400–500 mm | 500–800 mm | >800 mm |
| Temperature | 27–29 °C | 29–31 °C | 31–33 °C | <27 °C | >33 °C |

parameters are overlay as seasonal composite. The weighted value assigned for the physio-climatic suitability assessment based on the sensitivity analysis. Finally, a physio-climatically suitable rice area map was prepared for 2001, 2010 and 2020. Table 14.3 shows the physio-climatic parameter and their suitable range with three different sets of weighted value for sensitivity analysis.

The physical parameters and climatic parameters are combinedly used for physio-climatic suitability analysis. To fix the weighted value for the suitability model, we analyse the sensitivity of the parameters.

#### 14.2.4.3 Socio-Economic Suitability

The socio-economic parameters are selected and collected from different sources of data. The poverty dataset is collected from World Bank report published per head income count poverty situation in Bangladesh. The poverty dataset used the remote sensing, statistical and mobile detail call record (CDR) dataset to estimate the poverty in 1 km spatial resolution based on the income level in 2015 and census report of Bangladesh (Steele et al. 2017). The per pixel population (PPP) dataset also used remote sensing data and associated dataset to produce population density in 1 km grid scale (Stevens et al. 2015). The primary employment in agriculture (PEA) data collected from Bangladesh Bureau of Statistics (BBS) national household survey dataset. The statistical dataset in smallest administrative level (Upazilla) and converted it into 1 km spatial resolution geospatial dataset. The PPP and PEA dataset blended to prepared PPP and PEA ration dataset to use for the people engagement in agriculture assessment. The irrigation pump density dataset from minor irrigation survey data from BBS, 2001 to 2020 and the literacy dataset also collected from BBS census report. The irrigation pump density and literacy rate estimated in Upazilla level with geospatial techniques and produced 1 m spatial resolution dataset for socio-economic suitability assessment. All the parameters assigned appropriate suitability score based on literature review, expert opinion and prior knowledge. The weighted value was assigned after sensitivity and multi-collinearity analysis. Furthermore, the FAO land evaluation for rice crop is modified for the AWD irrigation suitability (Sys et al. 1993) (Table 14.4). In this study, we analysed the AWD suitability with physio-climatic parameters as four suitable classes: high suitable, moderately suitable, low suitable and very low suitable. Finally, the seasonal rice area map was used for suitable rice area extraction in 2001, 2010 and 2020.

### *14.2.5 Sensitivity and Multicollinearity Assessment*

The multicollinearity has been analysed to find out the similarities among the parameters. As the similar types of influencing factor responds same weighted value on the final suitability. The physical suitable parameters collinearity shows that the parameters are individual and negligibly correlated with each other. The highest correlation between soil inundation and permeability and other parameters is almost independent. In case of socio-economic parameters the highest correlation between irrigation pump density and farm holding size, and between PPP/PEA ratio and poverty among the parameters, and other parameters are almost independent (Table 14.5). The variance influence factors (VIF) analysis shows that the VIF value is less than 2. The VIF value of less than 4 is a very good independent variable for predicted variables (Murray et al. 2012). Furthermore, the sensitivity result of the physical parameter illustrates the less sensitive to the suitable classes and equal influence have been selected.

**Table 14.4** Selected socio-economic parameters and the assigned scores according to the importance of suitability for AWD rice irrigation based on the field experiment and estimation, sensitivity analysis and expert opinion

| Parameters and scored | Highly suitable (9) | Moderately suitable (7) | Low suitable (5) | Very low suitable (3) | Not suitable (1) |
|---|---|---|---|---|---|
| Pump density | >40 | 30–40 | 20–30 | 5–20 | 0–5 |
| Farm holding size | >1.30 | 1.20–1.30 | 1.15–1.20 | 1.10–1.15 | <1.15 |
| Literacy rate | 66–84 | 55–66 | 45–55 | 34–44 | 24–34 |
| Poverty | 0–120 | 120–150 | 150–180 | 180–210 | >210 |
| PPP/PEA | >10 | 2–10 | 0.15–2 | 0.08–0.15 | <0.08 |

**Table 14.5** The multicollinearity matrix for the (A) physical and (B) socio-economic parameters with the variance inflation factor (VIF)

| **Part (A)** | **S_Texture** | **S_Slope** | **S_pH** | **S_Permeability** | **S_Moisture** | **S_Inundation** | **S_Drainage** |
|---|---|---|---|---|---|---|---|
| S_Texture | 1 | | | | | | |
| S_Slope | −0.01615 | 1 | | | | | |
| S_pH | 0.16898 | 0.2501 | 1 | | | | |
| S_Permeability | 0.22472 | 0.1299 | 0.2006 | 1 | | | |
| S_Moisture | −0.04852 | 0.1009 | −0.094 | −0.05218 | 1 | | |
| S_Inundation | −0.08564 | −0.297 | −0.283 | 0.300281 | 0.215257 | 1 | |
| S_Drainage | 0.10060 | 0.1086 | 0.0969 | −0.0387 | −0.04342 | −0.09162 | 1 |
| **Part (B)** | **IPD** | **FHS** | | **Literacy rate** | **Poverty** | **PPP/PEA** | |
| IPD | 1 | | | | | | |
| FHS | 0.236082 | 1 | | | | | |
| Literacy rate | 0.089173 | 0.11066 | | 1 | | | |
| Poverty | −0.16322 | −0.13103 | | 0.142715 | 1 | | |
| PPP/PEA | −0.11076 | −0.00778 | | −0.14016 | −0.2435 | 1 | |

The sensitivity analysis is used to explain how the factors can influence the final output. The high sensitivity value means the high influencing capability. The physical suitability parameters sensitivity analysis result shows that the equally weighted value determined a very highly suitable area of 50.56%, whereas the 70% weight-imposed parameters determined highly suitable area fluctuated**.** The soil slope and drainage parameters are higher sensitive for highly suitable areas, 50.48% and 49.63%, respectively. Soil inundation and soil permeability exposed the less sensitive in highly suitable area case, 40.02% and 47.38%, respectively. The socio-economic parameters sensitivity result shows the various influenced result on the suitability classes. The suitable class influence illustrated 4.76–8.72% on the highly suitability class. The farm holding size, PPP/PEA ratio and the irrigation density showed the higher influenced on the highly suitable class (Fig. 14.3). The equally weighted parameter demonstrated the sensitivity on highly, moderately and low suitable classes are 9.21%, 47.64% and 43.1%, respectively. In this study, we used the equally weighted parameters for suitability analysis.

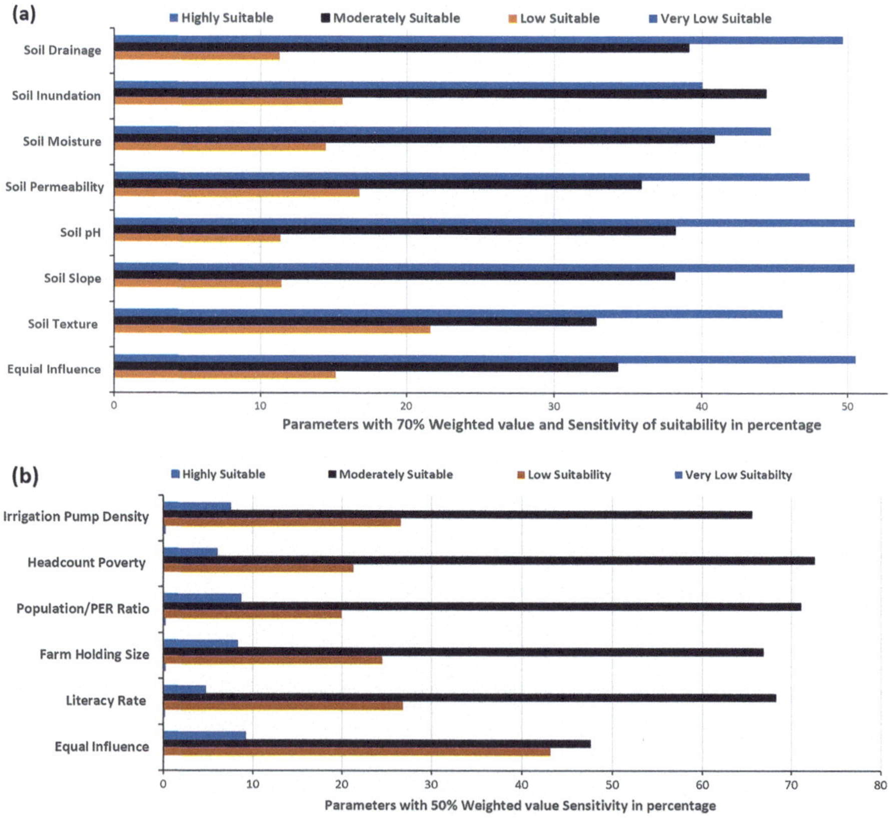


**Fig. 14.3** Sensitivity analysis result of the (**a**) physical factors with equal influence values and 70% weighted values; (**b**) socio-economic factors with equal influence values and 50% weighted values to compared sensitivity for AWD suitability assessment

### 14.2.6 AWD Suitability Validation with Field Data

The result from the AWD suitability analysis compared with the filed data of different AWD irrigation practices field. The AWD irrigation implemented 87 rice paddy site data over the study area were collected from Rural Development Academy (RDA) reported data. The field location and the water use dataset used for site specific water saving dataset. The AWD field location with water saving data was extracted from the AWD suitable class map. The AWD water saving and suitable class are compared to assess the model performance.

## 14.3 Results

### 14.3.1 Physical Suitable Rice Area

The physically suitable AWD irrigated result showed that Rajshahi, Rangpur and the northern part of Dhaka division of the country are highly suitable for AWD irrigated rice area. The comparatively high elevation, dominant silty clay soil, imperfect drainage soil and comparatively lower moisture induced the region to be physically highly suitable for AWD irrigated rice area. The southern part of the administratively Khulna division and the eastern part of the country Sylhet division are low suitable for AWD irrigated rice area. The southern coastal area with very low inundation and fine sandy soil types influenced the suitability. The southwestern part of the country is the mountain region with highly drainage soil, and soil texture makes the region low suitable rice area (Fig. 14.4).

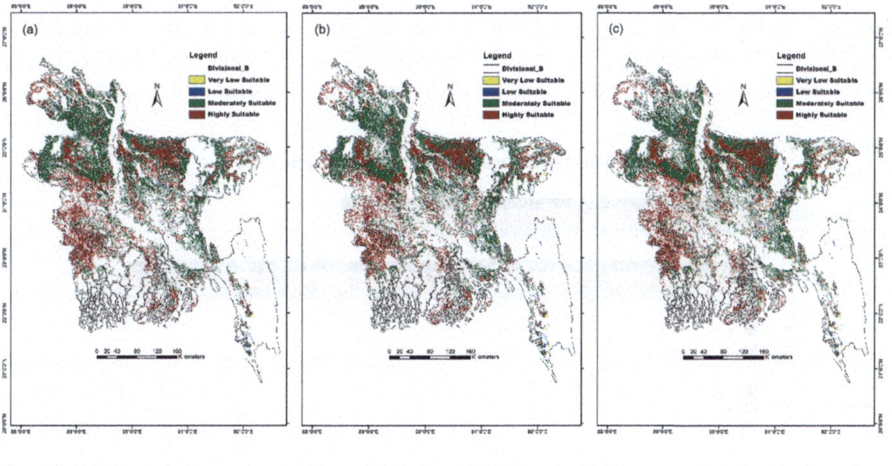


**Fig. 14.4** Physical alternative wetting and drying (AWD) suitable rice area distribution based on the physical properties over Bangladesh; (**a**) 2001, (**b**) 2010 and (**c**) 2020

Among the three different rice growing seasons, there were no significant difference of suitability. The most suitable for AWD irrigation was boro rice season (36.79%), whereas the Amon and Aus rice area were 38.21% and 39.18%, respectively. Similar trend in moderately, low and very low suitable rice area case. The physical condition of the country is almost similar all through the year and there are very few annual fluctuations over the time. In 2001, the highly suitable class for AWD irrigation was 39.18% and 38.21% in 2010 and 36.79% of total rice area in 2020. A similar trend was observed in the case of moderately, low and very low suitable classes for AWD irrigated rice area (Fig. 14.5). The probable reason behind the similar suitability trend is there is very limited seasonal variation of the physical parameters.

### 14.3.2 *Physio-Climatic Suitable Rice Area*

The sensitivity of the three different parameters for physio-climatic suitability showed that the precipitation was the most influential parameter for physio-climatic suitability (**S1**). In the set A and C case, the precipitation weighted value was constant and in case B the value decreased, and the suitability class also changed in B case. The physical parameter is moderately sensitive, and temperature is the least sensitive for the physio-climatic suitability. As a result, we used the set A value for the physio-climatic suitability determination for the study (Table 14.6).

The physical and climatic suitability parameters were overlayed and prepared for the physio-climatically suitable area. As the country is physically suitable for AWD irrigation from this model. But the seasonal weather condition changed the suitability classes over the area. The central-north and western part of the country especially known as Rajshahi and Rangpur divisions is high and moderately suitable for boro rice growing season. Some parts of this region exposed comparatively low suitable as the region is dominant wet land (Chalan beel) in the country. The southern and eastern parts of the country, especially Khulna and Sylhet division, are comparatively low suitable for AWD irrigated rice area. In the case of amon rice shows the moderately suitable and the distribution demonstrated the same as boro

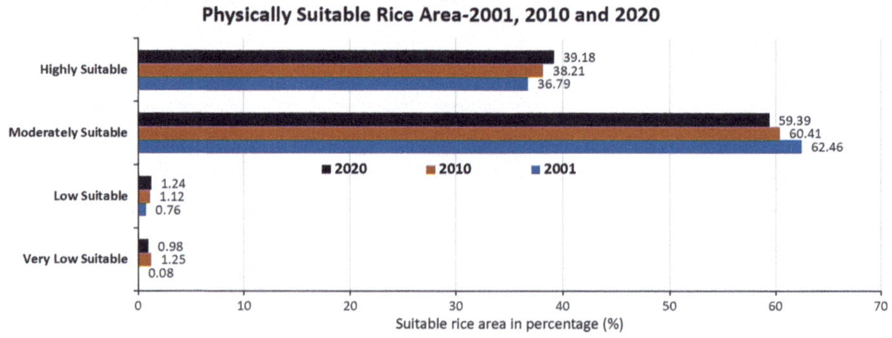


**Fig. 14.5** Physically suitable seasonal rice area of Bangladesh in 2001, 2010 and 2020

**Table 14.6** Physio-climatic parameters sensitivity result of three different set (S1) weighted value and suitability of Boro, Aus and Amon rice growing season

| Rice season | Boro rice area (%) | | | Aus rice area (%) | | | Amon rice area (%) | | |
|---|---|---|---|---|---|---|---|---|---|
| Suitability/cases | A | B | C | A | B | C | A | B | C |
| Highly suitable | 79.312 | 69.069 | 87.375 | 0.327 | 8.471 | 0.218 | 0.997 | 0.966 | 0.023 |
| Moderately suitable | 20.625 | 30.723 | 12.443 | 15.571 | 23.680 | 8.300 | 50.444 | 62.040 | 38.882 |
| Low suitable | 0.061 | 0.206 | 0.181 | 71.465 | 55.457 | 64.407 | 47.364 | 35.407 | 56.420 |
| Very low suitable | 0 | 0 | 0 | 12.635 | 12.389 | 27.073 | 1.193 | 1.585 | 4.674 |

rice season. The Aus rice season exposed as the lower suitable AWD irrigated rice area. Figure 14.6 shows the physio-climatically suitable rice area distribution of three different rice seasons over the country in 2001, 2010 and 2020.

There are very few temporal changes of suitable rice area among the different years. Statistically, 78.98% of total boro rice growing area in 2001 was highly suitable for the AWD irrigation and 82.88% in 2010 and become 79.21% in 2020 in terms of physio-climatic suitability. An almost similar trend assessed in case of moderately and low suitable for boro rice growing season. The dry season with less rainfall (BMD 2017) and totally irrigated condition (Portmann et al. 2010) made the boro rice season highly suitable for AWD irrigation. In the amon rice growing season, 1.23% was highly suitable, 66.32% moderately suitable, 28.55% low suitable and 3.9% of total amon rice area in 2001. In 2010 and 2020, the suitable amon rice area for AWD irrigation was almost the same. The early growing stage of amon rice paddy experienced huge rainfall but in the mature and flowering stage was almost rainless as a result the season is moderately and low suitable for AWD irrigation. The Aus rice season is the least suitable rice season for the AWD irrigation application. Only 0.34% of total Aus growing area is highly suitable and 15.77%, 71.46% and 12.63% of the total Aus rice area are moderately, low and very low physio-climatic suitable, respectively, in 2020. In 2010, the Aus suitable rice area was comparatively high (4.46%) in comparison with 2001 and 2020. The Aus rice season contributed very few (9.86%) to the country's total rice production. The growing period is a flood prone season and the plantation date shifting to cope the flood impacts. In terms of temporal changes, the boro rice area suitability slightly increased from 2001 to 2020, whereas the amon rice suitable area decreased in the same period. The suitability of Aus rice area gradually increased from 2001 to 2020. The detailed suitable area results for boro, amon and Aus rice area are shown in Fig. 14.7.

### 14.3.3 *Socio-Economically Suitable Area*

The socio-economic parameter determined suitable area for AWD irrigation in Bangladesh showed variation from physical and physio-climatic suitable rice area. Among the three different rice growing seasons, there is almost similar proportion of suitability. As the socio-economic parameters have no seasonal variation, the seasonal rice growing season also showed slight variation. Only 1.91% of total boro rice area was highly suitable, whereas the moderately, low and very low suitable

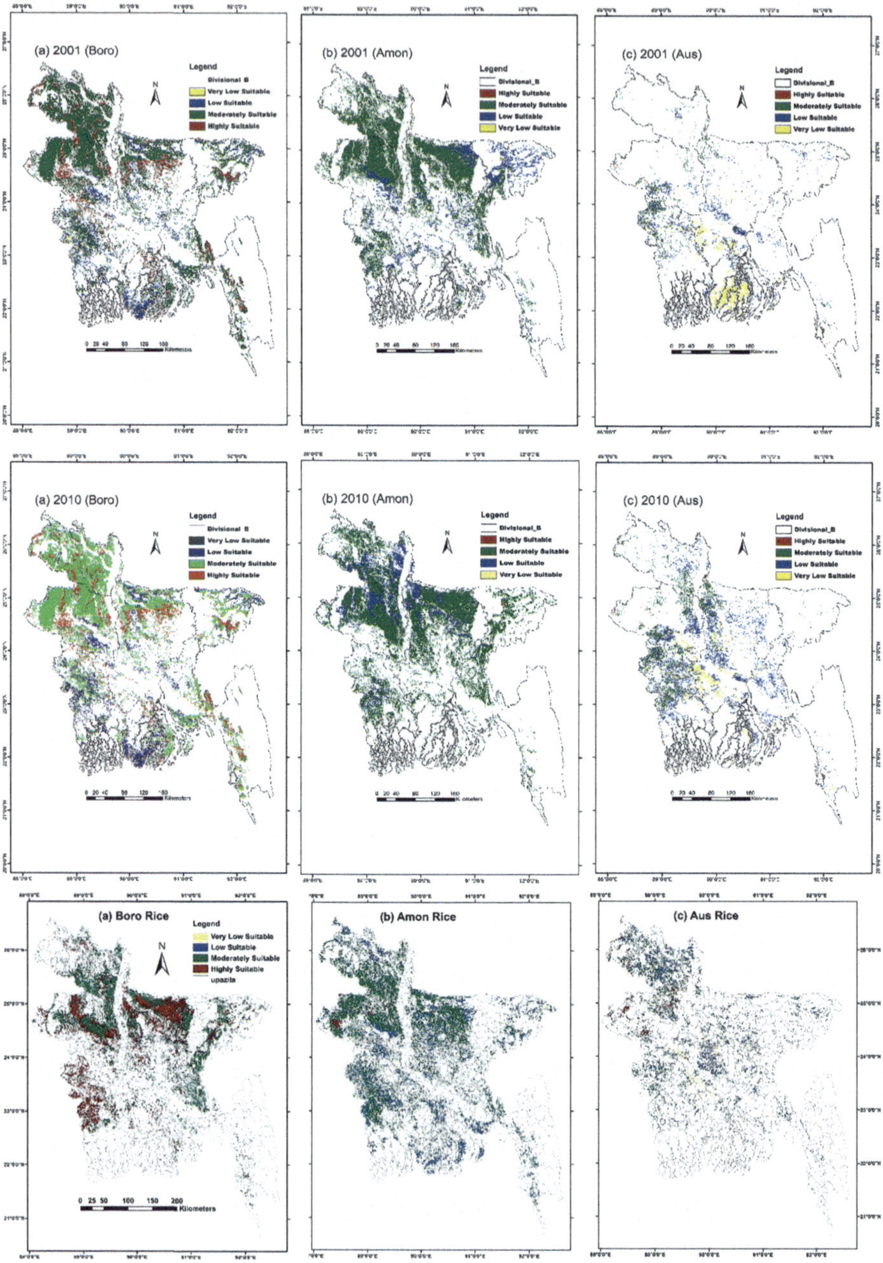


**Fig. 14.6** The physio-climatic suitability classes for AWD irrigated rice area (**a**) boro rice season in 2001, 2010 and 2020, (**b**) amon rice season in 2001, 2010 and 2020 and (**c**) Aus rice season in 2001, 2010 and 2020

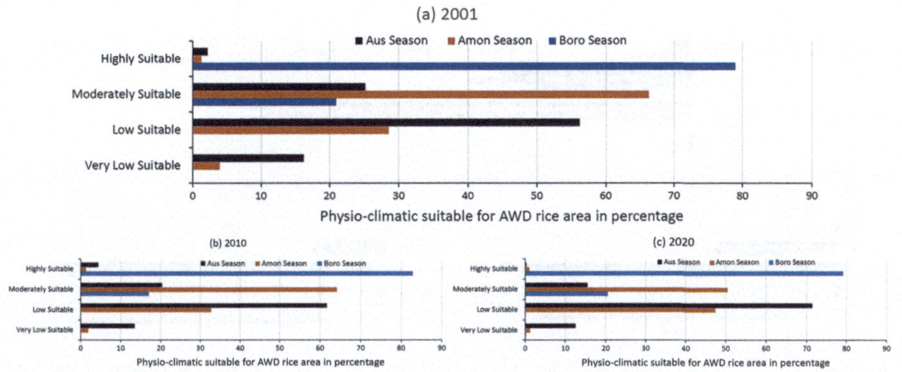


**Fig. 14.7** The physio-climatic suitability class statistics of AWD irrigated (**a**) boro, amon and Aus rice season in 2001, (**b**) boro, amon and Aus rice season in 2010 and (**c**) boro, amon and Aus rice season in 2020

boro rice area are 79.16%, 17.91% and 1.0%, respectively, in 2010. Almost similar trend was observed in 2001 and 2020 (Fig. 14.8). But physically and physio-climatically the boro rice area showed very highly suitable for AWD irrigation. The amon and Aus rice season also showed a similar percentage of suitability as boro rice season. The socio-economically suitability of AWD rice growing area gradually increased from 2001 to 2020. Increased irrigation facilities, poverty rate declination and growing awareness are the probable causes of this trend.

The central north-western and central-northern part of the country exposed comparatively higher suitability for AWD irrigation implementation. The area as boro-amon double rice pattern is the dominant rice cropping pattern. A very small area in central northern part is highly suitable for AWD irrigation. The probable causes of the suitability area are comparatively rice farmer, literacy rate and the irrigation facilities. The others part of this region exposed comparatively moderate suitability. Although the similar cases but the poverty rate is higher in this region makes as comparatively moderate suitable area for AWD application. The most northern part and the south-eastern part were the comparatively low suitable for the AWD irrigation implementation. The low irrigation facility, PPP/PEA ratio induced the region as such suitability rate. Figure 14.9 shows the socio-economically AWD irrigation suitable area distribution over the country.

### *14.3.4 Suitability Classes Validation*

This study determines AWD suitable area has been compared with the field data. The AWD irrigated rice data on water use, crop height and yield have been collected from Rural Development Academy (RDA). All together 86 field sites under AWD irrigation all over the country have been randomly selected for comparison. In Bangladesh, the AWD irrigation applied only in the Boro rice season and these data were collected in 2019, Boro rice cultivation season. The data were collected from

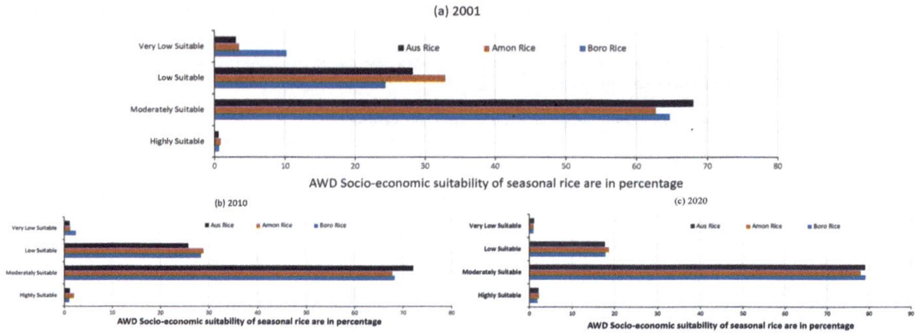


**Fig. 14.8** The socio-economic suitability class statistics of AWD irrigated (**a**) boro, amon and Aus rice season in 2001; (**b**) boro, amon and Aus rice season in 2010 and (**c**) boro, amon and Aus rice season in 2020

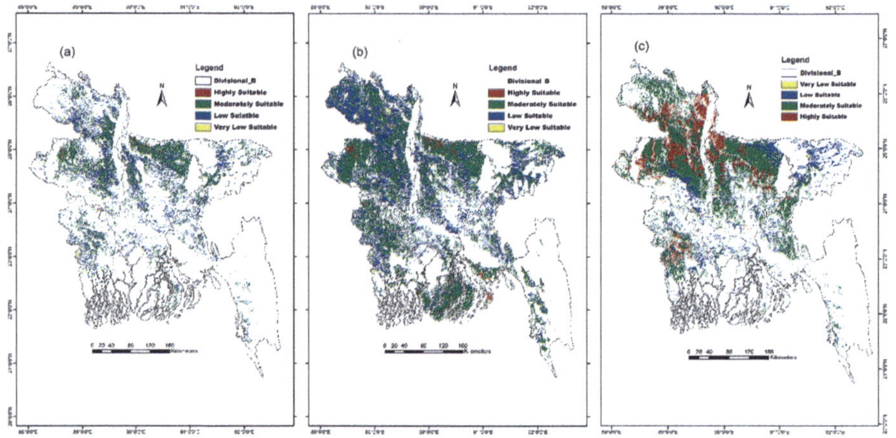


**Fig. 14.9** Socio-economically suitable rice area for AWD irrigation in Bangladesh; (**a**) 2001, (**b**) 2010 and (**c**) 2020

the AWD and non-AWD irrigated rice field. The water used in AWD field and non-AWD irrigated field compared and calculated the water saving from the AWD irrigation application. Most of the AWD field with higher water saving rate are s in the central and northern part of the country (Fig. 14.10). The water saving result distribution showed the harmonic trend with physio-climatic and socio-economic suitable classes from this study. Among the 86 fields, the 20 fields saved more than 30% water, 58 field saved 30–20% water and 8 field saved less than 20% water compared with non-AWD irrigated rice field. Among the >30% water saving field (20), 19 field area within the highly suitable AWD irrigated area and 1 field in moderately suitable area. The 20–30% water saving fields (58) also followed the suitability pattern, only 5 fields were in moderately suitable area and rest of them were in highly suitable area. There were eight less water saving (<20%) fields, six fields were in moderately suitable area and two fields in highly suitable area. Figure 14.10 illustrated the AWD water saving field with the AWD suitable classes all over the country. The highly suitable and water saving field are mostly located in the north-western

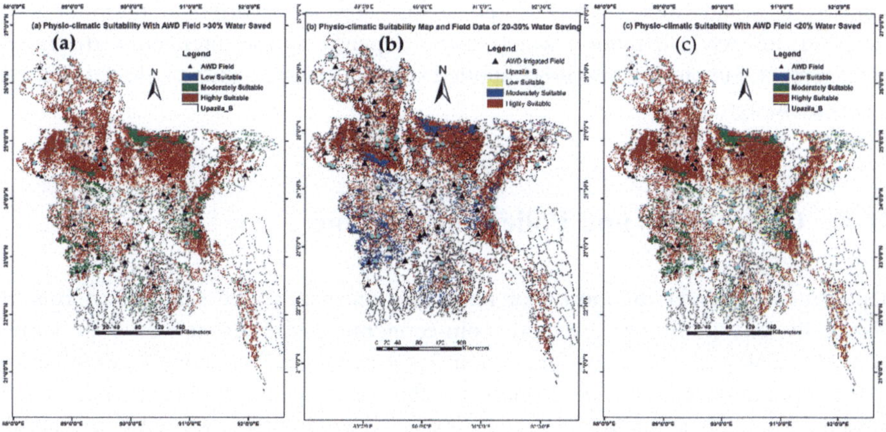


**Fig. 14.10** Estimated physio-climatically AWD suitable Boro rice area validation with field data, the water saving amount and AWD irrigated site in study area; (**a**) AWD field with >30% water saving, (**b**) AWD field with 20–30% water saving, (**c**) AWD field with <20% water saving in the study area

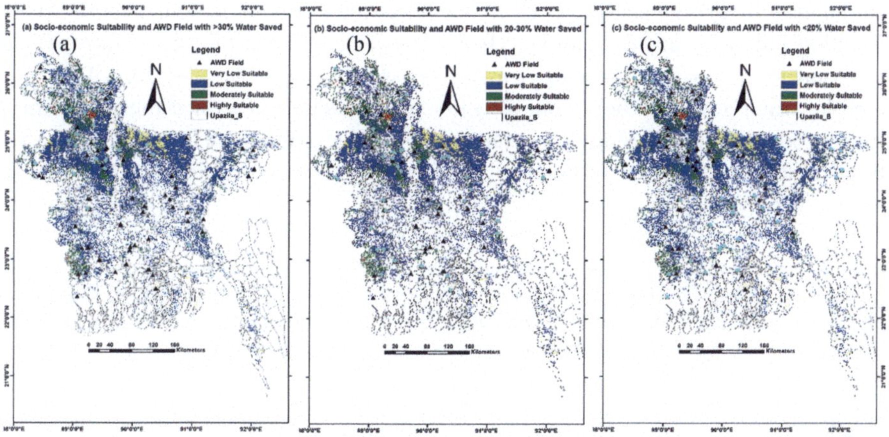


**Fig. 14.11** Estimated socio-economically AWD suitable Boro rice area validation with field data, the water saving amount and AWD irrigated site in study area; (**a**) AWD field with >30% water saving, (**b**) AWD field with 20–30% water saving, (**c**) AWD field with <20% water saving in the study area

part of the country. The less water saving field is mostly located in moderately suitable area and south-eastern part of the country.

In case of estimated socio-economic suitability classes and field data comparison shows good agreement. The socio-economic parameter especially the poverty in wealth index and suitability classes distribution result case, the north-western part of the country is poverty prone and has less capacity to adopt to AWD irrigation. However, the north-western part of the country is highly suitable in terms of physical and physio-climatic suitability. Figure 14.11 showed that there are very few regions that are socio-economically suitable for AWD irrigation, but the moderately suitable area and higher AWD water saving sites are distributed harmonically. However, the AWD irrigation water saving depends on the intellectual irrigation management, but our methodology applied suitable classes and the field data shows a very good agreement.

## 14.4 Conclusions and Policy Implications

The estimated AWD suitability for boro, Aus and amon rice season in terms of physically, physio-climatically and socio-economically result showed huge variation. Physically, the boro, Aus and amon rice growing season is highly suitable for AWD application. But, physio-climatically the boro season is highly suitable, amon season moderately suitable and Aus season is the least suitable for AWD irrigation. On the other hand, socio-economically very few rice growing areas are highly suitable but large areas are moderately and low suitable for AWD application. As the natural phenomena, it's difficult to change the physio-climatic suitability, but there is great opportunity for the country to increase the socio-economic suitability to adopt AWD irrigation. Moreover, the boro and amon rice growing seasons are the dominant rice growing season, but only the boro rice season is currently under AWD irrigation practices in the country. There is also huge opportunity to adopt AWD irrigation in Amon rice growing season. The AWD irrigation adoption offers not only the water saving benefits but also it's very effective to reduce methane emissions, production increase and cost reduction for rice growing. The result from this study could be helpful for the country's policy maker for sustainable rice production management.

**Authors' Contributions** Corresponding author Md Rahedul Islam was involved in conceptualization, methodology development, data analysis and paper writing. Second author Dr. Kim edited and corrected the paper and revised it again and again. Third author Professor Takeuchi guided to throughout the project, provided data, analysis, revised the methodology, and writing the manuscript.

**Funding** Please add: 'This research received no external funding'.

**Conflicts of Interest** 'The authors declare no conflict of interest'.

## References

Abatzoglou JT, Dobrowski SZ, Parks SA, Hegewisch KC (2018) TerraClimate, a high-resolution global dataset of monthly climate and climatic water balance from 1958-2015. Sci Data 5:1–12. https://doi.org/10.1038/sdata.2017.191

Adnan S (2007) Departures from everyday resistance and flexible strategies of domination: the making and unmaking of a poor peasant mobilization in Bangladesh. J Agrar Chang 7(2):183–224. https://doi.org/10.1111/j.1471-0366.2007.00144.x

Ahmed A, Ahmad K, Chou V, Hernandez R, Menon P, Naeem F, Naher F, Quabili W, Sraboni E, Yu B (2013) The status of food security in the feed the future zone and other regions of Bangladesh: results from the 2011–2012 Bangladesh integrated household survey, April, pp 1–255. http://www.ifpri.org/sites/default/files/publications/bihstr.pdf

Alam MS, Islam M, Salam M, Islam M (2010) Economics of alternate wetting and drying method of irrigation: evidences from farm level study. Agriculturists 7:82–89. https://doi.org/10.3329/agric.v7i1.5258

Alexandridis TK, Cherif I, Chemin Y, Silleos GN, Stavrinos E, Zalidis GC (2009) Integrated methodology for estimating water use in mediterranean agricultural areas. Remote Sens 1(3):445–465. https://doi.org/10.3390/rs1030445

Arunrat N, Pumijumnong N (2017) Practices for reducing greenhouse gas emissions from rice production in northeast Thailand. Agriculture (Switzerland) 7(1):1–20. https://doi.org/10.3390/agriculture7010004

BADC (2018) Minor irrigation survey report 2017-2018. Survey and monitoring project

BARC (Bangladesh Agriculture Research Council) (1998) Land resource database

Basak R (2016) Benefits and costs of climate change mitigation technologies in paddy rice focus on Bangladesh and Vietnam CGIAR research program on climate change, Agriculture and Food Security (CCAFS) 160

BBS (2022) Statistical Yearbook Bangladesh 2022, 42nd ed. BANGLADESH BUREAU OF STATISTICS (BBS). http://bbs.portal.gov.bd/sites/default/files/files/bbs.portal.gov.bd/page/b2db8758_8497_412c_a9ec_6bb299f8b3ab/2023-06-26-09-19-2edf60824b00a7114d8a51ef5d8ddbce.pdf. www.bbs.gov.bd

Bishwajit G, Barmon R, Ghosh S (2013) Reviewing the status of agricultural production in Bangladesh from. Russ J Agric Soc Econ Sci 1(25):19–27

BMD (Bangladesh Meteorological Department) (2017) Bangladesh Meteorological Department

Cancelliere A, Di Mauro G, Bonaccorso B, Rossi G (2007) Drought forecasting using the standardized precipitation index. Water Resour Manag 21(5):801–819. https://doi.org/10.1007/s11269-006-9062-y

Carrijo DR, Lundy ME, Linquist BA (2017) Rice yields and water use under alternate wetting and drying irrigation: a meta-analysis. Field Crop Res 203:173–180. https://doi.org/10.1016/j.fcr.2016.12.002

Carrijo DR, Akbar N, Reis AFB, Li C, Gaudin ACM, Parikh SJ, Green PG, Linquist BA (2018a) Impacts of variable soil drying in alternate wetting and drying rice systems on yields, grain arsenic concentration and soil moisture dynamics. Field Crop Res 222:101–110. https://doi.org/10.1016/j.fcr.2018.02.026

Carrijo DR, Lundy ME, Linquist BA, Sander BO, Wassmann R, Palao LK, Nelson A, Government TU, Hunger G, The Future Security Initiative, USAID, van der Hoek W, Sakthivadivel R, Renshaw M, Silver JB, Birley MH, Konradsen F, Nelson A et al (2018b) Rice yields and water use under alternate wetting and drying irrigation: a meta-analysis. Field Crop Res 19(1):1–10. https://doi.org/10.1016/j.fcr.2016.12.002

Chang Y, Tan T, Chen T, Chuah JH, Chang L, Wu M, Tatini NB, Ma S, Alkhaleefah M (2022) Spatial-temporal neural network for rice field classification from SAR images. Remote Sens 14(8):1929

Climate Watch (n.d.) Data for climate action—Bangladesh. https://www.climatewatchdata.org/countries/BGD. Accessed 4 Apr 2020

Demarez V, Helen F, Marais-Sicre C, Baup F (2019) In-season mapping of irrigated crops using Landsat 8 and Sentinel-1 time series. Remote Sens 11(2):118. https://doi.org/10.3390/rs11020118

Duan D, Sun X, Liang S, Sun J, Fan L, Chen H, Xia L, Zhao F, Yang W, Yang P (2022) Spatiotemporal patterns of cultivated land quality integrated with multi-source remote sensing: a case study of Guangzhou, China. Remote Sens 14(5):1250. https://doi.org/10.3390/rs14051250

Education E, Bangabandhu A, Mujibur S, Agric R, Fishing H, View F, View B (2016) Impact of alternate wetting and drying technique on rice production in the drought prone areas of Bangladesh

FAO (2014) Bangladesh AQUASTAT data, pp 1–16. http://www.fao.org/nr/water/aquastat/countries_regions/BGD/BGD-CP_eng.pdf
FAOSTAT (2015). http://www.fao.org/faostat/en/#data/GR. Accessed 4 Apr 2020
Fields R (2009) Estimation of water loss from irrigated. J Agric 7(1):29–42
Hossain I, Chamhuri S, Mokhtar M, Dey MM (2013a) Water productivity for Boro Rice production: study on floodplain Beels in Rajshahi, Bangladesh. J Bio Sci 21:1–7. https://doi.org/10.3329/jbs.v21i0.22526
Hossain MS, Islam MR, Bari MA (2013b) Trend in the availability of agricultural land in Bangladesh. Soil Resource Development Institute (SRDI), Dhaka
Hossen B, Yabar H, Mizunoya T (2021) Land suitability assessment for pulse (Green gram) production through remote sensing, GIS and multicriteria analysis in the coastal region of Bangladesh. Sustainability (Switzerland) 13(22):12360. https://doi.org/10.3390/su132212360
Islam R, Takeuchi W, Oyoshi K (2023) Mapping rice paddy cropping patterns with MODIS and ALOS-2 scan SAR over Bangladesh from 2001 to 2018s dataset uses for this study. In methodology, firstly we use multidate ALOS-2 ScanSAR and GPS field data 2022(1), 32–50
JAXA (2020) Mendeley-Desktop-1. https://sharaku.eorc.jaxa.jp/GSMaP/guide.html
Kirby M, Ahmad M, Mainuddin M, Palash W, Qadir E, Shah SM (2014) Bangladesh integrated water resources assessment supplementary report: land use, crop production, and irrigation demand. CSIRO, Canberra
Kubota T, Hashizume H, Shige S, Okamoto K, Aonashi K, Takahashi N, Ushio T, Kachi M (2006) Global precipitation map using satelliteborne microwave radiometers by the GSMaP project: production and validation. Int Geosci Remote Sens Symp 45(7):2584–2587. https://doi.org/10.1109/IGARSS.2006.668
Luo GJ, Kiese R, Wolf B, Butterbach-Bahl K (2013) Effects of soil temperature and moisture on methane uptake and nitrous oxide emissions across three different ecosystem types. Biogeosciences 10(5):3205–3219. https://doi.org/10.5194/bg-10-3205-2013
Mainuddin M, Kirby M (2015) National food security in Bangladesh to 2050. Food Secur 7(3):633–646. https://doi.org/10.1007/s12571-015-0465-6
Majumder S, Bala BK, Arshad FM, Haque MA, Hossain MA (2016) Food security through increasing technical efficiency and reducing postharvest losses of rice production systems in Bangladesh. Food Secur 8:361. https://doi.org/10.1007/s12571-016-0558-x
Mesgaran MB, Madani K, Hashemi H, Azadi P (2017) Iran's land suitability for agriculture. Sci Rep 7:7670. https://doi.org/10.1038/s41598-017-08066-y
Minamikawa K (2014) Greenhouse gas mitigation by alternate wetting and drying water management in irrigated rice paddies in Southeast Asia, pp 2–3
Mottaleb KA, Krupnik TJ, Erenstein O (2016) Factors associated with small-scale agricultural machinery adoption in Bangladesh: census findings. J Rural Stud 46:155–168. https://doi.org/10.1016/j.jrurstud.2016.06.012
Murray L, Nguyen H, Lee Y, Remmenga MD, Smith DW (2012) Variance inflation factors in regression models with dummy variables. In: Conference on applied statistics in agriculture. New Prairie Press. https://doi.org/10.4148/2475-7772.1034
Nelson A, Wassmann R, Sander BO, Palao LK (2015) Climate-determined suitability of the water saving technology "alternate wetting and drying" in rice systems: a scalable methodology demonstrated for a province in The Philippines. PLoS One 10(12):1–19. https://doi.org/10.1371/journal.pone.0145268
Neteler M (2010) Estimating daily land surface temperatures in mountainous environments by reconstructed MODIS LST data. Remote Sens 2(1):333–351. https://doi.org/10.3390/rs1020333
Pearson KA, Millar GM, Norton GJ, Price AH (2018) Alternate wetting and drying in Bangladesh: water-saving farming practice and the socioeconomic barriers to its adoption. Food Energy Secur 7(4):1–12. https://doi.org/10.1002/fes3.149

Portmann FT, Siebert S, Döll P (2010) MIRCA2000-global monthly irrigated and rainfed crop areas around the year 2000: a new high-resolution data set for agricultural and hydrological modeling. Glob Biogeochem Cycles 24(1):GB1011. https://doi.org/10.1029/2008gb003435

Rabia AH, Figueredo H, Huong TL, Lopez BAA, Solomon HW, Alessandro V (2013) Land suitability analysis for policy making assistance: a GIS based land suitability comparison between surface and drip irrigation systems. Int J Environ Sci Dev 2014:1–6. https://doi.org/10.7763/IJESD.2013.V4.292

Rahman MW, Parvin L (2009) Impact of irrigation on food security in Bangladesh for the past three decades. J Water Resour Prot 1(3):216–225. https://doi.org/10.4236/jwarp.2009.13027

Rahman MS, Sander BO (2017) Application of alternate wetting and drying (AWD) in Bangladesh agriculture: findings from case studies, pp 310–332

Runkle BRK, Suvočarev K, Reba ML, Reavis CW, Smith SF, Chiu Y-L, Fong B (2019) Methane emission reductions from the alternate wetting and drying of rice fields detected using the Eddy covariance method. Environ Sci Technol 53(2):671–681. https://doi.org/10.1021/acs.est.8b05535

Sander BO, Wassmann R, Palao LK, Nelson A (2017) Climate-based suitability assessment for alternate wetting and drying water management in The Philippines: a novel approach for mapping methane mitigation potential in rice production. Carbon Manag 8(4):331–342. https://doi.org/10.1080/17583004.2017.1362945

Seo D, Lakhankar T, Khanbilvardi R (2010) Sensitivity analysis of b-factor in microwave emission model for soil moisture retrieval: a case study for SMAP mission. Remote Sens 2(5):1273–1286. https://doi.org/10.3390/rs2051273

Steele JE, Sundsøy PR, Pezzulo C, Alegana VA, Bird TJ, Blumenstock J, Bjelland J, Engø-Monsen K, De Montjoye YA, Iqbal AM, Hadiuzzaman KN, Lu X, Wetter E, Tatem AJ, Bengtsson L (2017) Mapping poverty using mobile phone and satellite data. J R Soc Interface 14(127):20160690. https://doi.org/10.1098/rsif.2016.0690

Stevens FR, Gaughan AE, Linard C, Tatem AJ (2015) Disaggregating census data for population mapping using random forests with remotely-sensed and ancillary data. PLoS One 10(2):1–22. https://doi.org/10.1371/journal.pone.0107042

Sys C, Van Ranst E, Debaveye J (1991) Land evaluation part I (principles in land evaluation and crop production calculations). Agricultural Publications

Sys C, Ranst EV, Debveye J, Beernaert F (1993) Land evaluation part 3.pdf. International Trainning Center for Post-graduate Soil Scientists. University of Ghent, Ghent

Takaku J, Tadono T, Tsutsui K, Ichikawa M (2016) Validation of "AW3D" global DSM generated from ALOS PRISM. ISPRS Ann Photogramm Remote Sens Spatial Inf Sci 3:25–31. https://doi.org/10.5194/isprs-annals-III-4-25-2016

Thenkabail PS, Hanjra MA, Dheeravath V, Gumma M (2010) A holistic view of global croplands and their water use for ensuring global food security in the 21st century through advanced remote sensing and non-remote sensing approaches. Remote Sens 2(1):211–261. https://doi.org/10.3390/rs2010211

Wang ZP, DeLaune RD, Patrick WH, Masscheleyn PH (2010) Soil redox and pH effects on methane production in a flooded rice soil. Soil Sci Soc Am J 57(2):382. https://doi.org/10.2136/sssaj1993.03615995005700020016x

World Bank (2020) Interactive poverty map of Bangladesh. https://issuu.com/world.bank.publications/docs/poverty_maps_of_bangladesh_2010

Zhang M, Lin H, Wang G, Sun H, Fu J (2018a) Mapping paddy rice using a convolutional neural network (CNN) with Landsat 8 datasets in the Dongting Lake Area, China. Remote Sens 10(11):1840. https://doi.org/10.3390/rs10111840

Zhang X, Wu B, Ponce-Campos GE, Zhang M, Chang S, Tian F (2018b) Mapping up-to-date paddy rice extent at 10 M resolution in China through the integration of optical and synthetic aperture radar images. Remote Sens 10(8):1–26. https://doi.org/10.3390/rs10081200

# Chapter 15
# Remote Sensing Application and Machine Learning Approach to Estimate the Availability of Local Food: A Case Study of Urban Area in Indonesia

**Riska Ayu Purnamasari, Marwan Setiawan, Wardah Wardah, Parwa Oryzanti, Dian Oktaviani, and Yohanes Purwanto**

**Abstract** Remote sensing technology has many applications in agriculture, including monitoring and analyzing local food systems in urban areas. In urban environments, space is often limited, and food is often produced on small plots of land; this makes it challenging to monitor and assess the condition of the crops using traditional methods. Traditional methods of predicting local food availability may not provide as much detail nor consider factors such as the specific weather conditions in a particular location or the health and growth of individual crops. Remote sensing technology can provide more detailed and specific information about these factors, allowing for more accurate predictions about local food availability. One of the critical applications of remote sensing for local food systems in urban areas is detecting and monitoring farmland and crops. These methods are an essential part of land use planning and policy recommendations. Crop yield prediction can be measured by applying satellite remote sensing to determine physiological conditions during the growing season. Machine learning has shown to be effective in a variety of data-driven applications. This study used Sentinel-2 images from March to October 2018 to map the normalized difference vegetation index (NDVI) and leaf area index (LAI) via satellite. Different growth stages were observed. The generated models were validated by linear regression and random forest algorithms. The results showed that the random forest method had good accuracy. NDVI had the highest accuracy ($R^2$ = 0.90) in September 2018 and October 2018; however, the

R. A. Purnamasari (✉)
Research Center for Ecology and Ethnobiology, National Research and Innovation Agency (BRIN), Bogor Regency, West Java, Indonesia

Innovation Center for Tropical Sciences, Bogor City, West Java, Indonesia
e-mail: risk013@brin.go.id

M. Setiawan · W. Wardah · P. Oryzanti · D. Oktaviani · Y. Purwanto
Research Center for Ecology and Ethnobiology, National Research and Innovation Agency (BRIN), Bogor Regency, West Java, Indonesia



T. Ahamed (ed.), *Remote Sensing Application II*, New Frontiers in Regional Science: Asian Perspectives 77, https://doi.org/10.1007/978-981-97-1188-8_15

accuracy of LAI was very high ($R^2 = 0.91$) as of July 2018. These results presented the late growth stage had good accuracy for estimating local crop production as crops reached the peak vegetative periods. Thus, machine learning provides reliable tracking of local food production at local and regional levels. The results show that vegetation indices can be used to calculate site-specific local crop management and predict yield. These integrated models can be used for logistics and decision-making related to local agricultural production.

**Keywords** Linear regression · Machine learning · Local food · Random forest · Yield prediction

## 15.1 Introduction

The term local food does not possess a standardized definition and varies among countries (Meyerding et al. 2019). Local food can refer to food grown, processed, and distributed within a defined geographic area (Feenstra 1997). The exact boundaries of this area can vary, but it is typically defined as being within a certain radius of the point of consumption, such as a city or town. Local food systems can vary in size and complexity, from small-scale community gardens and farmers' markets to larger-scale networks of local food producers (Meyerding et al. 2019). A local food system aims to provide communities with access to fresh, healthy, locally grown foods while supporting local food producers and promoting sustainable food production and consumption practices (Granvik et al. 2017). There are several reasons why local food systems are essential in urban areas (Pothukuchi and Kaufman 1999). First, local food systems can improve food security by making fresh, healthy foods more accessible to residents of urban areas. This is especially important in low-income communities, where access to affordable and healthy foods can be limited (Siegner et al. 2018). Second, local food systems can support the local economy by providing opportunities for smallholder farmers and creating possibilities for small-scale producers. This can help to create jobs and stimulate economic development in urban areas (Deller et al. 2017). Third, local food systems can contribute to mitigating the environmental consequences of industrial agriculture by decreasing the necessity for transporting food over long distances and advocating for sustainable farming methods (Pradhan et al. 2020). Finally, local food systems can help to build stronger, more connected communities by bringing people together around the production, processing, and distribution of food (Azunre et al. 2019; Granvik et al. 2017).

However, the problem with local food systems in urban areas is the uncertainty of crop yields (Payen et al. 2022). Urban agriculture often relies on a mix of conventional and innovative farming techniques, making it challenging to predict crop yields accurately. Traditional methods of crop prediction often rely on manual observation and measurements, which can be time-consuming and labor-intensive. In contrast, remote sensing technology allows for the automatic collection of data on the health and growth of crops using satellite imagery and sensors (Wang et al. 2018). This can offer a more extensive and intricate perspective on the crops, enabling more precise forecasts. Remote sensing can be used to monitor crop development and conditions in urban agricultural areas, providing valuable information that can be used to improve crop management and predict yields with greater precision (Brown and McCarty 2017). Yield data could generate realistic assessments of food production on a smaller, localized scale (Morrison et al. 2011). This can help local food producers make more informed decisions about managing their crops and increasing their productivity, which can support the development of sustainable local food systems in urban areas (Siegner et al. 2018).

Satellite information captures crop development conditions through a variety of spectral bands, including optical (visible and near-infrared), thermal, and microwave bands. Information about crop growth conditions can be directly obtained from satellite data by examining different spectral bands across the electromagnetic spectrum. Through the intensity measuring of light in different spectral bands, it is possible to determine crops' health and growth stages. For example, healthy crops typically have a high level of chlorophyll, which is responsible for photosynthesis. By measuring the amount of light absorbed in the blue and red spectral bands, it is possible to estimate the chlorophyll content of crops and determine their health (Wang et al. 2021). A wide range of spectral and optical bands have been extensively employed to track crop growth conditions, predict crop yield, and distinguish different crop areas, often using various vegetation indices. A frequently employed vegetation metric is the normalized difference vegetation index (NDVI), often integrated with biophysical factors such as leaf area index (LAI) (Purnamasari et al. 2019).

The forecast of local crop yields can be achieved by utilizing satellite remote sensing techniques, as demonstrated by Guan et al. (2017). Many research efforts involve the integration of satellite imagery with statistical models to estimate crop yields (Chlingaryan et al. 2018; Purnamasari et al. 2019; Shiu and Chuang 2019). Statistical models rely on mathematical connections between various variables and can be employed to forecast yield using NDVI and LAI as inputs. These models might be based on superficial linear relationships to cover the complexity of large datasets of historical data depending on the monthly growing season. However, the linear equations have limitations in capturing complex relationships and patterns in the data. To address these limitations, the machine learning (ML) approach has the potential to be a helpful tool for crop yield prediction (Aghighi et al. 2018; Camps-Valls et al. 2006; Tuia et al. 2011). Machine learning techniques, which belong to the class of algorithmic approaches, are capable of capturing intricate and nonlinear relationships in the data, overcoming some of the limitations of traditional regression methods. Machine learning algorithms can significantly enhance accuracy in this context (Ali et al. 2016; Gopal and Bhargavi 2019; Liakos et al. 2018). Since the early stages of spectral reflectance and remote sensing technology development, researchers have utilized a range of models to predict crop yield through remote sensing methodologies. Satellite imagery has been harnessed to monitor the optical properties of canopies, monitor crop conditions, and predict yield and production in numerous countries across the globe, utilizing a variety of sensors.

Recent advancements in high-resolution satellite sensors offer promising capabilities for monitoring field crop production and conditions. This study explores the potential of the publicly accessible Sentinel-2 satellite in predicting crop yields within smallholder farm areas. Sentinel-2 boasts a high resolution, with imagery available at resolutions of up to 10 m, and it possesses unique spectral features, including 3 red-edge bands, with a revisit time of approximately 5 days. These characteristics render it a valuable instrument for tracking vegetation in agricultural and forestry activities. Sentinel-2's red-edge bands at 705, 740, and 783 nm are crucial for capturing reflective qualities associated with vegetation characteristics, making it especially beneficial in precision agriculture applications (Castillo et al. 2017).

In this study, satellite imagery was used to predict the local crop yield, such as cassava and sweet potato, in urban areas in the western area of Java Island, Indonesia. By analyzing data from satellite images, it is possible to identify areas where crops are being grown and estimate the crops' biomass and yield using linear regression (LR) and random forest (RF) regression. The statistical and machine learning model was used to predict the crop yield based on LAI and NDVI within the monthly growing season. The combination of remote sensing technology and machine learning to estimate the availability of local food in urban areas is a relatively new approach. The study also highlights the challenges of monitoring small plots of land and the benefits of using remote sensing to provide more detailed and accurate information.

## 15.2 Materials and Methods

### *15.2.1 Framework of Research*

As depicted in Fig. 15.1, the initial phase of the study involved conducting a ground truth data survey. This survey involved the collection of 12 georeferenced locations representing potential local crops. During a field survey, data points regarding the crop yield of cassava and sweet potato were obtained. The locations were subsequently linked with the respective vegetation indices (NDVI) and biophysical attributes (LAI) to create a machine learning-driven regression model for the early projection of potential crop yield in the specified fields. Sentinel-2 satellite datasets were obtained for the analysis of NDVI and LAI.

### *15.2.2 Study Area*

Located in Banten province, Indonesia, Cilegon City is one of Indonesia's most important industrial cities. It has high population growth, dynamic land use development, and high demand for public service facilities. Based on its

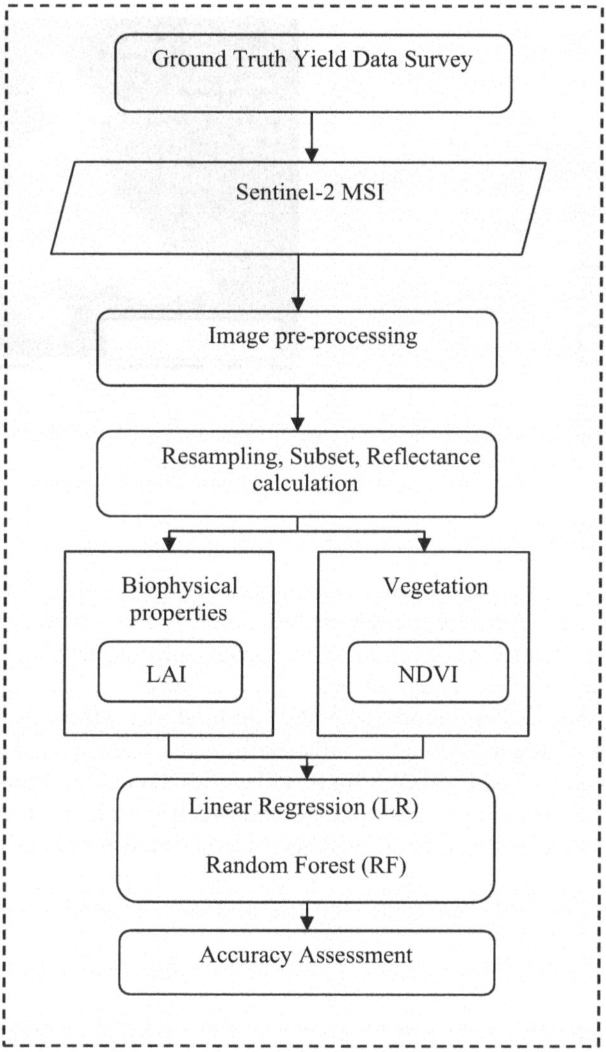


**Fig. 15.1** Framework of research

geographical location, the City of Cilegon is located in the western part of Java Island, in the position 5°52′24″ to 6°4′07″ South and 105°54′05″ to 106°05′11″ East (Fig. 15.2). The city encompasses a coastal area to the north, with rural regions both to the south and north, as well as an urban center situated in the middle of the area. Residences are concentrated in the central part of the region, where most infrastructures and facilities that support socioeconomic activities are developed; local crop cultivation, such as cassava and sweet potato, is found in the southern area.

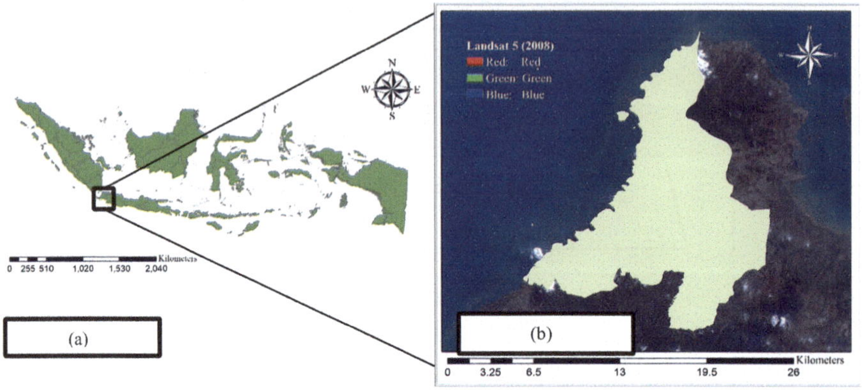


**Fig. 15.2** Geographical extent of the study area: (**a**) Indonesia and (**b**) Cilegon City in Banten province

#### 15.2.2.1 Elevation

In the city of Cilegon, the elevation varies between sea level, at 0 m, and 500 m above sea level. This diverse topography encompasses a wide range of altitudes. While much of the city's land is suitable for agricultural activities, it is important to note that there exists an optimal elevation range for agricultural production within this varied landscape. The optimal elevation for agricultural production in Cilegon typically falls within the range of approximately 62.5–137.5 m above sea level. Within this elevation band, environmental conditions, including temperature, soil quality, and precipitation, are often most conducive to successful crop cultivation. This information is crucial for farmers and land planners in Cilegon as it helps identify the most fertile and productive areas for agriculture, ensuring efficient land use and sustainable food production in the region (Fig. 15.3a).

#### 15.2.2.2 Road

Due to economic growth, the city has seen a rise in the number of automobiles, although the expansion of road networks has not kept pace at the same rate. As a result, traffic congestion in the city has worsened. In terms of socioeconomic issues, key roads are required to market fresh agricultural produce at any distance from growing areas. These key roadways are vital arteries for the local economy, ensuring that agricultural goods reach consumers and markets on time and at a reasonable cost. The convergence of these economic and transportation issues emphasizes the importance of managing traffic congestion in order to ensure a smooth flow of commodities and sustain the city's continued economic growth (Fig. 15.3b).

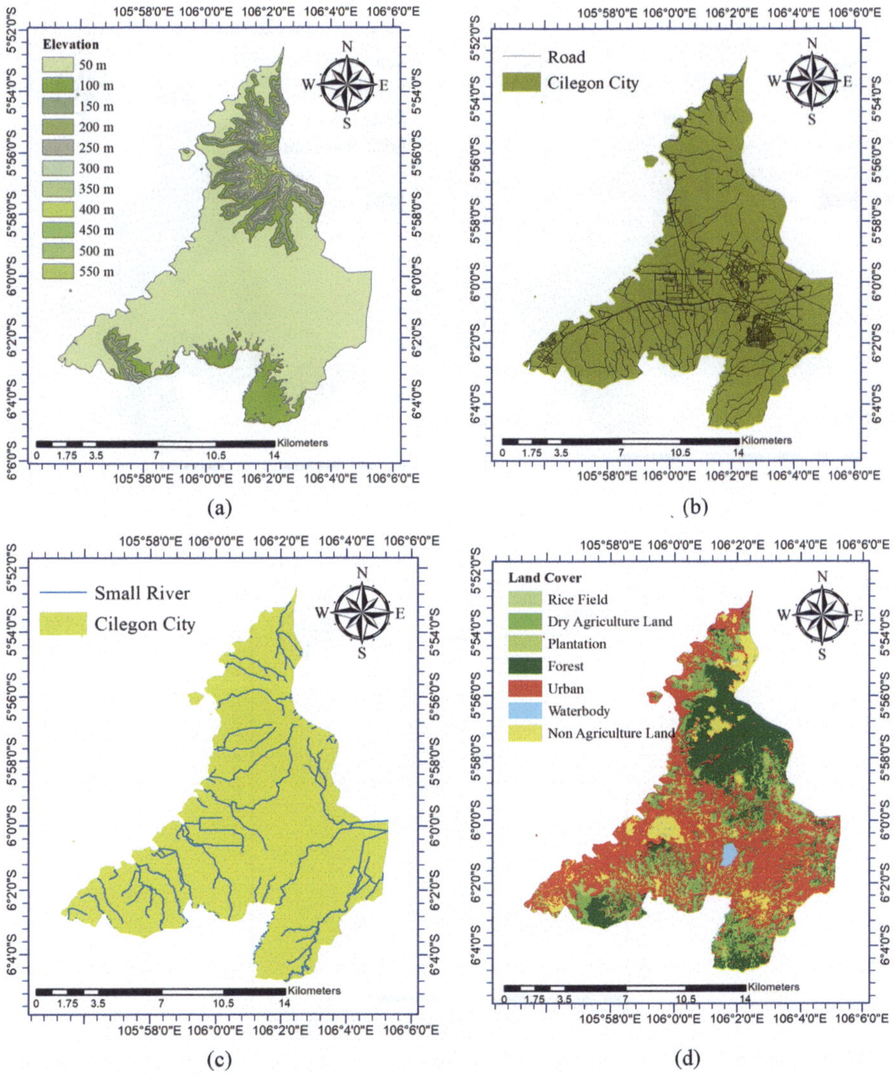


**Fig. 15.3** Study area ecological condition: (**a**) elevation, (**b**) road, (**c**) river, (**d**) land use and land cover (LULC), (**e**) slope, (**f**) soil, and (**g**) rainfall

#### 15.2.2.3 River

The city of Cilegon currently relies on the Cidanau River, which serves as the primary water source for the Krenceng Reservoir, to meet all of its water needs. The availability of water resources is limited. These limitations can pose significant challenges for ensuring a reliable and sustainable water supply for the city's residents and industries. Factors such as increased demand, potential environmental

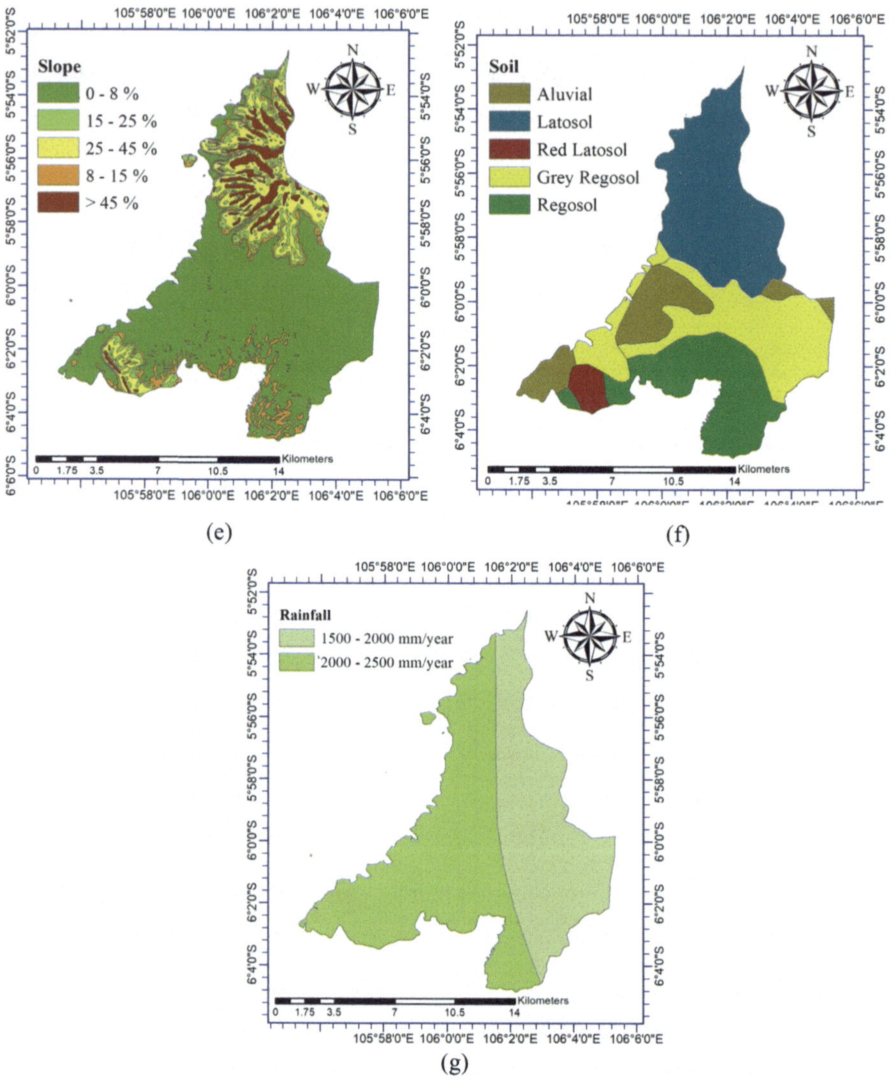


**Fig. 15.3** (continued)

stressors, and the effects of climate change can further exacerbate the vulnerability of Cilegon's water supply system (Fig. 15.3c).

#### 15.2.2.4 Land Cover and Land Use

Cilegon city's land had coastal land in the north, rural land in the south and north, and urban land in the center (Fig. 15.3d). The urban region had infrastructure-related facilities to assist the socioeconomic aspect. Residence was also concentrated in the city's center. The major land use in the north is rice agriculture, while the southern section is characterized by fields and dry ground topography.

#### 15.2.2.5 Slope

The city of Cilegon has primarily flat territory with an elevation of 500 m above sea level. The majority of the terrain contour in Cilegon City is classified as flat, with slopes ranging from 0% to 45% steepness (Fig. 15.3e). This means that most of the city's terrain is relatively level and not subject to significant changes in elevation. In terms of topographical contours, the majority of the city's landscape falls within the classification of being "flat," with slopes that vary from gentle inclines, measuring 0%, to moderately steep gradients, reaching up to 45% in steepness. This topographical description indicates that Cilegon's geography is relatively uniform, with variations in slope that are conducive to various types of development and land use within the city.

#### 15.2.2.6 Soil Type

Cilegon contains a diverse range of soil types, comprising alluvial, red regosol, red yellow podzolic, and latosol. Alluvial soils are primarily utilized in rice-focused agricultural practices, while red regosol soils are employed for upland rice cultivation and crop production in dry climates. These red regosol soils are often prevalent in hilly areas and at the foothills of mountains (Fig. 15.3f).

#### 15.2.2.7 Rainfall

The rainy season in Cilegon covers most of the year, lasting from October to May. The remaining 4 months are the dry season of the city. Located in the westernmost part of Java, Cilegon's highest wet season rainfall is in January with an average monthly rainfall of 319 mm (12.6 in.) and the lowest point in the dry season is August with monthly average rainfall 71 mm (2.8 in.) (Fig. 15.3g).

### *15.2.3 Local Crop Location Survey*

The Local Food Ground Truth Survey assessed the cultivation and yield of local crops, specifically cassava and sweet potato, in 12 different locations in Cilegon City, Indonesia. The survey aimed to gather crop production data, identify yield variations across different locations, and provide insights into the local food system. The survey team visited each of the 12 locations and conducted field surveys to collect relevant data (Table 15.1). Cassava and sweet potato are all important local

**Table 15.1** Survey location with latitude and longitude points in the study area

| No. | Village name | Latitude | Longitude | Yield (ton/ha) |
|---|---|---|---|---|
| 1 | Deringo | −6.03584 | 106.028566 | 6 |
| 2 | Kotasari | −5.987695 | 106.031169 | 4.5 |
| 3 | Mekarsari | −5.934035 | 106.007005 | 3 |
| 4 | Ciwedus | 6.0414853 | 106.0086726 | 4 |
| 5 | Cibeber | −6.037040 | 106.063654 | 1 |
| 6 | Deringo | −6.03584 | 106.028863 | 2.4 |
| 7 | Kotasari | −5.987403 | 106.021525 | 1 |
| 8 | Mekarsari | −5.987695 | 106.031169 | 2 |
| 9 | Tegal ratu | −6.035632 | 105.980338 | 4 |
| 10 | Deringo | −6.03584 | 106.028566 | 30 |
| 11 | Purwakarta | −5.998685 | 106.007005 | 13 |
| 12 | Kotasari | −5.987403 | 160.021525 | 8 |

foods in many parts of the world. Cassava and sweet potato are starchy root vegetables that are a staple food for millions of people in the tropics and subtropics. These crops have a wide variety of uses, including as food for humans and animals, as a source of biofuel, and as an industrial raw material.

These plants are highly nutritious and provide a valuable source of calories, carbohydrates, fiber, and other essential nutrients. Although these crops have distinct characteristics, they share some similarities, particularly in their reflectance properties that can be observed from satellite imagery. One of the primary similarities lies in the leaf structure of these crops. Cassava and sweet potato plants all have broad, green leaves that reflect a significant amount of visible light. This leaf structure and color similarity contribute to comparable reflectance patterns in satellite imagery. Another shared characteristic is the presence of chlorophyll in their leaves, which is responsible for the absorption of sunlight for photosynthesis.

### *15.2.4 Normalized Difference Vegetation Index (NDVI)*

NDVI, initially introduced by Rouse et al. (1973), serves as a metric for assessing the vitality and productivity of vegetation. It is computed by taking the difference between plants' near-infrared and red reflectance. The specific NDVI value for a cassava plant will depend on various factors, such as the age of the plant, its growth stage, and environmental conditions. Generally, a healthy cassava and sweet potato plant will have an NDVI value between 0.2 and 0.8 (Panganiban et al. 2019; Tedesco et al. 2021). This range has similar requirements for optimal growth and health. The NDVI was calculated as follows (NIR is near-infrared):

$$\mathrm{NDVI} = \frac{\mathrm{NIR} - \mathrm{RED}}{\mathrm{NIR} + \mathrm{RED}} \tag{15.1}$$

### 15.2.5 Leaf Area Index (LAI)

LAI, or leaf area index, quantifies the quantity of green leaf surface area within a plant canopy per unit of ground area. It is a valuable indicator for evaluating the productivity of vegetation. LAI is determined by dividing the total leaf area within a plant canopy by the corresponding land area where these plants are cultivated. The specific LAI value for a cassava and sweet potato plant depends on various factors, such as the age of the plant, its growth stage, and environmental conditions. A healthy cassava plant will usually have an LAI value between 3 and 5 (Fagbola et al. 1998; Phosaengsri et al. 2019; De Souza et al. 2017). This indicates that the combined leaf area within the plant canopy is approximately three to five times larger than the land area where the plants are cultivated. LAI is calculated as:

$$\mathrm{LAI} = \frac{\mathrm{Leaf\ Area}\left(\mathrm{m}^2\right)}{\mathrm{Ground\ Area}\left(\mathrm{m}^2\right)} \tag{15.2}$$

### 15.2.6 Sentinel-2A Imagery

The Sentinel-2 satellite is a part of the Copernicus program, an earth observation initiative led by the European Space Agency (ESA). It was launched in June 2015 and is equipped with an advanced multispectral imaging system designed to deliver high-resolution optical images of the Earth's surface. One of the key novelties of the Sentinel-2 satellite is its capability to predict multiple crops. The satellite's multispectral instrument (MSI) sensor captures data in 13 spectral bands, including visible, near-infrared, and shortwave infrared wavelengths. Sentinel-2 can observe subtle differences in vegetation, soil, and water content by collecting data across a wide spectral range. This information is crucial for monitoring and managing agricultural resources. The multispectral imagery provided by the Sentinel-2 satellite allows for identifying and mapping various crop types. Different crops reflect and absorb light in distinct ways, which can be detected by analyzing satellite data.

Sixteen satellite images from Sentinel-2A, covering Cilegon City, were captured from March 2018 to November 2018. These images were obtained from the United States Geological Survey (USGS) Earth Explorer online resource. Cloud masking and buffering processes were carried out to prepare the data for analysis using the IdePix toolbox, part of the Sentinel Application Platform (SNAP) version 5.0. Subsequently, the NDVI was computed from the processed imagery (Fig. 15.4). In this research, NDVI was employed as one of the predictors to estimate crop yield. Another significant predictor was the LAI, a biophysical indicator for monitoring essential plant properties. LAI is important in agricultural land monitoring and modeling studies due to its pivotal role in ecological processes. To derive LAI values from Sentinel-2 imagery, a dedicated algorithm from the SNAP biophysical operations toolbox was applied to compute LAI retrieval maps. These maps have

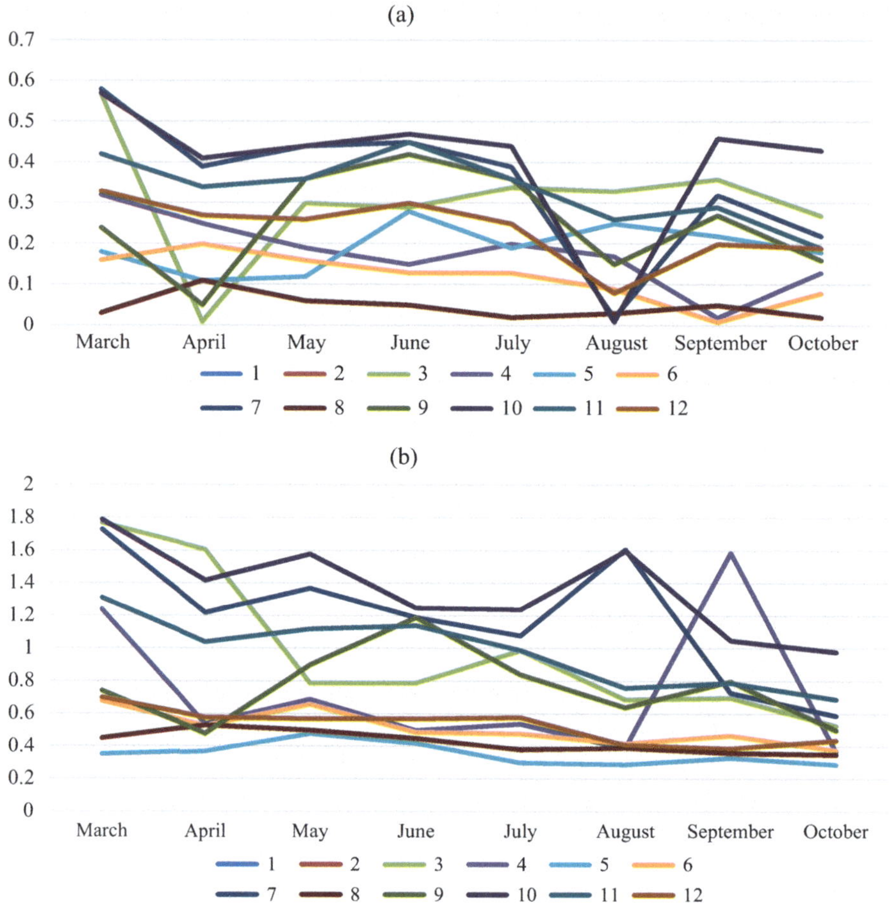


**Fig. 15.4** The monthly value of NDVI (**a**) and LAI (**b**) at the 12 locations of the study

found extensive use in numerous land observation research endeavors. During the optimal growing season, data for both NDVI and LAI were extracted by averaging the values obtained from each surveyed field within Cilegon City. The ideal growing season was determined by evaluating the vegetation indices and biophysical characteristics extracted from Sentinel-2 images (Fig. 15.5).

### *15.2.7 Machine Learning Approach*

Machine learning models can potentially be valuable tools for crop yield prediction using satellite imagery. These more suitable approaches will depend on the data's specific characteristics and the desired accuracy and complexity level.

Linear regression (LR) is a statistical technique used to establish a relationship between a dependent variable (in this context, crop yield) and one or multiple

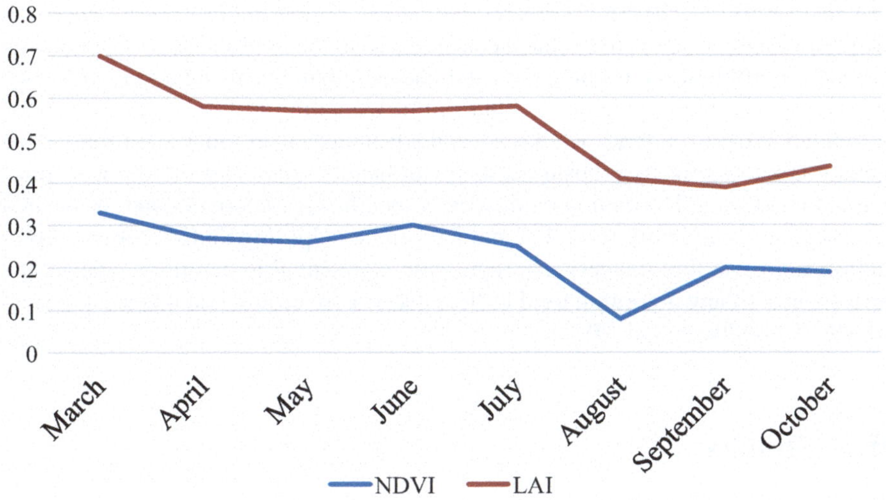


**Fig. 15.5** The optimum growing season based on NDVI and LAI

independent variables (like satellite imagery and other data sources). The essence of linear regression lies in finding a straight line that optimally aligns with the data points in a scatter plot. This is accomplished by minimizing the sum of the squared disparities between the anticipated values projected by the line and the actual observed values. Once this line is determined, it can be employed to make forecasts about the dependent variable's value based on the values of the independent variables. The linear regression (LR) algorithm employed a linear predictor function to model the data, incorporating the parameters. These parameters were calculated using measured crop growth variables and computed backscattering coefficients at various crop growth stages. The algorithm establishes this linear relationship by creating a linear equation that relates the explanatory variables. In machine learning, linear regression is a foundational algorithm used for regression tasks, where the goal is to predict a constant numeric value. It assumes a linear relationship between the variables, suggesting that the dependent variable can be forecasted based on the values of the independent variables.

In simple linear regression, there is just one independent variable, and a straight line represents the relationship between that independent variable and the dependent variable. The common equation for simple linear regression is:

$$y = mx + b \tag{15.3}$$

where

$y$ is the dependent variable (also known as the response variable or target variable).
$x$ is the independent variable (the predictor variable or feature).
$m$ is the slope of the line, representing the change in $y$ for a unit change in $X$.
$b$ is the $y$-intercept, representing the value of $y$ when $x$ is 0.

The random forest (RF) technique has proven the effectiveness and superiority over other machine learning approach for classification and regression tasks

(Amankulova et al. 2023). To create predictions, an ensemble learning approach integrates many decision trees. Each decision tree in the random forest is trained on a random portion of the training data and makes judgments using a random subset of features. The final prediction is obtained by averaging or voting the predictions of individual decision trees. RF has several advantages over other algorithms. This method is known for its robustness, as it can handle noisy and irrelevant features without overfitting. Depending on the data's specific characteristics and the desired accuracy and complexity level, LR and RF can be helpful tools for crop yield prediction using satellite imagery. RF regression is essential to carefully evaluate the performance of any algorithm used for this purpose, to ensure that it is accurate and reliable (Choudhary et al. 2022).

## 15.3 Results

### *15.3.1 Local Food Ground Truth Survey*

The surveys included information such as the crop type being cultivated, local farmers' cultivation practices, planting density, growth duration, and yield. This comprehensive data collection process allowed a thorough analysis of each location's local crop cultivation practices and yield levels. The survey focused on two critical local crops: cassava and sweet potato. The yield per unit area for each crop was recorded by collecting this data across multiple locations. The analysis of the survey data revealed significant variations in crop yield across the 12 locations (Fig. 15.6).

### *15.3.2 Local Crop Yield Predictions*

This section provides the analysis outcomes aimed at exploring the correlation between crop yield, NDVI, and LAI during each growth stage. Data from Sentinel-2 imagery were compared with yield data collected through field surveys. The aim was to understand how the vegetation indices, specifically NDVI and LAI, can be used as predictors of crop yield. The analysis examined the correlation between crop yield and the corresponding NDVI and LAI values obtained at different growing stages. The NDVI and LAI data were extracted from the Sentinel-2 imagery.

The NDVI and LAI values were computed during the yield forecast stage. The scatter plot is used to compare the performance of the LR algorithm (Figs. 15.7 and 15.9) with RF approach (Figs. 15.8 and 15.10) that have used different methods to predict the crop yield. The RF algorithm outperformed in terms of accuracy and robustness. The findings suggest that the random forest (RF) model demonstrates superior performance and exhibits higher accuracy compared to linear regression (LR) in predicting yield. The figures show the best results presented by the RF model

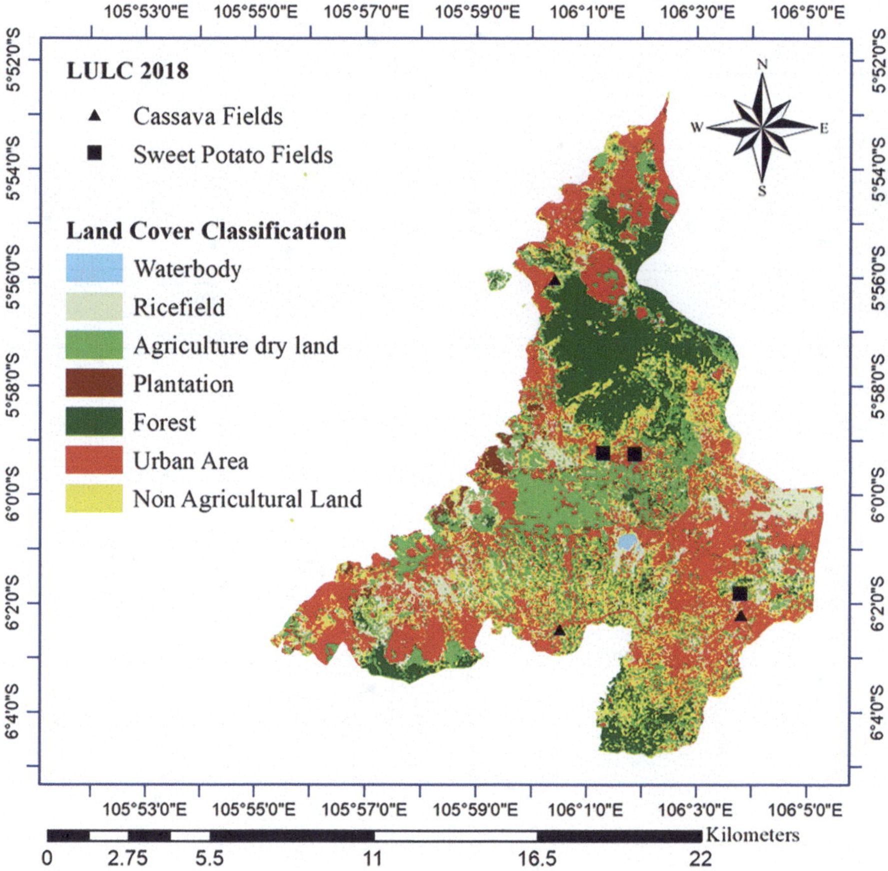


**Fig. 15.6** Survey location for local crops

for LAI with 0.91 $R^2$ and 0.21 RMSE values in July 2018 (Fig. 15.10). Also, high result was presented by RF model for NDVI with 0.91 $R^2$ and 0.23 RMSE value in September and October 2018 (Fig. 15.8). These results demonstrate that during the late growing stage, there was a notably high level of accuracy in estimating local crop yield, particularly as the crops approached their peak vegetative phases.

Understanding plant development and health depends significantly on the NDVI and LAI. The leaf area per unit ground area is measured by LAI, whereas the difference between red light, which vegetation absorbs, and near-infrared light, which vegetation strongly reflects, is measured by NDVI. Because these indicators show biomass, photosynthetic capability, and plant health, they are useful for forecasting crop production. The results were particularly noteworthy in the later developing stages, suggesting that the RF model's forecast accuracy utilizing NDVI and LAI data increased as crops come closer to peak vegetative phases. The RF model seems to perform very well, given the high $R^2$ value. However, there is a tendency to

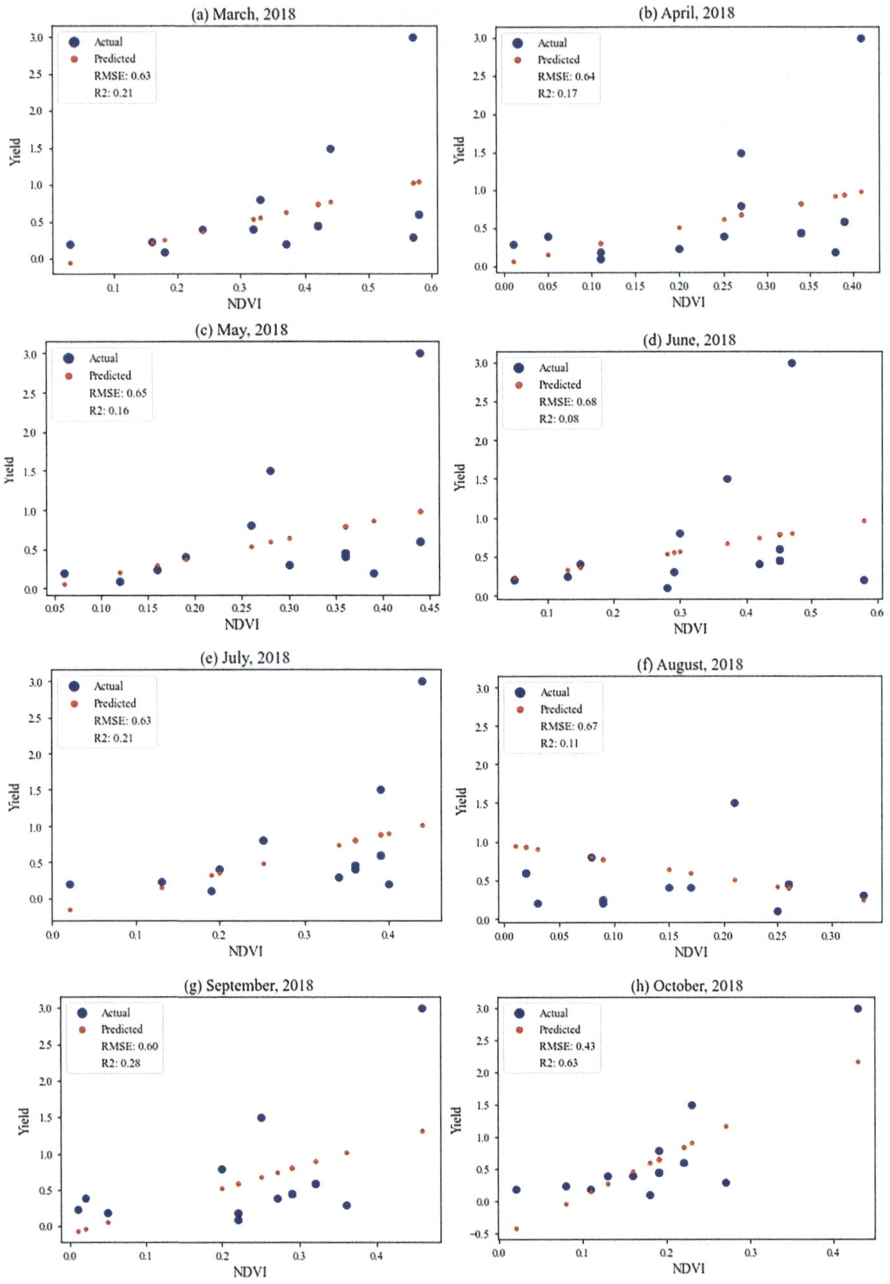


**Fig. 15.7** Scatter plots of linear regression algorithms for local crop yield (×10 ton/ha) by normalized difference vegetation index (NDVI) from (**a**) March to (**h**) October 2018

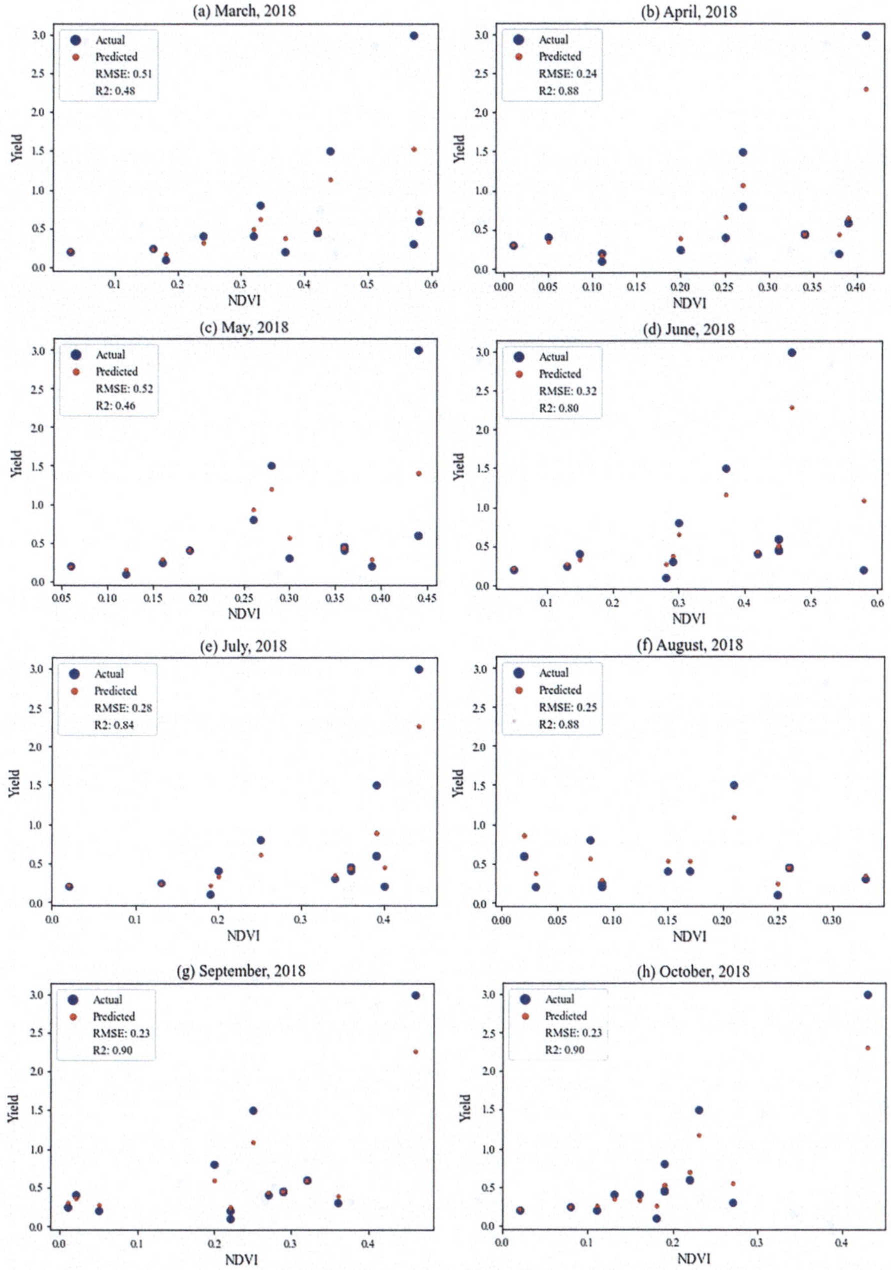


**Fig. 15.8** Scatter plots of random forest algorithms for local crop yield (×10 ton/ha) by normalized difference vegetation index (NDVI) from **(a)** March to **(h)** October 2018

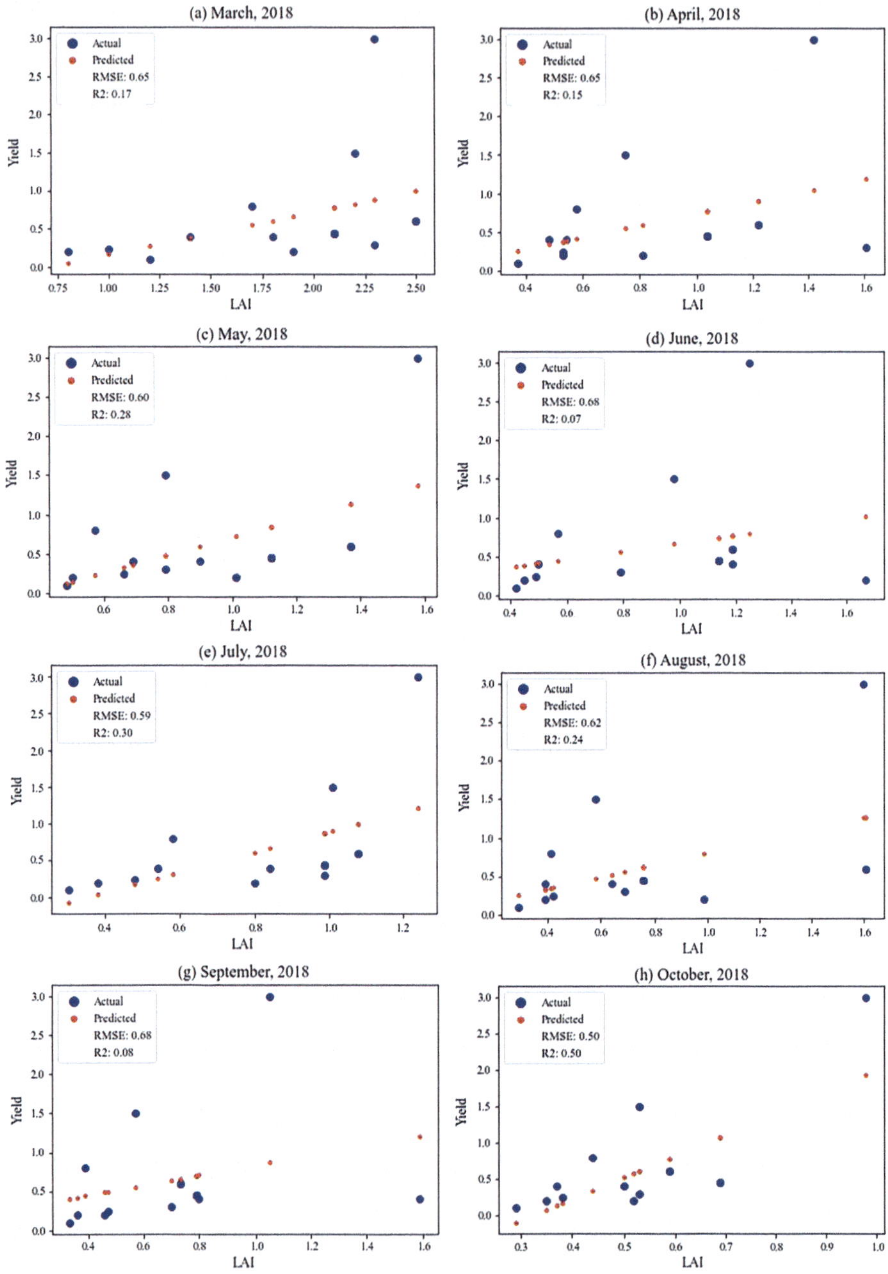


**Fig. 15.9** Scatter plots of linear regression algorithms for local crop yield (×10 ton/ha) by leaf area index (LAI) from (**a**) March to (**h**) October 2018

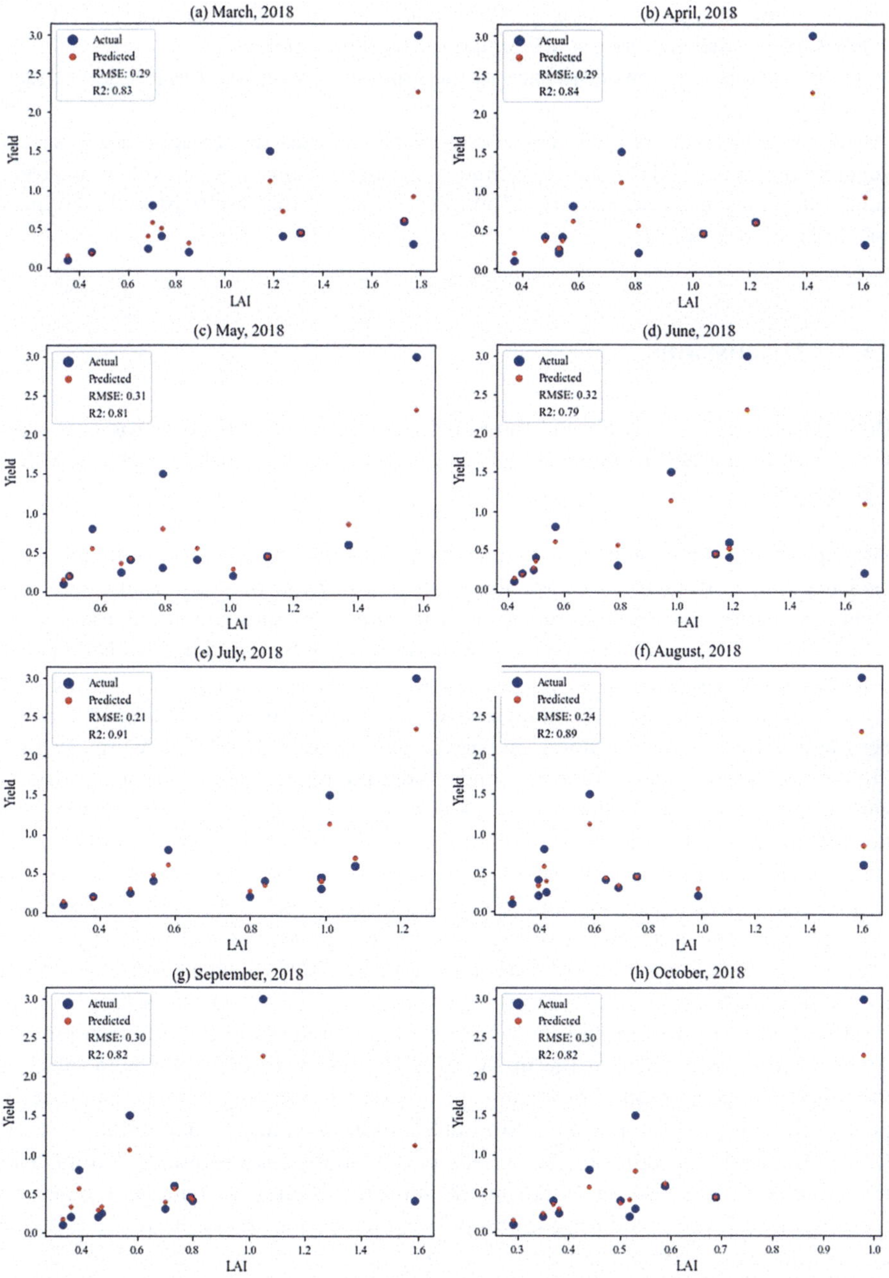


**Fig. 15.10** Scatter plots of random forest algorithms for local crop yield (×10 ton/ha) by leaf area index (LAI) from (**a**) March to (**h**) October 2018

underestimate the yield at higher NDVI and LAI values, which could be due to overfitting or other factors not accounted for by the model.

Overfitting is a serious drawback of the random forest model that might reduce its ability to predict outcomes on novel, untested datasets. This is especially true if the hyperparameters are not calibrated to a sufficient degree. Despite being useful vegetation indices, NDVI and LAI may not fully capture crop health indicators, hence they may miss more subdued markers of stress that multispectral or hyperspectral data may show.

## 15.4 Discussions

Both linear regression (LR) and random forest (RF) are utilized to forecast crop yield based on satellite imagery. LR is a conventional statistical approach that assesses the connection between a dependent variable and one or more independent variables. RF, on the other hand, is a well-known machine learning approach that integrates many decision trees to create predictions. In a random forest model, each decision tree is trained on a randomly selected portion of the training data and makes decisions using a random subset of features. The ultimate prediction is then derived by averaging or voting on the predictions made by individual decision trees. This ensemble approach enhances the overall predictive accuracy and robustness of the model. RF has several advantages over other algorithms, including its robustness and ability to handle noisy and irrelevant features. It also measures feature importance, allowing researchers to understand the relative contributions of different features in the prediction process. Depending on the data's specific characteristics and the desired accuracy and complexity level, LR and RF can be helpful tools for crop yield prediction using satellite imagery.

Previous studies have demonstrated the effectiveness of combining remote sensing data and machine learning techniques for crop yield prediction. For example, a study conducted by Amankulova et al. (2023) used a similar approach to predict soybean yields in Southeast Hungary. The researchers employed a random forest algorithm and multispectral data to develop a predictive model. The findings indicated that the most suitable time to forecast soybean yield at the field level was around 60–70 days prior to the harvesting period, specifically during the initial bloom stage. Similarly, another study by Choudhary et al. (2022) focused on predicting rice yields in the Guangdong province, China. The researchers utilized a combination of Sentinel-2 satellite data and machine learning algorithms, including random forest as well as decision tree and linear regression. The findings demonstrated the potential of this approach for accurately estimating rice yields, which is crucial for agricultural planning and food security in the region.

These studies collectively highlight the broad applicability and success of integrating remote sensing data with machine learning algorithms for crop yield prediction. By leveraging the rich information provided by satellite imagery and advanced modeling techniques, it is possible to achieve accurate and reliable predictions for various crops, including cassava and sweet potato. This knowledge can significantly benefit agricultural practices, resource allocation, and food security efforts at both regional and global scales.

## 15.5 Conclusions

The combination of remote sensing and machine learning techniques holds excellent potential for predicting and managing different crop types, including cassava and sweet potato. However, accurate predictions can be made with remote sensing data from satellites like Sentinel-2 and advanced machine learning algorithms. The multispectral data captured by satellites like Sentinel-2 provide valuable information about the reflectance properties of different crops, including cassava and sweet potato. Once the machine learning model is trained, it can be applied to large-scale satellite datasets to predict the distribution of cassava and sweet potato at regional and country levels. This information is invaluable for land use management and decision-making by providing insights into the spatial distribution of these crops, enabling policymakers to allocate resources, plan agricultural infrastructure, and develop targeted interventions to support farmers. Moreover, the predictive capabilities of remote sensing and machine learning can be extended to estimate crop yields for inventory planning and regional food security. By integrating satellite data with ground-based measurements, weather data, and historical yield records, machine learning models can be developed to accurately predict the potential yield of cassava and sweet potato crops.

The study's methodology, while innovative, encounters several limitations. The results may not be easily generalizable to other regions with different climates, crops, or agricultural practices. Hence, application of the models to a wider range of geographic locations with varying climatic zones and farming practices is required to test the robustness and adaptability of the predictive relationships. These predictions enable farmers and policymakers to plan and optimize production, manage post-harvest activities, and ensure an adequate food supply for local communities. Furthermore, the created model can be used to anticipate inventory planning yields to assure regional food security. These values are coupled with other elements such as plant growth stage, accessible water and nutrients, and environmental conditions to predict the crop's potential yield. Farmers, local food producers, and policymakers can use this information to optimize the local food system and assure a consistent supply of fresh, healthful food to the local community in urban areas.

**Acknowledgments** We want to express our sincere thanks to the Indonesian Geospatial Agency (BIG), the United States Geological Survey (USGS), and European Space Agency (ESA) for providing geographical and satellite data information for this research.

**Conflict of Interest** The authors declare no conflict of interest regarding the research.

## References


Aghighi H, Azadbakht M, Ashourloo D, Shahrabi HS, Radiom S (2018) Machine learning regression techniques for the silage maize yield prediction using time-series images of Landsat 8 OLI. IEEE J Sel Top Appl Earth Obs Remote Sens 11(12):4563–4577

Ali I, Cawkwell F, Dwyer E, Green S (2016) Modeling managed grassland biomass estimation by using multitemporal remote sensing data—a machine learning approach. IEEE J Sel Top Appl Earth Obs Remote Sens 10(7):3254–3264

Amankulova K, Farmonov N, Akramova P, Tursunov I, Mucsi L (2023) Comparison of PlanetScope, Sentinel-2, and Landsat 8 data in soybean yield estimation within-field variability with random forest regression. Heliyon 9:e17432

Azunre GA, Amponsah O, Peprah C, Takyi SA, Braimah I (2019) A review of the role of urban agriculture in the sustainable city discourse. Cities 93:104–119

Brown ME, McCarty JL (2017) Is remote sensing useful for finding and monitoring urban farms? Appl Geogr 80:23–33

Camps-Valls G, Bruzzone L, Rojo-Alvarez JL, Melgani F (2006) Robust support vector regression for biophysical variable estimation from remotely sensed images. IEEE Geosci Remote Sens Lett 3(3):339–343

Castillo JAA, Apan AA, Maraseni TN, Salmo SG III (2017) Estimation and mapping of above-ground biomass of mangrove forests and their replacement land uses in the Philippines using Sentinel imagery. ISPRS J Photogramm Remote Sens 134:70–85

Chlingaryan A, Sukkarieh S, Whelan B (2018) Machine learning approaches for crop yield prediction and nitrogen status estimation in precision agriculture: a review. Comput Electron Agric 151:61–69

Choudhary K, Shi W, Dong Y, Paringer R (2022) Random Forest for rice yield mapping and prediction using Sentinel-2 data with Google Earth Engine. Adv Space Res 70(8):2443–2457

De Souza AP, Massenburg LN, Jaiswal D, Cheng S, Shekar R, Long SP (2017) Rooting for cassava: insights into photosynthesis and associated physiology as a route to improve yield potential. New Phytol 213(1):50–65

Deller SC, Lamie D, Stickel M (2017) Local food systems and community economic development. Community Dev 48(5):612–638

Fagbola O, Osonubi O, Mulongoy K (1998) Growth of cassava cultivar TMS 30572 as affected by alley-cropping and mycorrhizal inoculation. Biol Fertil Soils 27:9–14

Feenstra GW (1997) Local food systems and sustainable communities. Am J Altern Agric 12(1):28–36

Gopal PM, Bhargavi R (2019) A novel approach for efficient crop yield prediction. Comput Electron Agric 165:104968

Granvik M, Joosse S, Hunt A, Hallberg I (2017) Confusion and misunderstanding—interpretations and definitions of local food. Sustainability 9(11):1981

Guan K, Wu J, Kimball JS, Anderson MC, Frolking S, Li B et al (2017) The shared and unique values of optical, fluorescence, thermal and microwave satellite data for estimating large-scale crop yields. Remote Sens Environ 199:333–349

Liakos KG, Busato P, Moshou D, Pearson S, Bochtis D (2018) Machine learning in agriculture: a review. Sensors 18(8):2674

Meyerding SG, Trajer N, Lehberger M (2019) What is local food? The case of consumer preferences for local food labeling of tomatoes in Germany. J Clean Prod 207:30–43

Morrison KT, Nelson TA, Ostry AS (2011) Methods for mapping local food production capacity from agricultural statistics. Agric Syst 104(6):491–499

Panganiban E, Plata IT, Bartolome BB, Taracatac AC, Labuanan FRE (2019) Cassava leaf NDVI-artificial neural network (CaNDVI-ANN): a low cost, portable and non-invasive cassava plant health monitoring device. Int J Adv Trends Comput Sci Eng 8:3555–3559

Payen FT, Evans DL, Falagán N, Hardman CA, Kourmpetli S, Liu L et al (2022) How much food can we grow in urban areas? Food production and crop yields of urban agriculture: a meta-analysis. Earth's Future 10(8):e2022EF002748

Phosaengsri W, Banterng P, Vorasoot N, Jogloy S, Theerakulpisut P (2019) Leaf performances of cassava genotypes in different seasons and its relationship with biomass. Turk J Field Crops 24(1):54–64

Pothukuchi K, Kaufman JL (1999) Placing the food system on the urban agenda: the role of municipal institutions in food systems planning. Agric Hum Values 16(2):213–224

Pradhan P, Kriewald S, Costa L, Rybski D, Benton TG, Fischer G, Kropp JP (2020) Urban food systems: how regionalization can contribute to climate change mitigation. Environ Sci Technol 54(17):10551–10560

Purnamasari RA, Noguchi R, Ahamed T (2019) Land suitability assessments for yield prediction of cassava using geospatial fuzzy expert systems and remote sensing. Comput Electron Agric 166:105018

Rouse JW, Haas RH, Schell JA, Deering DW (1973) Monitoring vegetation systems in the Great Plains with ERTS. In: Third ERTS symposium, NASA SP-351, vol I, pp 309–317

Shiu YS, Chuang YC (2019) Yield estimation of paddy rice based on satellite imagery: comparison of global and local regression models. Remote Sens 11(2):111

Siegner A, Sowerwine J, Acey C (2018) Does urban agriculture improve food security? Examining the nexus of food access and distribution of urban produced foods in the United States: a systematic review. Sustainability 10(9):2988

Tedesco D, de Oliveira MF, dos Santos AF, Silva EHC, de Souza Rolim G, da Silva RP (2021) Use of remote sensing to characterize the phenological development and to predict sweet potato yield in two growing seasons. Eur J Agron 129:126337

Tuia D, Verrelst J, Alonso L, Pérez-Cruz F, Camps-Valls G (2011) Multioutput support vector regression for remote sensing biophysical parameter estimation. IEEE Geosci Remote Sens Lett 8(4):804–808

Wang AX, Tran C, Desai N, Lobell D, Ermon S (2018) Deep transfer learning for crop yield prediction with remote sensing data. In: Proceedings of the 1st ACM SIGCAS conference on computing and sustainable societies, pp 1–5

Wang S, Guan K, Wang Z, Ainsworth EA, Zheng T, Townsend PA et al (2021) Unique contributions of chlorophyll and nitrogen to predict crop photosynthetic capacity from leaf spectroscopy. J Exp Bot 72(2):341–354

# Chapter 16
# Drought Assessment over Northern Africa Using Multi-source Satellite Product

**Emmanuel Owoicho Abah, Adigun Paul Ayodele, Ebiendele Precious, Ryozo Noguchi, and Paul Abuh Omale**

**Abstract** Drought represents a significant threat to human and ecological systems across Africa. Historically, drought assessment relied on limited ground-based meteorological precipitation and temperature measurements. However, these cannot capture the full complexity of drought impacts on vegetation, soil moisture, agriculture, and water resources. The study focused on assessing the skill of different satellite remote sensing products for comprehensive drought monitoring in providing extensive spatial coverage of multiple drought-related variables. We utilized multiple satellite precipitation products to analyze meteorological drought patterns across Northern Africa between 2001 and 2020. Three datasets—Climate Hazards Group InfraRed Precipitation with Station (CHIRPS) data, Climate Prediction Center morphing method (CMORPH) precipitation data, and Integrated Multi-Satellite Retrievals for Global precipitation measurement (IMERG)—were used for this study. Recent advancements have expanded monitoring capabilities through new drought indicators derived from multi-spectral sensors. However, challenges remain in maximizing the use of Earth observation data for early warning and mitigation. This chapter discussed the need for integrated drought monitoring to support evidence-based adaptation in Northern Africa's drought-prone environments. It reviewed different satellite-based approaches for characterizing agricultural and hydrological drought, including precipitation anomalies, vegetation indices, evapotranspiration deficits, and surface water declines. While precipitation datasets like

E. O. Abah (✉)
Federal Ministry of Agriculture and Food Security, Abuja, Nigeria

A. P. Ayodele · E. Precious
Department of Engineering Mechanics and Energy, University of Tsukuba, Tsukuba, Ibaraki, Japan

R. Noguchi
Graduate School of Agriculture, Kyoto University, Kyoto, Japan

P. A. Omale
Department of Agricultural and Environmental Engineering, Joseph Sarwuan Tarka University, Makurdi, Benue State, Nigeria



T. Ahamed (ed.), *Remote Sensing Application II*, New Frontiers in Regional Science: Asian Perspectives 77, https://doi.org/10.1007/978-981-97-1188-8_16

CHIRPS and Tropical Rainfall Measuring Mission (TRMM) have enabled large-scale meteorological drought monitoring, vegetation indices track delayed impacts on crop health. Operational evapotranspiration models and surface water mappings provide further insights into moisture stress. However, no indicator gave a complete view of drought propagation through climatic, ecological, and human systems. In this regard, an integrated monitoring system combining multiple indicators and datasets is required. The chapter emphasized the importance of rigorous validation and uncertainty characterization using independent ground-based data. Overall, satellite data offer new drought monitoring capabilities to help vulnerable regions anticipate and prepare for droughts exacerbated by climate change. Realizing this potential requires continued algorithm refinements, capacity-building, and seamless integration with local information to maximize early warning utility.

**Keywords** Drought · Evapotranspiration · Precipitation · Moisture stress · Satellite data · Earth observation

## 16.1 Introduction

Natural disasters like drought have the potential to have far-reaching effects on human civilizations and sensitive ecology. Climate change and its precursors (Abah et al. 2018, 2020, 2021, 2022) have further altered the natural pattern of droughts, causing it to be more prevalent, lengthier, and acute. Drought has an ominous presence in the African environment, threatening vital areas such as food supplies, water reserves, and the delicate fabric of economic development. Historically, meteorological measurements, especially precipitation and temperature trends, have been used to assess drought (Aziz et al. 2018). However, the limitations of meteorological data have prompted a compelling need to go beyond them, as it has become clear that meteorological data alone cannot capture the full complexity of drought, including crucial aspects like soil moisture levels and the health of vegetative cover (Jiao et al. 2021). Recently, there has been an undeniable upward trend in anticipation of a shift to making the most of satellite-derived data for drought assessment. Satellite data provide an unrivaled perspective, providing analysts with a birds-eye view of the many factors that go into a comprehensive drought study (Ahmadalipour et al. 2019). By combining so many sources of information, we could create new drought indices, ushering in an age in which traditional meteorological indices can be superseded and providing a detailed picture of the dire drought situation.

Drought, a persistent occurrence, can inflict harm on communities and various aspects of human life, including food security, agriculture, and the environment (Behrangi et al. 2015; Bryan et al. 2020). Over 50% of the 2.8 billion individuals worldwide who experienced natural disasters during the 1960s were affected by droughts (Behrangi et al. 2015). Drought, a natural phenomenon, can become disastrous when prolonged and cumulative. The duration of a drought is essential for categorizing it into distinct forms, identifying its regional distribution, effects on land, and severity. Drought's impact on hydrology, soil, agriculture, and water sources categorizes them as hydrological, ecological, meteorological, and socioeconomic (Erfurt et al. 2020). In various parts of South Africa, recent studies indicate drought's influence on industries like agriculture and water resources (Orimoloye et al. 2021). Frequent recurrence led the Free State Province Government to declare drought a disaster due to its harmful effects on farmers and water-dependent industries. Traditional assessment methods rely on ground observations of temperature and rainfall, but satellite sensors provide real-time data for forecasting onset, duration, and severity (Wang et al. 2016). Managing the stages of a drought crisis necessitates evaluating and recording incidents. Remote sensing, utilizing satellite photos, allows evaluation of drought conditions using Moderate Resolution Imaging Spectroradiometer (MODIS) data. Drought's effects on forest plantations in Zululand, South Africa, were assessed using MODIS time series and climate data, revealing patterns related to precipitation shortfall (Xulu et al. 2018; Botai et al. 2018).

Northern Africa's vulnerability to drought endangers communities, ecosystems, and economies. To manage this, understanding drought dynamics and impacts is essential. Advancements in Earth observation facilitate drought assessment using multi-source satellite data. Integrating information from multiple platforms offers a holistic view, aiding decision-making. As Northern Africa faces climate change effects, remote sensing provides a comprehensive understanding. Drought indices like Normalized Difference Vegetation Index (NDVI), Vegetation Condition Index (VCI), and Standardized Precipitation Index (SPI) quantify severity, and data assimilation techniques, and improve forecasting (Page and Zribi 2019). Despite transformative benefits, challenges like data accessibility and uncertainties persist. Collaboration and capacity-building are vital. Integrating multi-source satellite data shifts drought assessment, offering insights into impacts and enabling evidence-based policies. This study assesses drought in Northern Africa using multiple satellite sources to improve accuracy for water management, agricultural planning, and disaster preparedness. This region experiences persistent and severe drought throughout the year, indicating its high susceptibility to desertification (Neto et al. 2021); hence high temporal and spatial resolutions' detailed assessment of drought dynamics, including vegetation health and soil moisture changes, would provide added value in climate adaptation for stakeholders.

## 16.2 Advancing Drought Assessment in Africa: Leveraging Satellite-Merged Precipitation Datasets

In the context of drought assessment across Africa, satellite-integrated precipitation databases provide significant advantages. Precipitation Estimation from Remotely Sensed Information Using Artificial Neural Networks (PERSIANN) and the Tropical Rainfall Measuring Mission (TRMM) are two examples of such datasets that offer high-resolution precipitation estimates by combining observations from multiple satellite sensors. These datasets employ advanced techniques to process and integrate satellite data, enabling accurate and spatially continuous monitoring of precipitation patterns across the African continent, which is crucial for effective drought assessment and management. These datasets have been further refined by other approaches, such as the Climate Prediction Center (CPC) Morphing Technique (CMORPH), the Global Satellite Mapping of Precipitation (GSMaP), and the Global Precipitation Measurement (GPM) Satellite-merged precipitation datasets provide the advantage of including gauge information with correction methods, among other benefits. Combining the two data sources improves the accuracy of satellite-based estimations and gauge readings on the ground. Without gauge adjustments, satellite-based estimates might overestimate measured rainfall due to factors including sample uncertainties, human algorithm mistakes, and atmospheric impacts (Rossi et al. 2017). Furthermore, imperfections within these datasets might negatively affect their hydrological value because of the non-linearities within hydrological processes. Some satellite-merged datasets, such as CMORPH and TRMM3B42, perform better than others regarding hydrological applications. Compared to conventional sources like rain gauges, such datasets have shown their worth in streamflow modeling models. This is especially important in places like the Yarlung Zangbo River Basin (YZRB), where monsoon impacts and a lack of in-person observations make accurate forecasting difficult (Liu et al. 2019). Satellite-derived rainfall estimates are a viable alternative to manually collected and often unreliable rain gauge readings. This quality is paramount for early warning systems to fulfill the urgent requirement for accurate and timely information in far-flung places.

The water cycle is primarily driven by precipitation (Ayodele et al. 2019), which also plays a vital role in the global water balance over many different periods. The dynamics of rainfall are essential to grasp on a variety of geographical scales (Ayodele et al. 2019), from watersheds to regions and even nations (Alijanian et al. 2019; Duan et al. 2016; Gu et al. 2018; Moisa et al. 2022). Precipitation information gathered using traditional methods, such as rain gauges, is prone to errors and has poor geographical coverage (Gemeda et al. 2022; Sippel et al. 2018). Understanding precipitation's geographical and temporal variability requires reliable, spatially dense data over long periods, contributing to the popularity of satellite remote sensing datasets. The resolution of satellite data is far higher than that of historical meteorological records, which is a significant benefit (Bayat et al. 2021). Three different satellite remote sensing datasets—the Climate Hazards InfraRed Precipitation with Stations (CHIRPS), the CMORPH, and the Integrated Multi-satellite Retrievals for GPM (IMERG)—were compared (Fig. 16.1) to show the annual consecutive dry days exceeding 5 days in Africa. Consistent across all three datasets, the Sahara and the Sahel see the driest days in a row. However, there are variations in the severity of the dry conditions throughout the available datasets. For instance, the IMERG dataset reveals that the Sahel area has the most significant number of consecutive dry days, whereas the CHIRPS dataset shows that the Sahara Desert has the highest number of consecutive dry days. This shows why combining data from many

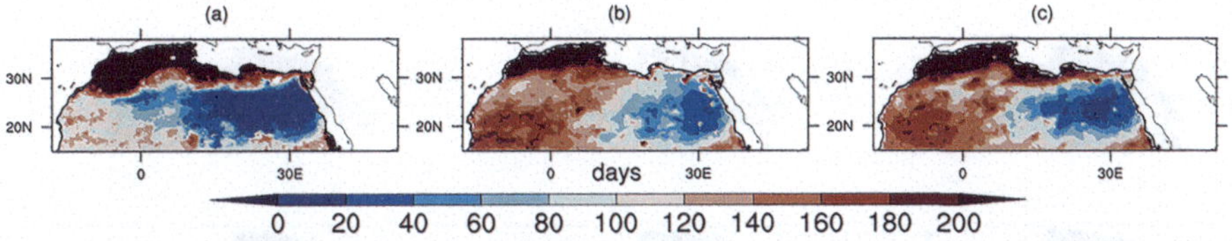


**Fig. 16.1** Annual spatial distribution of consecutive dry days over Africa: a comparative analysis between CHIRPS, CMORPH, and IMERG from 2001 to 2020. (**a**) ANN CHIRPS, (**b**) ANN CMORPH, (**c**) ANN IMERG

satellite and remote sensing datasets is crucial when accurately tracking drought in Africa. Data locations with high uncertainty may be found by comparing the outcomes of many datasets. Afterward, researchers and watchdogs will know precisely where to direct their attention. They were given the unreliability and limited spatial coverage of historical precipitation data collected through technologies like rain gauges. These restrictions may be addressed using satellite remote sensing datasets, which collect consistent, geographically rich data over extended periods. It graphically illustrates the use of satellite remote sensing data in tracking drought in Africa.

Figure 16.2 shows the seasonal spatial distribution of numbers of dry days across Northern Africa. The number of dry days is noted to be at its maximum during September, October, and November (SON), while it is at its lowest during the wet season in March, April, and May (MAM) (Fig. 16.2). It is important to note that some places, such as the Sahara Desert and the Sahel region, often have high numbers of consecutive dry days regardless of the season. When evaluating the severity of the drought in Africa, it is essential to consider the prevailing seasonal trends. It is possible to improve one's ability to foresee the likelihood of drought by first understanding the variations in the number of consecutive dry days that occur throughout the year. It is emphasized that conventional approaches to data collection on precipitation, such as rain gauges, are prone to mistakes and frequently need to provide more coverage throughout the whole geographical region.

In contrast, datasets derived from satellite remote sensing provide a solution to these restrictions by supplying accurate and densely packed data across prolonged times. This makes them an attractive alternative. The benefits of exploiting satellite remote sensing data to monitor drought conditions in the African environment are graphically emphasized by Fig. 16.2. Incorporating the insights provided by Fig. 16.2 into the larger background, it is essential to observe that precipitation, as the driving force of the water cycle, strongly impacts the global water balance throughout various periods. Dependable and comprehensive datasets are required to understand the spatial and temporal changes in the amount of precipitation that occur. Satellite remote sensing datasets are in a class of their own when it comes to being able to meet this need. These datasets not only make it possible to monitor the cyclical nature of droughts, but they also make it possible to forecast probable instances of drought based on the number of consecutive dry days. Evaluation of droughts is essential for comprehending their social, economic, and ecological repercussions because future deficits are expected to be more frequent and severe

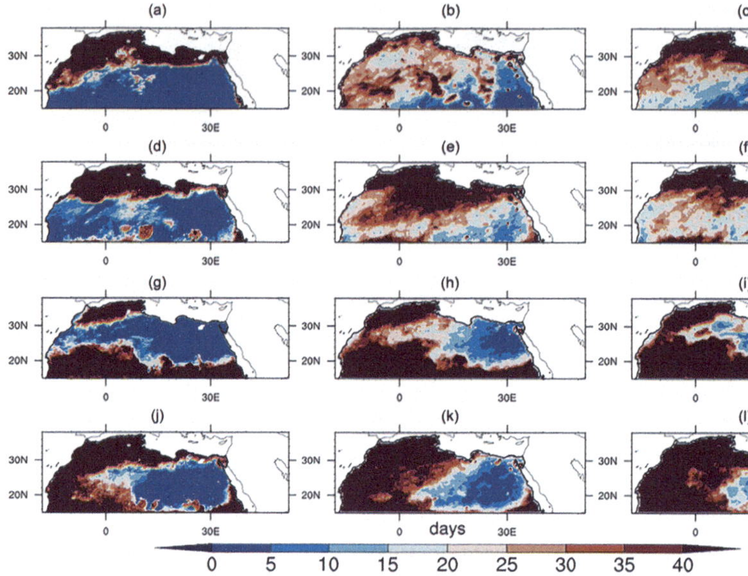


**Fig. 16.2** Spatial distribution of consecutive dry days across different seasons in Africa, for December, January, and February (DJF), March, April, and May (MAM), June, July, and August (JJA), and September, October, and November (SON): a comparative analysis between CHIRPS, CMORPH, and IMERG from 2001 to 2020. (**a**) DJF CHIRPS, (**b**) DJF CMORPH, (**c**) DJF IMERG, (**d**) MAM CHIRPS, (**e**) MAM CMORPH, (**f**) MAM IMERG, (**g**) JJA CHIRPS (**h**) JJA CMORPH, (**i**) JJA IMERG, (**j**) SON CHIRPS, (**k**) SON CMORPH, (**l**) SON IMERG

owing to climate change (Miller et al. 2001). Despite their occasional inadequate distribution in underdeveloped countries, drought research often uses meteorological stations. However, because of their limited geographic reach, they are frequently unable to record the ebb and flow of climatic variables adequately. Here, advancements in satellite technology have been beneficial, increasing the value of distant sensing information (IPCC 2014).

Drought evaluations, as well as other hydro-climatological applications, make substantial use of satellite-based precipitation estimates (Thiemig et al. 2012; Pona et al. 2021). Different satellite-based precipitation prediction methods have evolved. Some of them are the SPI (Yang et al. 2022; Hinge et al. 2021; Horton et al. 2022), the Microwave/Infrared Rain Rate Algorithm (Gemeda et al. 2020; de Oliveira-Júnior et al. 2021), and the CHIRPS dataset. One such system is the CHIRPS, which combines long-term infrared remote sensing data with ground-based observations, making it possible to study climate change and keep tabs on drought conditions over more than 30 years (Funk et al. 2015; Theon 1994; Nashwan et al. 2019). Abay Basin (Li et al. 2022), Rift Valley Lake Basin (Mianabadi et al. 2022), and Didessa Basin (Aksu et al. 2021) are only a few of the places in Ethiopia where the CHIRPS dataset's efficacy has been measured. Furthermore, CHIRPS is compatible and superior to other satellite-based precipitation products (Bayissa et al. 2017) by comparisons with ground data. Typically, meteorological or ground station data are used for estimating the possibility of a drought. However, obstacles include the generally short record duration and the scarcity/irregular distribution of ground stations within research regions. Due to their dependence on rain-fed crops, emerging countries' agricultural economies suffer greatly when agricultural droughts occur. To this end, satellite-derived precipitation databases that are both precise and extensive are essential.

## 16.3 Standardized Precipitation Index (SPI)

The SPI, a standardized drought indicator, is easier to use for determining the severity and length of a drought. The indicator can be used to determine the severity by normalizing rainfall anomalies and providing a probabilistic knowledge of the relevance of observed or anticipated rainfall. The SPI helps analyze drought severity over various timeframes, from short-term present circumstances to long-term trends, and can do for both short- and long-term situations (Kurnik et al. 2011; Pazhanivelan et al. 2023). McKee et al. (1993) developed the SPI to determine the length and severity of drought-like conditions (McKee et al. 1993). The initial step is by first normalizing rainfall anomalies, which reveals the probability importance of observed or anticipated rainfall amounts. The SPI measures the amount of wetness and dryness in the atmosphere by contrasting the total precipitation over various periods with the historical rainfall patterns. At a resolution of 10 km, the SPI may be estimated for various accumulation periods (Funk et al. 2003). The 10-day to 1-month SPI analyzes the present circumstances and how they relate to the soil moisture levels. The SPI for 3 months offers an estimate of the seasonal precipitation. In contrast, the SPI for 6 months and the SPI for 12 months reflect long-term precipitation patterns that often correlate with stream flows, reservoir levels, and occasionally groundwater levels. When the SPI number is positive, it indicates circumstances that are drier than the median, and when SPI number is negative, it indicates conditions that are wetter than the median. If the SPI is less than 1.0, this indicates a moderate drought, which occurs typically once every six dry episodes. If the SPI is below 1.5, there is a "severe" drought, translating to a 1 in 15 chance of dryness. Values less than 2.0 show the presence of "extreme" drought, which occurs in the driest 2% of all possible occurrences (Huffman et al. 2019).

The SPI-3, a drought index, quantifies precipitation deviations from the historical average over 3 months. A negative SPI-3 denotes drought conditions, with more negative values signifying heightened severity. Figure 16.3 visually demonstrates widespread drought occurrences across Africa between 2001 and 2022. The Sahel region, the Horn of Africa, and southern Africa have borne the brunt of the most intense droughts (Kurnik et al. 2011). The deficits exhibit a distinct spatial pattern, with the most severe instances concentrated in the continent's driest regions. Notably, the Sahel region is vulnerable due to its transitional position between the Sahara Desert and the Sudanian Savanna. Similarly, the Horn of Africa's

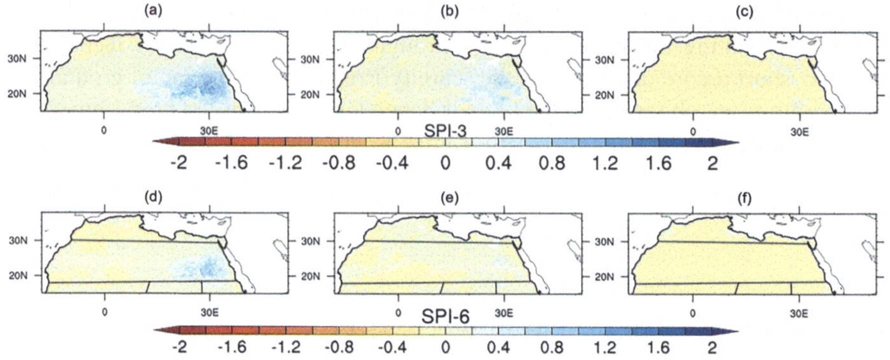


**Fig. 16.3** Spatial distribution of drought severity (months) from three satellite products over Africa for SPI-3 and SPI-6 during 2001–2022 from CHIRPS, CMORPH, and IMERG. (**a**) SPI-3 CHIRPS, (**b**) SPI-3 CMORPH, (**c**) SPI-3 IMERG, (**d**) SPI-6 CHIRPS, (**e**) SPI-6 CMORPH, (**f**) SPI-6 IMERG

susceptibility to drought is tied to its exposure to El Niño-Southern Oscillation (ENSO) events. At the same time, the Indian Ocean Dipole (IOD) puts southern Africa at risk. Recent years have witnessed an escalation in drought frequency and severity, likely attributed to climate change's role in triggering extreme weather occurrences, including droughts. This escalation is putting strain on the region's water resources and agricultural output. Thus, it becomes imperative to devise strategies to mitigate drought impacts and enhance resilience against impending drought episodes. As we delve into the outcomes, certain aspects warrant discussion. The current study exclusively leveraged three satellite products for SPI-3 computation. Incorporating additional satellite datasets for confirmation would be advantageous to bolster result validity. Moreover, the investigation's temporal scope spanned 2001–2020. Extending this range to encompass a more extended timeframe would provide insight into the consistency of drought severity trends over time.

The SPI-6 is a crucial drought measure that monitors precipitation patterns over 6 months and compares them to the average rainfall for that period. A negative SPI-6 indicates that drought conditions are present, and more negative readings indicate an increased drought severity. Figure 16.4d, e shows a significant increase in the number of instances of lack throughout Africa between 2001 and 2020, demonstrating that the three different satellite products' estimations of SPI-6 are only partially consistent. Those from CHIRPS and CMORPH indicate a closer level of agreement than those from IMERG. This disparity is probably due to the different approaches used to generate the estimations, as well as the use of various other data sources. When analyzing the results of the research, it is necessary to give due consideration to the implications of the uncertainty estimations. Certain regions being reported as being amid drought conditions are not, in fact, during a drought, and vice versa, which highlight the critical need for careful drought monitoring across Africa. The frequency and severity of droughts are expected to increase because of climate change, which poses a significant risk to the area. It is still of the utmost

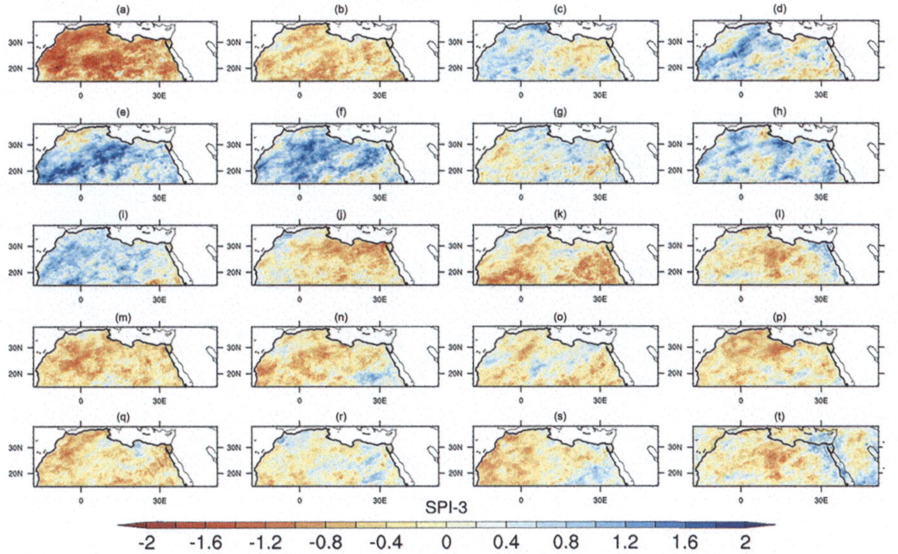


**Fig. 16.4** Spatial distribution of the 3-month SPI over Northern Africa for the period 2001–2020 from IMERG. (**a**) 2001 IMERG, (**b**) 2002 IMERG, (**c**) 2003 IMERG, (**d**) 2004 IMERG, (**e**) 2005 IMERG, (**f**) 2006 IMERG, (**g**) 2007 IMERG, (**h**) 2008 IMERG, (**i**) 2009 IMERG, (**j**) 2010 IMERG, (**k**) 2011 IMERG, (**l**) 2012 IMERG, (**m**) 2013 IMERG, (**n**) 2014 IMERG, (**o**) 2015 IMERG, (**p**) 2016 IMERG, (**q**) 2017 IMERG, (**r**) 2018 IMERG, (**s**) 2019 IMERG, (**t**) 2020 IMERG

importance to develop plans that may reduce the adverse effects of drought and strengthen resistance to the occurrence of future droughts.

Three-month SPI reveals considerable variability in meteorological drought across Northern Africa from 2001 to 2020 (Fig. 16.5). Several severe drought episodes stand out that brought hardship across the region. The years 2001, 2002, and 2011 saw extreme drought conditions emerge, with SPI values below −2 over vast areas, indicating acute rainfall deficits. These major droughts triggered food insecurity, water scarcity, and economic losses across Northern Africa. The regions most affected were the Sahel, the Horn of Africa, and northwestern Africa along the Mediterranean coast. These extreme events and moderate drought conditions, with SPI values from −1 to −1.5, have become more prevalent in recent decades. Central Africa and the Sahel seem most prone to persistent aridity, indicating an increasingly drought-vulnerable climate. The human and ecological impacts of repeated moderate droughts accumulate over time, undermining resilience. However, the precise timing and spatial pattern of lack has varied, and some periods, like 2001, 2002, 2011, and 2013, experienced regional-scale drought spanning most of Northern Africa (Gidey et al. 2018). During other episodes, like 2003–2009, drought was concentrated in specific subregions while the rest of Northern Africa saw normal or wet conditions. These highlight the urgency of targeted drought monitoring and early warning systems to assist the most vulnerable communities. Enhanced

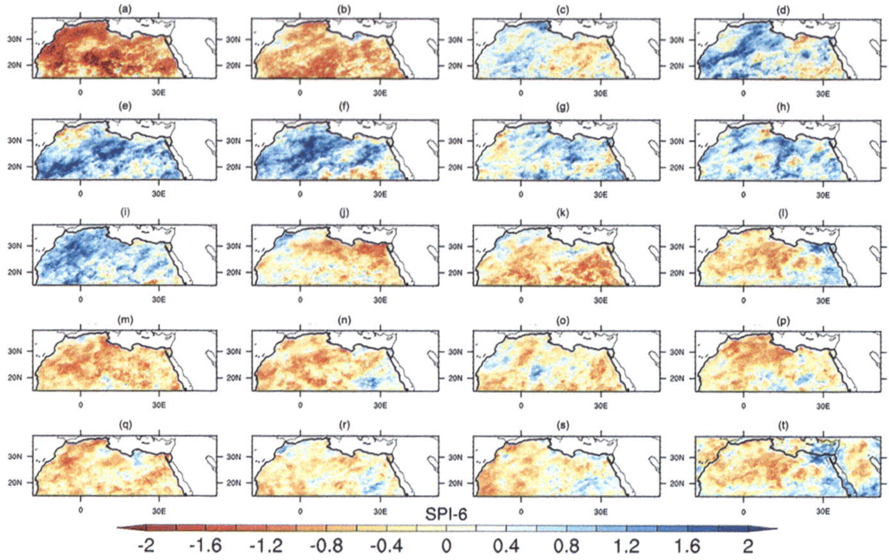


**Fig. 16.5** Spatial distribution of the 6-month SPI over Northern Africa for the period 2001–2020 from IMERG. (**a**) 2001 IMERG, (**b**) 2002 IMERG, (**c**) 2003 IMERG, (**d**) 2004 IMERG, (**e**) 2005 IMERG, (**f**) 2006 IMERG, (**g**) 2007 IMERG, (**h**) 2008 IMERG, (**i**) 2009 IMERG, (**j**) 2010 IMERG, (**k**) 2011 IMERG, (**l**) 2012 IMERG, (**m**) 2013 IMERG, (**n**) 2014 IMERG, (**o**) 2015 IMERG, (**p**) 2016 IMERG, (**q**) 2017 IMERG, (**r**) 2018 IMERG, (**s**) 2019 IMERG, (**t**) 2020 IMERG

water management and climate-smart agriculture will be essential to adapt to projected increases in aridity. While the analysis provides valuable insights, incorporating multiple datasets and linking meteorological drought to impacts would further strengthen understanding. Overall, the results underscore Northern Africa's susceptibility to droughts likely to increase with climate change.

The occurrence of meteorological drought, as characterized by negative Standardized Precipitation Index (SPI) values, has exhibited substantial temporal and spatial variability across Northern Africa over the past two decades (2001–2021). The most intense and widespread droughts, with SPI values below −2 indicating extremely dry conditions, impacted broad areas of Northern Africa in 2001, 2002, 2011, and 2013. During these major drought years, aridity stretched across much of the region, bringing hardship to agriculture, pastoralism, and water resources. In other years, such as 2010 and 2015, drought conditions were less intense and more geographically confined, with SPI values predominantly between −1 and −0.5 in certain subregions. The Sahel zone stands out as most prone to recurring droughts, given its semi-arid climate. Several periods of exceptionally low rainfall since 2001 have disrupted livelihoods and food security in Sahelian countries. The geographic extent of droughts has also shifted over time. Some years, like 2001 and 2013, saw drought encompass most of Northern Africa. In other periods, like 2003–2009, arid conditions were concentrated in specific subregions while the rest of the region experienced near-normal rainfall. This highlights the need for targeted early warning and adaptation efforts directed at the most vulnerable areas. Comparison of 3-month and 6-month SPI indicates general agreement, but also greater small-scale spatial variability evident in the 6-month index. This reflects how longer-term precipitation deficits may emerge over locations. Overall, the characterized drought variability points to intensifying risk across Northern Africa due to climate change and volatile rainfall. Improved monitoring and drought preparedness will be essential to building resilience. This analysis provides a useful overview of recent drought patterns but could be strengthened by incorporating additional precipitation datasets and linking meteorological drought to agricultural and hydrological impacts. The discussion highlights the importance of understanding where, when, and with what intensity droughts have historically occurred to manage risk and support adaptation.

## 16.4 Degree of Agreement Between CHIRPS, CMORPH, and IMERG Satellite Precipitation Products

The correlation analysis between different precipitation products reveals several valuable insights. A strong correlation exists between the 3-month and 6-month SPI calculated using CHIRPS, CMORPH, and IMERG across most of the study region (Fig. 16.6). Correlation coefficients exceeding 0.8 dominate for most grid cells, indicating these products show similar effectiveness in capturing meteorological drought patterns on seasonal timescales. This convergence lends confidence in the ability of all three satellite-based precipitation datasets to monitor developing drought conditions for early warning. It suggests interpolation methods and input satellite observations lead to consistent SPI outputs, the most widely used drought index. Areas of divergence emerge, particularly for 6-month SPI in topographically complex regions where orographic effects likely influence precipitation totals differently across products. But generally, the three products display enough agreement to be used interchangeably for drought monitoring and other applications requiring precipitation data at a 3- to 6-month SPI. By validating multiple precipitation products against each other, the analysis provides valuable guidance for selecting appropriate data for research or operational needs. Knowing where products converge, or exhibit discrepancies aids in more informed usage of the expanding precipitation datasets for climate modeling, water management, and agricultural decision-making. Overall, the strong correlations highlight the maturing of satellite-based precipitation estimation and processing methods that now lend confidence in their ability to effectively monitor precipitation variability and drought early warning across complex terrains like Northern Africa.

The interclass correlation analysis (Fig. 16.7) reveals strong agreement in captured meteorological drought patterns between essential precipitation products across Northern Africa. For the widely used 3-month SPI, interclass correlation coefficients between 0.5 and 0.55 emerge across most of the study region when

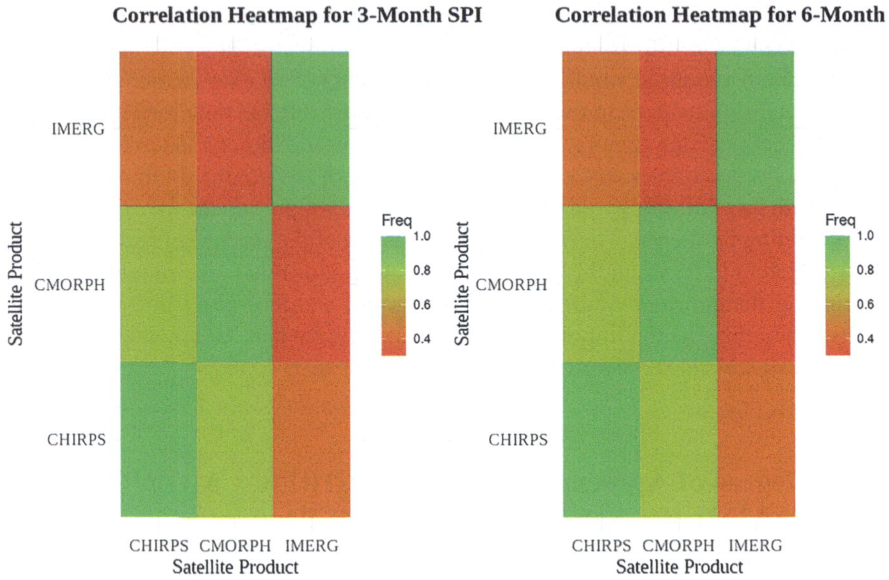


**Fig. 16.6** Correlation heatmap for 3-month and 6-month SPI between CHIRPS, CMORPH, and IMERG

comparing CHIRPS, CMORPH, and IMERG. This indicates these datasets similarly identify emerging short-term precipitation deficits associated with agricultural and hydrological drought onset. The 6-month SPI produces an interclass correlation of around 0.49 for most of the study area. This suggests satisfactory convergence in estimated longer-term precipitation anomalies that drive drawn-out drought episodes. Overall, the high correlation lends confidence in the interchangeable use of these products for drought monitoring, early warning, and other applications requiring precipitation data. However, some divergence does arise, indicating the continued value of an ensemble approach. In topographically complex areas, orographic effects likely influence precipitation totals differently across products, leading to slightly weaker correlations for the 6-month SPI measure in some zones.

Each product also has relative strengths that users must consider. The strong agreement suggests flexibility in selecting different validated products for drought analyses based on the needed temporal scale, location of interest, and desired accuracy versus accessibility. This allows tailoring outputs to specific stakeholder needs, whether for policymaking, agricultural management, or hydrological planning. Using any single precipitation product risks potential hidden biases. Checking multiple datasets remains advisable to obtain the most accurate picture of meteorological drought and minimize uncertainty for impact projections. An ensemble approach combining validated products provides greater confidence in determining where, when, and how intensely lack unfolds. The convergence of these widely used datasets increases the credibility of remote sensing and model-based precipitation estimates across the data-sparse Northern African domain. The high interclass

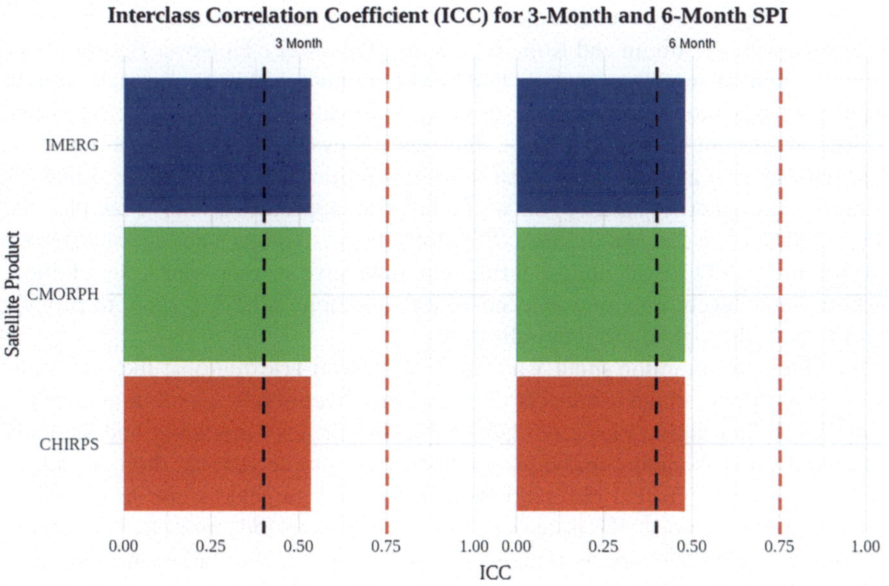


**Fig. 16.7** Interclass correlation coefficient (ICC) for 3-month and 6-month SPI between CHIRPS, CMORPH, and IMERG

correlations indicate sophistication in algorithms and blending methods for this complex region. However, continued ground validation and product improvements are still required. Overall, this analysis provides a critical validation baseline and suggests flexibility in selecting between quality-controlled datasets for diverse Northern Africa applications.

## 16.5 Drought Monitoring Based on Vegetation Index

Vegetation health monitoring provides useful agricultural drought information, capturing delayed impacts relative to rainfall deficits. Plants initially respond to moisture stress through water conservation without outward changes detectable by satellites. However, prolonged precipitation shortfalls can decrease photosynthesis and worsen vegetation conditions, evident through remote sensing. The Normalized Difference Vegetation Index (NDVI) widely monitors vegetation stress from water scarcity. It exploits the near-infrared and red bands to isolate moisture effects on plant health. Each pixel's 10-day maximum NDVI composite reduces atmospheric contamination when tracking drought emergence. Accelerated MODIS NDVI at 10-day intervals enables comparison to prior years to identify crop stress. Time series smoothing refines the NDVI record by minimizing noise from non-moisture factors over 2001–2010. Negative NDVI anomalies signaling below-normal vegetation activity provide straightforward drought identification once validated (Fan and

Liu 2016). Minor land cover changes are unlikely to influence anomaly performance at broader scales (Adigun and Koji 2022; Seun et al. 2022). However rigorous validation remains essential to confirm that NDVI anomalies effectively track drought impacts on agriculture. Evapotranspiration (ET) monitoring also supplies important ecological drought information. ET comprises soil evaporation and plant transpiration, representing atmospheric moisture loss. Satellite data can estimate actual ET using surface energy balance or water balance approaches. For example, the Simplified Surface Energy Balance (SSEBop) model exploits satellite land surface temperature (LST) to calculate ET fractions indicative of crop stress. In addition, surface water levels integrate antecedent precipitation and ET signals to provide short-term hydrological drought indications.

Satellites can monitor small water bodies in semi-arid regions, like programs evaluating climate change impacts on pastoralist livelihoods. Combining satellite rainfall, ET, and digital elevation data enables water level modeling and anomaly tracking. Rigorous validation builds confidence in remote sensing drought indicators. For example, ground observations have verified modeled water levels, while flux tower data support ET accuracy. However, linking indicators to agricultural yields and livelihood impacts remains an ongoing need. Rainfall monitoring provides the earliest meteorological drought identification before propagation through soils and vegetation. Africa has extensive gauge networks, but interpolation between stations using satellite data expands coverage. For instance, the FEWS NET RFE integrates station and satellite rainfall data to create gridded precipitation estimates for drought analysis (Funk et al. 2015). Soil moisture data also supply agriculturally relevant hydrological drought information. Microwave and thermal sensors can estimate moisture in different soil depths, capturing precipitation deficits. NASA's SMAP satellite provides global surface and root zone soil moisture data to identify emerging drought conditions.

The SPI composites precipitation anomalies over different timescales to characterize drought severity (McKee et al. 1993). Positive SPI indicates greater than median precipitation, while negative SPI signals emerging drought. FEWS NET uses 1–36-month SPI from satellite and ground observation data to monitor drought onset in Africa (Funk et al. 2015). Besides rainfall deficits, high temperatures causing increased ET can induce or exacerbate droughts. Land surface temperature (LST) anomalies from thermal satellites track elevated ET loss that dries soils and vegetation. LST data improve drought detection skills, capturing ET-related processes missed by precipitation alone. However, only some indicators provide a complete picture of complex drought manifestations across climatic, ecological, and agricultural spheres. An integrated monitoring approach combining multiple indicators at different temporal scales provides robust monitoring to identify onset and track drought impacts. For instance, the U.S. Drought Monitor assimilates precipitation, temperature, soil moisture, ET, and vegetation data into a converged depiction of emerging drought (Mark et al. 2002). Rigorous validation of indicators and warning systems remains critical to improve monitoring capabilities and avoid false alarms. Evaluating diverse drought events using independent ground data builds confidence in detection accuracy. Ongoing monitoring system refinement to address known limitations further enhances operational reliability. Overall, remote sensing provides expanded coverage for drought monitoring, but ground-truthing and integration with field data are essential to maximize value for agricultural and water management.

## 16.6 Discussion

This study evaluated three widely used satellite-based precipitation products—CHIRPS, CMORPH, and IMERG—for drought monitoring across Northern Africa. Several insightful conclusions emerge from these datasets' correlation, interclass correlation, and error analyses. A robust spatial correlation exists between the 3-month and 6-month SPI calculated from CHIRPS, CMORPH, and IMERG over most Northern Africa. Correlation coefficients exceeding 0.8 for most grid cells highlight convergence in the ability of these precipitation products to capture emerging meteorological drought patterns on seasonal timescales. This lends confidence in their interchangeable use for drought early warning systems, water management, and agricultural decision-making requiring precipitation data. Interclass correlation analysis reveals satisfactory product agreement in identifying precipitation deficits associated with agricultural and hydrological drought onset. Interclass coefficients between 0.5 and 0.55 for 3-month SPI and around 0.49 for 6-month SPI demonstrate these datasets similarly characterize short- and long-term precipitation anomalies. This further supports flexibility in selecting between validated products based on user needs. However, some divergence does exist, particularly for topographically complex regions where orographic effects likely influence precipitation totals differently across products. This highlights that relying solely on any single dataset risks potential hidden biases. An ensemble approach combining products is advisable to minimize uncertainty for drought projections. While this analysis demonstrates the maturation of satellite precipitation estimation for an intricate region like Northern Africa, continued refinement through ground validation is required. Relative product strengths should guide tailored usage for specific applications. Overall, the convergence between CHIRPS, CMORPH, and IMERG increases confidence in remote sensing capabilities for data-sparse areas. However, integrating these datasets with other drought indicators and field data remains essential for robust agricultural and hydrological monitoring.

Satellite remote sensing data have emerged as a powerful tool to address the limitations of conventional ground-based data collection methods. The benefits of satellite data lie in their ability to provide accurate, comprehensive, and densely packed information across large geographical areas and extended periods. This attribute is particularly crucial for monitoring drought conditions in regions like Northern Africa, where data coverage is often sparse, and ground observations alone need to be more comprehensive. The 3-month (SPI-3) and 6-month (SPI-6) Standardized Precipitation Index using CHIRPS, CMORPH, and IMERG (Fig. 16.8) reveals substantial convergence in capturing meteorological drought patterns. High correlation coefficients and low mean squared errors indicate that these products effectively identify precipitation anomalies associated with drought conditions. This convergence provides confidence in the interchangeable use of these products for drought monitoring and early warning systems. The study also recognizes specific contexts where product discrepancies arise, particularly in topographically complex regions, underlining the importance of cautious interpretation. Traditional data collection methods, such as rain gauges, have limitations in providing comprehensive coverage across vast geographical regions. In contrast, satellite remote sensing offers a solution by giving accurate and densely packed data over extended periods. This analysis demonstrated that satellite-derived precipitation datasets, like CHIRPS, CMORPH, and IMERG, are crucial in monitoring and forecasting drought conditions. These datasets not only track the cyclical nature of droughts but also enable predicting potential drought events based on consecutive dry days. The importance of an integrated monitoring approach that combines multiple drought indicators, including vegetation health, soil moisture, and temperature anomalies, lies in the fact that such an approach provides a comprehensive understanding of drought manifestations across different spheres, allowing for more accurate drought detection and impact assessment.

Incorporating vegetation health monitoring through indicators like the Normalized Difference Vegetation Index (NDVI) and evapotranspiration (ET) further enhances drought assessment. These indicators offer valuable insights into the delayed impacts of drought on agriculture and ecosystems. A holistic picture of drought emergence and progression emerges by integrating multiple indicators,

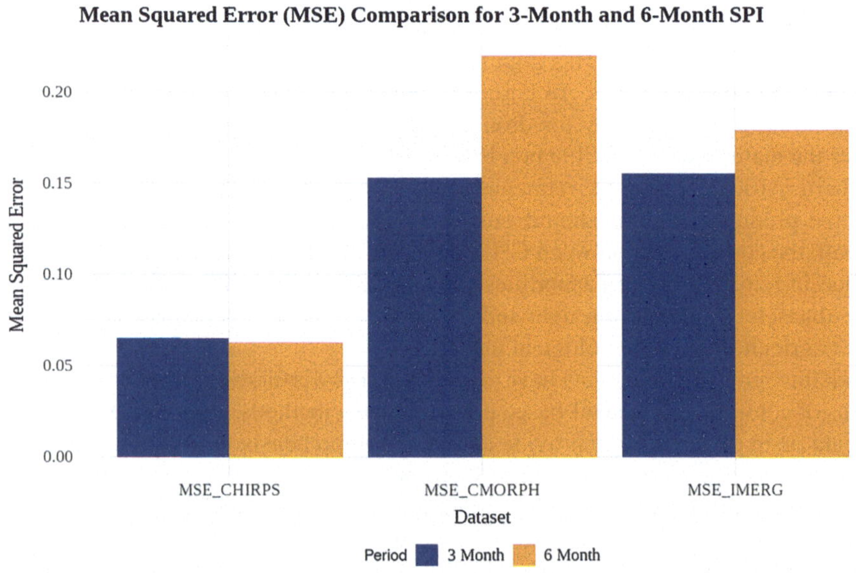


**Fig. 16.8** Mean squared error (MSE) comparison for 3-month and 6-month SPI between CHIRPS, CMORPH, and IMERG

such as rainfall monitoring, soil moisture data, and land surface temperature anomalies. Such comprehensive monitoring enables the identification of drought onset and the tracking of its impacts across various sectors. In conclusion, the analysis presented in this chapter underscores the critical role of satellite-based precipitation products in effectively monitoring meteorological drought conditions in Northern Africa. The degree of agreement among CHIRPS, CMORPH, and IMERG demonstrates their reliability and suitability for drought monitoring and early warning systems. However, it is essential to consider each product's specific strengths and limitations and integrate multiple indicators for a more comprehensive understanding of drought dynamics. As the impacts of climate change intensify the frequency and severity of drought events, the insights from this study will serve as a valuable resource for policymakers, researchers, and stakeholders seeking to mitigate the adverse effects of drought and enhance resilience in the Northern Africa region. Ongoing validation, continuous refinement of monitoring systems, and seamless integration with ground-truth data will propel our capabilities to confront the challenges of meteorological drought effectively. The imperative for robust and dependable drought monitoring systems assumes heightened significance given the climate change-induced drought frequency and severity amplification. Anticipating an upsurge in drought occurrences due to climate change, the necessity for accurate and reliable early warning systems becomes even more pronounced. The insights gleaned from this analysis not only enhance our comprehension of the evolving nature of droughts but also serve as a compass to steer adaptation strategies to mitigate the harmful impacts of drought on agriculture, water resources, and communities. The comparison of the CHIRPS, CMORPH, and IMERG datasets distinctly underscores their credibility and appropriateness in monitoring drought conditions. As climate change continues to influence precipitation patterns, the meticulous evaluation and timely alerting of drought occurrences become paramount for informed decision-making and the fostering of resilient development. The conclusions drawn from this analysis contribute substantively to the broader arena of drought research and emphasize the potential of satellite remote sensing in enriching our comprehension of drought dynamics and their cascading effects on vulnerable regions.

## 16.7 Conclusion

This chapter delves into the degree of agreement between three prominent satellite precipitation products—CHIRPS, CMORPH, and IMERG—across different seasons (MAM, JJA [June, July, and August], SON) from 2001 to 2020 over Northern Africa. The study primarily focuses on meteorological drought monitoring over the Northern Africa region, which is particularly susceptible to drought impacts due to its semi-arid to arid climate. The findings provide valuable insights into the efficacy of these satellite products in tracking drought patterns and assessing their agreement, thus enhancing our understanding of their potential for accurate drought monitoring. Meteorological drought, a critical concern for agriculture, water resources, and society, is intricately linked to precipitation deficits. By evaluating the number of consecutive dry days and utilizing well-established drought indices such as the SPI, this study has shed light on the variations in drought severity across different seasons. It is evident that the choice of satellite product plays a significant role in capturing these variations, and each product exhibits specific strengths in other contexts. The Sahara Desert and the Sahel region, known for their aridity, consistently experience high numbers of consecutive dry days regardless of the season, underlining the importance of monitoring drought in these regions. The study revealed significant variation in drought patterns across different regions and seasons in Northern Africa. December through February (DJF) and June through August (JJA) exhibited distinct patterns of drought occurrence. The number of dry days and drought severity varied across different seasons, highlighting the importance of understanding seasonal trends for effective drought prediction and management.

## References

Abah EO, Iwai K, Sakurai K, Noguchi R (2018) Comparative real-time assessment of particulate matter emissions in rice husk combustion. J Jpn Inst Energy 97:222–225. https://doi.org/10.3775/jie.97.222. (Peer reviewed article)

Abah EO, Ahamed T, Noguchi R (2020) Investigative study for low particulate matter emission in rice husk combustion. J Jpn Inst Energy 99:220–229. https://doi.org/10.3775/jie.99.220. (Peer reviewed article)

Abah EO, Ahamed T, Noguchi R (2021) Catalytic temperature effects on conversion efficiency of PM2.5 and gaseous emissions from rice husk combustion. Energies 14:6131. https://doi.org/10.3390/en14196131. (Peer reviewed article)

Abah EO, Inalegwu J, Ariyawansha KT, Ahamed T, Noguchi R (2022) Rice husk handling and combustion in rural areas: assessment of household proximity to exposure risk in Nigeria. Agric Mech Asia Afr Latin Am 53(3):15–23. ISSN 0084-5841

Adigun P, Koji D (2022) Potential impact of increased urbanization on surface temperature in southwestern Nigeria. In: 水文・水資源学会研究発表会要旨集 水文・水資源学会/日本水文科学会 2022 年度研究発表会 (p 41). 水文・水資源学会

Ahmadalipour A, Moradkhani H, Castelletti A, Magliocca N (2019) Future drought risk in Africa: integrating vulnerability, climate change, and population growth. Sci Total Environ 662:672–686. https://doi.org/10.1016/j.scitotenv.2019.01.278

Aksu H, Cavus Y, Eris E, Aksoy H, Akgul MA, Turker S (2021) Spatiotemporal analysis of drought by CHIRPS precipitation estimates. Theor Appl Climatol 148:517–529

Alijanian M, Rakhshandehroo GR, Mishra A, Dehghani M (2019) Evaluation of remotely sensed precipitation estimates using PERSIANN-CDR and MSWEP for spatio-temporal drought assessment over Iran. J Hydrol 579:124189

Aziz A, Umar M, Mansha M, Khan M, Javed M, Gao H, Farhan S, Iqbal I, Abdullah S (2018) Assessment of drought conditions using HJ-1A/1B data: a case study of Potohar region, Pakistan. Geomat Nat Haz Risk 9:1019–1036. https://doi.org/10.1080/19475705.2018.1499558

Bayat B, Nasseri M, Delmelle E (2021) Uncertainty-based rainfall network design using a fuzzy spatial interpolation method. Appl Softw Comput 106:107296. https://doi.org/10.1016/j.asoc.2021.107296

Bayissa Y, Tadesse T, Demisse G, Shiferaw A (2017) Evaluation of satellite-based rainfall estimates and application to monitor meteorological drought for the Upper Blue Nile Basin, Ethiopia. Remote Sens 9(7):669. https://doi.org/10.3390/rs9070669

Behrangi A, Loikith P, Fetzer E, Nguyen H, Granger S (2015) Utilizing humidity and temperature data to advance monitoring and prediction of meteorological drought. Climate 3:999–1017. https://doi.org/10.3390/cli304099970628%

Botai C, Botai J, Adeola A (2018) Spatial distribution of temporal precipitation contrasts in South Africa. S Afr J Sci 114:70–78. https://doi.org/10.17159/SAJS.2018/20170391

Bryan K, Ward S, Roberts L, White M, Landeg O, Taylor T, McEwen L (2020) The health and well-being effects of drought: assessing multi-stakeholder perspectives through narratives from the UK. Clim Chang 163:2073–2095. https://doi.org/10.1007/s10584-020-02916-x

de Oliveira-Júnior JF, de Gois G, de Lima Silva IJ, de Oliveira Souza E, Jardim AMRF, Da Silva MV, Shah M, Jamjareegulgarn P (2021) Wet and dry periods in the state of Alagoas (Northeast Brazil) via standardized precipitation index. J Atmos Sol Terr Phys 224:105746. https://doi.org/10.1016/j.jastp.2021.105746

Duan Z, Liu J, Tuo Y, Chiogna G, Disse M (2016) Evaluation of eight high spatial resolution gridded precipitation products in Adige Basin (Italy) at multiple temporal and spatial scales. Sci Total Environ 573:1536–1553

Erfurt M, Skiadaresis G, Tijdeman E, Blauhut V, Bauhus J, Glaser R, Schwarz J, Tegel W, Stahl K (2020) A multidisciplinary drought catalogue for southwestern Germany dating back to 1801. Nat Hazards Earth Syst Sci 20:2979–2995. https://doi.org/10.5194/nhess-20-2979-2020

Fan X, Liu Y (2016) A global study of NDVI difference among moderate-resolution satellite sensors. ISPRS J Photogramm Remote Sens 121:177–191. https://doi.org/10.1016/J.ISPRSJPRS.2016.09.008

Funk C, Michaelsen J, Verdin J, Artan G, Husak G, Senay G, Gadain H, Magadazire T (2003) The collaborative historical African rainfall model: description and evaluation. Int J Climatol 23:47. https://doi.org/10.1002/JOC.866

Funk C, Peterson P, Landsfeld M, Pedreros D, Verdin J, Shukla S, Husak G, Rowland J, Harrison L, Hoell A, Michaelsen J (2015) The climate hazards infrared precipitation with stations—a new environmental record for monitoring extremes. Sci Data 2:150066. https://doi.org/10.1038/sdata.2015.66

Gemeda DO, Feyssa DH, Garedew W (2020) Meteorological data trend analysis and local community perception towards climate change: a case study of Jimma City, southwestern Ethiopia. Environ Dev Sustain 23:5885–5903

Gemeda DO, Korecha D, Garedew W (2022) Monitoring climate extremes using standardized evapotranspiration index and future projection of rainfall and temperature in the wettest parts of southwest Ethiopia. Environ Chall 7:100517. https://doi.org/10.1016/j.envc.2022.100517

Gidey E, Dikinya O, Sebego R, Segosebe E, Zenebe A (2018) Modeling the spatio-temporal meteorological drought characteristics using the standardized precipitation index (SPI) in Raya and its environs, Northern Ethiopia. Earth Syst Environ 2:281–292. https://doi.org/10.1007/s41748-018-0057-7

Gu Z, Duan X, Shi Y, Li Y, Pan X (2018) Spatiotemporal variation in vegetation coverage and its response to climatic factors in the Red River Basin, China. Ecol Indic 93:54–64

Hinge G, Mohamed MM, Long D, Hamouda MA (2021) Meta-analysis in using satellite precipitation products for drought monitoring: lessons learnt and the way forward. Remote Sens 13:4353. https://doi.org/10.3390/rs13214353

Horton P, Schaefli B, Kauzlaric M (2022) Why do we have so many different hydrological models? A review based on the case of Switzerland. Wiley Interdiscip Rev Water 9(1):e1574

Huffman GJ, Bolvin DT, Nelkin EJ, Tan J (2019) Integrated Multi-satellitE Retrievals for GPM (IMERG) technical documentation. IMERG Tech 9. https://gpm.nasa.gov/sites/default/files/document_files/IMERG_doc_190909.pdf

IPCC (2014) Impacts, adaptations and vulnerability. In: Working group II contribution to the fifth assessment report of intergovernmental panel on climate change. Cambridge University Press, Cambridge

Jiao W, Wang L, Mccabe M (2021) Multi-sensor remote sensing for drought characterization: current status, opportunities and a roadmap for the future. Remote Sens Environ 256:112313. https://doi.org/10.1016/J.RSE.2021.112313

Kurnik B, Barbosa P, Vogt J (2011) Testing two different precipitation datasets to compute the standardized precipitation index over the Horn of Africa. Int J Remote Sens 32:5947–5964. https://doi.org/10.1080/01431161.2010.499380

Li M, Lv X, Zhu L, Ochege FU, Guo H (2022) Evaluation, and application of MSWEP in drought monitoring in central Asia. Atmosphere 13:1053. https://doi.org/10.3390/atmos13071053

Liu L, Xu Y, Pan S, Bai Z (2019) Potential application of hydrological ensemble prediction in forecasting floods and its components over the Yarlung Zangbo River basin, China. Hydrol Earth Syst Sci 23(8):3335–3352. https://doi.org/10.5194/HESS-23-3335-2019

McKee TB, Doesken NJ, Kleist J (1993) The relationship of drought frequency and duration to time scales. In: 8th conference on applied climatology. American Meteorological Society, Boston, pp 179–184

Mianabadi A, Salari K, Pourmohamad Y (2022) Drought monitoring using the long-term CHIRPS precipitation over Southwestern Iran. Appl Water Sci 12:183. https://doi.org/10.1007/s13201-022-011705-4

Miller SW, Arkin PA, Joyce R (2001) A combined microwave/infrared rain rate algorithm. Int J Remote Sens 22(17):3285–3307

Moisa MB, Merga BB, Gemeda DO (2022) Multiple indices-based assessment of agricultural drought: a case study in Gilgel Gibe Sub-basin, Southern Ethiopia. Theor Appl Climatol 148:455–464

Nashwan MS, Shahid S, Wang X (2019) Assessment of satellite-based precipitation measurement products over the hot desert climate of Egypt. Remote Sens 11(5):555. https://doi.org/10.3390/rs11050555

Neto R, Santos C, Silva J, Silva R, Santos C, Mishra M (2021) Evaluation of the TRMM product for monitoring drought over Paraíba State, northeastern Brazil: a trend analysis. Sci Rep 11:1097. https://doi.org/10.1038/s41598-020-80026-5

Orimoloye I, Zhou L, Kalumba A (2021) Drought disaster risk adaptation through ecosystem services-based solutions: way forward for South Africa. Sustainability 13:4132. https://doi.org/10.3390/SU13084132

Page M, Zribi M (2019) Analysis and predictability of drought in Northwest Africa using optical and microwave satellite remote sensing products. Sci Rep 9:1466. https://doi.org/10.1038/s41598-018-37911-x

Pazhanivelan S, Geethalakshmi V, Samykannu V, Kumaraperumal R, Kancheti M, Kaliaperumal R, Raju M, Yadav MK (2023) Evaluation of SPI and rainfall departure based on multi-satellite precipitation products for meteorological drought monitoring in Tamil Nadu. Water 15(7):1435. https://doi.org/10.3390/w15071435

Pona HT, Xiaoli D, Ayantobo OO, Tetteh ND (2021) Environmental health situation in Nigeria: status and future needs. Heliyon 7(3):e06330. https://doi.org/10.1016/j.heliyon.2021.e06330

Rossi M, Luciani S, Valigi D, Kirschbaum D, Brunetti M, Peruccacci S, Guzzetti F (2017) Statistical approaches for the definition of landslide rainfall thresholds and their uncertainty using rain gauge and satellite data. Geomorphology 285:16–27. https://doi.org/10.1016/J.GEOMORPH.2017.02.001

Seun AI, Ayodele AP, Koji D, Akande SO (2022) The potential impact of increased urbanization on land surface temperature over South-West Nigeria. Curr Res Environ Sustain 4:100142

Sippel S, Reichstein M, Ma X, Mahecha MD, Lange H, Flach M, Frank D (2018) Drought, heat, and the carbon cycle: a review. Curr Clim Chang Rep 4:266–286

Theon JS (1994) The tropical rainfall measuring mission (TRMM). Adv Space Res 14(3):159–165

Thiemig V, Rojas R, Zambrano-Bigiarini M, Levizzani V, De Roo A (2012) Validation of satellite-based precipitation products over sparsely gauged African river basins. J Hydrometeorol 13(6):1760–1783

Wang J, Wang H, Hong Y (2016) Comparison of satellite-estimated and model-forecasted rainfall data during a deadly debris-flow event in Zhouqu, Northwest China. Atmos Ocean Sci Lett 9:139–145. https://doi.org/10.1080/16742834.2016.1142825

Xulu S, Peerbhay K, Gebreslasie M, Ismail R (2018) Drought influence on forest plantations in Zululand, South Africa, using MODIS time series and climate data. Forests 9(9):528. https://doi.org/10.3390/F9090528

Yang L, Lin L, Fan L, Liu N, Huang L, Xu Y, Mertikas SP, Jia Y, Lin M (2022) Satellite altimetry: achievements and future trends by a scientometrics analysis. Remote Sens 14:3332. https://doi.org/10.3390/rs14143332

# Chapter 17
# Remote Sensing Applications II: Climate Change Perspectives in Agriculture

**Tofael Ahamed**

**Abstract** This chapter concludes the book covering remote sensing applications with different analytical approaches for assessing the vulnerability and impacts of climate extremes such as riverbank erosion, flash floods, tropical cyclones, and drought. Different coping strategies, such as land suitability analysis, identifying renewable energy sources, microclimate characterization, climate risk assessment, and damage-based crop insurance models, are presented in this book. To perform the analysis, different remote sensing methods and analytical approaches were adopted. In riverbank erosion, flood, and drought vulnerability analysis, the multicriteria decision-making (MCDM) method with a fuzzy expert system was used. The same approach was used for land suitability analysis. To facilitate the analysis, different criteria indices, such as the normalized difference water index (NDWI), distance from the river, soil adjusted vegetation index (SAVI), land use and land cover (LULC), normalized difference vegetation index (NDVI), modified vegetation condition index (mVCI), disaster vegetation damage index (DVDI), and standardized precipitation index (SPI), were considered. An inventory of knowledge from spatially variable communities was also presented as a source of indigenous knowledge for mitigating climate change impacts. These combined approaches may help policy-makers and researchers with subsequent studies related to the effects of climate change and the impact on the development of proper adaptation policies.

**Keywords** Climate change perspectives · Climate impact assessment · Climate risk mitigation · Climate adaptive crop selection

---

T. Ahamed (✉)
Institute of Life and Environmental Sciences, University of Tsukuba, Tsukuba, Ibaraki, Japan
e-mail: tofael.ahamed.gp@u.tsukuba.ac.jp



T. Ahamed (ed.), *Remote Sensing Application II*, New Frontiers in Regional Science: Asian Perspectives 77, https://doi.org/10.1007/978-981-97-1188-8_17

## 17.1 Introduction

Recently, climate change and its impacts have been a notable concern among scientists worldwide. The countries in Southeast Asia and Africa located in low-lying deltaic regions are very vulnerable to climate change and often face natural disasters such as floods, flash floods, tidal surges, cyclones, tsunamis, westerlies, onshore coastal bank erosion, riverbank erosion, droughts, and saline water intrusion. These climate change impacts result in increased vulnerability of the lives of people through damage to land, agricultural property, and crops, which has become a notable challenge and threat to food security. To mitigate recent climate extremes, numerous adaptation strategies are needed to ensure food, life, and property security. These strategies may include vulnerability analysis, land suitability analysis, drought security assessment, microclimate analysis, identifying and using renewable energy sources, and knowledge of affected people. Analytical approaches may include multicriteria analysis and the use of different machine learning approaches considering remote sensing. Recently, the application of satellite remote sensing methods with different geographical information systems (GISs) and cloud computation systems, such as the Google Earth Engine (GEE), has attracted considerable attention from researchers and policy-makers because of the spatial and temporal coverage levels and rapid computational and decision-making capabilities of these methods. In this book, different approaches to remote sensing applications and adaptation policies are presented to resolve the adverse effects of climate change in Asia and Africa.

## 17.2 Climate Change Perspective: Droughts and Floods

In the introductory chapter, several remote sensing methods with different analytical approaches are introduced for assessing the vulnerability and impacts of climate extremes, such as riverbank erosion, floods, cyclones, and droughts. At the first stage, an erosion vulnerability analysis was performed, in which four criteria, namely, the normalized difference water index (NDWI), distance from the river, soil adjusted vegetation index (SAVI), and land use and land cover (LULC), were used (Fig. 17.1). At the second stage, a flood vulnerability map was prepared from multiple criteria (NDWI). The climatic factor of drought is one of the critical challenges during the transition periods of agriculture. Several examples are given in the literature, with a focus on the impact of drought on orchard and grain production. Several researchers have proposed drought stress assessment and land suitability models to avoid severe impacts on crop production. Das et al. (2021) used remote sensing phenological and geospatial soil–plant biophysical property databases to assess the effect of drought on tea production in Bangladesh. A remote sensing approach in which the standardized precipitation index (SPI) was used to assess drought stress on tea farms was reported. The land surface temperature (LST) and soil moisture index (SMI) were estimated from Landsat-8 Operational

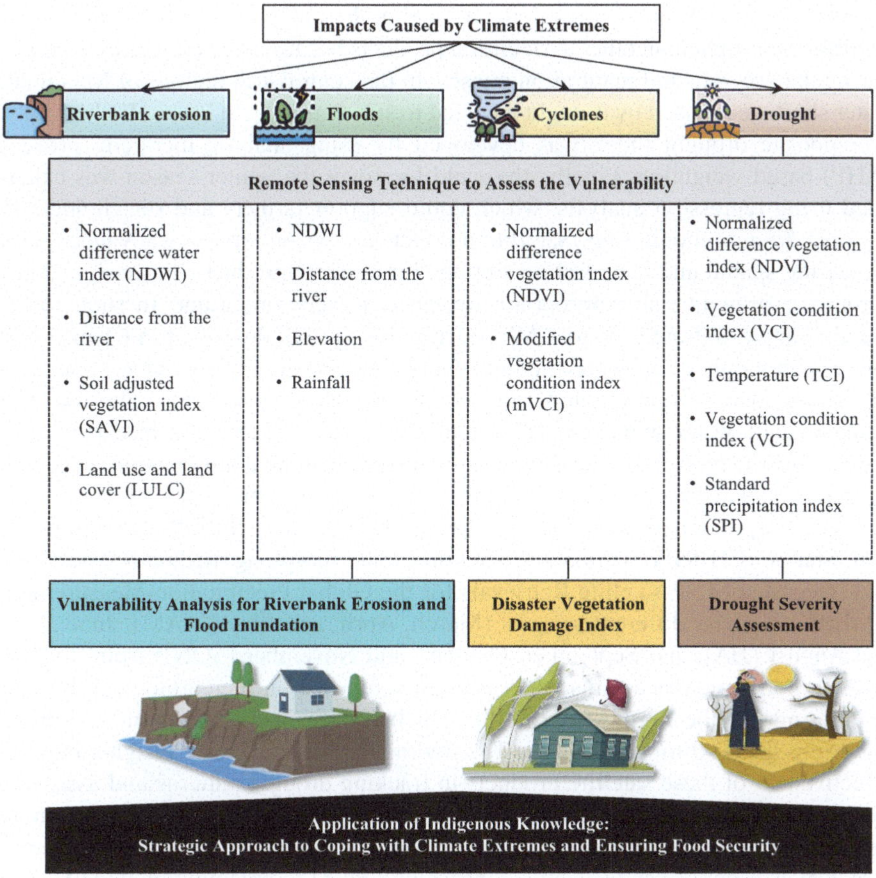


**Fig. 17.1** Vulnerability assessment of the impacts of climate extremes using remote sensing methods

Land Imager (OLI)/Thermal Infrared Sensor (TIRS) images. In addition, Sentinel-2 satellite scenes were used for generating normalized difference moisture index (NDMI), normalized difference vegetation index (NDVI), and leaf area index (LAI) maps.

Standardized precipitation index and optical and thermal infrared remote sensing methods are proposed in this book as a means of assessing and classifying drought stress for tea farms located in the northwestern part of Bangladesh. The drought stress assessment and drought classification results for tea plantations in Bangladesh could be useful for policy-makers, land use planners, and scientists in making decisions regarding how to ensure early warning of drought conditions for planning against yield loss due to drought.

In Chap. 11 of this book, biophysical, meteorological, and moisture indices were utilized to quantify the drought severity in each drought-affected region of the central dry zone of Myanmar. Persistent water limitations in this region have resulted in severe drought and fluctuations in the agricultural yield. Rice plants are more susceptible to drought than other agricultural products because they require more water during the growing and maturation phases. In the central arid regions of Myanmar, water shortages caused by drought have led to substantial yield losses. In this study, a composite drought index was developed by using analytic hierarchy process (AHP)-based weighting. Finally, the yield loss over the winter season was calculated using regression analysis, which could support farmers and stakeholders in regional drought monitoring. Satellite products are becoming more advanced with increasing spatial and temporal resolutions for climatic products. In terms of mitigating the drought risk, appropriate irrigation is very important. In such cases, the alternative wetting and dry (AWD) irrigation system described in Chap. 14 of this book not only offers water-saving benefits but also effectively reduces methane emissions, increases production, and reduces rice production costs. The use of a climate-smart water management system and sensible water use based on site-specific information on the land type and irrigation potential are recommended.

In this book, Chap. 16 delves into the degree of agreement between three prominent satellite precipitation products—Climate Hazards Group InfraRed Precipitation with Station (CHIRPS), Climate Prediction Center morphing method (CMORPH), and Integrated Multi-satellite Retrievals for the Global Precipitation Measurement (IMERG)—across different seasons (March, April, and May [MAM], June, July, and August [JJA], and September, October, and November [SON]) from 2001 to 2020 in northern Africa. This study primarily focused on meteorological drought monitoring over the North Africa region, which is particularly susceptible to drought due to its semiarid to arid climate. The findings provide valuable insights into the effectiveness of these satellite products in tracking drought patterns and assessing their agreement, thus enhancing our understanding of their potential for accurate drought monitoring. Meteorological drought, a critical concern in terms of agriculture, water resources, and society, is intricately linked to precipitation deficits. The Sahara Desert and Sahel region, known for their aridity, consistently experience large numbers of consecutive dry days regardless of the season, underlining the importance of drought monitoring in these regions. In this study, significant variations in drought patterns were revealed across different regions and seasons in North Africa. December, January, and February (DJF) and June, July, and August (JJA) exhibited distinct patterns of drought occurrence. The number of dry days and drought severity varied across the different seasons, highlighting the importance of understanding seasonal trends for effective drought prediction and management.

## 17.3 Climate Impact Assessment: Land Use and Land Cover Changes

Climate change is the most important and powerful driving force behind the changing land use and land cover (LULC) patterns. Due to climate change, land use patterns have changed; for example, plain lands are converted into flooded areas, crop lands are converted into orchards, and water bodies are filled for extra land. One of the significant studies in this book reported LULC changes from 1995 to 2021 for the major river systems in Bangladesh within a 25-km region to determine the most notably changing land use conditions (Alam et al. 2022). Bangladesh is one of most notable climate impact-affected countries due to its location in a delta. In this research, permanent water bodies were identified using the NDWI, and LULC maps were created in the Google Earth Engine (GEE) using supervised classification of the five primary land use groups (water bodies, bare land, vegetation, agricultural land, and urban areas). Furthermore, fuzzy expert-based multicriteria evaluation analysis provided an agricultural land use vulnerability map that might be helpful for future land use planning to fulfill the rising need for food production and guarantee livelihoods (Fig. 17.2).

In land use planning, land erosion and accretion in the coastal area of Bangladesh have also been reported as climatic change impacts that could be helpful for coastal boundary delineation and future land use planning (Shamsuzzoha et al. 2023). Land use planning faces additional challenges due to climate change and the sensitivity of crops, which requires careful attention to land selection. In land use planning, perineal crops such as orchards are more sensitive to land selection than other crops. In orchard production, grapes are grown in more than 100 countries worldwide, and poor vine growth due to variety selection has reduced yields worldwide.

In this book, land suitability analysis and yield prediction at the early stage of crop development can be helpful for making the right decisions. In this regard, Arab et al. (2022) performed land suitability analysis to assess the potential of extending vineyards in Afghanistan using remote sensing datasets and field data. This study may help vineyard owners and managers in Afghanistan identify more suitable and appropriate areas for long-term perineal grape orchards. Regarding plantation crops,

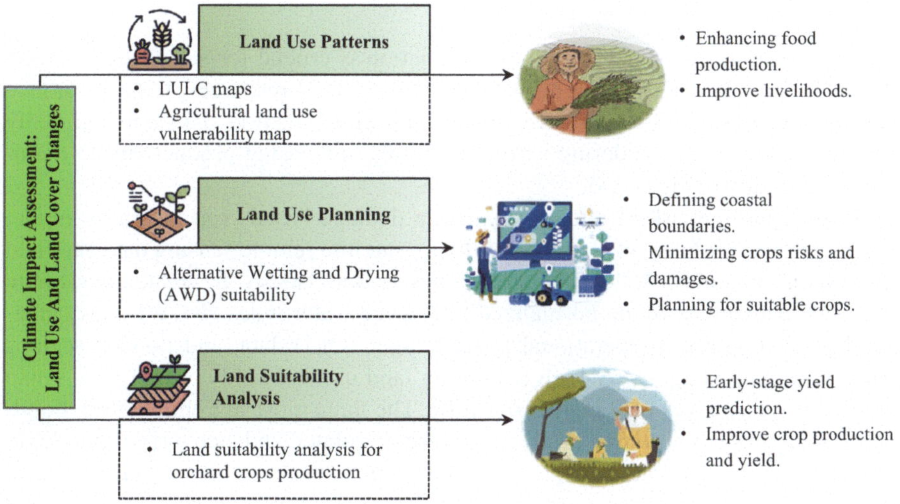


**Fig. 17.2** Assessing climate impact based on changes in land use and land cover

Das et al. (2020) also performed land suitability analysis for realizing sustainable tea production in Bangladesh, which could be helpful for agricultural land use planning and land policy-making in selecting suitable areas for increasing tea production.

## 17.4 Microclimatic Adaptation: Crop Selection and Watering

In Chap. 12 of this book, microclimate characterization and climate risk assessment of agroforestry-based Arabica coffee production were presented. Shaded plants regulate the microclimate of coffee plants when integrated with or under pine, alnus, mango, and chayote trees. Significant reductions in the minimum and maximum air temperatures, soil temperature, wind velocity, and light intensity were noted, while no differences in the relative humidity were observed. Microclimate changes are favorable for the growth and development of coffee trees. Arabica coffee production in Atok (Caliking and Topdac) experiences high exposure to typhoons and strong winds, erosion/landslides, and monsoons or persistent rains, and these areas are more sensitive to the effects of typhoons and pests and diseases than other regions, as these hazards greatly impact the yield. To attain high suitability and sustainability in coffee production areas, appropriate climate assessments were performed. Among the agroforestry systems assessed, coffee plants grown under Alnus trees as shade plants exhibited a more favorable microclimate environment favoring higher production potential and protection from climate hazard exposure.

## 17.5 Climate Risk Mitigation: Crop Insurance System

Land appropriateness for agricultural production is crucial for satisfying the food demand of a growing population with decreasing land resources. Crop insurance based on land suitability may effectively control production risks and rationally reward landowners considering various qualities, increasing productivity to maintain food security (Fig. 17.3).

To mitigate the risk of climate change, a damage-based crop insurance model was developed by Islam et al. (2022) utilizing satellite remote sensing data and field assessment data of flash flood-affected zones in Bangladesh. To create an accurate boro rice damage map, the normalized difference water index (NDWI) was calculated using Landsat-8 Operational Land Imager (OLI) data, and flooded regions were delimited using a very high-resolution land use and land cover (LULC) map derived from a Survey of Bangladesh (SoB). The three most susceptible flash flood-prone locations for boro rice production were calculated and categorized into high, moderate, and negligible damage zones. Farmers and socioeconomic factors influence the decision to adopt a damage-based crop insurance system. An empirical model for flash flood incidence in both the temporal and geographical domains

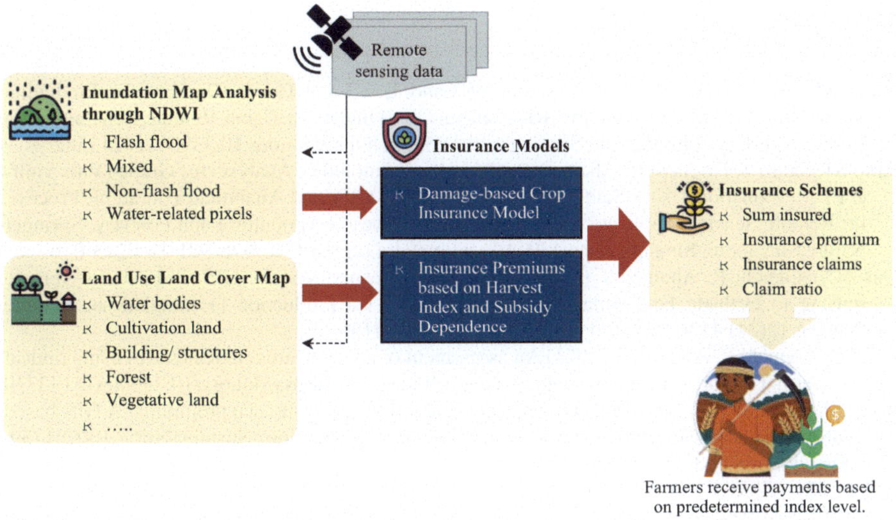


**Fig. 17.3** Mitigating climate risks through a crop insurance system for achieving resilient agriculture and safeguarding livelihoods

provides an appropriate metric for adopting a damage-based crop insurance model. In another study reported in this book, a land suitability model for rice production was developed based on suitability levels, and insurance premiums were proposed to obtain maximum returns based on the harvest index and subsidy dependence factor for the marginally and moderately suitable areas in the northern part of Bangladesh (Islam et al. 2018). Climate risk mitigation, crop insurance, and subsidence factors can play key roles in motivating farmers to adopt intensive production practices in climate risk regions in South Asian, Southeast Asian, and African countries.

## 17.6 Conclusions

This book focuses on four main thematic points: climate change effects (drought and flood); climatic impact assessment (land use and land cover changes, erosion, and vulnerability); microclimatic adaptation (climate-smart crop selection); and climate risk mitigation (crop insurance system). In this regard, the application of remote sensing methods along with machine learning models fulfills a significant role in the adaptation of policies to mitigate the adverse effects of climate change through vulnerability analysis, land cover/land use change analysis on large temporal and spatial scales, and suitable crop and orchard land selection analysis to increase production. The integrated outcomes reported in this book are recommended for food security planners, regional policy-makers, and researchers for achieving sustainability in addressing climate change effects to mitigate the risk of declining agricultural production for ensuring food security.

## References

Alam KF, Ahamed T (2022) Assessment of Land Use Land Cover Changes for Predicting Vulnerable Agricultural Lands in River Basins of Bangladesh Using Remote Sensing and a Fuzzy Expert System. Remote Sensing 14(21):5582. https://doi.org/10.3390/rs14215582

Arab ST, Salari T, Noguchi R, Ahamed T (2022) Land Suitability Analysis for Grape (Vitis vinifera L.) Production Using Satellite Remote Sensing, GIS, and Analytical Hierarchy Process. In: Remote Sensing Application: Regional Perspectives in Agriculture and Forestry. Springer Nature Singapore, Singapore, pp 149–184. https://doi.org/10.1007/978-981-19-0213-0_6

Das AC, Noguchi R, Ahamed T (2020) Integrating an expert system, GIS, and satellite remote sensing to evaluate land suitability for sustainable tea production in Bangladesh. Remote Sensing 12(24):4136. https://doi.org/10.3390/rs12244136

Das AC, Noguchi R, Ahamed T (2021) An assessment of drought stress in tea estates using optical and thermal remote sensing. Remote Sensing 13(14):2730. https://doi.org/10.3390/rs13142730

Islam MM, Ahamed T, Noguchi R (2018) Land suitability and insurance premiums: A GIS-based multicriteria analysis approach for sustainable rice production. Sustainability 10(6):1759. https://doi.org/10.3390/su10061759

Islam MM, Ujiie K, Noguchi R, Ahamed T (2022) Flash flood-induced vulnerability and need assessment of wetlands using remote sensing, GIS, and econometric models. Remote Sensing Applications: Society and Environment 25:100692. https://doi.org/10.1016/j.rsase.2021.100692

Shamsuzzoha M, Ahamed T (2023) Shoreline Change Assessment in the Coastal Region of Bangladesh Delta Using Tasseled Cap Transformation from Satellite Remote Sensing Dataset. Remote Sensing 15(2):295. https://doi.org/10.3390/rs15020295

# Index







T. Ahamed (ed.), *Remote Sensing Application II*, New Frontiers in Regional Science: Asian Perspectives 77, https://doi.org/10.1007/978-981-97-1188-8

SPRINGER NATURE

# GPSR Compliance

*The European Union's (EU) General Product Safety Regulation (GPSR) is a set of rules that requires consumer products to be safe and our obligations to ensure this.*

*If you have any concerns about our products, you can contact us on ProductSafety@springernature.com*

In case Publisher is established outside the EU, the EU authorized representative is:

Springer Nature Customer Service Center GmbH
Europaplatz 3
69115 Heidelberg, Germany

The manufacturer's authorised representative in the EU is Springer
Nature Customer Service Centre GmbH, Europaplatz 3, 69115 Heidelberg,
Germany. If you have any concerns regarding our products, please
contact ProductSafety@springernature.com

Printed and bound by CPI Group (UK) Ltd, Croydon, CR0 4YY
01/10/2026
02225863-0017